Ex dono Dr RD Hutchins
olein commensalis
2000.

GEOPHYSICS AND ASTROPHYSICS MONOGRAPHS

AN INTERNATIONAL SERIES OF FUNDAMENTAL TEXTBOOKS

Editor

B. M. McCormac, *Lockheed Palo Alto Research Laboratory, Palo Alto, Calif., U.S.A.*

Editorial Board

VOLUME 19

THE
BRIGHTEST
STARS

CORNELIS DE JAGER

The Astronomical Institute at Utrecht, Holland

D. REIDEL PUBLISHING COMPANY

DORDRECHT : HOLLAND / BOSTON : U.S.A.
LONDON : ENGLAND

Library of Congress Cataloging in Publication Data

Jager, Cornelis de
    The brightest stars.
    (Geophysics and astrophysics monographs; v. 19)
    Bibliography: p.
    Includes indexes
    1. Stars, Brightest. I. Title. II. Series.
QB843.B75J33   523.8′2   80–36702
ISBN 90–277–1109–7
ISBN 90–277–1110–0 (pbk.)

Published by D. Reidel Publishing Company,
P.O. Box 17, 3300 AA Dordrecht, Holland.

Sold and distributed in the U.S.A. and Canada
by Kluwer Boston Inc.,
190 Old Derby Street, Hingham, MA 02043, U.S.A.

In all other countries, sold and distributed
by Kluwer Academic Publishers Group,
P.O. Box 322, 3300 AH Dordrecht, Holland.

D. Reidel Publishing Company is a member of the Kluwer Group.

Printed in The Netherlands

TABLE OF CONTENTS

One of the brightest and most enigmatic stars known, Eta Carinae, is hidden in luminous clouds in the central part of this photograph, obtained with the 1 m ESO-Schmidt telescope at La Silla (Chile). The figure shows an area of ∼4° diameter. The object and its remarkable history and properties are described in some detail in Section 6.17 of this volume. Figure 99 there shows Eta Carinae and its surroundings as observed in soft X-rays by the Einstein Observatory.

(By courtesy of R. M. West, European Southern Observatory, Genève.)

FOREWORD

No part of the Hertzsprung–Russell diagram shows a more pronounced diversity of stellar types than the upper part, which contains the most luminous stars. Can one visualize a larger difference than between a luminous, young and extremely hot Of star, and a cool, evolved pulsating giant of the Mira type, or an S-type supergiant, or – again at the other side of the diagram – the compact nucleus of a planetary nebula?

But there is order and unity in this apparent disorder! Virtually all types of bright stars are evolutionally related, in one way or the other. Evolution links bright stars. In many cases the evolution is speeded up by, or at least intimately related to various signs of stellar instability. Bright stars lose mass, either continuously or in dramatic sudden events, they vibrate or pulsate – and with these tenuous, gigantic objects this often happens in a most bizarre fashion. Sometimes the evolution goes so fast that fundamental changes are observable in the time span of a human's life – several of such cases have now been identified.

So, what I hope this book will bring forward is the *unity of bright stars*, notwithstanding seeming diversity. In ordering the material I have primarily attempted to collect observational data on the various types of bright stars and, while showing the specific properties, tried to stress the mutual relationships. This is done by many cross-references and also by describing in some detail a number of 'prototypes'. In addition I have tried to show how similar phenomena occur in stars of greatly differing types.

Naturally I have briefly reviewed the theoretical aspects necessary for understanding the evolutionary relations (Chapter 4), and for grasping the connection between the various parts of extended stellar atmospheres (Chapters 5 and 6).

In the nomenclature I have normally followed the rules given in the *IAU Style Book* (*Transactions of the IAU* **XIVB**, 1971, pp. 254-264). But there are a few exceptions. Throughout this book I have written T_e instead of T_{eff} for the effective temperature, and M_b for M_{bol} the bolometric absolute magnitude. This is less cumbersome and I have taken care to avoid confusion with T_{ex} or T_{el}.

Following Van Genderen I have assumed the name 'hypergiant' for the usual 'super-super-giant'. Finally, a personal idiosyncrasy: as a spectroscopist I find it impossible to call a volume filled with H^+ particles a H II region. The symbol H^+ denotes a type of particle; and H II denotes a spectrum (of a proton!), and whatever spectrum would be emitted by a H^+ cloud, certainly not a H II spectrum. Therefore: H^+ region.

I dedicate this book to two persons who have inspired my interests during different parts of my life: my father taught me, when I was eleven, why a red star is cooler than a blue one, and let me wonder why stars are different. I also dedicate it to my wife Duotsje, the brightest star above my horizon.

During the preparations of the manuscript I have greatly profited from lively and most inspiring discussions with so many of my colleagues. These are in the Utrecht Astronomical Institute A. G. Hearn, H. J. Lamers, and K. A. van der Hucht. I am particularly obliged for most rewarding discussions to J. P. Cassinelli, P. S. Conti, M. Friedjung, and D. Mihalas.

I acknowledge with thanks the many comments received from C. Andriesse, L. I. Antipova, R. Barbon, J. Castor, N. N. Chugaj, C. de Loore, M. W. Feast, H. J. Habing, R. Hammer, J. B. Hutchings, K. Kodaira, L. B. Lucy, A. Maeder, A. G. Massevitch, D. C. Morton, E. R. Mustel, Å. Nordlund, F. Praderie, D. Reimers, L. Rosino, T. P. Snow, C. Sterken, R. Stothers, T. Tsuji, P. Ulmschneider, G. Vaiana, A. M. van Genderen, M. S. Vardya, N. R. Walborn, G. N. Wallerstein, A. J. Willis, O. C. Wilson, A. Winnberg, and B. Wolf.

Several of my co-workers at the Astronomical Institute at Utrecht have greatly helped in producing the manuscript. I am sincerely obliged to Hans Braun for the drawings. The laborious typing of the manuscript was done by Henny van Egmond, Pepi Gelderman, Annemiek Grootenboer, Ursule van der Jagt, Lydia Swart, and Sofie van der Waaij, and guided by Lidy Negenborn. I should not fail to mention the efficient help of my daughter Els Groeskamp.

Finally, it is a pleasure to thank the Editor of the Series, Billy McCormac, and Mrs. N. Pols and Mr. B. Vance at Reidel, Dordrecht for their contribution in producing a good-looking volume.

Utrecht, 31 January 1980   CORNELIS DE JAGER

CHAPTER 1

THE UPPER BOUNDARIES
OF THE HERTZSPRUNG–RUSSELL DIAGRAM

1.1. Preamble

Why is it that there are no stars brighter than (roughly) a few million times the Sun, hotter than approximately a hundred-thousand degrees, or cooler than a few thousand degrees?

What processes occurs in stars close to these boundaries, objects that are presumably at the fringes of their stability?

Recent new stellar observations of various kind, such as high-resolution ultraviolet stellar spectroscopy, intensity-measurements in the infrared, microwave and X-ray ranges, as well as refined observations in the optical spectral range, have shed new light on the various features and processes that are characteristic for near-instable stars, features such as stellar chromospheres, coronae, extended shells, and processes like mass-loss and erratic stellar variability.

For this book we have chosen as a title 'The Brightest Stars', rather than something like 'Supergiants'. This was done so because there are important groups of stars that show many of the features described in the previous paragraph, while not being normally classified as supergiants: we just mention Oe, Be, and Wolf–Rayet stars. Also, the name 'supergiant' is confusing, since it applies to stars of greatly varying bolometric luminosity. An M-type supergiant has approximately the same absolute bolometric luminosity as a typical Be star. The F–G type 'super-supergiants' are less bright than Of stars or B-type supergiants. An M-type giant is larger than a B-type supergiant.

In brief, this book deals with stars brighter than about $10^3 L_\odot$, but with special emphasis on the *objects* close to the boundaries described in the first paragraph of this section, and on the *features* and *processes* relevant to such stars. The book may be timely – I hope! – because of the many new interesting observations of the last few years, and also because the last monograph on this subject appeared half a century ago, when Cecilia Payne (1930) published her fundamental monograph 'The Stars of High Luminosity'. Her book appeared at an important moment in the evolution of stellar physics, about ten years after the discovery of Saha's law, and during the earliest phase of quantitative stellar spectrophotometry. During these vital developments her book served as a fundamental source for many years after its publication. In the time passed since, basic astrophysics has greatly developed. In addition, approximately 10^{40} more photons of stellar origin fell on Earth. A very minor fraction of them has been collected by stellar telescopes and still less, approximately 10^{22}, were

1

collected by stellar spectrographs. Of these about one percent originated from intrinsically luminous stars. In this book we wish to summarize the main conclusions
these 10^{20} photons have enabled the astrophysicists to draw.

1.2. The Most Luminous Stars

Figure 1 shows a Hertzsprung–Russell diagram for bright stars in the Magellanic
Clouds, as published by Hutchings (1976a). It agrees reasonably with similar diagrams
established for the Large Magellanic Cloud (LMC) by Brunet and Prévot (1972),

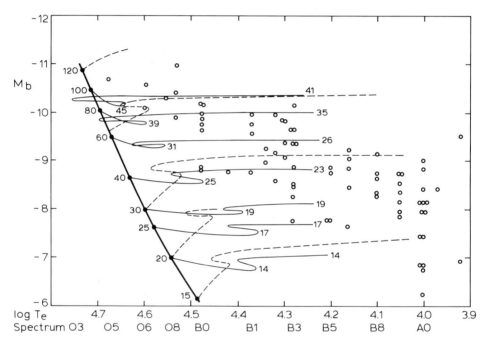

Fig. 1.   HR-diagram of bright stars in the Magellanic Clouds after Hutchings (1976a). The dashed
and thin solid lines give the evolutionary tracks computed by Stothers (1972a) without mass-loss,
and by De Loore *et al.* (1977, 1978a, b) with mass-loss. Initial masses in solar units are given along
the main sequence, and intermediate and final masses (for the computations with mass-loss) are
labeled alongside the evolutionary tracks.

Osmer (1973), and Humphreys (1978). Since the whole material refers to stars from
two extragalactic systems there are only two distances involved, and the material thus
forms a fairly homogeneous collection. The ordinate of the diagram gives the bolometric absolute magnitude, which is a measure for the total luminosity of the star
expressed in the magnitude scale. The relation between magnitude (M) and luminosity
(L) is given by

$$- dM = 2.5 \, d \log L.$$

The bolometric absolute magnitude of the stars can be compared with that of the Sun by taking for the absolute bolometric magnitude of the Sun $M_b=4.76$ (Code *et al.*, 1976) and the total flux $L_\odot=3.83\times10^{33}$ erg s^{-1}. These data lead to the relations

$$L = 3.07 \times 10^{35} \times 10^{-0.4M_b} \text{ erg s}^{-1}, \tag{1.2; 1}$$

$$\log L/L_\odot = 1.90 - 0.4M_b.$$

(See also the discussion of the relevant constants by Lang (1974); p. 562; use of his data would yield only slightly different values.)

The effective temperature T_e is related to the radiation flux πF per cm^2 and second at the stellar surface by the relation

$$\pi F = \sigma T^4_e$$

and is related to the luminosity L and radius R by

$$4\pi R^2 \pi F = L.$$

Taking $(T_e)_\odot=5784$ K, one obtains

$$\log(R/R_\odot) = \tfrac{1}{2} \log(L/L_\odot) - 2\log T_e + 7.52. \tag{1.2; 2}$$

Other useful expressions may be derived for stars of which blue (B) and V-magnitudes are known. Assuming black body emission functions one then derives, with Arp (1961):

$$B - V = 6900/T - 0.56 + \Delta X, \tag{1.2; 3}$$

$$M_V = -0.06 - 5\log R + 28400/T + X_V, \tag{1.2; 4}$$

where X is a small correction term of decreasing importance as $\lambda T \to \infty$, and $\lim \Delta X = 2.5 \log(\lambda_V/\lambda_B)$ for $T \to \infty$; R is in solar units.

Figure 1 shows that the brightest stars in the Magellanic Clouds have a bolometric magnitude of approx. -11 which yields a luminosity of about 2×10^6 as expressed in solar units.

An uncertainty in this limit may be due to the errors in the distances of the Magellanic Clouds, which affects the scale of absolute magnitudes. Therefore, we show in Figure 2 a Hertzsprung–Russell diagram for galactic stars as composed from several sources. For the early-type stars we used mainly compilations by Snow and Jenkins (1977), and by Conti and Frost (1977). For early-type stars not figuring in these compilations, as well as for later-type stars we used data given elsewhere in this volume. Attention is drawn to the stars HD 93 129 A, HD 15570, and η Car, the brightest known stars in our Galaxy.

If we apply the conventional 'rule of caution' not to include the three brightest objects in the statistics, we arrive again at a largest luminosity of $2\times10^6\,L_\odot$, cor-

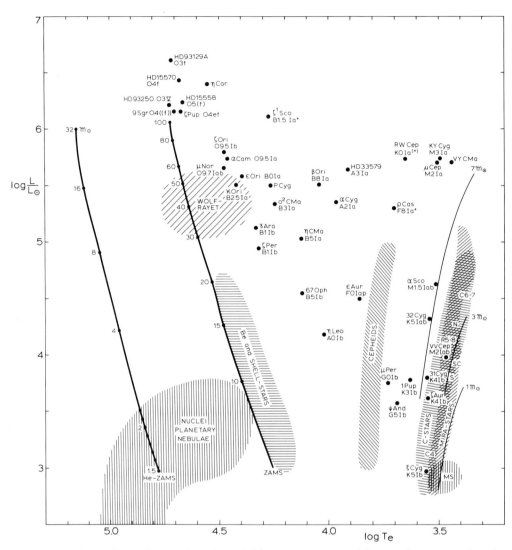

Fig. 2. $\log L - \log T_e$ diagram for galactic bright stars as composed from various sources (see the text). The ZAMS is after Stothers (1972), and De Loore *et al.* (1977, 1978a, b). The He–ZAMS is from Paczynski (1971a). The position of the Wolf–Rayet stars is uncertain.

responding to 8.6×10^{39} erg s^{-1}. Humphreys (1978), who made an extensive study of some 1000 supergiants and bright O-type stars in the Galaxy confirmed this result by reporting that the brightest O stars have M_b between -10 and -12. Why do brighter stars not exist?

1.3. Gravitational Acceleration Versus Radiation Forces

Eddington (1921) suggested that the upper limit for stellar luminosity is reached for those stars for which the inward gravitational acceleration would be balanced by the outward acceleration due to radiation forces. For the former we have

$$g_{\text{grav}} = G\mathfrak{M}/R^2. \tag{1.3; 1}$$

The acceleration by radiation forces is

$$-g_{\text{rad}} = \frac{1}{\varrho}\frac{dP_{\text{rad}}}{dz} = k_F \pi F/c,$$

where k_F is the flux-mean-opacity – cf. our expression (5.4; 17) further in this volume. With $\pi F = L/4\pi R^2$, one obtains, defining Γ_F:

$$\Gamma_F \equiv -g_{\text{rad}}/g_{\text{grav}} = k_F L/4\pi c G\mathfrak{M}. \tag{1.3; 2}$$

The upper limit of stellar brightness is assumed to be reached for $\Gamma_F = 1$, hence

$$L_{\text{lim}} = 4\pi G\mathfrak{M}c/k_F = 2.52 \times 10^4 \, \mathfrak{M}/k_F. \tag{1.3; 3}$$

We introduce the dimensionless parameters

$$l = L/L_\odot \quad \text{and} \quad m = \mathfrak{M}/\mathfrak{M}_\odot,$$

and with $\mathfrak{M}_\odot = 1.99 \times 10^{33}$ g, $L_\odot = 3.83 \times 10^{33}$ erg s^{-1}, one obtains

$$l_{\text{lim}} = 1.31 \times 10^4 \, m/k_F. \tag{1.3; 4}$$

It is easy to draw some conclusions from the expression (1.3; 4) for hot supergiant atmospheres, since for these the opacity k_F is nearly solely due to scattering by free electrons. Assuming a solar-type composition the scattering coefficient is

$$\sigma_e \approx 0.3 \text{ cm}^2 \text{ g}^{-1}. \tag{1.3; 5}$$

The significance of Equation (1.3; 4) is that it allows us to find out whether the brightest stars are radiating at, above, or below their Eddington-limit. It does not unambigously answer the question why there are no stars brighter than $\sim 2 \times 10^6 \, L_\odot$, since any observed luminosity corresponds to a mass via Equation (1.3; 4). The question is only whether that mass exists. As an example: substitution of σ_e from Equation (1.3; 5) for k_F in Equation (1.3; 4) shows that the assumed maximum luminosity, $l = 2 \times 10^6$, would correspond with a star of $46\mathfrak{M}_\odot$. This mass is about a factor 2 smaller than what would be inferred from the computed masses and evolutionary tracks as shown in Figure 1. It is also smaller than the observed masses of the most massive stars, as given in Section 2.4 and Table VIII there.

This leads to the conclusion that most of the brightest stars seem to radiate below the Eddington-limit. Hence, they would have stable atmospheres, a conclucion that sounds trivial but that is not because precisely in the very upper part of the HR-diagram there are a number of fairly unstable stars, with strong mass-loss and erratic variability.

To pursue the matter of atmospheric stability we have to go to individual stars. Therefore we have collected in Table I basic data for a few of the brightest stars of greatly different effective temperatures. We even included Nova V1500 Cyg 1975, the brightest and fastest nova on record; the parameters in the table refer to the phase of maximum brightness. The Eddington-limit for these stars is calculated for hot stars in the way described above and for cool stars we used data from the models of very tenuous stellar atmospheres as given by De Jager and Neven (1975).

TABLE I

Stellar parameters and Eddington limits for a few well-studied stars

	HD 93 250	ζ Pup	η Car	Nova Cyg 1975 (max)	ϱ Cas	α Sco	VY CMa
Spectral type	O3 V((f))	O4 ef	–	Q1 (A0 Ia⁺)	F8 Ia⁺	M1.5 Iab	M3–5e I
T_e (K)	52 500	46 000	29 000	9000	5000	3600	2700
$\log(L/L_\odot)$	6.4	6.2	6.5	5.9	5.3	4.7	5.7
$\mathfrak{M}/\mathfrak{M}_\odot$	120	80	115	1	25	20	?
$R/R_\odot$	20	20	60	360	600	700	3000
$\dot{\mathfrak{M}}(10^{-6}\,\mathfrak{M}_\odot\,\mathrm{yr}^{-1})$	0.2	8	50000	*	25	3	200
$\log(L/L_\odot)_{\mathrm{Edd}}$	6.7	6.5	6.7	4.6	8.3	9.5	–
Ref. this book, Section:	3.3	3.2	6.17	9.5; 9.6	3.8	3.11	6.16

* $\int -\dot{\mathfrak{M}}\,dt = 6\times10^{-5}\,\mathfrak{M}_\odot$ in $\sim 10d$.

Table I shows, perhaps surprisingly, that only Nova Cyg 1975 had at maximum a luminosity exceeding the Eddington-limit. All other non-eruptive stars, even η Car, *seem* to radiate below that limit. But is this really so? In the present calculations we considered the atmospheric stability assuming that only radiation forces are counteracting gravitation. But there are other forces: in Section 1.6 we will show that for most of the brightest stars the acceleration caused by dissipation of turbulent energy may be the dominant mechanism in defining the structure of the outer atmospheres of unstable stars. In Chapter 5 we will show that strong fraunhofer lines may add by a factor up to ~ 10 to the k_F-value, which will reduce the Eddington-limit.

We note further that the Eddington limits for helium stars (for which $\sigma_e \approx 0.15$) with $10\mathfrak{M}_\odot$ (being the mass of a typical Wolf–Rayet star) or $0.6\mathfrak{M}_\odot$ (a typical nucleus of a planetary nebula) are found at $\log l = 5.94$ and 4.7 respectively. This means that nuclei of planetary nebula are stable – from this point of view – but Wolf–Rayet stars are hardly, or not – depending on the precise value of their luminosity (see the discussions in Section 3.4).

1.4. Inclusion of the Mass-Luminosity Relation

The question why there are no stars brighter than $\sim 2\times10^6\,L_\odot$ has not yet been answered. It is clear that Equation (1.3; 4) would have an unambiguous solution when a relation between L and $\mathfrak{M}$ would be known. Then, both $L_{\max}$ and $\mathfrak{M}_{\max}$ could be derived. Observations of masses and luminosities of main-sequence stars and theoretical studies have shown the existence of a definite mass-luminosity relation (Figure 14 in Chapter 2). In a limited range of mass or luminosity this law can be approximated by

$$l = \text{const.}\ m^n. \qquad (1.4; 1)$$

For $m \lesssim 7$ Heintze (1973) finds $n=3.7$, and McCluskey and Kondo (1972, see also Section 2.4) found in a range extending to slightly larger luminosities: $n=3.88$. But the exponent n decreases with increasing mass. From our Figure 13 we derive for $4.6 < \log l < 6.7$: $n=3.0$, (cf. Equation (2.4; 2)), but when including only the very brightest stars, also the evolved object BD$+40°4220$, we get $n=2.7$.

The introduction of Equation (1.4; 1) into Equation (1.3; 4) gives

$$l_{\text{lim}} = (1.28 \times 10^4/k_F)^{n/(n-1)}. \qquad (1.4; 2)$$

With $n \approx 3.8$ and $k_F = 0.3$ this would yield

$$\log l_{\text{lim}} = 6.3; \qquad (M_b)_{\text{lim}} = -11,$$

but $n = 2.7$, the value that applies to the region of more extreme masses and luminosities, yields:

$$\log l_{\text{lim}} = 7.4; \qquad (M_b)_{\text{lim}} = -14!$$

For $n=1$ the limiting luminosity would be ∞.

This result shows that actually the Eddington-limit for main-sequence stars, calculated just on the basis of gravitational and radiation forces *does not exist!* This is so, because for increasing mass and luminosity the exponent n tends to decrease and in the extreme case – for hypothetical very massive stars – it would approach unity. That this happens is a consequence of the Eddington relation (1.3; 3) and can be understood qualitatively: For stars for which $\Gamma_F \approx 1$ a hypothetical addition of extra mass would be related to a proportional increase of luminosity with $n=\mathrm{d}L/\mathrm{d}\mathfrak{M} \approx 1$.

The reason why there *is* an apparent limit to atmospheric stability will be discussed in Section 1.6: dissipation of mechanical energy contained in non-thermal stochastic motions sets an upper limit to the mass and luminosity of atmospheres of stars, as it introduces a pressure gradient and hence a force which can compete with gravity in supergiants.

1.5. The Effective Value of the Acceleration of Gravity; Influence of Radiation Pressure

An important aspect of the foregoing discussion is that the effective value of the acceleration of gravity g_{eff} takes values close to zero near the limit of stellar atmospheric stability: From Equation (1.3; 2) one derives

$$g_{\text{eff}} = g_{\text{grav}} (1 - \Gamma_F). \qquad (1.5; 1)$$

With $r=R/R_\odot$, $t=T/T_\odot$, $g/g_\odot=mr^{-2}$, $l=r^2t_e^4$, and with Equation (1.3; 4) one obtains

$$g_{\text{lim}} = 8 \times 10^{-5} k_F t_e^4 g_\odot. \qquad (1.5; 2)$$

With Equation (1.3; 5), $(T_e)_\odot = 5784$ K and $g_\odot = 2.76 \times 10^4$ one thus obtains, for example for a star with $t_e = 2$ hence for a late B-type star: $g_{lim} = 10$, while for $t_e = 10$ hence for a very early O-type: $g_{lim} = 6.5 \times 10^3$. Only stellar atmospheres with g-values larger than g_{lim} can exist. But for stars with g_{grav} close to g_{lim} the *effective* value of g will be *very close to zero*, and may change considerably with height in the stellar atmosphere. Occasionally it may even obtain negative values in those parts of the atmosphere, where k_F assumes a large value. How close is g_{eff} to zero?

To answer this question Figure 3 gives for fourteen A- and B-type supergiants and hypergiants their position in the $\theta_e(=5040/T_e)$-log g diagram, and the locus of the

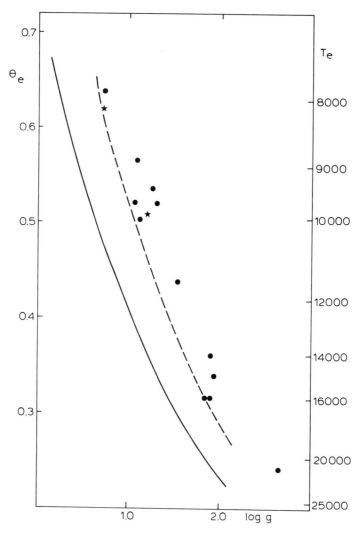

Fig. 3. Locus of the g_{lim}-curve in a $(\theta_e;$ log $g)$-diagram after data from Sterken (1976a, b). Dots denote galactic objects, asterisks are LMC objects. The lower envelope of the observed points (dashed) does not coincide with the theoretical limit.

g_{lim}-values, as determined from Equation (1.5; 2). The diagram shows that all stars examined are at the right-hand side of the limiting curve – as should be the case! It shows moreover that the effective limit – the dashed line – lies at an approximate distance $\Delta \log g = 0.4$ to the right of the theoretical limit.

The above discussion is sufficient for a first review. A more detailed treatment of the influence of radiation pressure is needed for near-unstable stars. Such a discussion was given by Kurucz and Schild (1976) on the basis of detailed atmospheric models by Kurucz (1979). The radiative acceleration g_{rad} is determined by

$$-g_{\text{rad}} = \frac{4\pi}{c} \int_0^\infty \kappa_\nu \pi F_\nu \, d\nu. \tag{1.5; 3}$$

Since $\int_0^\infty \pi F_\nu \, d\nu$ is proportional to T_e^4, g_{rad} is roughly so. The dependence of κ_ν on T and P produces deviations from this proportionality.

Figure 4 gives $\kappa_\nu H_\nu$, where $H_\nu = \frac{1}{2} \int_{-1}^{+1} d\mu \, \mu I_\nu(\mu)$, for early-type stars with $\log g = 4$ and shows the contribution of the various spectral line complexes for three different T_e-values.

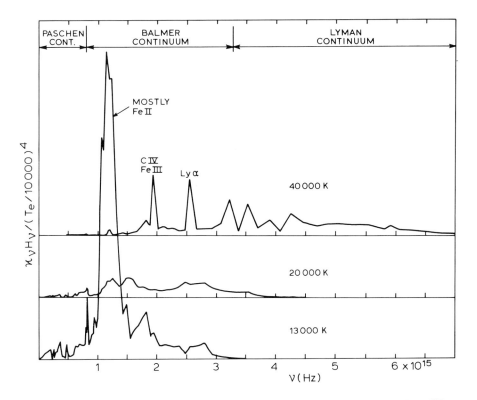

Fig. 4.   The monochromatic flux-term $\kappa_\nu H_\nu$ as a function of the frequency ν for stars of three different T_e-values and $\log g = 4$. The diagram shows the main contributors to the radiative acceleration. Important is the large peak due to the singly ionized iron group elements in the near ultraviolet for late B-type stars. (Kurucz and Schild, 1976).

The consequent radiative acceleration at an optical depth $\tau_{Ross} = 10^{-4}$, as computed from Equation (1.5; 3) for different T_e and log g-values is given in Figure 5, where the straight line is proportional to T_e^4. Also this figure makes clear that early-type stars can only be stable for fairly large g-values. The deviations from the straight line show the influence on g_{rad} of the various strong spectral lines for stars with different effective temperatures.

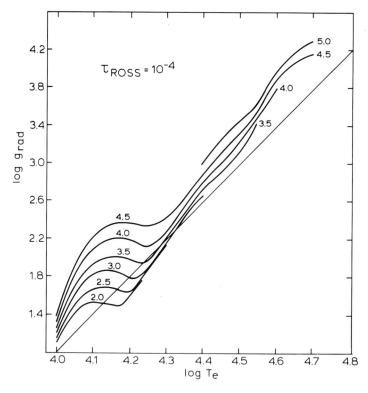

Fig. 5.   Radiative acceleration $\log g_r$ as a function of T_e for different values of $\log g_{grav}$. The straight line is proportional to T_e^4. (Kurucz and Schild, 1976).

1.6. Influence of Rotation and of Turbulent Motions

In previous sections we assumed with Eddington that the stability of a stellar atmosphere is solely determined by the balance between the gravitational force and that due to the radiation-pressure-gradient.

   In addition, there is another ouward force which is due to *rotation*. For a rigidly rotating stellar surface

$$g_{rot} = -\omega^2 R \cos\beta, \tag{1.6; 1}$$

where β is the stellar latitude. This effect is particularly important in Be-type stars.

Further, *turbulent motions* with a small characteristic length (so-called micro-turbulence – see Section 2.7) exert a pressure

$$P_\mu = a' \varrho v_\mu^2,$$

where v_μ us the velocity amplitude of the micro-turbulent motions. The factor a' is of the order unity; it is $\frac{1}{2}$ if the velocity distribution is gaussian. The *dissipation* of turbulent pressure produces an acceleration

$$g_\mu = \frac{1}{\varrho} \frac{d}{dz} \ (a' \varrho v_\mu^2). \qquad (1.6; 2)$$

But also macro-turbulent motions exert a pressure, P_M, being the average value of the relevant part of the momentum flux density tensor $\varrho v_i v_k$, and related to the loss of momentum from the rising to the descending components.

The macro-turbulent pressure P_M is estimated from such calculations (cf. for details: Nordlund, 1976):

$$P_M \approx a'' \varrho v_M^2,$$

where a'' is dimensionless and of order unity. Hence, in the *general case*, where the random velocity field is described by *a spectrum of turbulence*, one has

$$P_t = a \varrho \langle v_t^2 \rangle, \qquad (1.6; 3)$$

and

$$g_t = \frac{1}{\varrho} \frac{d}{dz} \ (a \varrho \langle v_t^2 \rangle) \qquad (1.6; 4)$$

where $a[F(k)]$ is dimensionless but depends on the spectrum $F(k)$ of turbulence – see Section 2.9. It is assumed, but not yet proven, that a is of the order unity.

Hence, in general

$$g_{eff} = g_{grav} + g_{rad} + g_t + g_{rot}, \qquad (1.6; 5)$$

where the various accelerations are defined by Equations (1.3; 1), (1.5; 3), (1.6; 4), and (1.6; 1).

It is known that in those cool stars that have fairly extended convection regions just below the photospheric surface, strong turbulent motions can develop at smaller scales; these motions may propagate outwards. Hence in cool stars strong turbulent motions of various scales may be expected in the outer parts of the atmospheres.

However, also in hot stars turbulent motions have been reported to occur, often with velocities, close to, or in cases even superceding that of sound, s. It is evident that such motions must have a large influence on the stability of an atmosphere since for $v_t \approx s$ we have $a \varrho v_t^2 \approx 2aNkT$, so that the turbulent pressure then approximates the thermal pressure. Moreover, one may expect considerable dissipation of turbulent energy in atmospheres with (super-)sonic macro-turbulent velocities. These aspects, therefore, completely change the atmospheric structure as compared with a similar atmosphere without turbulent motions. With a view to Equation (1.6; 4) we may conclude that *in atmospheres with turbulent motions $\gtrsim s$ the effective acceleration of gravity is principally determined by the contribution of turbulence.*

In order to obtain a feeling for the relative importance of the three different accelerations we make a rough comparison of the conditions in two supergiant atmospheres; see Table II. In this table we have compared two model atmospheres, with characteristic temperatures of 42 000 and 4500 K, both for $g_{eff} = 10$. The comparisons

TABLE II

Rough comparison of g_{eff}, g_r, and g_{turb} in two model supergiant atmospheres with $g_{eff} = 10$; the values of v_t and Δv_t in the two last lines are chosen as *examples* and are estimated from $v_t \, \Delta v_t$ which was found assuming $g_t = -g_{eff}$ and $\Delta z = H$

Symbol	Values		Dimensions
T	42 000	4500	K
g_{eff}	10	10	cm s^{-2}
τ_{5000}	0.3	0.3	
κ_{5000}	0.3	5.9×10^{-3}	cm^2 g^{-1}
P_g	1.1	2.4×10^3	g cm^{-1} s^{-2}
ϱ	1.7×10^{-13}	6.4×10^{-9}	g cm^{-3}
H	7×10^{11}	3.7×10^{10}	cm
g_{rad}	-1.7×10^3	-5×10^{-3}	cm s^{-2}
$v_t \Delta v_t \approx$	7×10^{12}	4×10^{11}	cm^2 s^{-2}
$\begin{cases} v_t \\ (dv_t/dz) \, H \end{cases}$	45 / 15	10 / 4	km s^{-1} / km s^{-1}

are made at $\tau_{0.5} = 0.3$. The quantity H denotes the density scale height of the atmosphere. In the *hot* atmosphere the radiative acceleration is in absolute value much larger than the assumed effective acceleration so that the gravitational acceleration should also be of the order of 2×10^3: hence $-g_{rad} \approx g_{grav}$. The effective g-value is nearly reduced to zero by the influence of radiation forces. In the *cool atmosphere* radiation forces are virtually absent, and $g_{eff} \approx g_{grav}$. In order to estimate the value, necessary for the gradient of the turbulent pressure to compete with the other accelerations, we have, for simplicity but obviously not correctly rewritten Equation (1.6; 4) as

$$v_t \, dv_t = -g_t \, dz.$$

Assuming $g_t = -g_{eff}$ and taking $\Delta z = H$ one obtains values for $v_t \Delta v_t$ from which combinations of v_t and Δv_t can be derived, as done, *as an example*, in the lower part of Table II. (Obviously other combinations of values of v_t and of (dv_t/dz). H could have been selected just as well.) This shows again that, for the turbulent acceleration to compete with the effective acceleration of gravity (super-)sonic velocities may be needed. This result becomes immediately clear when one realizes that the gas-pressure $P_g = \frac{1}{2} \varrho v_{th}^2$ and the turbulent pressure $P_t = \alpha \varrho \langle v_t^2 \rangle$ are comparable when the Mach number v_t / v_{th} is of the order unity (cf. the discussion by Gustafsson *et al.*

(1975)). The velocities required, expressed in Mach numbers, are roughly equal for hot and cool supergiants and have indeed been observed, but v_t/v_{th} seems to be slightly larger in early-type supergiants than in those of late-type.

The outcome of this brief comparison is that for a more thorough discussion of the stability of the brightest stars a critical discussion of their velocity fields is necessary. This is the more so because in stellar atmospheres at the limit of their stability, outward streaming motions occur, often involving large energies and energy fluxes. These outward motions, in turn, can again provoke strong turbulence, which may enhance the instability. In that connection it seems important to make the following remark (de Jager, 1978).

We refer to Equation (1.6; 4). It is easy to show that as soon as there is a gradient in the 'turbulent' pressure there must be dissipation of a mechanical energy flux,

$$F_m = \theta \varrho \langle v_z^2 \rangle s, \tag{1.6; 6}$$

where θ is a factor of the order unity, and s the velocity of sound. Equation (1.6; 6) is only valid when the velocities v_z are of a stochastic nature, and relate to *pressure disturbances*. Streaming motions contribute also to the turbulent pressure but with energy fluxes of the order of $\varrho(\varDelta v)^3$. A comparison of (1.6; 4) and (1.6; 6) yields

$$g_t \approx -\frac{1}{\varrho}\frac{\mathrm{d}}{\mathrm{d}z} \ (F_m/s). \tag{1.6; 7}$$

Since s changes only slowly with height (because s is practically proportional to $T^{1/2}$ which changes slowly in the upper stellar photospheres), the fact that g_t does not vanish in many stars – also in hot supergiants – shows that such stars must have an outward flux of mechanical energy $F_m \neq 0$ which, moreover, dissipates. This may, consequently, lead to the formation of warm chromospheric or hot coronal regions around such stars, as are indeed observed.

The foregoing discussion may also be useful in a determination of the upper limit of the region of existence of atmospheres of cool stars. Since g_{rad} is negligible in such stars, in contrast to the acceleration due to dissipation of mechanical energy, we have for the cool atmospheres at the limit of their existence:

$$g_{grav} + g_t = 0. \tag{1.6; 8}$$

With Equations (1.3; 1) and (1.6; 7), and assuming the velocity of sound s independent of height, which is an acceptable approximation in a discussion in which only orders of magnitude are relevant, condition (1.6; 8) transforms into

$$-\mathrm{d}F_m/\mathrm{d}z = G\mathfrak{M}\varrho s/R^2. \tag{1.6; 9}$$

We now write $s = (\gamma P/\varrho)^{1/2}$ with $\gamma = c_p/c_v$, and $\mathrm{d}F_m/\mathrm{d}z = fF_m/H$. Here H is the density scale height $H = \mathfrak{R}T/\mu g$, and f is the fraction of the mechanical energy dissipated over a vertical path-length of one scale height. Introducing further $g = G\mathfrak{M}/R^2$, condition (1.6; 9) transforms into

$$fF_m = (\gamma \mathfrak{R}/\mu)^{1/2} PT^{1/2}. \tag{1.6; 10}$$

With $\gamma=\frac{5}{3}$, $\mu=1.44$ (a value for cool tenuous atmospheres), the gas constant $\mathfrak{R}=$ 8.3×10^7, and knowing that the limiting value for P/g at small optical depths in cool and very tenuous atmospheres is approx. 300 (see e.g. De Jager and Neven, 1975), the condition (1.6; 10) becomes

$$fF_m \approx 3 \times 10^5 \, gT^{1/2}. \qquad (1.6; 11)$$

To illustrate the significance of this expression we consider the observed upper limit of luminosity of cool stars which runs roughly from $(\log T_e; \log(L/L_\odot))=(4.0; 6.05)$ to $(3.5; 5.5)$ in the Hertzsprung–Russell diagram. For the stars on this line we assume the masses as given in Table III, and since R is found from T_e and L with Equation (1.2; 2) the values of g and limiting fF_m values result as given in the table. To under-

TABLE III

The upper brightness limit of cool stars; consequences for the dissipation of mechanical energy flux

$\log(L/L_\odot)$	6.05	5.8	5.6	5.5
$\log T_e$	4.0	3.8	3.6	3.5
$\log(\mathfrak{M}/\mathfrak{M}_\odot)$	1.65	1.48	1.16	1.0
$g(\text{cm s}^{-2})$	9.0	1.9	0.23	0.08
fF_m	2.7×10^8	4.5×10^7	4.3×10^6	1.3×10^6
$f(\text{approx.})$	2×10^{-3}	3×10^{-3}	0.2	0.5

stand the scope of these values we assume $\langle v_z^2\rangle^{1/2}=s$, and derive F_m with Equation (1.6; 6) taking $\theta=1$, which yields f. Apparently $f\approx10^{-3}$, over the greater part of the area, but f rapidly approaches unity for the coolest stars. This indicates that very moderate values for the dissipation of mechanical energy (or stated otherwise: for the decrease of turbulent pressure with height) should be sufficient to explain the observed brightness limit of medium-type stars, but that fast dissipation should be needed to account for the absence of very luminous cool stars, if there are no other mechanisms, e.g. of an evolutionary character.

1.7. 'Boundaries' to the Hertzsprung–Russell Diagram

Although the upper part of the $(\log L, \log T_e)$-diagram is fairly homogeneously filled with stars, a diagram like Figure 2 suggests that there are boundaries at the three sides: at large luminosities, as well as at high and low temperatures. The upper boundary is determined by various causes. The balance between inward acceleration of gravity and the outward acceleration caused by dissipation of turbulent energy plays a role over the whole range of T_e-values. In addition radiation forces contribute in the region of high temperatures. Figure 6 shows in the low T_e-region the line assumed there for the condition $g_{\text{grav}}+g_t=0$, and the consequent fF_m-values. We note that this condition necessarily leads to the conclusion that in such atmospheres mechanical energy is dissipated (Equation (1.6; 7)), causing outer-atmosphere heating and thus contributing to mass-loss. Translated in other notions this conclusion would

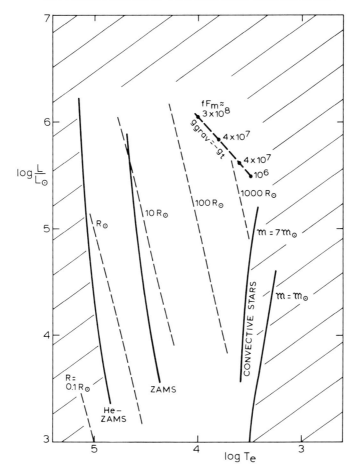

Fig. 6. Boundaries to the upper part of the Hertzsprung–Russel diagram; see text. Dashed lines are loci of constant R-values.

mean that the stellar evolution in this part of the HR-diagram can be accompanied by appreciable mass-loss. However, one should not exclude other possible evolutionary aspects to explain the lack of massive red giants – evolutionary tracks might not reach that regions for other causes than just mass-loss (cf. Chapter 4).

The right-hand boundary of the HR-diagram – for low temperatures – is described in Section 4.1, where it is explained that the 'Hayashi-track' marks the positions of fully convective stars in hydrostatic equilibrium. Stars that are only partly in convective equilibrium are found to the left of the Hayashi-track, while stars in more than complete convective equilibrium do obviously not exist. Objects that are not in hydrostatic equilibrium, such as collapsing proto-stars, can exist to the right of the Hayashi-track. In passing we have to add that the precise position of the Hayashi-track depends slightly on the stellar mass, and further, that the Hayashi-tracks mark the loci of stars in hydrostatic equilibrium in part of their pre-main-sequence contraction phase, as well as in very late phases of stellar evolution. The figure shows the tracks for $\mathfrak{M}=7\mathfrak{M}_\odot$ (Iben, 1974), and $\mathfrak{M}_\odot$.

For the left-hand side of the diagram – the high-temperature region – we have to consider two cases.

For *non-degenerate stars* we realize that most normal stars are virtually *hydrogen stars*, and are found at the main sequence or to its right. However there are stars to the left of the main sequence. These contain more He. We assume that the high-temperature boundary is *practically* marked by the Zero Age Main Sequence (ZAMS) for *pure helium stars*. The locus of these stars is discussed in Section 4.2, and the position of this He–ZAMS is also indicated in Figure 6. Stars that consist only for a part of pure helium are found to the right-hand side of that curve.

As a next step one could also consider ZAMS's for pure carbon, oxygen, ..., etc. stars. The C–ZAMS is situated at the high-temperature side of the He–ZAMS, as is shown indeed by computations of pure He stars: when the star develops a considerable C-core (Paczynski, 1971a) it may move towards the left of the He–ZAMS. The C–ZAMS was calculated by Salpeter (1961) and is shown in Figure 60 (Chapter 4).

Degenerate stars of the high luminosities and the temperatures considered here do not exist: bright stars of temperatures around 10^5 K would have radii of 0.1 to $10R_\odot$ (as indicated by the loci of constant R in Figure 6), and extrapolation of the $\mathfrak{M}$–R law for degenerate objects shows that for such radii the masses would be so small that the matter cannot be degenerated (see e.g. Schatzman, 1958, p. 98). Could one think of a hypothetical white dwarf with $R=10^{-2} R_\odot$, $L=10^4 L_\odot$, which would then have (Equation (1.2; 2): $T_e=6\times10^5$ K? Such a star, with an assumed mass of $1\mathfrak{M}_\odot$ would have a total particle kinetic energy $\int NkT \, dV$ of about $2\times10^{41} T$ erg, where T is the internal temperature of the star. Since it radiates 4×10^{37} erg s^{-1} its lifetime would be $5\times10^3 T$ s. Even for a high internal temperature of 10^8 K such a star would only exist for about 10^4 yr. In addition its life-time would be considerably shortened by energy-loss caused by neutrino-emission. Therefore such objects would be very rare.

1.8. Terminology Related to Extended Atmospheres

A. OUTER STELLAR LAYERS

Since many aspects and features of the stars, discussed in this volume, are related to extended atmospheres, stellar winds, mass-loss, warm and hot stellar envelopes (chromospheres and coronae) it seems useful to define the relevant terminology. Thereby we shall partly follow Conti (1976).

Atmosphere: the region where the absorption lines are formed. Normally, the atmosphere comprises the photosphere.

Photosphere: the region where the continuous radiation is formed.

Chromosphere: the definition is given in Section 6.1: normally a region with temperature $>T_e$ with small (<1) continuous optical depths, and large (>1) optical depths in strong lines.

Envelope: a region surrounding a star, which produces the emission lines. The temperature should not necessarily be higher than the photospheric value, but the density should be sufficient in order that the emission lines become visible. In most cases an envelope is expanding; it then becomes a stellar wind (see below).

Stellar wind: outward material motion in the atmosphere or envelope with $v(r) > 0$, and normally $dv/dr > 0$.

Corona: an envelope in which $T \gg T_{\mathrm{phot}}$.

Shell: an envelope in which locally $d\varrho/dr > 0$. (Note that in normal envelopes $d\varrho/dr < 0$.)

B. TEMPERATURES

The notion temperature, symbol T, is in many cases just a thermodynamic *parameter*, not more. It is used for describing the emergent stellar *radiation flux*, the *velocity distribution* of the particles, the state of atomic *excitation and ionization*.

The *effective temperature*, T_e is the temperature of a black-body with the same radius as the star and the same integrated (over ν or λ) radiation flux.

The *brightness temperature* T_{br} is the temperature of a black-body with the same radius as the star that has in one wavelength interval the same radiation flux as the star.

The *radiation temperature* $T_r(\nu)$ at frequency ν is the T-value which, inserted in Planck's law $B(\nu, T)$, satisfies Kirchhoff's relation $\varepsilon_\nu / \kappa_\nu = B(\nu, T)$, where ε_ν and κ_ν are the emission- and absorption-coefficients respectively.

The *colour temperature* T_c is the temperature of a black-body with in a certain wavelength interval the same shape or gradient of the energy-wavelength- (or: energy-frequency-)curve as the star.

The *kinetic temperature* T_k is a parameter describing the velocity distribution of the atomic particles (not necessarily in the range $v = 0$ to $v = \infty$), through Maxwell's law.

The *electron temperature* T_{el} is the kinetic temperature of the electrons.

The *excitation temperature* T_{ex} is a parameter describing the relative population of two (or more) atomic or ionic levels, through Boltzmann's equation.

The *ionization temperature* T_i is a parameter describing the relative degree of ionization of two (or more) states of ionization of an element, through Saha's equation.

THE MAIN OBSERVATIONAL CHARACTERISTICS
OF THE MOST LUMINOUS STARS

2.1. Spectral and Luminosity Classification

While the Harvard classification (O, B, A, ..., M) classifies the stars according to (decreasing) photospheric temperature, the Yerkes (or MK: Morgan–Keenan) classification arranges the stars in classes I ... V according to decreasing intrinsic luminosity as well (Morgan and Keenan, 1973). Thus, a two-dimensional classification originates. Initially all supergiants were labeled with I, but it turned out fairly soon that this class consists of too wide a variety of absolute brightnesses, thus asking for a further subdivision. Furthermore, another class, that of the 'super-supergiants' (Feast and Thackeray, 1956) was added, initially consisting of the four reddest of the brightest stars in the Large Magellanic Cloud. These are of luminosity class 0 (zero) but are also called Ia–0, or Ia$^+$ (Keenan, 1971). The reason for not initially adopting the 'plus-notation' is because the plus sign is used among giant stars to denote 'less luminous than' (cf. Feast and Tackeray, 1956). However at present the plus-notation is frequently used. We shall use it throughout this book. As a rule we will use the name 'hypergiants' for super-supergiants (following Van Genderen, 1979).

So we have the luminosity classes

Ia$^+$ the most luminous supergiants (hypergiants)
Ia
Iab
Ib   least luminous supergiants
II   bright giants
...

Lists of standard stars for the MK-classification were published by Morgan and Keenan (1973, pp. 33 and 39ff). These lists are partly revisions of earlier published standards. Standards for O-type stars were given by Walborn (1973d) and Conti and Leep (1974). Morgan *et al.* (1978) published a 'Revised MK Spectral Atlas for Stars Earlier than the Sun'. An atlas of spectra of the coolest stars (type G and later) was published by Keenan and Mc Neil (1976).

Table IV gives the Yerkes classification of bright standard supergiant stars as defined in the fundamental work of Morgan and coworkers.

Since the MK-classification is an empirical and qualitative one, based on the aspect of the spectra it needs to be calibrated. Figure 7 gives the result of such a calibration for supergiants of spectral types A0 to G0, after Bouw and Parsons (1972). Note the

TABLE IV

Yerkes classification of bright standard supergiant stars (Morgan *et al.*, 1943; Keenan and Morgan, 1951; as revised by Morgan and Keenan, 1973). The revised types are marked with a dot. LC=luminosity classification.

	Star	Spectr.	LC		Star	Spectr.	LC	Star	Spectr.	LC		Star	Spectr.	LC
.	HD 46223	O4		.	ε Ori	B0	Ia	χ Aur	B5	Iab		ζ Ori	O9.5	Ib
.	HD 46150	O5		.	κ Ori	B0.5	Ia	σ Cyg	B9	Iab	.	ζ Per	B1	Ib
.	ζ Pup	O5	f		χ² Ori	B2	Ia	44 Cyg	F5	Iab		η Leo	A0	Ib
.	15 Mon	O7			o² CMa	B3	Ia	o¹ CMa	K3	Iab	.	α Lep	F0	Ib
.	τ CMa	O9	Ib		55 Cyg	B3	Ia	α Ori	M2	Iab	.	α Per	F5	Ib
					η CMa	B5	Ia					β Cam	G0	Ib
				.	β Ori	B8	Ia					ζ Cep	K1	Ib
				.	α Cyg	A2	Ia					119 Tau	M2	Ib
					φ Cas	F0	Ia							
					δ CMa	F8	Ia							
					μ Cep	M2	Ia							

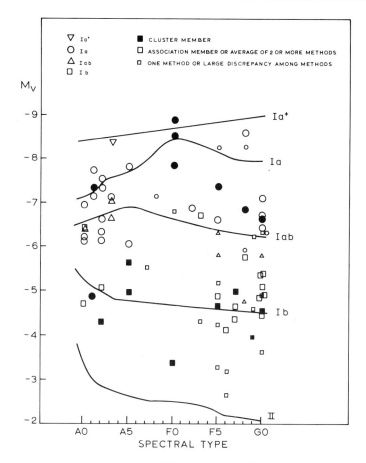

Fig. 7.   Absolute visual magnitudes of *A* and *F* supergiants; calibration of the Yerkes classification
After Bouw and Parsons (1972). Note the large dispersion of the points. The Ia⁺ class is in this
diagram defined by one point only!

very large scatter of the points, which is due to errors in the assumed distances of the stars, classification errors, and to 'cosmic dispersion' (the degree to which the relation between the spectral criteria and the stellar luminosities departs from a single-valued relation).

With increasing stellar temperature the luminosity classes gradually loose their meaning, particularly for spectral types earlier than approx. O5 since the lines connecting stars of the same luminosity classes in the Hertzsprung–Russell-diagram tend to converge for early O-type stars. For instance, the cluster Trumpler 14 contains both an O3 If$_*$ star* (HD 93 129 A) and an O3 V star (HD 93 129 B), with a magnitude difference of only 1.5 mag. (Walborn, 1973a)!

The Harvard classification system has a decimal subdivision (O3, O4 ... B0, B1 ... B9, A0 ... etc.) It appeared, however, suitable to introduce intermediate types O5.5, O6.5 ... O9.5, B0.5, and Walborn (1971b) even introduced the types O9.7, B0.2, and B0.7.

The earliest stars have spectral type O3.

The original Harvard classification was based on ratios of line strength estimated from visual inspections. A refinement would be to *measure* ratios of equivalent widths of lines of different excitation or ionization of the same element, and thus to (a) obtain a quantitatively calibrated spectral classification, which (b) is related to excitation or ionization temperatures. Thus Morgan *et al.* (1973), Conti and Alschuler (1971), and Conti and Frost (1977) used $\lambda\lambda 4471/4541$ He I/He II equivalent-width ratios (see Table 3 of Conti and Alschuler and our Figure 8 and Table V) for spectral classification of O-type stars.

For spectral classifications of stars between K5 and M0 one uses the ratio of Ca I $\lambda 4226$/Fe I $\lambda 4144$, $\lambda 4325$.

Beyond M2 the TiO bands are used. Bands of VO define the subclass M7. White and Wing (1978) developed a photoelectric two-dimensional classification system of M supergiants, based on observations in 8 narrow bands, essentially on those in a TiO (0.71 μm) and in CN bands (0.81 and 1.10 μm). See further Section 3.11.

From Chapter 1 it is clear that extremely bright stars are also stars that are at the fringe of instability. Atmospheric instability is shown by *variability* (Chapter 8) and by the occurrence of *emission lines* in the spectrum. As a rule emission lines are markers of extended atmospheres. Emission line stars are mostly brighter than non-emission stars of the same spectral type (Feast *et al.*, 1960). Conti and Leep (1974) have investigated the relation between luminosity and the occurrence of an extended atmosphere, as shown by the emission-line strength of Hα. Figure 9 gives the Hα/Hγ line ratio as a function of M_v for cluster members. For a comparison it is important that the predicted values for this ratio, based on the non-LTE plane-parallel models of Auer and Mihalas (1972) are close to unity, and but slightly dependent on spectral type (right hand ordinate of the diagram). The predicted value is 1.3 for models with small g-values. The diagram shows that for all stars fainter than $M_v = -6$ the observed

---

* See later for the f notation.

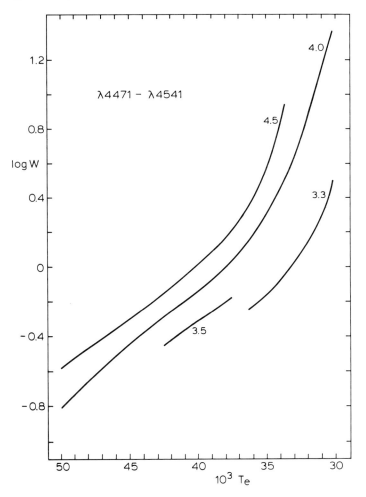

Fig. 8. Predicted $\log W$ for $\lambda 4471$ He I *minus* $\lambda 4541$ He II, for different $\log g$-values, from the non-LTE models of Auer and Mihalas (1972). (After Conti, 1973b).

TABLE V

Classification criteria of early O spectral types based on the $\lambda 4471/\lambda 4541$ ratio $\log W$ (after Conti and Frost, 1977); the $\log W$-values refer to difference $^{10}\log W(\lambda 4471) - {}^{10}\log W(\lambda 4541)$

$\log W$ limit		Spectral type		$\log W$ limit
-0.45	$\leq$	O5.5	$<$	-0.30
-0.60		O5	$<$	-0.45
		O4	$<$	-0.60
		O3		He I $\lambda 4471$ absent

ratio is close to the predicted value, but that for brigther stars the ratio becomes smaller *viz.* Hα tends to vanish, or Hα even occurs in emission (this observation should be related to the fact that Hα emission is characteristic for an envelope; see Section 3.2).

It is also interesting to see that for the brightest early-type stars there is very little difference in the luminosity. even for large differences in the Hα/Hγ ratio, indicating that for such stars the optical thickness of the envelope can take very different values for the same luminosity.

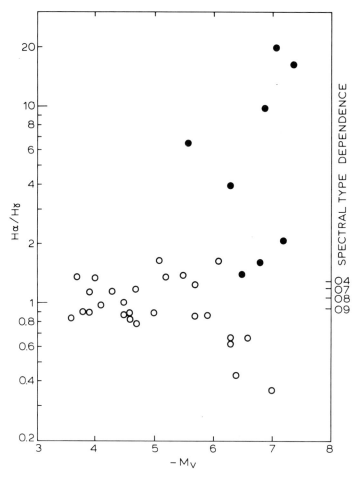

Fig. 9. Relation between M_v and the observed ratio of Hα/Hγ, a measure of the extent of the envelope. *Open circles*, Hα in absorption; *filled circles*, Hα in emission. The predicted ratio is near unity according to the non-LTE plane-parallel models. The exact spectral-type dependence of this ratio for main-sequence stars is shown along the right-hand ordinate; that for lower-gravity models is about 1.3. For stars with M_v fainter than −6, the line ratio is near that predicted; these lines are formed in plane-parallel geometry. For stars brighter than this, the Hα line weakens or goes into emission, implying a greater and greater extension of the envelope. There is a tendency for the envelope to be stronger with brighter M_v, but the scatter is such that the Hα strength cannot be used to predict M_v very accurately. (From Conti and Leep, 1974).

Stellar spectra with Balmer emission lines are normally denoted by the addition of the letter e to the spectral type. Mild effects are denoted by using parenthesis: (e). The letter p refers to stars whose spectrum cannot be understood with 'normal' or expanding envelopes. Further, Of stars are O stars with the N III lines near $\lambda 4640$ and $\lambda 4686$ He II in emission, while O(f) and O((f)) stars are those where the He II line is weak, or in absorption respectively. The ((f)), (f), f classification of O stars is a luminosity criterium (see Section 3.2).

2.2. Absolute Magnitudes and Luminosity Criteria

The royal way to determine absolute magnitudes of stars is – of course – through parallactic or other distance determinations. Thus Conti and Alschuler (1971) determined luminosities for a fifty O-type stars in clusters and associations. When absolute magnitudes have been determined for a sufficient number of stars, one may use these in setting up relations between certain spectral characteristics and absolute magnitudes and thus establish luminosity criteria.

We start with O- and B-type stars. While, as shows Figure 8, the difference between the He-equivalent widths may serve as a suitable criterium for $\log g$, hence for luminosity, the better luminosity criterium is the ratio Si IV/He I. In this connection we refer to the list of equivalent widths of the Si IV (1400 Å) resonance doublets as measured with the OAO instrument in 118 O- and B-type stars by Panek and Savage (1976).

For B- and A-stars Rosendhal (1973b) determined relations between the 'net' equivalent width W_{net} of Hα (i.e. the sum of absorption and emission expressed in the width of the adjacent continuum) and the absolute visual magnitude M_v for B0–B6, and for B8–A5 stars, as well as between the 'emission equivalent width' and M_v for these two groups of stars. In the same way Rosendhal (1974) found a linear relation between the mean equivalent width W(Si II) of the Si II lines at 6347 and 6371 Å and the absolute visual magnitude M_v for B9–A2 supergiants (Figure 10): $M_v = -6.56W$(Si II)-2.52, with considerable scatter (0.4 mag). The equivalent width varies from 0.38 Å at $M_v = -5$ to 0.84 Å at $M_v = -7.9$. Feast (1976) has noted that this relation – slightly extrapolated – is not satisfied for one of the brightest hypergiants in the Small Magellanic Cloud, i.e., SMC HD 7583 (A0Ia$^+$) (with $M_v = -9.1$ mag where W(Si II)$=0.40$ only). The same discrepancy, by a factor of about 2, applies to less bright supergiants in the SMC. This result may either mean that Rosendhal's relation is not satisfied for stars with $M_v < -8$ mag, or that silicon is underabundant by a factor of more than 2 in the Small Magellanic Cloud supergiants. There are other indications for metal-defiencies in SMC stars (see e.g. Hutchings et al., 1977). Luminosity criteria for F5 to G2 supergiants are the product E of the equivalent widths of O I $\lambda 7774$ and Fe I $\lambda 7748$, according to: $M_v = -10.98E - 1.79$ (Rao and Mallik, 1978). For cooler stars luminosity criteria are the ratios Sr II $\lambda 4077$/Fe I $\lambda 4063$ and others, and the intensities in the CN and CO bands (Keenan, 1963; Yamashita, 1966, 1967; Spinrad and Wing, 1969).

An interesting attempt to calibrate the absolute magnitudes of early type very

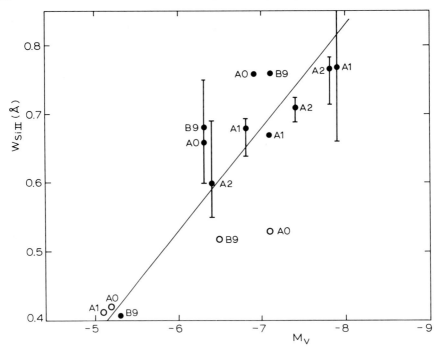

Fig. 10.   Relation between M_v and the silicon strength index for B9–A2 stars. For stars for which the silicon line strength may be variable the full range of observed variation is shown.
(After Rosendhal, 1974).

luminous stars was made by Burnichon (1976) who used multiple stars. Statistical laws relating colours and absolute magnitudes, stellar reddening .. etc. are better known for the dwarf stars than for giants or supergiants. By using such properties for the faint stars of a binary – and assuming these not to be evolved stars – the absolute magnitudes for the bright objects could be determined with a fair precision.

2.3. The Effective Temperature Scale and the Bolometric Correction

In a few cases only has the effective temperature been determined straightforwardly, i.e. from absolute photometry, and measured apparent radii.

    If the stellar radius and the distance from Earth are R and d respectively, and the observed flux at Earth's distance – corrected for interstellar extinction – f(erg cm^{-2} s^{-1}) then the flux πF and effective temperature T_e of the star's photosphere are found with

$$\sigma T_e{}^4 = \pi F = \frac{d^2}{R^2} f = 4f/\theta^2, \tag{2.3; 1a}$$

where $\theta = 2R/d$(rad) is the angular diameter of the star. If, in addition, the star's luminous stars was made by Burnichon (1976) who used multiple stars. Statistical distance d is known, the radius and luminosity are known, hence M_b. Similarly for 'mono'-chromatic fluxes:

$$\pi F_\lambda = 4 f_\lambda / \theta^2. \tag{2.3; 1b}$$

Thus resulted for some of the hottest stars remarkably low T_e-values: 32 500 K for the O4ef-star ζ Pup – see Section 3.2, and 30 200 K for the WC8 star γ^2 Vel.

However, for stars with very extended atmospheres with stellar winds that are optically thick in certain wavelength ranges the radii may take different values at different wavelengths, thus introducing too low a value for T_e. This may explain why in several stars photospheric temperatures derived from spectral lines give temperatures, higher than T_e deduced from L and R.

At the other end of the T-axis temperatures of 3900 ± 150 K and 3580 K were found for α Ori (Betelgeuse, M2 Iab), on the basis of measured fluxes and model atmosphere analyses (Tsuji, 1976b; Scargle and Strecker, 1979). This fairly high value for the effective temperature of a late type star may also be related to sphericity effects, as seems indicated by diameter measurements of the Mira stars o Cet (Mira) and R Leo by Labeyrie *et al.* (1977), who found that the stellar diameter in strong TiO absorption features is twice that in the 'continuous' spectrum (whatever that may be!), suggesting smaller 'continuous' radii, and a higher T_e-value.

Table VI lists a number of 'direct' determinations of T_e, based on measured values of fluxes and angular diameters. These results apply to medium- and late-type stars. The main imput to this table is the work of Code *et al.* (1976).

In most of the other cases the effective temperature was found along indirect methods. Several of such methods have been used:

(a) An elegant iterative method for the simultaneous determination of T_e and θ is due to Blackwell and Shallis (1977). The method assumes that f_λ is measured over a broad λ domain so that by integration over λ the value of f can be determined, if necessary after extrapolation over unobserved parts of the spectra. Assume an approximated value for T_e known. Model calculations then yield for a given wavelength λ the stellar flux πF_λ. With Equation (2.3; 1b) one thus finds a first approximation of θ which, inserted in Equation (2.3; 1a), gives a new value of T_e. If this value disagrees with the first approximation the iteration loop is repeated. The method converges rapidly, as was also shown by Tsuji (1979a) in an application to α Ori. It was also applied by Underhill (1978) – who used one iteration only and found as a check that calculated angular diameters agreed well with those of Hanbury Brown *et al.* (1974). This work was extended to 170 stars in a later paper (Underhill *et al.*, 1979). The results, presented in Figure 11, show the well-known fact that supergiant T_e-values are smaller than those of main-sequence stars of the same spectral type. The fact that they are also smaller than those derived from line spectra may suggest the influence of an extended atmosphere.

(b) A method, comparable to the one described in (a) is to fit the observed intensity-wavelength variation in the stellar continuum to *model predictions* (Morrison, 1975; Luck, 1977a). Thus Luck, following Van Paradijs (1973b) established a T_e-

TABLE VI

Relation between spectral type and T_e for luminous stars

(a) Determinations from measured fluxes at Earth's distance, and angular diameters

Star	Spectral type	T_e (K)	±	Ref.
γ^2 Vel	{ WC8+O9 I { WC8	{ 32510 { 30200	2520	Code *et al.* (1976) Willis and Wilson (1978)
ζ Pup	O4ef	32510	1930	Code *et al.* (1976)
ζ Ori	O9.5 Ib	29910	2110	,,
ε Ori	B0 Ia	24820	920	,,
κ Ori	B0.5 Ia	26390	1270	,,
β CMa	B1 II-III	25180	1130	,,
ε CMa	B2 II	20990	760	,,
η CMa	B5 Ia	13310	560	,,
β Ori	B8 Ia	11550	170	,,
α Car	F0 Ib-II	7460	460	,,
δ CMa	F8 Ia	6110	430	,,
α Ori	M1.5 Iab	3900	150	Tsuji (1976b)
19 Psc	C	3050	200	Scalo (1976)
χ Cnc	N	2700	100	,,

(b) T_e-determinations from infrared fluxes by Blackwell and Shallis (1977), and Scargle and Strecker (1979)

Star	spectral type	T_e (K)	±	Ref.
δ CMa	F8 Ia	5880	390	}
α Car	F0 Ib-II	7210	170	} B. and S.
β Ori	B8 Ia	11180	210	}
α UMa	K0 III	4730		
β Gem	K0 III	5360		
α Boo	K2 III	4060		
γ And	K3 II	4200		
α Tau	K5 III	3490		
β And	M0 III	3520		
α Cet	M1.5 III	3560		
α Sco	M1.5 III	3560		
α Ori	M1-M2 Ia-Ib	3580		
TU CVn	M2 III	3530		
μ Cep	M2 Ia	3600		
β Peg	M2.5 II-III	3530		
S Per	M4 Ia	2700		
α Her	M5 Ib-II	3400		
VY CMa	M5 Ia	2740		
g Her	M6 III	3450		
RS Cnc	M6 III	3350		
RX Boo	M8 III	3150		
Y CVn	C5.4	3440		
TX Psc	C6.2	3790		

spectral-type relation for Ib supergiants of medium spectral types (G, M). Scargle and Strecker (1979) applied it to late-type stars, by using 1 to 5 μm photometry. The weak side of this method is the unknown influence of interstellar extinction. Furthermore, for stars with very small g_{eff}-values, such methods may yield unreliable results since the models for such stars are unreliable.

(c) It is also possible to compare measured *equivalent widths* of lines with values calculated for model atmospheres to derive T_e-values. Thus, Conti (1973a, 1975) used the measured ratios of equivalent widths $\lambda 4471/\lambda 4542$ (He I/He II), coupled with predictions from model atmospheres to determine effective temperatures for early-type stars. His calibration curve is given in Figure 8, and the resulting T_e-values for hot stars (earlier than B0) in Figure 11.

(d) A group of determinations is based on the Zanstra-method, in which essentially fluxes in two greatly differing wavelength bands are compared, so that no effective but rather colour temperatures are determined. Morton (1969a) has derived a colour temperature scale for early-type stars imbedded in gaseous nebulae by applying the Zanstra-method, and using as basic data the nebular intensity of $H\alpha$, and the apparent magnitude of the stars. Zanstra's method is based on the assumption that all Lyman continuum photons from the star are eventually converted into Balmer photons of the nebula, line and continuum. The number of Lyman continuum photons is

$$N_L = \int_{\nu_L}^{\infty} \frac{\pi F_\nu \, d\nu}{h\nu} \ \text{cm}^{-2} \, \text{s}^{-1},$$

where ν_L is the frequency of the Lyman limit. Then, for a nebula at distance d to Earth, around a star with radius R, the $H\alpha$-flux to be expected at the Earth would be

$$f_a = \frac{4\pi R^2 N_L hc}{4\pi d^2 \lambda_a q_a(T)} \ 10^{-0.4A} \ \text{erg cm}^{-2} \, \text{s}^{-1},$$

with

λ_a = wavelength of $H\alpha$;

q_a^{-1} = fraction of Balmer photons emitted in $H\alpha$;

A_a = interstellair absorption at $H\alpha$.

Since Equation (2.3; 1b) gives essentially a relation between the ratio R^2/d^2 and the apparent magnitude in a colour band, one may combine the intensity flux in the stellar Lyman continuum as deduced from the observed nebular $H\alpha$ intensity, and the apparent stellar brightness, to derive the colour temperature. A further comparison with the energy-frequency distribution as derived from photospheric models, then yields information on the stellar effective temperature.

A variant of the Zanstra method is to use the He II Balmer α line at 1640 Å to estimate the flux in the He II continuum at $\lambda < 228$ Å. We will call this the He–Zanstra method.

A method, somewhat similar to the one described here was used by Pottasch *et al.* (1977) for the determination of effective temperatures of stars imbedded in nebulosities, among which O stars, nuclei of planetary nebulae and Wolf–Rayet stars. Basic data was the absolute UV photometry obtained by the Astronomical Netherlands Satellite. The measurements, where necessary corrected for the (estimated) influence of line emission, gave the continuous spectrum which appeared fairly well to fit to black-body curves. In one case where this could be tested this 'black-body temperature' agreed well with T_e. In addition Zanstra-temperatures were determined. These, mostly, agreed well with T_e found from continuum photometry. Absolute luminosities were found from a comparison with photometry in the visible and radii from the observed luminosity at 4500 Å, and the assumption that the wavelength distribution of the radiation can be described by that of a black-body at the Zanstra-temperature T_z. This *seems* a circle-argumentation, but since at 4500 Å R^2 depends only on T_z, T_e depends to the $\frac{1}{4}$ power on the Zanstra temperature, so that the method converges. The derived T_e-spectrum relation for O-type stars is compared with other data in Figure 11.

The observed radioflux of a nebula can be used in a similar way to obtain information on the colour temperature of the exciting star (Hjellming, 1968), because the

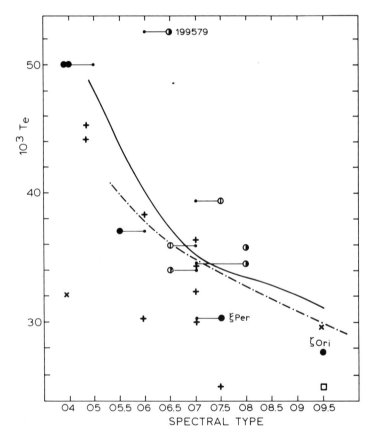

Fig. 11.   The temperature scale for early-type giants and supergiants. The circles and dots are from Conti (1973b). The luminosity types are shown by the symbols (*circle with vertical line*, type III; *half-filled circles*, IIIf; *filled circles*, I of If), connected with a horizontal line with the spectral type used by Morton in those cases where there is a difference. This is compared with Morton's (1969a) Zanstra method determination (*solid line*), data from Pottasch *et al.* (1977) for O stars embedded in nebulosity (*plus signs*), Code *et al.* (1976; *crosses*), and Underhill *et al.*'s value for a supergiant (1979, *square*). The *dash-dotted line* is the synthesis for supergiants by Flower (1977); see also our Table VII.

microwave emission is related to the ionization and excitation of the nebula by ultra-violet stellar radiation.

The large scatter of the data in Figure 11 is certainly depressing, if it does not indicate physical differences between stars of the same spectral subclasses. There is a tendency for the Zanstra temperatures, which are essentially colour temperatures, to be lower than other determinations. May this be a clue to explaining part of the differences? The fact that the greater part of the emission of very early-type stars is in the invisible extreme ultraviolet spectral range ($\lambda < 912$ Å, the Lyman jump) may play a role.

Figure 12 gives a summary of what we consider to be the best T_e-spectral-type relation for stars of medium and late spectral type; the figure is compiled from the

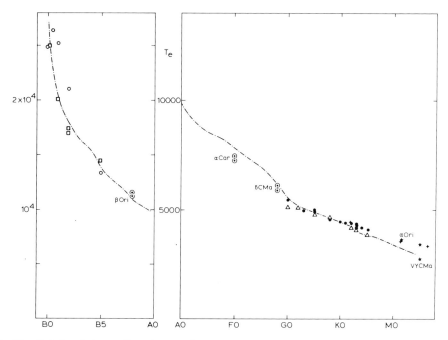

Fig. 12.   Relation between T_e and spectral type.
$\bigcirc$ direct measurements, Table VI.
$\triangle$ results from Van Paradijs (1973b).
$\bullet$ data from Luck (1977a).
$\square$ hot supergiants (Underhill *et al.*, 1979).
*   cool supergiants
$+$ cool giants (averaged)$\}$Scargle and Strecker (1979).
.—.—   synthesis by Flower (1977) for supergiants; Table VII.

data of Table VI, those of Luck, Van Paradijs, and Scargle and Strecker; we also give Flowers's (1977) synthesis – Table VII.

For very late-type giants and supergiants, mainly of the spectral types M, R, N, S, C, MS, Scalo (1977) has established a relation between the colour-indices, mainly the red and infrared ones of the stars, and the socalled 'black-body colour temper-

TABLE VII

Temperature classification and bolometric correction for supergiants and giants. After Flower (1977).

Supergiants					Giants				
B-V	MK	T_e	$\log T_e$	BC	B-V	MK	T_e	$\log T_e$	BC
−0.33	O5.5	40495	4.607	−3.73	−0.33	O6	39280	4.594	−3.50
−0.31		37135	4.570	−3.46	−0.31	O7.5	35410	4.549	−3.25
−0.29	O7.5	34060	4.532	−3.20	−0.30	O8.5	33620	4.527	−3.12
−0.28	O9	32615	4.513	−3.06	−0.29	O9.5	31920	4.504	−2.99
−0.26		29910	4.476	−2.79	−0.28		30310	4.482	−2.87
−0.25	B0	28640	4.457	−2.68	−0.26		27320	4.437	−2.60
−0.23		26265	4.419	−2.40	−0.25		25940	4.414	−2.48
−0.20		23065	4.363	−2.02	−0.23		23385	4.369	−2.22
−0.18		21155	4.325	−1.76	−0.21	B1	21080	4.324	−1.97
−0.17	B1	20260	4.307	−1.67	−0.20		20015	4.301	−1.84
−0.16		19400	4.288	−1.56	−0.18		18045	4.256	−1.58
−0.12	B3	16315	4.213	−1.16	−0.17	B3	17135	4.234	−1.46
−0.11	B4	15625	4.194	−1.05	−0.16		16265	4.211	−1.34
−0.08	B5	13720	4.137	−0.82	−0.12	B7	13220	4.121	−1.07
−0.05	B7	12050	4.081	−0.64	−0.11	B8	12550	4.099	−0.72
−0.025	B8	10915	4.038	−0.51	−0.08	B9	11400	4.057	−0.46
0.00	B9	10255	4.011	−0.38	−0.05	B9.5	10545	4.023	−0.32
0.05	A2	9120	3.960	−0.17	−0.025	A0	10000	4.000	−0.24
0.07	A4	8790	3.944	−0.10	0.00	A1	9570	3.981	−0.19
0.10	A5	8510	3.930	0.00	0.05	A2	8995	3.954	−0.09
0.14	A8	8205	3.914	0.09	0.07		8790	3.944	−0.06
0.20	F0	7800	3.892	0.14	0.10	A3	8510	3.930	−0.03
0.33	F5	7000	3.845	0.13	0.14	A5	8205	3.914	−0.20
0.50	F8	6210	3.793	0.08	0.20		7800	3.892	0.00
0.60		5835	3.766	0.04	0.33	F1	7000	3.845	0.01
0.70		5525	3.743	−0.01	0.43	F5	6500	3.813	−0.01
0.88	G2	5070	3.705	−0.12	0.50		6210	3.793	−0.03
0.92	G3	4990	3.698	−0.14	0.60		5875	3.769	−0.06
1.04	G6	4755	3.677	−0.22	0.70	G1	5585	3.747	−0.10
1.16	G9	4530	3.656	−0.35	0.80		5300	3.724	−0.16
1.30	K2	4280	3.632	−0.46	0.88	G5	5085	3.706	−0.22
1.45	K3	4030	3.605	−0.67	0.92	G6	4990	3.698	−0.26
1.54	K4	3875	3.588	−0.84	1.04		4720	3.674	−0.37
1.61	K7	3750	3.574	−1.00	1.16	K2	4465	3.650	−0.49
1.70		3500	3.544	−1.43	1.30		4185	3.622	−0.66
1.72		3250	3.512	−1.72	1.45	K4	3905	3.592	−0.92
1.76	M4	3000	3.477	−2.50	1.54	K8	3750	3.574	−1.19
1.80		2800	3.447	−3.36	1.61		3500	3.544	−1.66

atures', which is a notion with some degree of superfluity: the colour temperature is normally defined as the temperature parameter of the black-body curve that fits best the observations of the continuous radiation in a given spectral range. Thus, Scalo published a relation between the infrared colour index R-I, and the colour-temperature, T_{R-I} for this wavelength range. It may still be too large a step from this relation to that between T_e and spectral type. In any case, for the few late-type stars for which T_e could be found from 'direct' T_e-determinations (Scalo, 1977; Scargle and Strecker, 1979) the difference between T_{R-I} and T_e is still fairly large. The T_e-values for late-type stars in Table VI are the few 'direct' determinations available. The values given are the average between those for the T_e-values derived for fully darkened disks and uniform ones.

A determination of T_e-values and of the bolometric corrections for M-type supergiants by Lee (1970) was based on observed spectral energy distributions from 0.36 to 3.4 μm, with estimated flux-values for longer wavelengths. Flower (1975) published a relation for M-type giants. T_e-values for a few N-type stars were determined from measured angular diameters and fall in the range 2500–3000 K (Walker *et al.*, 1979) – cf. also Section 3.11.

Effective temperatures and *bolometric corrections*, $BC = M_V - M_B$, are closely related. Flower (1977) established a synthesizing relation between B-V, the spectral type, the bolometric correction and the effective temperature, based on temperatures, colours, and bolometric corrections of Code *et al.* (1974, 1976) for stars hotter than the Sun, of Lee (1970) and Flower (1975) for M-type supergiants and giants, and a temperature-colour relation of Van Paradijs (1973b) for G and K supergiants. His relation for giants and supergiants is copied here as Table VII. For the hottest stars the values of BC are practically only known with some certainty for the main-sequence.

2.4. Masses of Massive Stars

The truly reliable way to find stellar masses is the direct determination via the observed orbital motions of the components of a binary. Indirect methods are based on stellar model computations, in actual practice on a comparison of calculated evolutionary tracks for stars of various masses with the observed positions in the Hertzsprung–Russell diagram, or on the determination of photospheric parameters: a determination of g and R would make it possible to determine the stellar mass $\mathfrak{M}$.

We first summarize the results from direct methods which are thereafter compared with results from indirect methods. For those stars for which the bolometric absolute magnitude is known (Section 2.3) this information then enables one to determine a *mass-luminosity relation*. In this section we restrict ourselves to a discussion of masses of massive early-type stars, *on* or *close to* the main sequence. The most recent values are summarized in Table VIII. The stars given there may be considered as not or slightly evolved (the latter applies to the Of stars) and are therefore best suited to establish the mass-luminosity relation. Later in this book we also discuss masses of various types of *evolved* stars. For these we refer to:

- Table XIV (Section 3.4) for Wolf–Rayet stars;
- Table XXI (Section 3.10) for Zeta Aurigae stars;
- Table XXXV (Section 4.11) for massive X-ray binaries;
- Table XLIV (Section 9.3) for post-novae.

TABLE VIII

Masses of early-type, not or slightly evolved massive stars ($\mathfrak{M} \gtrsim 15\mathfrak{M}_\odot$) determined from binary motions

Designation	Name	Spectral types	$\mathfrak{M}\sin^3 i/\mathfrak{M}_\odot$	M_b	References
HD 93205 A		O3 V, O8 V	>39		Conti and Walborn (1976), Conti and Frost (1977)
HD 93403 A		O5.5((f))	$\gtrsim 5.2/\sin^3 i$		Thackeray and Emerson (1969), Conti and Frost (1977)
HD 206267		O6, O9	18.7, 6.5		Crampton and Redman (1975)
BD+40°4220	V 729 Cyg	O7f Ia, O6f Ia	47±7, 11±5	−11.4	Bohannan and Conti (1976)
			59±9, 14±6	−7.9	Leung and Schneider (1978b)
HD 57060	UW CMa (29 CMa)	O7f–O8.5 If, O7	24, 29	−8.7	Stothers (1972a), review
			19.3, 23.2	−6.6	McCluskey and Kondo (1972), Parthasarathy (1978), Hutchings (1977)
HDE 228854	V 382 Cyg	O7, O7	23, 30	−9.3	Leung and Schneider (1979), Stothers (1972a), review
			46, 34		
HD 152248		O7f, O8	37, 33	−10	Hill et al. (1974), De Loore (1980), review
			24.4, 22.5		
HD 166734		O7 If, O9 I	29, 31		Conti and Massey (1980)
HD 159176		O7, O7	38, 38	−8.8	Conti et al. (1975), Lloyd Evans (1979), De Loore (1980), review
HDE 228766		O7.5, O5.5f	34±10, 23±5	−8.8	Massey and Conti (1977)
HD 47129	Plaskett's star	O8 V, O8(f)	~51, ~64	−8.5	Stothers (1972a), review, Hutchings and Cowley (1976)
			58–100, 64–90		Batten (1976)
HD 19820		O8, O8	18.9, 9.2		Stothers (1973a), review
HD 1337	AO Cas	O9 III, O9 III	19, 26	−8.3	Hutchings and Hill (1971)
			18, 23	−7.3	Leung and Schneider (1978a; 1979)
			25, 29		

HD	Name	Sp. (1)	Sp. (2)					Reference
HD 36486	δ Ori A	O9.5 II	B	28	10	−9.1	−5.5	Stothers (1972a), review; De Loore (1980), review
HD 198846	Y Cyg	O9.5 IV	O9.5 IV	18	18	−7.6		Hutchings (1976f), review
HD 25638	SZ Cam	B0		21	6		−4.3	McCluskey and Kondo (1972)
HD 193611	V 478 Cyg	B0 V	B0 V	15	15	−6.6	−6.4	Stothers (1972a)
HD 35652	IU Aur	B0p	B1p	17.4	11.8	−6.1	−6.3	McCluskey and Kondo (1972); Mammano et al. (1978)
HD 93206	QZ Car	B0 Ib	O9 V	16.7	28	−6.6	−6.4	Eaton (1978)
HD 216014	AH Cep	B0.5 IV	B0.5 IV	16	14	−7.3	−6.9	Leung et al. (1979); Hutchings (1976f)
HD 163181	V 453 Sco	B0.5 Ia	B0 V ≥	14.3 ± 0.6; 13	24.9 ± 0.6; 22	−9.9	−6.5	McCluskey and Kondo (1972); Woodward and Koch (1975)
HD 42933	δ Pic	B0.5 III	B0.5–3	16	7			Hutchings (1975a); Stothers (1972a), review
HDE 227696	V 453 Cyg	B0.5 IV	B0.5 IV	16.8 ± 0.9	12.9 ± 0.7	−6.4	−5.9	Lacy (1978)
HD 218066	CW Cep	B0.5 V	B0.5 V	11.9 ± 0.1	11.2 ± 0.1			Lacy (1978)
HD 65818	V Pup	B1 V	B1.5	17.4	9.6	−5.1	−3.8	Stothers (1973a); Eaton (1978); Schneider et al. (1979)
HD 10516	φ Per	B1 III-V pe	O9.5 V	29 ± 14	3.8 ± 1.7			Poeckert (1979)
HD 190967	V 448 Cyg	B IIb-II		18	22	−6.8	−7.6	Stothers (1973a)
HD 151890	μ¹ Sco	B1.5 V	B3	12.8	8.4	−4.5	−3.2	Stothers (1973a); McCluskey and Kondo (1972)
HD 34333	EO Aur	B3 III	B3 III	20	20	−5.8	−5.3	Schneider et al. (1979); Stothers (1972a) review
HD 33357	SX Aur	B3.5	B6	7.8	4.7	−3.9	−2.2	De Loore (1980), review; Chambliss and Leung (1979)

The first modern systematic survey of masses and luminosities of visual binaries is due to Harris *et al.* (1963). It refers mainly to stars of medium masses. In later investigations masses of more massive stars have been communicated. Stothers (1972a, 1973a) has compiled masses of early-type massive stars as derived from a fifty well studied eclipsing binaries, among which are ten supergiants.

McCluskey and Kondo (1972) summarized masses and luminosities of fourty visual and eclipsing binaries, as well as of some 100 for which only total masses of the system can be derived. Hutchings (1976f) reviewed masses for massive double-lined spectroscopic binaries. Masses of un-evolved and evolved massive binaries were summarized by Vanbeveren and De Loore (1980). The most massive couples known are (a) HD 47 129, Plaskett's star, for which the component masses were determined by Hutchings and Cowley (1976) at $58 < \mathfrak{M}_1/\mathfrak{M}_\odot < 100$ and $64 < \mathfrak{M}_2/\mathfrak{M}_\odot < 90$, and (b) V729 Cyg ($=\mathrm{BD}+40°4220$) a contact binary for which Leung and Schneider (1978) found $\mathfrak{M}_1/\mathfrak{M}_\odot = 58.7 \pm 9.1$, and $\mathfrak{M}_2/\mathfrak{M}_\odot = 13.7 \pm 6.3$. Earlier, Bohannan and Conti (1977) found smaller values. For VV Cep the mass of the primary was initially given as 80 to 84$\mathfrak{M}_\odot$ (Perry, 1966; Batten, 1968) but these values were too large (Hutchings and Wright, 1971); see Table XXI. For the M2 ep Iab star Boss 1985 a mass is derived $< 34\mathfrak{M}_\odot$ (Cowley, 1969).

In Table VIII we collect the best values from these sets as well as from a number of other recent determinations, thereby restricting ourselves to fairly massive stars with $\mathfrak{M} \gtrsim 15\mathfrak{M}_\odot$.

For those binary members for which bolometric luminosities are known the data permit one to establish a mass-luminosity diagram. Figure 13 is based on the work of Harris *et al.* (1963) with the above mentioned modern additions for the massive stars (Table VIII) and from results given in Table IX.

McCluskey and Kondo (1972) derived from their compilation for stars with $-8 \le M_b \le +10.5$ the empirical relation

$$\log(\mathfrak{M}/\mathfrak{M}_\odot) = 0.504 - 0.103 M_b,$$

from which we deduce, with Equation (1.2; 1):

$$\log(\mathfrak{M}/\mathfrak{M}_\odot) = 0.014 + 0.258 \log(L/L_\odot). \qquad (2.4; 1)$$

From Figure 13 we would derive

$$M_b = 1.6 - 7.5 \log(\mathfrak{M}/\mathfrak{M}_\odot) \quad \text{for} \quad -7 > M_b > -12,$$

hence with Equation (1.2; 1):

$$\log(L/L_\odot) = 1.26 + 3.0 \log(\mathfrak{M}/\mathfrak{M}_\odot) \text{ for } 4.7 < \log(L/L_\odot) < 6.7. \quad (2.4; 2)$$

For the most massive stars we would find a coefficient of 2.7 (cf. Figure 13).

It should next be possible to compare the results from 'direct' determination, particularly Table VIII and Figure 13, with theoretical computations, and with results from the 'indirect' determinations.

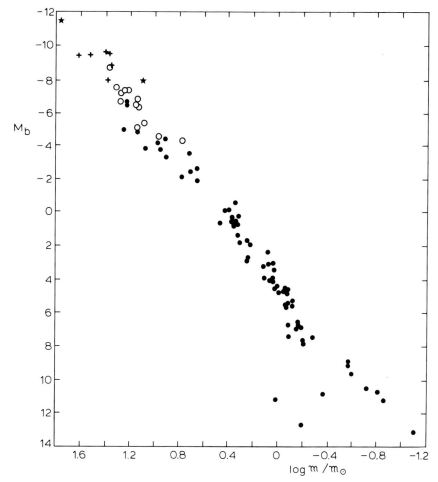

Fig. 13. Mass-luminosity law, based for the fairly massive stars on Harris *et al.* (1963, *dots*), with for the large masses: binary data from Table VIII (*open circles*); photospheric data from De Loore *et al.* (1977; *crosses*, see Table IX); *asterisks* denote the components of the massive contact binary V729 Cyg (=BD+40°4220).

To that end we compare in Figure 14 the mass-spectral-type relation as derived from the available sources with theoretically computed relations, derived assuming no rotation and chemical abundances defined by $(X; Z)=(0.639; 0.021)$ (*solid lines*) or $(0.739; 0.044)$ (*dashed lines.*) The lower and upper lines refer to the Zero Age Main Sequence (ZAMS) and the Temperature Minimum Main Sequence (TAMS), respectively; the latter curve being defined as the locus of the points where, during the evolution, the stellar effective temperatures reach a minimum.

There is a fair agreement between computations and theoretical predictions: there are no stars to the left-hand side of the lower curve; stars occurring in the upper right part of the diagram are apparently evolved objects. In view, however, of the

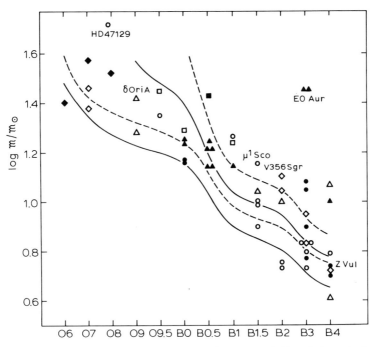

Fig. 14. (Mass, spectral type)-diagram for observed stars (*symbols*) and theoretical stellar models (*lines*). Empirical masses have been derived from eclipsing binary orbits; the most reliable masses are represented by filled symbols. The shape of a symbol refers to the stellar luminosity class: *squares*, I–II; *triangles*, III–IV; *circles*, V; *diamonds*, unknown luminosity class. The theoretical curves are based on models with no rotation and $(X, Z) = (0.739, 0.021)$ (*solid lines*) or (0.739, 0.044) (*dashed lines*). The lower and upper lines of each type refer to the ZAMS and TAMS, respectively. From Stothers (1972a).

nearly linear relation between $\log \mathfrak{M}$ and spectral type the diagram does not allow conclusions related to possible shifts of the points in a diagonal (upper left-lower right) direction. For that purpose one has to compare observed positions of individual stars in the Hertzsprung–Russell diagram with their theoretical evolutionary tracks. Along these lines Conti and Burnichon – see Figure 15 – from a comparison with Stothers's (1972a) calculations, without mass-loss, concluded that the most massive stars should have masses even exceeding $120 \mathfrak{M}_\odot$!

However these large mass-values are not confirmed observationally. For instance, Lamers and Morton (1976) compared the mass, thus derived for the binary V729 Cyg=BD+40°4220 (O7f+O6f), being $\sim 80 \mathfrak{M}_\odot$, with the values found from the binary orbit, being $\mathfrak{M}(O7) = 50 \pm 10 \mathfrak{M}_\odot$ and $\mathfrak{M}(O6f) = 13 \pm 5 \mathfrak{M}_\odot$; see Table VIII. This single comparison already suggests that the masses derived from model computations could be overestimated by a factor close to 2. The most probable explanation of this difference is immediately clear from a glance at Figure 1: mass-loss tends to reduce the stellar mass of massive stars to about $\frac{2}{3}$ or $\frac{1}{2}$ the initial value during the first phase of stellar evolution. The example given above does indeed refer to (slightly) evolved stars. An alternative, but less probable, explanation that should not

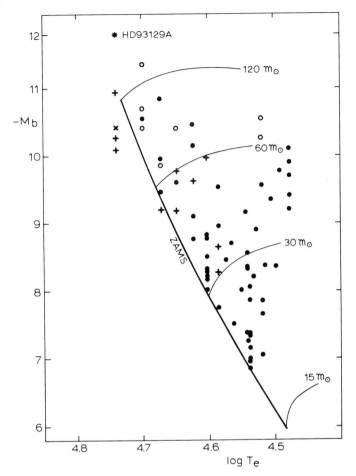

Fig. 15. $\text{Log} T_e$-M_b plot of observed O stars super-imposed on the theoretical evolutionary tracks during core-hydrogen burning phase of the $15\mathfrak{M}_\odot$, $30\mathfrak{M}_\odot$, $60\mathfrak{M}_\odot$, and $120\mathfrak{M}_\odot$ models by Stothers (1972a) with composition ($X=0.75$; $Z=0.02$). *Open circles:* Of stars. *Filled circles:* all other O stars $=$O, O((f)), O(f), Oe... *Crosses:* stars in Tr 14. *Plus signs:* stars in Tr 16/Cr 228. (From Conti and Burnichon, 1975).

be forgotten yet is that there might be an error in the model calculations. If we accept Stother's (1976) models calculated on the basis of Carson's (1976) opacities then one has to admit lower T_e-values for the upper main sequence, hence smaller BC-values, lower luminosities and smaller masses.

Finally, we describe a few stellar mass-determinations based on *photospheric studies*. For the three O–B supergiants ζ Ori, ε Ori and η CMa De Loore *et al.* (1977) determined the radii from measured angular diameters and distances; the g-value was found from spectral investigations. If one is certain that the spectroscopically determined $g_{\text{eff}}=g_{\text{grav}}$, and that R is not wavelength-dependent. then $\mathfrak{M}$ can be derived (Table IX) with the relation (Osborn, 1975; see for the limitations: Vardya, 1976):

$$\log(\mathfrak{M}/\mathfrak{M}_\odot)=\log g_{\text{grav}}-4 \log T_e-0.4M_b+12.49.$$

TABLE IX

Masses and luminosities for selected supergiants (part(a)) after De Loore et al., 1977)

(a)

Star	Type	M_v^1	V_0^2	BC^2	M_b	θ_{ld}^2 (10^{-3} arc sec)	d (pc)	$R_*/R_\odot$	$\log g$ (cm s^{-2})	$\log \mathfrak{M}_*/\mathfrak{M}_\odot$
ζ Ori	O9.5 Ib	−6.4	1.95	−2.9±0.3	−9.3±0.3	0.48±0.04	468	24±3	3.2±0.2[3]	1.52±0.21
ε Ori	B0 Ia	−6.8	1.50	−2.5±0.1	−9.3±0.3	0.69±0.04	457	34±4	3.0±0.1[4]	1.62±0.12
η CMa	B5 Ia	−7.0	2.44	−0.9±0.1	−7.9±0.1	0.75±0.06	772	62±8	2.2±0.2[5]	1.35±0.21

(b)

Star	Type	T_e (K)	M_b	$\log g$ (cm s^{-2})	$\log \mathfrak{M}_*/\mathfrak{M}_\odot$
o² CMa	B3 Ia	20000±2000	−8.9±0.3	2.5 ±0.2	1.35±0.29[6]
HD 7583	A0 Ia⁺	8960±300	−9.5±0.3	0.87±0.10	1.37±0.19[7]
HD 33579	A3 Ia⁺	8130±300	−9.6±0.3	0.7 ±0.2	1.40±0.28[8]

[1] Snow and Morton (1976).
[2] Code et al. (1976).
[3] Auer and Mihalas (1972).
[4] Lamers (1975).
[5] Underhill and Fahey (1973).
[6] Van Helden (1972).
[7] Wolf (1973).
[8] Wolf (1974).

For stars of which the angular diameter is not known, but with known distances and M_v-values (o² CMa, HD 7583, and HD 33579 in Table IX), T_e, hence BC, and g_{eff} are determined from the spectra; this yields R with Equation (1.2; 2) and with g one determines $\mathfrak{M}$. Early data from atmospheric studies were summarized by Harris *et al.* (1963); they found for ζ Per (B1 Ib): $37\mathfrak{M}_\odot$; α Cyg (A2 Ia): $14\mathfrak{M}_\odot$; HD 33579 (A5 Ia⁺): $45\mathfrak{M}_\odot$; and δ CMa (F8 Ia): $34\mathfrak{M}_\odot$.

Also for these cases a plot of the data in a theoretical Hertzsprung–Russell diagram would show the derived masses to be smaller than the theoretical ones by factors up to two. One may, therefore, assume this factor in the interpretation of evolutionary tracks for massive stars calculated without mass-loss.

2.5. Radii of Luminous Stars

Particularly the fundamental work by Hanbury Brown and Twiss (review by Hanbury Brown *et al.*, 1974), has opened the way to interferometric determination of the angular diameter θ. Another method is to use stellar absolute fluxes or T_e-values, because then the value of θ can be derived with Equation (2.3; 1). Additional direct determinations of θ are obtained by making use of stellar occultations (summarized by Barnes and Evans, 1976). If the stellar distance is known, this enables one the 'direct' determination of radii of single stars (no binaries).

There are also more indirect methods to determine R; these make use of the M_b and T_e-values through Equation (1.2; 2), or through

$$\log R/R_\odot = 0.2M_b - 2\log T_e + 8.47. \tag{2.5; 1}$$

By using the relation between radius, absolute and apparent magnitude and the stellar angular diameter one obtains a related formula, derived by Wesselink (1969) and somewhat modified by Barnes and Evans (1976):

$$\log T_e + 0.1C = 4.221 - 0.1V_0 - 0.5\log\theta, \tag{2.5; 2}$$

where C is the bolometric correction; V_0 the unreddened magnitude in the *UBV* system; and θ the stellar angular diameter in milli-arc sec.

The right-hand side of Equation (2.5; 2) is often denoted by F_V; the significance of this definition is that F_V is linearly related to the ^{10}log of the visual surface brightness. Barnes and Evans (1976) and Barnes *et al.* (1976, 1978) give graphs showing the relation between F_V and various kinds of colour-indices. The relation between F_V and the colour index $(V-R)_0$ appears not to depend on luminosity class for all spectral types between O4 and M8; and for S- and C-stars. Photometry in the *VR*-system would hence yield F_V, with

$$F_V = 3.84 - 0.32\,(V-R)_0 \quad \text{for} \quad (V-R)_0 \le 0.8 \text{ mag.}$$

A difficulty in the application of this method is that T_e is often only poorly determined. However Blackwell and Shallis (1977) made use of the fact that the *infrared* flux is fairly insensitive to the effective temperature and used measurements of the infrared fluxes (range: 2 to 13 μm) to simultaneously determine, in successive approximations, the angular diameter θ with the method described in Section 2.3, under (a). It seems that this method allows accuracies of 5% in the angular diameter, in the best conditions.

Underhill's (1978) determinations of T_e, described in Section 2.3, is very similar, and yields results that agree well with those of Hanbury Brown except for ζ Pup (HR 3165) and α Eri, for which smaller values are found. These discrepancies are certainly related to the ambiguity in defining T_e for stars with extended atmospheres.

TABLE X

Apparent angular diameter, θ, expressed in milli-arc sec for supergiants, as found from absolute fluxes (F), lunar occultations (O), interferometrically (I), or by speckle interferometry (S). From Hanbury Brown *et al.* (1974), Barnes and Evans (1976), Barnes *et al.* (1976), Lynds *et al.* (1976), Blackwell and Shallis (1977), Labeyrie *et al.* (1977), Balega and Tikhonov (1977), Underhill (1978), Tsuji (1978a), Ridgway *et al.* (1977b, 1979), Scargle and Strecker (1979), Walker *et al.* (1979).

Star	Spectral type	Uniform disc		Complete darkening		F_V	Method	Remarks
		θ	$\pm$	θ	$\pm$			
γ^2 Vel	WC8+O9 I	0.43	0.05	0.44	0.05		I	multiple star
ζ Pup	O4ef	0.41	0.03	0.42	0.03	4.20	I	
				0.33	0.02		F	
ζ Ori	O9.5 Ib	0.47	0.04	0.48	0.04			
ε Ori	B0 Ia	0.67	0.04	0.69	0.04	4.03	I	
κ Ori	B0.5 Ia	0.44	0.03	0.45	0.03	4.20	I	
η CMa	B5 Ia	0.72	0.06	0.75	0.06	4.04	I	
β Ori	B8 Ia	2.43	0.05	2.55	0.05	4.00	I	
				2.67	0.08		F	
α Car	F0 Ib-II	6.1	0.7	6.6	0.8	3.89	I	
				7.08	0.19		F	
δ CMa	F8 Ia	3.29	0.46	3.60	0.50	3.80	I	
				3.89	0.5		F	
α Sco	M1.5 Iab			41	2	3.32	O	
				42	2		I	
		42					F	
α Ori	M1.5 Iab			54	3	3.31	I	
		49	1	74	1		S	
		53	4				S	
		37	3				F	
		47					F	
α Her	M5 Ib-II	40					F	
U Ari	M4-6e	6.11	0.34				O	variable
S Psc	M5-7e	3.84	0.37				O	variable
R Leo	M6.5e-M9e			76	5		O	
		30	7				S	⎫
		54	9				S	⎬ phase-dependent
		40-56					F	⎭
V Cnc	S3-9e			2.8	0.8		O	
o Cet	M6e	34	7				S	⎫
		103	20				S	⎬ phase-dependent
		60					F	⎭
VV Sgr	M8e	5.21	0.42				O	variable
AQ Sgr	C7 (N3)	5.48	0.45				O	

Table X summarizes apparent diameters of supergiants obtained by the various methods described here. The uncertainty in the stellar limb-darkening makes the interpretation of any observation of angular diameters somewhat uncertain. Therefore one either assumes complete limb-darkening, i.e. assuming the surface brightness $I(\vartheta)$ to vary as $I(0)\cos\vartheta$, or one assumes a uniform disk. The differences in the found θ-values are usually small (see the Table).

Another uncertainty, already mentioned in Section 2.3, and particularly applying to the coolest and brightest stars is related to the existence of extended atmospheres, and hence to sphericity of the atmosphere, causing wavelength dependence of the stellar radius. The 'diameters' determined *in* the molecular absorption bands can be much larger (factors 2 are indicated!) than those measured in the continuous spectrum (Labeyrie *et al.*, 1977). In addition, the existence of scattering dust halos around such stars may simulate a larger diameter than just the photospheric one, even when the radial optical depth of such halos is only small. Tsuji (1978a) thus explained the apparent discrepancy between the angular diameter of α Ori determined interferometrically and from stellar fluxes (see the relevant data in Table X). This also explains why the largest discrepancies between I, S, and F diameters occur for short wavelengths – for increasing λ these groups of values seem to converge (see also the discussion of α Ori in Section 3.11). Vardya (1977) has drawn attention to the fact that the colour index of stars is approximately proportional to five times the logarithm of the ratio of radii at the two wavelength bands.

With θ and the trigonometric parallaxes known one can find the radii; to give an example, one finds for α Sco and α Ori radii of $700R_{\odot}$ and $600R_{\odot}$ respectively.

Particularly powerful becomes the method when photometric and spectral observations of eclipsing binaries are combined with stellar distances known from other sources, and with the surface brightness relation of Barnes and Evans. Thus, Lacy (1978) determined very accurate parameters for the three eclipsing binaries CW Cep, V453 Cyg, and AG Per (see also Table XI).

Another suitable way to determine stellar radii is by making use of photometric (eclipsing) binaries. If the spectra of both components are moreover observable the method allows one to compare the spectral-type-radius relation with predictions on the basis of an evolutionary scenario.

This is notably important for those binaries in which mass exchange between components plays a part. A comparison of the radius with the calculated size of the Roche lobe (Section 4.10; 1) is then an important means to find whether mass exchange is likely to occur in the binary, or not.

Table XI lists a few modern determinations of radii of components of photometric binaries. This list is certainly not exhaustive. For the close (near-contact) binary 29 CMa the radius of the primary is close to the theoretical Roche radius (Hutchings, 1977). The same remark applies to the components of V729 Cyg.

TABLE XI

Some determinations of radii of important eclipsing binaries

Binary	Spectral type		$R/R_\odot$		Reference
V 729 Cyg*	O7f Ia	O6f Ia	30±2	28±2	Bohannan and Conti (1976)
(=BD+40°4220)					Massey and Conti (1977)
			33±9	17±4	Leung and Schneider (1978b)
29 CMa	O8 Ia	<O7	12.3±1.2		Hutchings (1975a)
AI Aur	B0p	B1p	5.0	8.5	Mammano et al. (1978)
V 453 Sco	B0.5 Ia	B	26±3	9.0±0.7	Woodward and Koch (1975)
			23	≥9	Hutchings (1975a)
CW Cep	B0.5 V	B0.5 V	5.40±0.10	4.95±0.10	Lacy (1978)
V 453 Cyg	B0.5 IV	B0.5 V	9.33±0.30	5.74±0.18	Lacy (1978)
V Pup	B1 V	B1.5	6.1 ±0.5	5.2 ±0.6	Schneider et al. (1979)
μ^1 Sco	B1.5 V	B3	5.3 ±0.2	5.0 ±0.2	Schneider et al. (1979)
AG Per	B4 V	B5 V	2.87±0.10	2.60±0.10	Lacy (1978)

* Contact system; cf. Table VIII.

2.6. g_{eff}-Values in Extreme Objects

Knowledge of masses and radii of stars allows one to determine the acceleration of gravity in a stellar atmosphere. For unstable stars the effective acceleration g_{eff} may differ considerably from that value.

Emission line stars are apparently objects at the fringe of instability. This was very clearly shown by Th. and J. Walraven (1971) who compared multicolour observations of supergiant stars in the Large Magellanic Cloud (LMC) with theoretical computations based on stellar atmospheric models by Mihalas (1966) and De Jager and Neven (1967). In Figure 16, the $(B-U)$ and V-B values are plotted for the stars, as well as lines of equal apparent B-magnitude (thin lines). The theoretical computations allow one to draw lines of equal effective g-values. The straight line at the bottom of the diagram indicates the position of black-body colours, which would correspond to those of stars with extremely small values of g. Actually the authors assume that black-body colours would correspond to the theoretical case of effective gravitation zero (see also Chapter 5 of this volume).

Figure 17 shows essentially the same diagram as Figure 16 but here the spectral types are indicated as well as the positions of most of the emission line stars. The position of the most luminous supergiant of the LMC, HD 33579, is also marked. It is interesting to note that all emission line stars fall virtually on the black-body line, which supports the above statement that these objects are essentially stars with nearly zero values for the effective acceleration of gravity in their atmospheres.

It is clear from Figure 17 that for the supergiants without emission lines the *effective* log g-values are roughly of the order of 1 and that they may even go up to 2 for the earlier B-type stars. For later type supergiants Parsons, on the basis of 6-colour photometry of 57 supergiants in the range F3–G3 found that for Ib-stars log g_{eff} decreases from 2.0 at F3 to about 1.2 at G3. For early A-type supergiants Aydin (1971) gives log g_{eff} between 1 and 2.

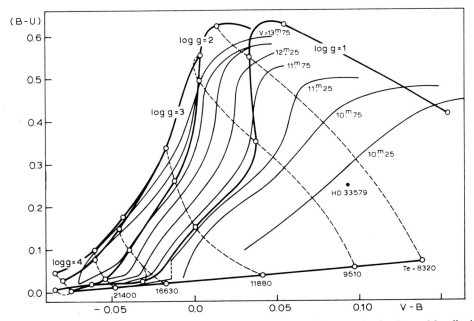

Fig. 16. Observed and calculated stellar colour-colour values for stars in the Large Magellanic Cloud. *Thick lines*: g_{eff}=constant (model computations); *thin lines*: V=constant for the observed stars; *dashed*: lines of constant effective temperatures (models); *straight line at bottom of diagram*: black-body colours. From Walraven and Walraven (1971).

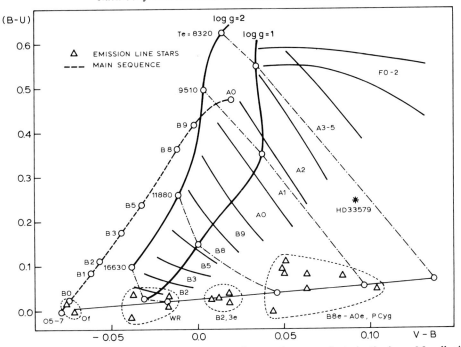

Fig. 17. Spectral types in the $(B-U)$ versus $V-B$ diagram for supergiants in the large Magellanic Cloud. The main sequence is indicated by *circles on the dashed line*. For the supergiants the boundaries between the spectral types are indicated by *thin lines*. Emission line objects and similar stars fall on the black-body line. (After the Walravens, 1971).

For stars of later spectral types the g_{eff} values are mostly badly known but as a rule they are small. For example the giant Arcturus (α Boo, Ko III) has log $g_{eff}=$ 1.48$\pm$0.15 (Blackwell and Willis, 1977), while much smaller values are measured for more extreme stars such as red giants and supergiants.

2.7. The Distribution of Light over the Surface of Supergiants

The powerful method of speckle interferometry, developed by Labeyrie (1970, 1974) has enabled one – in principle – to reach the theoretical limit of resolution of a telescope, and thus to observe details on the surfaces of the stars with the largest angular diameters. So far the technique has been applied successfully to α Ori (M1.5 Iab) by Gezari et al. (1972), Bonneau and Labeyrie (1973), Lynds et al. (1976), and Wilkerson and Worden (1977), and to several binaries and stars with circumstellar shells (Blazit et al., 1977).

Essentially, the method is based on the assumption that for a source of sufficiently small angular size and for a wave-front of sufficiently small area, the influence of atmospheric turbulence for any time interval of sufficiently short duration, is to deviate the direction of the wavefront in the same way for rays arriving from all parts of the source. Hence, an image of a star ($=$a source of sufficiently small angular size) obtained at the focal plane of a telescope in an exposure time small as compared with the time scale of the variations of the optical transfer function of the Earth's atmosphere ($\gtrsim 10$ ms), will appear to be decomposed into a number of sharp images (speckles) displaced randomly with regard to each other, but each of them being essentially an image of the star taken with a resolution corresponding to the resolving power of the telescope. Combinations of the various speckle-images into one resulting picture may give the resulting image a resolution corresponding to the practical resolution of the telescope.

Lynds et al. (1976) have applied the technique to α Ori, in a part of the continuous spectrum (5100$\pm$50 Å) and in a TiO band (5180$\pm$50 Å) and found for this star diameters in m-arc sec of 49$\pm$1(clnt.) or 53$\pm$1(TiO) for an assumed uniform intensity disk, and 74$\pm$1 and 79$\pm$1 respectively for a core-shaped intensity profile. They even succeeded in obtaining an 'image' of α Ori, and although Lynds et al. took care to stress the preliminary character of these results it seems that α Ori might show some faint structures, with amplitudes slightly above the noise level and which persisted during the consecutive nights. Such structures were absent in the star γ Ori, which was observed as a comparison object. However, some doubt was casted to the reality of the inhomogeneities because in a subsequent investigation of Wilkerson and Worden (1977) they were absent. Betelgeuze is very highly limb darkened, with

$$I/I(0) = 1 + \mu(\cos\vartheta - 1),$$

where

$\mu = 0.75\pm0.13$ in the continuous spectrum at 5100 Å; and
$= 0.93\pm0.03$ in the TiO band at 5200 Å.

2.8. Magnetic Fields of Luminous Stars

The observed magnetic fields of stars are averages over the stellar surface, and are fairly weak for the few luminous stars for which such measurements have been reported. The measuring accuracy of the conventional photographic method, based on the measurement of the amount of Zeeman splitting in a longitudinal field, is as a rule not better than 100 G. A photoelectric measuring technique for the longitudinal component, developed by Severny *et al.* (1974) has so far been applied to a few supergiants only, and yields for γ Cyg (F8 Ip) field components fluctuating rather erratically between -170 and $+200$ G. Is this due to stellar spots? For β Ori (B8 Ip) the value of $H_{//}$ ranges from -100 to $+260$ G. A cross-over effect ($H_{//}$ changing sign during the observations) was observed on several occasions. For these cases the measuring error was given as ± 40 G for γ Cyg and ± 10 G for β Ori.

Another supergiant of intermediate spectral type for which a magnetic field was detected is Canopus (α Car; F0 Ib) for which Rakos *et al.* (1977), in observations covering the three years period 1973–1976, find a field that varies systematically between $+600$ and -100 G in a period of 22.35 days. Tentatively this period was identified with the star's period of rotation. This would then indicate that the magnetic structures were semi-permanently on the star, at least during the three-years period of observations.

The only late-type supergiant for which a magnetic field has been detected is VV Cep (M2 Iab; see Table XXI) for which the field component varies between $+850$ and -360 G (Ledoux and Renson, 1966).

2.9. Turbulent Velocities

The investigation of the random velocity field of the supergiant atmospheres is still completely open. First of all it is hard to distinguish between rotational broadening and macroturbulence. Next, only spectra taken with the highest possible resolution and of good photometric quality would enable one to really disentangle the various components of the motion field.

The *macroturbulent motions* should be distinguished from the rotational effects. This is difficult and leaves ambiguity. One method is to study the distribution function of the half-widths of the lines for stars of similar spectral types. For 64 supergiants of spectral types O9.5 Ia to A5 Ia studied by Rosendhal (1970) it was shown that large-scale turbulent motions are at least as important in broadening the spectral lines as rotation. The smallest observed linewidth for each spectral type can be assumed to reflect the 'macroturbulent velocity component, ζ_M. In that case the ratio of ζ_M to the velocity of sound s appears to be of the order $3 \ldots 4$ over the range of spectral types from O9.5 Ia through A5 Ia (Lamers and De Loore, 1976). Another method is to analyse the Fourier transform of profiles of various lines. The rotational profiles are bowl-shaped, while conventionally the macro-turbulent profiles are assumed gaussian – this may not be correct (Vermue and De Jager, 1979). Thus Ebbets

(1979) found for early-type supergiants that $\zeta_M \approx 30$ km s^{-1}; main-sequence stars seem to have only little macroturbulence.

It needs to be stressed that these methods can yield confusing results since also a microturbulent velocity component produces line broadening so that the procedures described here rather yield a value for the combined effect of macro- and micro-turbulence. This uncertainty is shown clearly in a study of α Cyg (A2 Ia) by Groth (1972) where the same set of line profiles can be mimiqued by greatly different combinations of the large-scale radial velocity field and the microturbulent ve-locities, the combinations ζ_M; $\zeta_\mu = (1; 15$ km s$^{-1})$ and $(15; 6$ km s$^{-1})$ being possi-ble.

The *microturbulent velocity component* ζ_μ is in most cases determined from curve-of-growth studies. Struve and Elvey (1934) found that the ζ_μ-value increases with luminosity; a similar correlation was found by Rosendhal and Wegner (1970) for A-type supergiants. We have to remark that reliable data can only be obtained if the investigation is based on good atmospheric models. On the average the ζ_μ-values are of the order of $0.5s$ to s, although exceptions occur: curve-of-growth analysis of κ Cas (B1 Ia) gave a microturbulent velocity component of 20 km s^{-1} (Kovachev and Duerbeck, 1976).

Groth (1961) and Lamers (1972) give data that suggest that the microturbulent velocity components in supergiants are of the order of 10 km s^{-1} in the deep layers, and increase to 20 km s^{-1} in the outermost layers. But the time is perhaps not yet ripe for an analysis of the depth-variation of ζ_μ.

Lamers and De Loore (1976) have summarized data on microturbulent velocities for a number of Ia supergiants, and have added to this listing a few Ib supergiants, as well as the O4ef star ζ Pup. Only such stars have been selected for which reliable model analyses are available. These results together with some data from others, collected in Table XII show, that in broad outline the microturbulent velocities have about half to one time the value of the velocity of sound s.

For less extreme luminosity class (Ib) supergiants of spectral types G0–K5, Luck (1977a) found average values of $\zeta_\mu = 2.5 \pm 0.5$ km s^{-1} and $\zeta_M = 5 \pm 3$ km s^{-1}. Roughly, these values are about $\frac{1}{2}s$ and s.

The expected turbulent velocities in late-type stars can be computed if one assumes that the turbulence is due to acoustic waves generated in the stellar convection zones. The predicted acoustic fluxes for stars of different T_e- and g-values by De Loore (1970), and stellar models from Carbon and Gingerich (1969) and Peytremann (1974b) were used by Edmunds (1978) to predict r.m.s. wave amplitudes for stars with T_e between 4000 and 8000 K, and log $g = 1$ to 5. Edmunds assumed the conventional relation $F_{ac} = \varrho v^2 s$. These predictions agree qualitatively with values for the few late-type stars for which turbulent velocities are known.

Micro- and macroturbulence and the notion 'spectrum of turbulence'. The unsatis-factory state of the interpretation of observations related to the stellar velocity field is partly of an observational nature – in most stellar spectra the spectral resolution and photometric accuracy are insufficient to allow one deducing much information

TABLE XII

(Micro-)turbulent velocity components ζ_μ in some supergiant atmospheres for which reliable model analyses are available, as collected mainly by Lamers and De Loore (1976). The values of T_e and $\log g$ are those assumed in the model analyses.

Star	Type	T_e (10³ K)	$\log g$	ζ_μ km s⁻¹	References
ζ Pup	O4ef	50	4.0	5	Snijders and Underhill (1975) Peterson and Scholz (1971)
HD 164 402	B0 Ib	33	4.1	10	Dufton (1972)
ε Ori	B0 Ia	29	3.0	15	Lamers (1974); Lamers and Snijders (1975)
HD 96 248	B1 Iab	26	3.3	10	Dufton (1972)
o² CMa	B3 Ia	20	2.5	24	Van Helden (1972); Lamers and Snijders (1975)
o² CMa	B3 Ia	20	2.5	24	Van Helden (1972); Lamers en Snijders (1975)
η CMa	B5 Ia	19	2.3	15	Underhill and Fahey (1973) Lamers and Snijders (1975)
β Ori	B8 Ia	14	1.9	5	Crivellari and Stalio (1975) Lamers and Snijders (1975)
HD 21 389	A0 Ia	11	1.0	9	Aydin (1972)
α Cyg	A2 Ia	9.9	1.0	10	Aydin (1972)
6 Cas	A3 Ia	9.3	1.0	10	Aydin (1972)
α Per	F5 Ib	6.4	1.6	5	Parsons (1967)
ϱ Cas	F8 Ia	–	–	10	Lamers and De Loore (1976)
β Aqr	G2 Ib	5.7	1.4	10	Parsons (1967)

from a spectral line. Partly, however, the unsatisfactory situation is also related to lack of clarity in using the terms micro- and macro-turbulence.

This situation can be improved by introducing the notion 'spectrum of turbulence'. When the components of the velocity field along the line of sight are decomposed in their Fourier components, the spectrum of turbulence is the function $K(k)\,dk$, giving the kinetic energy of the velocity field components along the line of sight contained between wave numbers k and $k+dk$. Here, as usual $k=2\pi/l$, where l is the wavelength of the Fourier component. Usually, it is assumed that the velocity field contains a smallest wave number k_0, corresponding to a longest wavelength l_0. Often, it is also assumed that at least approximately $l_0 \approx H$, the density scale height of the atmosphere. Furthermore (Van Bueren, 1973), the density scale height is mostly within a factor 2 equal to the optical scale height, θ, defined by

$$dz = \theta \, d\log\tau.$$

We consider next the influence of the spectrum of turbulence on an infinitely narrow spectral line, not widened by damping or thermal influences. It is furthermore assumed that the spectral line refers to a geometrical point x on the stellar surface. The resulting line will have a certain profile, describing the photospheric velocity distribution along the line of sight, as influenced by radiative transport in the stellar photosphere.

If we call this distribution function $\varphi(v)$ (with $\int_{-\infty}^{\infty} \varphi(v)\,dv=1$), then for one point x on the stellar disk, the macroturbulent velocity component is defined (according to De Jager, 1972) as the first moment of this distribution function:

$$\zeta_{M,\,x} = \int_{-\infty}^{\infty} v\varphi_x(v)\,dv$$

and the microturbulent velocity component as the second moment

$$\zeta_{\mu,\,x}^2 = \int_{-\infty}^{\infty} (v - \zeta_M)^2\,\varphi_x(v)\,dv.$$

For the whole disk $\zeta_M^2 = \int \zeta_{M,x}^2 I(x)\,d\Sigma$ and $\zeta_\mu^2 = \int \zeta_{\mu,x}^2 I(x)\,d\Sigma$, when $I(x)$ is the emergent intensity at x, and Σ the area.

It is, hence, clear that micro- and macro-turbulence are not simply asymptotic limiting cases of the velocity field, as is so often claimed. With a continuous spectrum of turbulence the motions with the smallest wavelengths contribute fully to the microturbulent line broadening component, but those with longer wavelengths contribute too, but in reduced measure. Reversely, the motions with very long wavelengths contribute fully to the macroturbulent displacement component, and those with smaller wavelengths contribute increasingly less. In this connection it is useful to introduce the notion 'micro- and macro-turbulent filters' (De Jager, 1972). In dimension-less form these filters are given as functions of $k\theta$, where θ is the optical scale height, already defined earlier. Figure 18 gives these filters for the stellar case (Vermue and De Jager, 1979), calculated for weak lines in a simplified model of a stellar atmosphere. They are given for two cases: isotropic turbulence and radial pulsations. The filter-functions for the case of isotropic turbulence were recomputed by Durrant (1979) for a more realistic model, that of an exponential, constant-property atmosphere. In the treatment the line is no longer assumed infinitely thin and radiation transfer is considered. Durrant's improved filter function for micro-turbulence is also shown in Figure 18.

2.10. Rotational Velocities in Atmospheres of Bright Stars

A difficulty in determining the rotational velocity of stars from observed line profiles is that lines broadened by macroturbulence and rotation are very much alike. Figure 19 shows the average equatorial rotational velocity for types later than O8, derived from the classical investigation of Boyarchuk and Kopylov (1958). The emission line stars and other peculiar stars are not included in this diagram. Supergiants have apparently fairly small rotational velocities, ranging from 70 km s^{-1} for O9–B0 stars down to 25 km s^{-1} for B through F-type stars. This is confirmed by Stothers (1972a) from a comparison of a theoretically computed HR diagram for the case of no rotation with an observed diagram for massive early-type stars.

For emission line stars the rotational velocities are larger. From Boyarchuk and Kopylov (1958) it appears e.g. that for B0e stars the average equatorial rotational

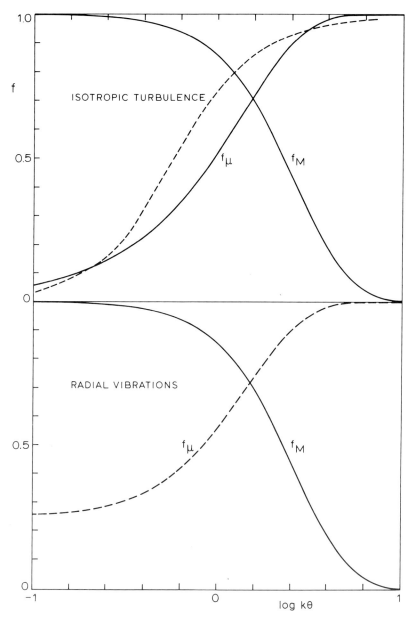

Fig. 18. Macro- and micro-turbulent filter functions for stellar atmospheres, plotted against $k\theta$. *Upper diagram:* case of isotropic turbulence; *solid lines:* filters after De Jager and Vermue (1979); *dashed:* improved calculations by Durrant (1979); *lower diagram:* case of radial waves (De Jager and Vermue, 1979).

velocity is about 415 km s^{-1}, and for later B-type stars it is approx. 345 km s^{-1}. As will also be shown in Section 3.3 there is a link between emission features of Oe and Be stars and the rotational velocities.

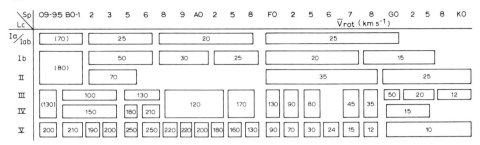

Fig. 19. Average equatorial rotational velocities for stars as summarized by Schmidt-Kahler (1965) from the publication of Boyarchuk and Kopylov (1958). The emission-line stars are not included in this diagram.

The material of Figure 19 does not include the O-type stars, but a group of 205 O-type stars were studied by Conti and Ebbets (1977) who found – see Figure 20 – that the distribution function of rotational velocities is bimodal for main sequence stars (luminosity class V) with a large maximum at approx. 100 km s^{-1} and a smaller near 300 km s^{-1} with a gap at 200 km s^{-1}, while the class I stars have only one maximum at approx. 100 km s^{-1}. The most rapid as well as the slowest rotators appear to be rarer towards earlier spectral types and disappear for types O3 and O4.

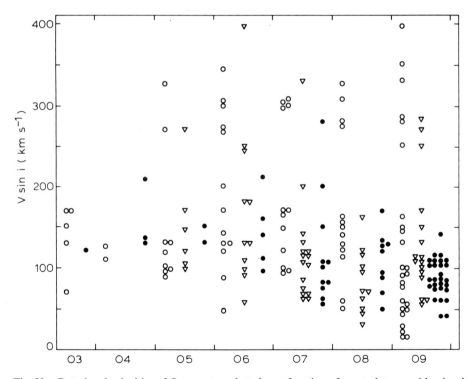

Fig. 20. Rotational velocities of O-type stars plotted as a function of spectral type and luminosity class. ○=main sequence (V), ▽=giants (III), ●=supergiants (I). (From Conti and Ebbets, 1977).

Similarly, for a sample of 256 late B-type stars Wolff and Preston (1978) found that about half of the stars have sharp lines and have apparently true v_{rot}-values with a maximum in the velocity interval 30–70 km s^{-1}. The others have larger velocities, up to approx. 350 km s^{-1}.

Obviously, not the real equatorial velocity v_{eq} is determined but the component $v_{eq} \sin i$, with i=inclination of polar axis to the celestial sphere. Hutchings (1976e) and Hutchings and Stoeckley (1977) have shown that the ratio of the Full Width at Half Maximum (FWHM) of a line in the UV to that of a line in the visual depends on wavelength according to the values of ω/ω_{crit} (the angular velocity as a fraction of the critical) and i – Figure 21. The explanation of this effect is simple, and related to the fact that the UV limb-darkening is always much stronger than that in the visual, due to the fact that for a given temperature change the (black-body) flux change in the UV is by far larger than in the visual. The results of Figure 21 were computed at the basis of simplified models, with as an essential assumption that the

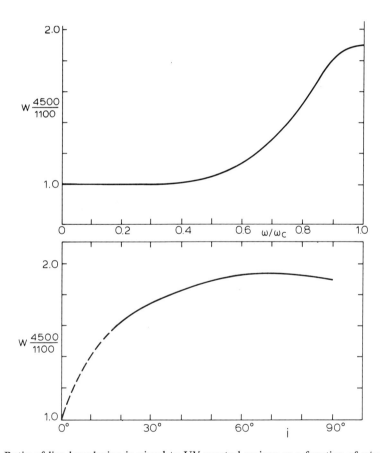

Fig. 21.   Ratio of line broadening in visual to UV spectral regions as a function of ω/ω_{crit} and i. Other parameters are T_e=21 000 K; limb darkening 0.7–1.0; gravity darkening coefficient 0.25. (After Hutchings and Stoeckley, 1977).

gravity darkening is according to Von Zeipel's darkening law $(1-K\sin\vartheta)$ in which K is the (variable) gravity darkening coefficient and ϑ the angle of emergency of the emitted radiation. Gravity darkening is essentially based on the fact that in a star in radiative and thermal equilibrium the local radiation flux is proportional to the local gradient of the gravitation potential. In a rapidly rotating star this gradient may take strongly different values between pole and equator. The assumption of gravity darkening gives the link between the star's rotation and the wavelength dependence of the line widths. Hence, by assuming certain photospheric parameters it is possible in principle to determine v_{eq} and i from line width measurements in the visual and ultraviolet ranges. Thus, Hutchings and Stoeckley find for ζ Per (B1 Ib) for which $v_{eq}\sin i=72$ km s^{-1}, values of $v_{eq}\approx 280$ km s^{-1}, $i\approx 15°$ and $\omega/\omega_{crit}\approx 0.9$. For the B0 IVe star γ Cas, with $v_{eq}\sin i=300$ km s^{-1} and $v_{crit}=550$ km s^{-1}, one finds $i=47°$, $v_{eq}=420$ km s^{-1}, and $\omega/\omega_{crit}=0.75$. If, however, $v_{eq}\sin i$ would be 400 km s^{-1} then $i\approx 65°$, $v_{eq}=450$ km s^{-1}, and $\omega/\omega_{crit}=0.82$.

For ζ Pup (O4ef), assuming $v_{eq}\sin i=175$ km s^{-1}, $v_{crit}=530$ km s^{-1} one derives $v_{eq}=175$–210 km s^{-1}, $i\geq 50°$, and $\omega/\omega_{crit}=0.33$–0.40. (See however Section 3.2, where we give $v\sin i\approx 300$ km s^{-1}.)

2.11. Polarization of Light of Stars with Extended Atmospheres

The continuous radiation of several stars with extended atmospheres appears to show some degree of linear polarization. Polarization has been found in red variables (Kruszewski et al., 1968; Kruszewski and Coyne, 1976) up to 8% and 13% in two cases, in Mira stars (Dyck and Sanford, 1971), in K and M giants and supergiants (Dyck and Jennings, 1971; Tinbergen et al., 1980; Koch and Pfeiffer, 1978), in Be stars – Section 3.3, γ Cas – (Serkowski, 1970; Elvius, 1974; Poeckert and Marlborough, 1977), an Oe star (Hayes, 1978), and Of stars (Hayes, 1975).

The largest degree of polarization measured is found in the peculiar M supergiant VY CMa, for which $22\pm2\%$ was measured at 0.36 μm (Shawl, 1969; Hashimoto et al., 1970) – cf. the discussion of this star in Section 6.16.

The linear polarization discovered by Hayes in the continuous spectra of four Of stars and the Oe star λ Cep is variable. However only two of the five stars are binaries. The variable polarization for such stars implies a photosphere that is extended, non-spherical, and variable. For λ Cep (Oe) the variations are on a time scale of a day (Hayes, 1978). Also, several of the very cool stars examined by Kruszewski and Coyne (1976) and by Tinbergen et al. (1980) appear to show variations of the polarization at time scales of days, to a year (for α Ori and α Sco).

We may distinguish between two cases. For certain stars the polarization, particularly if it is pronounced in the infrared, may be due to a circumstellar dust cloud. This case is discussed in Section 6.12.

For other stars like the Of and Be stars (Sections 3.2 and 3.3) but also late-type supergiants, one should rather think in terms of an extended and spherically asymmetric gaseous photosphere.

The spherical asymmetry could either be caused by stellar rotation or else by features like a disk-shaped envelope, or by rays extending outward, and the like. In the cases of gaseous atmospheres (no dust particles) the polarization is generally caused by scattering at electrons, free (Thomson scattering) or bound (Rayleigh scattering). Polarization is measured by determining the four Stokes parameters I, Q, U, and V, where I represents the intensity of the wave, Q and U the linear Stokes parameters, and V is a measure for the ellipticity of polarization (for $V=0$ the radiation is linearly polarized).

We refrain from repeating here the detailed expressions for the Stokes parameters, but refer, as well as for the detailed expressions for scattering, to a compendium like Lang's (1974, pp. 11, 68, and 72).

In dealing with stellar atmospheres the problem may be approximated by assuming single scattering (i.e. only considering the last scattering process). The more elaborate way of dealing with multiple scattering has been treated for a purely scattering atmosphere by Chandrasekhar (review: 1950) and has been applied to stars and the Sun by other authors. Generally the polarization to be expected for main sequence stars is very small, well below 1%. For very extended atmospheres one may expect larger values.

Approximations. For the cases discussed in this section (supergiants, or very extended atmospheres) we are allowed to simplify the treatment by dealing, for the cases of Thomson and Rayleigh scattering with *linear* polarization, and *single* scattering at particles *at large distance* to the photosphere, the 'coronal' case.

The angular distribution of radiation for single scattering is, in the case of Thomson scattering, per scattering electron:

$$\sigma_T(\varphi) = \tfrac{3}{4}\sigma_{el}(1 + \cos^2\varphi) \tag{2.11; 1}$$

with $\sigma_{el}=0.665$ cm^{-1}; and φ is the angle between directions of incident and scattered radiation.

For Rayleigh scattering, per scattering atom (see Unsöld, p. 180):

$$\sigma_R(\varphi) = \sigma_T(\varphi)\left(\frac{\langle\lambda_r\rangle}{\lambda}\right)^4 \Sigma f_r \tag{2.11; 2}$$

where $\langle\lambda_r\rangle$ is the weighted average of the wavelengths of the resonance lines of the atom (for hydrogen $\langle\lambda_r\rangle\approx 1026$ Å), and f_r their oscillator strength; Σf_r is equal to the number of electrons in the outer shell; $\lambda(\gg\langle\lambda_r\rangle)$ is the wavelength.

Consider the case of a scattering volume at a distance r – projected distance ϱ – to the star, as shown in Figure 22. The intensity of the electrical vibration tangential to the stellar limb is called $I_t(e)$, that of the radial vibrations is $I_r(e)$. The degree of polarization is

$$p = (I_t - I_r)/(I_r + I_r). \tag{2.11; 3}$$

Following Schuster (1879), Minnaert (1930), Baumbach (1937), Van de Hulst (1950), as reviewed by Unsöld (1955, pp. 644f) we decompose the average intensity of radia-

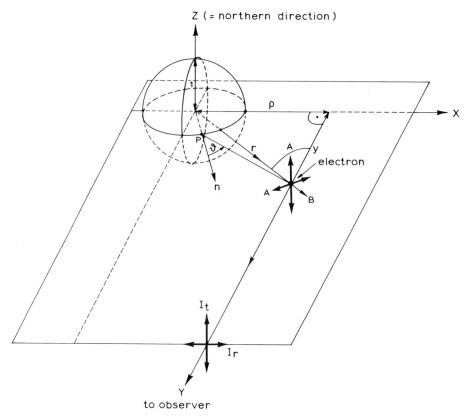

Fig. 22.   The computation of polarization in the 'coronal case'. Diagram after Unsöld (1955, p. 642).

tion received in the point **r** in a part $\pi F2A(r)$, where A is proportional to the mean square of the vector components of which the electrical vector vibrates in any transversal direction, hence at right angles to the radius vector, and a part $\pi FB(r)$ of which the electric vector vibrates radially to the star. The total average intensity of radiation in the point r is then

$$J(r) = \pi F(2A(r) + B(r)). \tag{2.11; 4}$$

Here F is the mean surface brightness of the star with

$$\pi F = \int\limits_0^\pi \int\limits_0^{2\pi} I \cos\vartheta \, \sin\vartheta \, d\vartheta \, d\varphi \approx 2\pi \int\limits_0^{\pi/2} I \cos\vartheta \, \sin\vartheta \, d\vartheta. \tag{2.11; 5}$$

The latter equality assumes a spherically symmetric star.

   For large values of r the star becomes a point source, hence in the limit for $r \to \infty$ $A(r) \to (2r^2)^{-1}$, and $B(r) \to 0$. Schuster (1879) and Minnaert (1930) have shown that $A(r)$ and $B(r)$ may be found by a straightforward computation; for a given r-

value their values depend on the star's limb-darkening. Tables to compute A and B are given by Van de Hulst (1950) for several limb-darkening coefficients. The vibrations of the electrons scattering the incident radiation are distributed in a vibration ellipsoid with axes A, A and B; this defines the degree of polarization of the light emitted with a certain direction.

Let n_e and n_R be the numbers of free and bound electrons per cm³ in the point $\mathbf{r}$, then the intensity of light scattered in all directions is

$$4\pi S = \pi F(2A + B)(\sigma_T n_e + \sigma_R n_R), \qquad (2.11; 6)$$

where S is the mean source function.

The source function for light vibrating at right angles to, respectively in, the plane through the star's center is then

$$\left.\begin{array}{l} S_t = \tfrac{3}{8}FA(r)(\sigma_T n_e + \sigma_R n_R), \\ S_r = \tfrac{3}{8}F\{A(r)\cos^2\varphi + B(r)\sin\varphi\}(\sigma_T n_e + \sigma_R n_R). \end{array}\right\} \qquad (2.11; 7)$$

In this expression φ is the angle between the $\mathbf{r}$ and y directions (Figure 22). Integration over the y-axis, taking self-absorption into account, yields the intensities I_t and I_r; they define the polarization vector at the point (x, z) of the celestial sphere.

This treatment has so far been restricted to scattering in the EW plane. Consider now a scattering element in a point making angles ψ and $(90\text{-}\varphi)$ with the NS line and the EW line respectively (Figure 23). This case can be brought back to the previous one by a rotation around an axis $//$ y-axis through the star's center. Calling

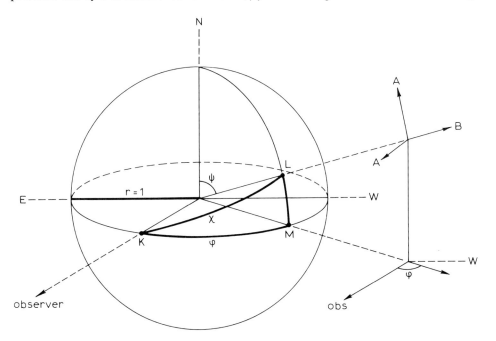

Fig. 23. Geomety for polarivation calculations.

χ the angle KL then $\cos\chi=\sin\psi\cos\varphi$, and the angle φ in Figure 22 should be replaced by this χ-value. The amount of rotation is the angle LKM, with

$$\sin LKM = (\sin\psi - \cos\chi\cos\psi)/\sin\chi\sin\psi.$$

The total polarization for the star as a whole is thus found after integration over the celestial sphere around the star, yielding the polarization vector p.

Conventionally, in observations of linear polarization the Stokes parameters Q and U are measured, where the $+Q$ axis ($\theta=0$) corresponds to the N–S direction in the equatorial system of coordinates. The polarization angle is then $\psi=\frac{1}{2}$ arc $\tan(U/Q)$, it increases in a counterclockwise manner.

Theoretical computations of the polarization vector were published by Poeckert and Marlborough (1977). In the frame of a rotating star the polarization angle is defined such that it is zero when the polarization vector is parallel to the rotation axis of the star. For a point in the envelope the amount of scattered radiation, and the local Stokes parameters q and u are determined (for the geometry of the problem I refer to the original publication) and the integrated Stokes parameters Q and U are determined from an integration over the whole volume.

The comparison of computations with observations of the Be star γ Cas confirms the idea that the star is surrounded by an axis-symmetric rotating gaseous disk (see Section 3.3).

To explain the polarization of red variables and Mira stars Peraiah (1976) has calculated the expected degree of linear polarization from stars with extended atmospheres, assuming the necessary asymmetry due to rotation (a non-rotating star would obviously be completely symmetric, and would not have any polarization). Furthermore, the inclination i between the stellar axis of rotation and the line of sight is a parameter.

Peraiah succeeded, under certain assumptions, to find theoretical results that are in fair agreement with the observations of Kruszewski *et al.* and Dyck and Sandford for reasonable assumptions on the ratio f between the equatorial centrifugal acceleration, and that of gravity. Values of the order $f\approx0.5$ yield in most cases degrees of polarization comparable to the observed values.

CHAPTER 3

SPECTRAL CHARACTERISTICS AND STELLAR PARAMETERS
FOR THE MAIN GROUPS OF LUMINOUS STARS;
SOME PROTOTYPES

3.1. O- and Bright B-Stars

Since the main part of the emission of these hot stars is in the ultraviolet part of the spectra the study of that spectral region is important. Instrumental for the progress in this field were various ultraviolet-sensitive spectrometers launched in spacecraft in the seventies, such as the Wisconsin instruments aboard the NASA OAO-2 (Code *et al.*, 1970), the Utrecht Orbiting Stellar Spectrometer S59 in the ESA TD-1A satellite (De Jager *et al.*, 1974), the Culham–Edinburgh–Liège instrument S2-68 in the ESA TD-1A satellite (Boksenberg *et al.*, 1973), the Princeton satellite instrument Copernicus (Rogerson, Spitzer *et al.*, 1973; Snow and Jenkins, 1977), the objective-prism instrument aboard Skylab (Henize *et al.*, 1976a), the Groningen ultraviolet photometer in the Astronomical Netherlands Satellite (Van Duinen, Aalders *et al.*, 1975), the high-resolution near-ultraviolet balloon spectrograph BUSS (Kondo, De Jager *et al.*, 1979), (De Jager, Kondo *et al.*, 1979); the International ultraviolet Explorer, IUE, (Wilson *et al.*, 1978; Willis, 1979). The discovery of X-rays from early-type stars was made principally by the Einstein Observatory – HEAO 2 (Giacconi *et al.*, 1979).

Van der Hucht *et al.* (1976) published line identifications of nine bright stars observed with S59, and they give in their Table 1 a list of all stars for which ultraviolet stellar identification lists were published. Snow and Jenkins (1977) give UV spectra for 60 O- and B-type stars, among which 13 supergiants. Code and Meade (1979) present low resolution (12–22 Å) spectra from 1160–3600 Å of 164 bright stars observed with Copernicus. Jamar *et al.* (1976) published a Catalogue with low resolution (~20 Å) spectra of 1356 stars observed with S2-68 in the ESA TD-1A satellite. A second S2-68 Catalogue (Thompson *et al.*, 1978) contains photometric data in the UV of over 30 000 stars.

The *classification of O-type stars* depends on the line strength ratio of He I $\lambda 4471$ to the He II $\lambda 4541$, and thus represents an ionization sequence. The O3 stars have the highest ionization temperature: in their spectra the He I line $\lambda 4471$ is no longer present (but we note that Kudritzki (1979) found the line as a weak absorption in the very luminous O3 star HD 93250 – see Section 3.3). Lists of O-type stars thus classified have been given by Walborn (1972b, 1973d) and by Conti and Leep (1974); for O-stars earlier than O6 such lists were given by Conti and Frost (1977).

Conti (1973a, 1974) has measured the equivalent widths of Hγ and of Hα in O-type stars, and has published diagrams of these values for various luminosity classes. Mihalas and Hummer ((1974b) have thereupon compared these diagrams with com-

puted values, according to the predictions from spherical model computations by
Mihalas and Hummer (1974a), and of non LTE-planar atmospheric models of Auer
and Mihalas (1972). These are reproduced here as Figure 24. It is clear from the figure
that the supergiants have, as a rule, the lines with the smallest equivalent widths. All

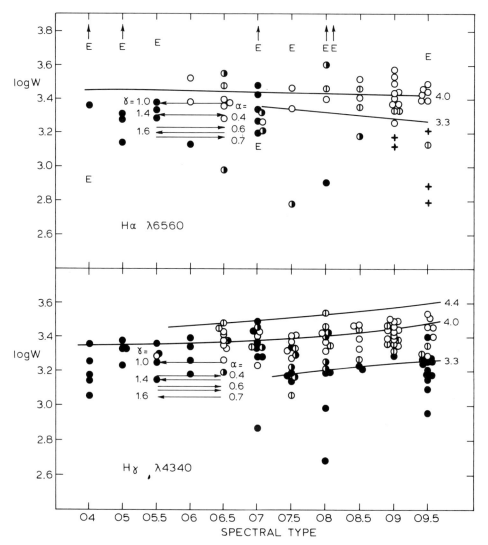

Fig. 24. Equivalent widths of Hα (upper) and Hγ versus spectral type (after Conti, 1974).
Spherical NLTE model results, represented by horizontal lines near O6, are labeled by γ or α, and
planar NLTE model results of Auer and Mihalas (1972), represented by solid lines, are labeled by
values of log *g*. (For the meaning of γ and α, see *List of Symbols*, and Chapter 5). The symbols represent values from individual stars: *open circles*, luminosity type V; *circle with vertical line*, luminosity type III; *half filled circles*, luminosity type IIIf; *filled circles*, luminosity type If or f; and *plus
sign*, luminosity type I. The symbol *E* denotes an emission line with a negative equivalent width of the
magnitude indicated by the symbol; the arrows indicate that the value lies off the figure (After
Mihalas and Hummer, 1974b).

stars earlier than O6.5 are similar to supergiants – see Section 3.2. The figure suggests that most supergiants later than O7 have $\log g$-values $\lesssim 3$. The $\log g$-values are the gravitational ones; the g_{eff}-values are expected to be smaller.

Similarly important for the interpretation of O and B stellar spectra in terms of model atmospheres are Kamp's (1978) computations of equivalent widths of 59 spectral lines of Si II, III and IV, for $T_e = 15000$–35000 K, $\log g = 3.0$–4.5. The ratio of equivalent widths Si IV 1394 Å (:) C IV 1548, 1551 Å is an excellent indicator for early B spectral type (B0–B2) (Henize et al., 1976b).

Many early-type stars – spectral types O, early B, WR – are *X-ray emitters*, as was discovered from observations made with the Einstein Observatory (Harnden et al., 1979; Seward et al., 1979). The X-ray luminosity L_x between 0.2 and 4.0 keV, is $\sim 5 \times 10^{33}$ erg s^{-1}. The ratio between L_x and the visual luminosity L_v is a few times 10^{-4}, which no clear dependence on spectral type in the range of early spectral types. Earlier already, Hearn (1975a) and Cassinelli et al. (1978b) – cf. Sections 6.11 and 7.3 – had found indications from visual and ultraviolet line studies that O- and early B-type supergiants should be surrounded by thin but fairly dense coronas. Whether these coronas have the shape of a spherical shell or are rather filamentary remains to be investigated; stability considerations would perhaps favour the latter model.

Most O, B and A type supergiants (as well as O and also the Oe stars) have an *infrared excess emission* detectable at $\gtrsim 10$ μm, but not at much shorter wavelengths (see Barlow and Cohen, 1977). This is ascribed to radiation emitted by an outflowing ionized gas, warmer than the star. Wright and Barlow (1975) and Panagia and Felli (1975) have shown that an outflow with constant velocity, hence yielding a density distribution $\sim r^{-2}$, gives rise to a free-free emission spectrum, proportional to $v^{2/3}$; – see Section 6.13. Observations of the infrared excess have been used by Barlow and Cohen (1977) to determine the stellar mass flux (Section 7.5). An interesting observation that needs verification is the detection of CO in the heavily reddened star 9 Cep (HD 206165; B2 Ib) by Tarafdar et al. (1980). If this CO is really circumstellar it must originate at fairly large distances to the star, through cooling of the stellar wind gas.

The fraction of O- and B-type stars that are *binaries* is hard to determine because radial velocity measurements are confused by the presence of stellar winds, sometimes even variable. Conti and Ebbets (1977) found 11% of their sample of 205 O-type stars certain to be binaries and 44% probable, and while Bohannan and Garmany (1978) found only 2 out of a sample of 18 bright O-type stars to be double, Garmany and Conti (1980) found that 50% of the O4–B5 and 40% of the O-stars are binaries. The mass-ratio of O-type binary components is as a rule close to unity. Table VIII in Section 2.4 gives masses, luminosities, and spectral types of the best studied O, Of and B-type binary stars. Interesting is the massive contact binary BD+40.4220 in the Cygnus OB2 association (cf. Tables VIII and XI). This binary appears to be an X-ray emitter with $L_x = 7 \times 10^{33}$ erg s^{-1} in the energy interval 0.2–4.0 keV (Harnden et al., 1979). This couple is to be ranked under the 'weak' X-ray binaries. In addition there are several 'strong' X-ray binaries known, with typically $L_x \approx 10^2$ to 10^4 times $L_\odot$.

They consist nearly all of an early-type supergiant and a degenerate companion. Their parameters are given in Table XXXV in Chapter 4.

In Table XIII we give the main characteristics of a number of the best studied early-type supergiants.

As *prototypes* of the stars described in this section we select two stars: the O9.5 Ib supergiant ζ Ori A, and the O9.5 Ia supergiant α Cam.

ζ *Ori A* (O9.7 Ib) has an apparent visual magnitude $m_v = +1.9$, it is at a distance of 468 pc, and hence has $M_v = -6.4$, see Tables IX and XIII. The – limb darkened – angular diameter is 0.48×10^{-3} arc sec (Hanbury Brown et al., 1974), hence the radius is $24 R_\odot$ ($= 16.8 \times 10^6$ km). Spectroscopically Auer and Mihalas (1972) determined log $g = 3.2$ and $T_e = 31\,000$ K, while Davis and Shobbrook (1977) derive $T_e = 29\,900 \pm \pm 2100$ K. With these T_e-values the bolometric correction is BC $= -2.93$ (Code et al., 1976); hence $M_{bol} = -9.3$. Snow and Jenkins (1977) give -9.35. With these values for R and g one would obtain $\mathfrak{M} = 33\mathfrak{M}_\odot$. To give an impression on the precision of this value: with the distance of 550 pc (sometimes also found in literature) one would obtain $M_b = -9.7$, and $\mathfrak{M} = 45\mathfrak{M}_\odot$ (Sterken, 1976b). Uesugi and Fukuda (1970) derived $v_{rot} \sin i = 127$ km s^{-1}, in fine agreement with Ebbets's (1979) later result: 120 km s^{-1}. The latter also derived $v_M = 30$ km s^{-1}.

Hearn (1975a), Cassinelli et al. (1978b), and Ebbets (1980) have used the Hα profile[*] to study the temperature variation with distance outside the photosphere, and found evidence for a 'coronal' temperature increase (to values of several times 10^5 K). The corona should have its basis close to the photosphere, and should not extend further than approximately $2.5 R_*$; thereafter the temperature should decrease again, presumably due to cooling by radiative recombination emission. At the same place the stellar wind should increase in velocity. A mass-loss rate of $\sim 10^{-6}\,\mathfrak{M}_\odot$ yr^{-1} is found by various authors. A detailed analysis of the temperatures and mass-loss in the outer layers of ζ Ori is given in Section 7.3; see also Table XXXIX.

The *photosphere of the supergiant α Cam* (O9.5 Ia) was investigated by Takada (1977) who compared the spectra (line and continuous, the latter down to 1100 Å) with predictions on the basis of planar LTE and non-LTE models. Slightly better agreement is found in the former case, for which the following photospheric parameters are derived: $T_e = 29\,000 \pm 2000$ K; the He/H abundance ratio is (He) $= 0.16$; $v_t = 20$ km s^{-1}; v_t must be fairly uncertain. Ebbets (1979) derived $v_{rot} \sin i = 100$ km s^{-1} and $v_M = 20$ km s^{-1}. Furthermore $M_v = -6.7 \pm 0.7$, hence $M_b = -9.6$. With the $(\mathfrak{M}, L)$ law of our Figure 13 one would find $\mathfrak{M} = (32 \pm 7)\mathfrak{M}_\odot$, and log $g_{grav} = 3.05 \pm 0.3$. Further, $g_{eff} = g_{grav} + g_{rad} \approx 2.3$. The profiles of the two strongest Balmer lines, Hα and Hβ cannot be described by planar atmospheric models, LTE nor non-LTE. These profiles are P Cygni-like and must be formed in the stellar wind. Hence mass-loss is of importance. Rapid variation, on time scales of days down to hours are observed in the stellar wind terminal velocity. The variations are largest in lines formed at the largest distances to the star (De Jager et al., 1979) – cf. Section 8.5.

---

[*] An error in the zeropoint of the wavelength scale in the first two investigations does not essentially influence the results – cf. Section 7.3.

TABLE XIII

Main characteristics of some well-studied early-type bright O and Of-stars and B-type supergiants. From Snow and Jenkins (1977), Abbott (1978), Ebbets (1979); and from data for some specific stars as given elsewhere in Chapter 3.

Star	HD	Spectral type		m_v	$\log T_e$	M_b	$\mathfrak{M}/\mathfrak{M}_\odot$	$R/R_\odot$	$\log g_{grav}$	$v \sin i$	v_{esc}	v_∞ (wind)
		Conti/Leep	Walborn									
ζ Pup	66811	O4 ef	O4 I(n)f	+2.26	4.70	−10.6	100	20	3.8	250	1050	2660 km s^{-1}
9 Sgr	164794	O4 ((f))	O4 V((f))	5.97	4.70	−10.55	100	15	4.1	168	1290	3440
α Cam	30614	O9.5 I	O9.5 Ia	+4.29	4.46	−9.6	32	32	3.0	100	590	1890
ζ Ori A	37742	O9.5 I	O9.7 Ib	1.90	4.48	−9.3	33	24	3.3	120	630	2290
μ Nor	149038		O9.7 Iab	4.89	4.47	−9.4	40	27	3.2	133	630	2190
ε Ori	37128	B0 Ia		1.70	4.43	−9.4	45	33	3.0	85	580	2010
κ Ori	38771	B0.5 Ia		2.09	4.42	−9.1	36	37	2.9	81	520	1870
139 Tau	40111	B1 Ib		4.83	4.32	−8.4	(24)	35	2.7	131	460	(640)
ϱ Leo	91316	B1 Iab		3.85	4.32	−8.3	(23)	33	2.8	69	470	1580
γ Ara	157246	B1 Ib		3.33	4.32	−8.1	(22)	30	2.8	230:	480	(1050)
o² CMa	53138	B3 Ia		3.04	4.18	−8.3	(24)	64	2.2	77	340	(580)
η CMa	58350	B5 Ia		2.46	4.12	−7.8	(21)	62	2.2	57	330	(590)
67 Oph	164353	B5 Ib		3.97	4.12	−6.6	(14)	36	2.5	17	370	(500)
β Ori A	34085	B8 Ia		0.13	4.07	−7.6	(21)	77	2.0	42	300	(530)

3.2. The Of and Oef Stars

These stars are the brightest early type O stars, they occur on or above the extreme high-luminosity high-temperature part of the main sequence.

Of stars were initially defined (Plaskett and Pearce, 1931) as stars with the N III lines at $\lambda\lambda$4634, 4640, 4642 and λ4686 He II in emission; often other non-Balmer line emissions occur too. Gradually the classification criteria have changed: Walborn (1971b) and Conti and Leep (1974) define Of stars as those that have emission in He II λ4686. Following these authors a further subdivision is introduced, denoting by ((f)) those O stars showing N III in emission but λ4686 He II clearly in absorption; while (f) denotes those O stars with N III in emission but λ4686 weakly in absorption or missing, and f those showing both N III and λ4686 strongly in emission. The early Of stars show N V 4604, 4620 Å in absorption. The late Of stars (O6 to O9.5) have C III 5696 Å in emission and C III 8500 Å in absorption. After the findings by Conti and Leep (1974) that there is a definite correlation between the presence of both λ4686 He II and Hα in emission, and that both lines are characteristic for the presence of an envelope, Conti (1976) proposed the term Of to denote O stars with envelopes, even though some may show only Hα emission. We will not follow this suggestion because of the confusion with Oe stars.

Walborn (see e.g. 1977) has introduced the notation f$_*$ for spectra in which N IV λ4058 emission is stronger than N III $\lambda\lambda$4634+40+42, while f+ signifies that Si IV $\lambda\lambda$4089+4116 are in emission in addition to He II λ4686 and the N III complex. It looks probable that these emissions are related to a rather dense wind, so that the $*$, + notation would describe a sequence of decreasing wind densities. There are three Of$_*$ stars known; they have all weak Si IV emission in their spectra. They have all spectral type O3f$_*$ and are: HD 93129 A in the Carina Nebula (Walborn, 1971c; Conti and Burnichon, 1975; Conti et al., 1979), VI Cyg no7 in the Cygnus Association OB2 (Walborn, 1973c; Humphreys, 1978) and HDE 269810 in the Large Magellanic Cloud (Walborn, 1977).

Oef stars. This is a class, introduced by Conti and Leep (1974); the stars show "an emission structure in He II λ4686, similar to that in Oe stars, namely central emission with a central absorption reversal, giving the appearance of double emission in this line. Usually there is no emission in the Balmer lines". The Oef and O(ef) stars have in addition to the Oe stars broad absorption widths in the other lines of the spectrum, suggesting large rotational velocities. It is very significant that the stars are all of early spectral type, not later than O6.5. Among the 130 O-type stars classified by Conti and Leep (1974, their Table 1) there are only four ef stars; viz. ζ Pup, O4ef; HD 14442, O6ef; λ Cep, O6ef; BD+60°2522, O6.5 IIIef. There is one (ef) star, viz. HD 192281, O5.5 (ef). The ef classification, certainly useful, is not generally used.

Figure 25 gives the early-type upper part of the Hertzsprung–Russell diagram, and shows locations of galactic Of, O(f), and O((f))-type stars. The Of and O(f) stars are generally the most luminous types, which indicates that λ4686 He II emission is related to the luminosity, and that the ((f)), (f), f sequence is a luminosity sequence, com-

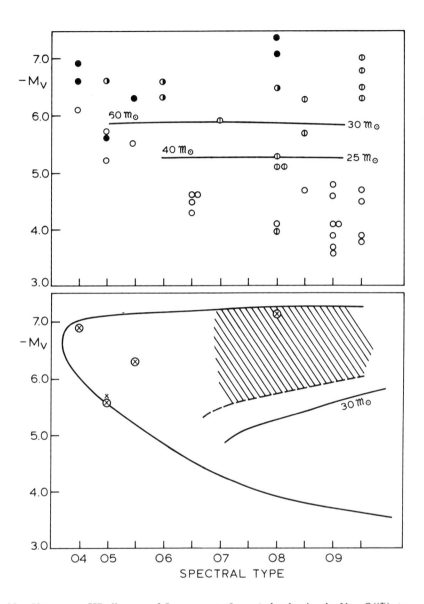

Fig. 25. *Upper part:* HR diagram of O-type stars. *Open circles:* luminosity V or O((f)) stars; *circles with vertical lines:* luminosity III or I; *half-filled circles:* O(f) stars; *filled circles:* Of stars. The lines show evolutionary tracks of stars of $60\mathfrak{M}_\odot$ and $40\mathfrak{M}_\odot$, calculated *with* mass-loss. *Lower part:* schematic version of upper part. The lower line is the 'Zero Age Main Sequence' for O-type stars; the upper line gives the upper M_v-boundary. In between that line and the broken one is the region of the Of stars. In the *hatched area* all stars are found with $\lambda 5696$ C III in emission as well as the lines $\lambda 4485$ and $\lambda 4503$. The *crosses* refer to stars with C III $\lambda 4647.50$ in emission; the *open circles* to Si IV $\lambda 4089$ emission. The line gives the evolution of a $30\mathfrak{M}_\odot$-star, *without* mass-loss. (Basically after Conti, 1973c, with modifications).

parable to v, III, I (Walborn, 1978a). Increasing strength of the N III emission lines in the $\lambda 4640$ group is related to decreasing g-value, hence increasing L-value (Mihalas et al., 1972; Mihalas and Hummer, 1973). Hutchings (1979b) finds for O6f–O9f stars M_b-values of $\sim -10.3 \pm 0.3$.

Figure 25 shows in addition that:
– most stars earlier that O6 are Of or O(f) stars;
– most stars more massive that $50\mathfrak{M}_\odot$ are Of or O(f) stars.
(This seems not fully true for O3 stars – Conti and Frost (1977) – but that may be due to a definition criterion). The earliest Of-stars have spectral type O3f, a type introduced by Walborn (1971c). In the Small Magellanic Cloud the earliest Of star has the type O4 III (n) (f) (Walborn, 1978b).

Figure 25 suggests that all Of stars are fairly massive. The masses in Figure 25 were derived from a comparison of positions in the HR-diagram with theoretical evolutionary paths computed by De Loore et al. (1978a, b) assuming mass-loss. Direct mass-determinations should be based on analysis of binary motion. Not all Of stars are binaries (Conti et al., 1977b); we refer to the few masses of Of stars collected in Table VIII.

That Of stars must lose mass follows from the emission lines in their spectra, and notably from the P Cygni profiles detected in the UV and the visible region. Conti and Leep (1974) gave typical line profiles for a few Of stars; these are reproduced here as Figure 26. The profiles refer to the stars HD 152408, O8 If; HD 108, O7 If; and 9 Sge, O8 If. Characteristic is the strong He emission, showing a P Cygni profile in 9 Sge and the P-Cygni type profiles of $\lambda 5876$ He I in HD 152408, and of $\lambda 4686$ He II

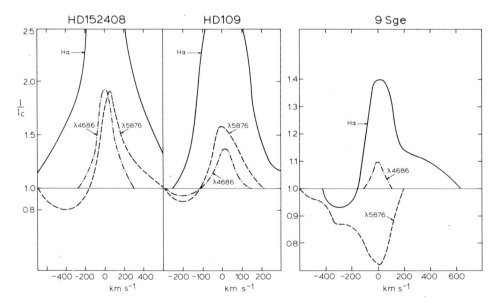

Fig. 26.   Normalized line profiles of Hα, $\lambda 5876$ He I, and $\lambda 4686$ He II for the Of supergiants HD 152408, HD 108 and 9 Sge. (After Conti and Leep, 1974).

in HD 108. In 9 Sge it is remarkable that $\lambda 5876$ has a highly asymmetric absorption line profile, whereas $\lambda 4686$ is symmetric. An illustrative comparison of the – emission line – profiles of Hα and He II 4686 Å in four O3–4f type stars was given by Conti (1976). The similarity between shape (and asymmetry) of the profiles indicates that the lines are formed in the same region of the envelope. The overall line width of Hα is $\approx 10^3$ km s^{-1}. The line profiles suggest a high rate of mass-loss. In the spectrum of λ Cep (O6ef) the profiles of He II $\lambda 4686$ and of Hα show night-to-night variations at the 1% level (Conti and Frost, 1974; Hutchings and Sanyal, 1976).

A statistical analysis of the average wavelengths in spectra of Of stars by Conti *et al*. (1977a) gave for these stars a K-term of -29 ± 3 km s^{-1}. This value, confirmed by Lynds (1979), who found -25 km s^{-1}, suggests that the wind in these stars takes already significant values in atmospheric levels, and is an independent indication of the strong outstreaming motions in these stars. This indication is further amplified by the clear relation between the (emission) equivalent width W of the He II line at $\lambda 4686$, and the peculiar velocity v_{pec} of the stars. Roughly $\log W (4686)=2.6+0.016 v_{pec}$, where W is in mÅ and v_{pec} in km s^{-1}. Along the same track Conti *et al*. (1977b) found in the spectra of Of stars indications of a photospheric velocity gradient which shows how the expansion builds up. The average velocity gradient for seven Of stars investigated is 7 km s^{-1} eV^{-1} with no systematic dependence on the spectral subtype.

A summary of the mass-loss of Of stars is given in Section 7.4 and Table XXXIV.

Some Of stars are surrounded by a dust nebula, as is shown from infrared measurements. Cohen and Barlow (1973) list two such cases: NGC 7635 (central star: BD+60°2522; O6.5 IIIf), and IC 1470 (central star has spectral type O7f). See also Section 6.15.

The line spectrum: apart from the lines mentioned before, other interesting lines in the visible part of the spectrum are the C III lines at 4647–51 and 5696 Å. The near infrared spectra of Of stars were investigated by Mihalas *et al*. (1975a), and by Y. Andrillat and Vreux (1975). On the enhanced near infrared continuous radiation there is only one emission line, that of He I $\lambda 10830$, which occurs in virtually all Of supergiants (investigated were all spectral subclasses from O3fI through O9fI), and in practically none of the O V – or O((f)) – stars. Furthermore there are the lines of C III $\lambda 8500$, 9701–15 Å. The Paschen lines of hydrogen, particularly Pδ, and the He II line at 10124 Å occur faintly in emission in some stars; in others they are neither absorption nor emission lines.

The interpretation of the line spectra with their remarkable profiles and intensities offers problems. There is similarity in the emission spectrum of Of-type stars and Wolf–Rayet stars but there are essential differences. For instance hydrogen does not occur with 'normal' abundance in the atmospheres of Wolf–Rayet stars whereas the Of stars appear to have the same hydrogen abundance as other O-type stars. Conti and Leep (1974) have given some qualitative indications how to explain the differences between the emission profiles with central, unshifted absorption, as observed in Oe or Be type stars, and the P Cygni type profiles as seen in some – but not all! – lines of Of type stars, – see Figure 27. In this simplified picture we assume an extended

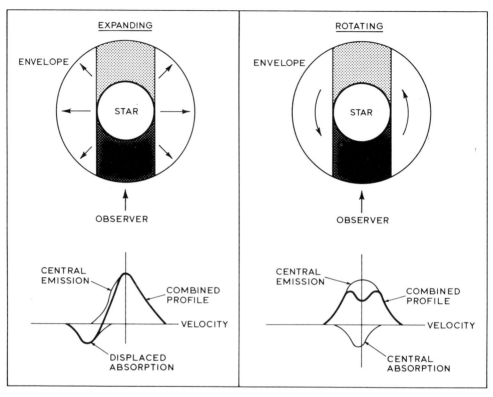

Fig. 27.   Idealized geometrical representations of a spherical envelope with expansion (left) or rotation (right). The shaded volume behind the star is not visible to the observer; the darker volume in
front produces the absorption line; and the volumes on either side of the star produce the emission
part of the line. In the expanding case, the absorption component is violet displaced, although the
emission is central. The combined profile is a P Cygni type. In the rotating case, both the absorption
and the emission are central. The combined profile gives the characteristic double emission appearance. (Conti and Leep, 1974).

envelope, which produces the emission lines; its thickness may be comparable to the
photospheric radius. Normally, in the envelope $dT/dr < 0$. That part of the envelope
which is in front of the photosphere should produce an absorption feature, whereas
the part on either side of it would produce an emission line. (In a more strict reasoning
this simplified picture cannot be correct – it is the source function of the line in the
atmosphere, and its distribution with height that defines the ultimate profile; if
$S_{env} > B_{phot}$, the part in front of the star does not produce an absorption line; furthermore, problems of redistribution in the line profile can play an important role; cf. for
a further discussion Section 6.2). Following this reasoning the P Cygni profiles of
Of stars should indicate the presence of an expanding envelope – according to the
left-hand part of Figure 27 – whereas the symmetric profiles with central absorption,
as they occur in the Oe- and Be-type stars, should be explained by the right-hand part
of Figure 27, which is a rapidly rotating star with negligible expansion. Complicated
situations arise when the outflow velocity and that of rotation are comparable.

Although this explanation is elementary and schematic it helps! Other geometrical structures may be considered with different consequences for the spectrum.

Let us next consider the interpretation of a few of the characteristic lines of Of stars! The C III lines at 5696 Å ($3p^1P^0$-$3d^1D$) and 8500 Å ($3s^1S$-$3p^1P^0$) are interesting because the former is an emission line while the latter occurs in absorption. In a first investigation Mihalas et al. (1975a) suggested that this could be explained with the assumption that they are formed in a relatively extended envelope, where the dilution factor is ~0.1. But Cardona–Núñez (1978), who discussed the non-LTE behaviour of these lines on the basis of a detailed solution of the coupled statistical-equilibrium and transfer equations, could reproduce the observed behaviour of the lines for a plane-parallel atmosphere! The explanation is found in the relative overpopulation of the $3d$-level by means of recombination from C^{4+} followed by downward cascades, while the – intermediate – $3p$-level is drained by the process of two-electron transitions which couples the $3p$-level to the (lower) $2p^2(^1S, {}^1D)$-levels. Thus, for $\lambda 5696$: $T_{ex} > T_{br}$ yielding an emission line, while the reverse is true for $\lambda 8500$. This behaviour is the case for any temperature above ~7000 K.

The N III emission, initially the main characteristic for the Of stars, can similarly be understood by a detailed computation of the whole ensemble of the levels in the N^{++}-ion, and including transitions to the N^{+++}-ion, as was done by Mihalas et al. (1972), and Mihalas and Hummer (1973). It appears that the upper ($3d$) levels of the transition are filled by di-electronic recombination from an auto-ionizing N IV level. The relative overexcitation of the $3d$ levels with respect to the lower ($3p$) level, yielding a line excitation temperature, higher than the brightness temperature of the (local) continuum and thus producing emission lines, is due to the arrangement of the transition probabilities from $3p$ downward. It thus appears that for the N III lines to occur in emission, the N^{+++}-ion should be sufficiently abundant. Mihalas et al. (1972) further show that the N III emission lines can also be formed in a plane atmosphere, and are therefore not typical for an envelope. We *conclude* that the behaviour of the C III and N III lines in spectra of Of-stars can be understood with the plane-parallel atmospheres, and just reflects special density and temperature combinations. However, emission in the Hα line (and in $\lambda 4686$ He II which is correlated to Hα emission) appears to be caused by an envelope (Mihalas, 1974).

Hence O((f)) stars which do not show He $\lambda 4686$ emission do not have an envelope and will have a small mass-loss rate, while Of stars, having He II emission, do have an envelope and a larger rate of mass-loss. This was the main reason for Conti (1976) to propose defining Of stars as O stars with envelopes, as shown by emission of He II $\lambda 4686$ and/or Hα.

A prototype of an Of star: Zeta Puppis. The hottest of the bright Of stars is ζ Pup, with spectral type O4 I(n)f (Walborn, 1972b), or O4ef (Conti and Leep, 1974). The visual spectrum was described by Baschek and Scholz (1971), Heap (1972), and Conti (1976); the UV spectrum by Morton (1976), Holm and Cassinelli (1977), and very detailed by Morton and Underhill (1977). An absolutely calibrated rocket spectrum (12 Å to 15 Å resolution) was obtained by Brune et al. (1979).

The *main parameters* of the star were summarized by Lamers and Morton (1976), from which we abstract: Distance $d=450\pm50$ pc (Brandt *et al.*, 1971); $M_v=-6\pm0.25$; BC$=-4.5$, hence $M_b=-10.6$; $L=1.44\times10^6 L_\odot$; angular diameter $\theta=(4\overset{\prime\prime}{.}2\pm0.3)\times$ $\times10^{-4}$ (Hanbury Brown *et al.*, 1974), hence $R_*=20.3\pm2.7R_\odot$; the photospheric radius seems to be smaller: $R_p=16R_\odot$; acceleration of gravity: log $g=3.8\pm0.2$ (cm s^{-2}) at R_*. These data would yield $\mathfrak{M}=60\mathfrak{M}_\odot$ to $90\mathfrak{M}_\odot$. In an other approach the mass is estimated from a comparison of the position in the HR-diagram with computed evolutionary tracks (Conti and Burnichon, 1975): $\mathfrak{M}=100\mathfrak{M}_\odot$ (see also Section 2.4). Such a determination depends on theoretically predicted evolutionary paths and may yield values that are too high as shown in Section 2.4.

The star is a fast rotator; several values were derived:

- Barnes *et al.* (1974)   385$\pm$30 km s^{-1};
- Snijders and Underhill (1975)   200 km s^{-1};
- Conti (1976)   300 km s^{-1}; from He I and II lines;
- Morton and Underhill (1977)   260$\pm$60 km s^{-1}, from weak ultraviolet lines.

The determination of the *effective temperature* yields large problems. The value, derived from L and R or from absolute photometry would be $T_e=32510\pm1930$ K (Code *et al.*, 1976) or 31900 ± 1800 K (Brune *et al.*, 1979). These values are similar to those of the later type stars ζ Oph (O9.5 V) and τ Sco (B0 V), and seem therefore much too low. To rescue this situation, Lamers and Morton (1976) suggested that the photospheric radius be smaller than the value measured interferometrically (viz. $R_{\mathrm{ph}}=16R_\odot$); then T_e could become 50000 K, a value that would fit the observed ratio of equivalent widths of He I $\lambda4471$ to He II $\lambda4541$. However, this would not solve the problem posed by UV spectrophotometry. The matter of the T_e-determination is therefore further examined on the basis of the continuous and line spectra: The continuous spectrum of the star seems flatter than should be expected from the high temperature suggested by the spectral line features. A plane-parallel atmosphere with an effective temperature of about 5×10^4 K would give a spectrum that is much too steep for the observations. Mihalas and Hummer (1974a) and Cassinelli and Hartmann (1975) gave arguments showing that this problem could be remedied by assuming a spherical extended outflowing atmosphere (see Sections 5.4 and 5.5). Their two solutions should certainly not be considered as the only possible ones, since the parameters of the models used there, have not been checked for consistency with regard to the main parameters of the star. This doubt is strengthened by an investigation by Castor *et al.* (1975) who computed a stellar outer atmospheric model, thought to be representative for ζ Pup (see also Section 5.9). The model included a curved atmosphere, *and* outward streaming motions (the radiation-driven stellar wind). The predicted energy-wavelength curve of the model did not show the ultraviolet flux deficiency, demanded by the observations. From a redetermination of the absolute spectral energy distribution, derived in the UV from OAO-2 observations (Bless *et al.*, 1976), Holm and Cassinelli (1977) could find agreement in two cases. The first was a model with $T_e=50000$ K, log$g=4.5$ (Kurucz *et al.*, 1974). It is to be

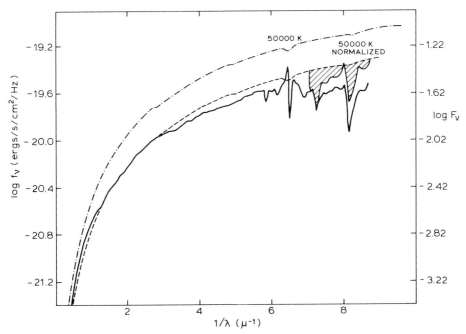

Fig. 28. The energy-frequency curve of ζ Pup corrected for $E(B-V)=0.04$ with the mean extinction curve. The angular diameter (Hanbury Brown *et al.*, 1974) was used to determine the intrinsic stellar surface flux. For comparison, the flux F_ν from a 50 000 K, $\log g=4.5$ plane-parallel model atmosphere (Kurucz *et al.*, 1974) is shown and in addition the flux from this model after adjustments made to approximate the appearance of a strong stellar wind model (Castor *et al.*, 1975): A 33% increase in the effective radius by the halo effect has been assumed to normalize the model flux to the observed flux at 2.257 μm^{-1}. An empirical correction for line blocking (illustrated by the hatched area) from high-resolution Copernicus scans is used to allow for the effects of line blanketing in an expanding envelope. (From Holm and Cassinelli, 1977).

remarked that a model with such a value of $\log g$ is plane-parallel – see our considerations in Sections 5.2 – and has $\Gamma \ll 1$, a result that does not contradict the results of our Figure 17 – a problem still to be solved. After the application of an approximate $\Delta f(\lambda)$ correction to account for a strong stellar wind, and after having reduced the flux at 4430 Å by a factor 0.57 ± 0.09, (such a reduction would mean that the radiation is not emitted by the photosphere proper but by a halo with radius $1.33R_*$), agreement between theory and observations was obtained. The second solution proposed was obtained by taking a model with $T_e=42\,000$ K, and assuming a flux correction at 4430 Å of 0.68 ± 0.10, needing a halo of $1.21R_*$. The thus 'corrected' energy distribution lies only 11% above the observed energy distribution in the ultraviolet which is fully in the brackets of the expected errors in the absolute energy calibration and in the determination of the star's radius, particularly if the uncertain far ultraviolet line blocking factor is also considered (Figure 28). Earlier attempts by Mihalas and Hummer (1974a, b) who assumed a *curved* atmosphere with $T_e \approx 40\,000$ K, and $\Gamma \approx 1$ did not lead to satisfactory agreement – see also Section 5.4.

A high T_e-value was found from a study of Copernicus observations of the helium spectrum by Snijders and Underhill (1975) who compared observations of the Balmer, Paschen, Bracket and $n=5$ series of He II with non-LTE predictions obtained for assumed non-LTE-plane-parallel model atmospheres by Auer and Mihalas (1972). The agreement between observations and theory appears to be good for the upper members of the $n=3$ and $n=5$ series for a model with $T_e=50000$ K and $\log g=4.0$. The higher members of the $n=4$ series (4–17 and higher), and the two members of the $n=2$ series appear stronger than is predicted by the model. Also the He II $\lambda 6683$ line is much stronger than in any other star (Conti and Frost, 1977). Tentatively, Conti and Frost (1977) suggest to explain these observations by assuming that ζ Pup has an abnormally small H/He ratio.

We *conclude* that observations of the continuous and of the line spectra of ζ Pup suggest $T_e \approx 46000$ K ± 4000 K.

No clear information is yet available on the *photospheric velocity field* or on the microturbulent velocity component of ζ Pup. Although an average $v_\mu=5$ km s^{-1} is quoted (Lamers and de Loore, 1976), actually there are no differences in the computed spectra with $v_\mu=0$ and 15 km s^{-1} (Snijders and Underhill, 1975). With the large rotational velocity of this star it seems difficult to detect such small turbulent velocities. Brucato (1971) reported large variations of the equivalent width of He II $\lambda 4686$ on a time scale of ten minutes, but in a critical discussion of the expected variations in the measured equivalent widths, Conti and Frost (1977) made it clear that there is no reliable support for this statement.

The *stellar wind* of ζ Pup was studied by various authors; cf. the description of model calculations of the radiation driven wind of an O4f type star in Section 5.9, and the review of the various discussions of the wind observations in Section 7.3. The conclusions reached there can be summarized as follows:
– there is a thin corona at the basis of the wind. In order to ionize the various ions of C, N, and O its temperature should be $\sim 5 \times 10^6$ K (Olson, 1978);
– the star has a rate of mass-loss of $\sim 6 \times 10^{-6}$ $\mathfrak{M}_\odot$ yr^{-1}.
– the stellar wind velocity can change visibly in a time span of a few hours only.

Other information on the stellar wind of ζ Pup is derived from the radiation emitted in the *microwave* region; at 14.7 GHz Morton and Wright (1978, 1979) measured a flux of 6.7 ± 1.5 mJy; at 5 GHz Abbott *et al.* (1980) measured 1.4 mJy. This is free-free emission of an ionized stellar wind; it yields a mass-loss of $(5 \pm 1.5) \times 10^{-6}$ $\mathfrak{M}_\odot$ yr^{-1}. From microwave and infrared observations of the continuous spectrum Castor (1979) found a temperature in the stellar wind $T_{el} \approx 42000$ K. This value refers to the most remote parts of the stellar envelope (see the discussion in Section 6.13).

We *conclude* that observations of the envelope and wind of ζ Pup lead to a consistent picture: the photospheric temperature is 46000 K; higher up there is a hot ('coronal') region fairly close to the star; the most remote parts of the wind have $T_{el} \approx 42000$ K. Hence the wind originates *in*, and extents *above* the 'corona' – a fundamental result! The corona itself must be fairly thin in order that the resonance-line-forming region is still sufficiently dense to produce lines as strong as they are actually observed. This model of a stellar wind, the *corona plus cool wind model* has been developed and elaborated mainly by Cassinelli and coworkers (review by Cassinelli, 1979), on the basis of a proprosal by Hearn (1975a).

3.3. The Most Luminous Galactic Star

This seems to be HD 93129A classified O5 by Smith (1968b) and later as O3 If* by Walborn (1971c). It is a member of the cluster Tr 14 in the η Car complex. Conti and Burnichon (1975) attribute to it a bolometric magnitude of approx. -11.5 mag, hence $\log(L/L_\odot) \approx 6.5$. The T_e-value has not yet been determined but if one would assume $T_e = 52\,000$ K (from a comparison with ζ Pup (Section 3.2) and with HD 93250, below) then its position in the HR diagram would be defined (Figure 2). The mass would then follow from a comparison with the theoretical main sequence, yielding $\mathfrak{M} > 120\ \mathfrak{M}_\odot$. Further $R \approx 20 R_\odot$ and $\log g > 3.9$. Its spectrum is shown in Figure 29. The star seems to be spectroscopically variable perhaps in the light of the different classifications given, but also because the He II $\lambda 4686$ emission has shown variations in wavelength and intensity (Walborn, 1971c). Another aspect of the spectrum is a faint indication of a P Cyg profile at N v $\lambda 4604$. Walborn (1971c) describes several similarities between the spectrum of this star and two bright WN stars (HD 92740–WN7, and HD 93131–WN6) found in the same region (Figure 29). The comparison suggests that the main difference between HD 93129 A and the WN star HD 93131 is the existence of a considerable amount of ejected material around the latter star.

Another bright object, one of the brightest known, is η Car a hot star embedded in gas and dust. The object, and its remarkable recent history are described in Section 6.17.

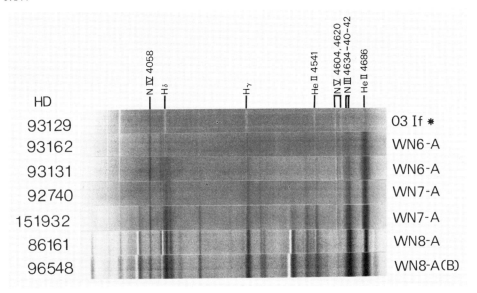

Fig. 29. Comparison of spectra of the O3 If* star HD 93 129 A and six southern narrow line WN stars. Cerro Tololo photographs. (By courtesy of Walborn, 1974; original print: *Astrophys.J.*).

Comparable to HD 93 129 A is the star HD 93 250 O3 V((f)) also in the η Carinae nebular complex. The continuous spectrum from 1500 Å to 50 000 Å was measured by Thé *et al.* (1980a). On the basis of these and spectral data Kudritzki (1979) determines $\log L/L_\odot = 6.4$, $R/R_\odot = 19$, $M_b = -11.2$, $T_e = 52000$ K, $\log g = 3.95$. Its mass should then be $\mathfrak{M} \approx 120 \mathfrak{M}_\odot$ (Conti and Burnichon, 1975), as determined from a comparison of these data with model calculations. The mass loss is astonishingly low for such a bright massive star: $-\dot{\mathfrak{M}} = 2 \times 10^{-7} \mathfrak{M}_\odot$ yr^{-1} (Conti, 1979), and this fact is apparently related to the ((f))-characteristics of the spectrum, which, as stated before, indicates the absence or weakness of a shell or stellar wind. It is remarkable that the two stars described in this section, which differ so little in absolute luminosity and in the other parameters ($\mathfrak{M}$, R) are so different as far as the wind-characteristics are concerned – a phenomenon that should be investigated!

3.4. The Wolf–Rayet Stars

These stars were first observed by Wolf and Rayet in 1867. A review was given by Underhill (1968). Symposium proceedings were edited by Gebbie and Thomas (1968) and by Bappu and Sahade (1973). A catalogue of WR-stars was published by Van der Hucht *et al.* (1980).

The *spectrum* contains broad emission lines of ions of He, C, N, and O, super-imposed on a continuous spectrum with an energy distribution like that of an O or a B-type star. The UV has also lines of Si IV and Al III. Many resonance lines and some others have P Cygni profiles (see definition in Section 3.5). The emission lines can have large widths, up to hundreds or even thousands of km s^{-1}.

There are two *sub-groups*: The WC stars in which the lines from ions of C (C III, C IV) and O dominate and with the NV resonance line at 1244 Å as the only conspicuous N-line, and the WN stars in which the lines from ions of N (N III, N IV) dominate. The subclasses are WN3–WN8, and WC5–WC9 based on line intensity ratios of the C, N, and O ions, analogous to the O-type stars. Near infrared spectra of these stars have been obtained by Kuhi (1966), Barnes *et al.* (1974), Andrillat (1976), Bappu *et al.* (1977), Bernat *et al.* (1977). There appears to be a marked difference between the line spectra of Carbon WR stars and those of the nitrogen sequence: the former showing a spectrum, rich in emission lines of He I, He II. The C III line at 9713 Å is the most intense in the infrared spectrum. The WR stars of the nitrogen sequence has only few near infrared emission lines: these of He I 10830 Å, N III 10430 Å, He II 8237 Å, and 10124 Å, the latter being the most intense. Spectra from 0.9 to 1.7 μm of HD 50896 (WN5) and HD 151 932 (WN7) (Bernat *et al.*, 1977) show only emission lines of H, He I, and He II.

Hiltner and Schild (1966) have introduced two parallel N sequences: WN–A stars have relatively narrow lines and an enhanced continuum; in WN–B stars the lines are wider and stronger.

There are similarities between WR stars and Of stars, both having broad emission lines, but these in WR stars are somewhat broader, see Figure 30. In both types of

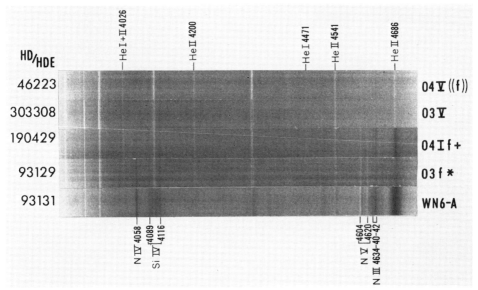

Fig. 30. Similarities between early O spectra and a WR spectrum suggest physical correspondence between these types. Cerro Tololo and Kitt Peak photographs. (By courtesy of Walborn, 1971c; Original print: *Astrophys. J.*).

stars the lines of the Balmer series show a marked v_{rad}-progression, indicating a photospheric velocity gradient (Conti *et al.*, 1979). There are differences too: the v_{rad} gradient is stronger in WR stars. Furthermore in the WR stars the He II 4686 Å line is markedly stronger than in Of stars. In WR stars the lines of the Balmer and Pickering series have P Cygni profiles. See the spectra in Figure 29. All this should be interpreted by assuming that WR stars have more developed envelopes than Of stars. It may be significant that a C–N dichotomy has also been claimed to exist in Of stars (Swings and Struve, 1941b). This suggestion has not been examined afterwards.

The similarity between these two groups of stars is further supported by Conti's finding (1976) that some WN7/WN8 stars contain appreciable amounts of hydrogen, and have N-emission-spectra, whose relative intensities are close to these of O6–7f stars (Leep, 1978). Their emission line widths and strengths are less than in most other WR stars. Other arguments that at least some of the WN7/WN8 stars form a somewhat separate class between Of and WN stars are (1) they seem slightly brighter than other WN stars (not very convincing: Table XVI); (2) they seem younger than the others – as follows from cluster membership data; (3) the ionization and excitation in the envelope differs from that in other WN stars (Moffat and Seggewiss, 1979). In some cases the upper Balmer lines are in absorption (Niemelä, 1973), and these WN7/WN8 stars are then called 'transition WN7/8 stars'. Note that not all WN7/WN8 stars are 'transition stars'. Interesting is the case of HD 15570, classified in the visible as O4f, but the ultraviolet spectrum shows N IV λ1486 and He II λ1640 strongly in emission – these spectral characteristics belong to WN stars rather than to Of. From a comparison with WN7 and Of spectra Willis and Stickland (1980) suggest that this star may be intermediate between Of and WN7 stars.

In the Carbon sequence there is a correlation between the width of the emission features and the spectral type. Actually, the width can be used as a criterion for the excitation. This however is not the case in the N sequence (Beals, 1938) and this was the main reason for the introduction of the narrow-line (WN–A) and broad line (WN–B) sequences. There are, however, intermediate widths and perhaps a continuous transition between the groups. This has brought Walborn (1974) to introduce the sequence of categories A, A(B), AB, (A)B, and B to denote increasing line width. See the difference in the spectra of HD 86161 and HD 96548 in Figure 29: all intensity ratios are approximately equal in the two stars but all lines are broader in the latter object.

The apparent *lack of hydrogen* in the atmospheres of WR stars has been noted by Kuhi (1973), Paczynski (1973), and others. This brought the suggestion that WR stars are hydrogen-poor stars, a conclusion with interesting consequences.

At the time of writing (1980) about 300 Wolf–Rayet stars are known, including 164 in our galactic system (Van der Hucht et al., 1980), 101 in the Large Magellanic Cloud (Fehrenbach et al., 1976; Melnick, 1978; Azzopardi and Breisacher, 1979), eight in the Small Magellanic Cloud (Breysacher and Azzopardi, 1979a) and 25 in M33. It is intriguing that the spectral types WC 6, 7, 8, and 9 have not been observed in the LMC. All but one of the WR stars in the SMC are WN stars. The narrow-line WN–A stars are usually found in massive young regions such as 30 Doradus in the LMC, the η Carinae Nebula, Scorpius OB1. It is remarkable that the 30 Dor complex contains a large number of WN7/8-transitory-type stars.

Many WR stars may be *members of a binary*; Kuhi (1973a) estimates the frequence of binaries among the WR stars at 73% but later data from Conti and Vanbeveren (1980) reduce this number to $\sim$36%. Burton et al. (1978) observed ultraviolet variability in several WR binaries and only one supposed 'single' star. Moffat and Haupt (1975) even given arguments, derived from photometric observations of short-term brightness fluctuations that most population I WR stars are members of binaries, but this is far from sure. In the large Magellanic Cloud 28% of the WR stars have been found to be binaries by Fehrenbach et al. (1976), and $\sim$40% by Conti and Vanbeveren (1980). Most WR stars in the SMC are binaries (Breysacher and Westerlund, 1978; Breysacher and Azzopardi, 1979b). Duplicity is hard to detect, and the above percentages are therefore lower limits. However, there are also clear indications that not all WR stars are binaries. Moffat (1978) showed that HD 93162 (WN7) in the Carina cluster Trumpler 16 is certainly single; the same applies to the WN7 star HD 93131 (Conti, 1978b).

A relation between the A–B subclassification and duplicity has not yet been established. Of those WR stars that are members of a binary, the companion is in nearly all cases an O- or B-type star (e.g., Cohen and Kuhi, 1977). An exception is the system HR 6392 which consists of a G5 Ia primary with $M_v=-9.0$, $M_b=-9.25$, and a secondary of spectral type WN6 with $M_v=-4.1$, $M_b=-9.8$ (Andrews, 1977; Moffat and Fitzgerald, 1977) – the components of this visual binary are sufficiently far apart that no mass exchange has taken place: therefore this couple is a difficult one for any theoretical scenario explaining WR stars by mass exchange.

Masses have been determined for ten WR stars in binaries. Table XIV list these data. From these values a – still uncertain – average mass of 10 $\mathfrak{M}_\odot$ follows. The WR masses are in most cases smaller than those of the companions. Table XIV contains one couple for which this is not true: the extraordinary case of HD 92 740 (WN7+O) for which the masses are much larger than all others. This may be another indication

TABLE XIV

Masses and periods of Wolf–Rayet binary stars after Smith (1968), Kuhi 1973a), Niemelä (1976, 1979), Moffat and Seggewiss (1977), Niemelä and Sahade (1979), Vanbeveren and De Loore (1980). All masses are in solar units.

HD	$\mathfrak{M}_1 \sin^3 i$	$\mathfrak{M}_2 \sin^3 i$	$P(d)$	Spectra	Remarks
113 904	11	30	18.34	WC6+O9.7 Iab	θ Mus
152 270	13	35	8.89	WC7+O8	
168 206	8.3	35	29.64	WC8+B0(?)	CV Ser
68 273	17	32	78.5	WC8+O9 I	γ^2 Vel
190 918	5	20	112.7*	WN4+O9 I	
186 943**	5	21	9.56	WN4+B	
193 576	11	26	4.21	WN5+O6	V 444 Cyg
90 657	6.8	13.6	6.46	WN5+O6	
211 853	6	20	6.69	WN6+O6 I	GP Cep
92 740	64	24	80.34	WN7+O	

* Fraquelli (1977).
** Ganesh and Bappu (1968).

that some WN7 stars are 'transition stars'. But there are a few more cases in which $\mathfrak{M}_w/\mathfrak{M}_2 > 1$. These are like HD 92 740 *all* W7 stars and are listed below:

HD 92 740 WN7+? $P = 80\overset{d}{.}3$ 4 $\mathfrak{M}_w/\mathfrak{M}_2 = 2.7$;   Conti *et al.* (1979), Niemelä
                                                                       (1979);

HD 192 641 WC7+B5–7e   4.5 > 1;   Bracher (1966);
HD 197 406 WN7+A7 (?)   4.3 $= 2$ to 4; Bracher (1966), Moffat (1979).

Abundances: The (H/He) ratio, compiled from results from various authors, is given in Table XV. The (H/He) ratio is very hard to determine for WC stars because of severe blending of carbon lines with the Pickering series. In both classes there are large differences between results from various authors, and this reflects the difficulty in the spectrophotometry as well as the uncertainty in the photospheric models, and additionally the difficulty of the interpretation of the emission lines. Nevertheless the certain outcome is that (He/H) > 1, and even close to 10 in several stars. So, the atmospheres of WR stars are strongly H-deficient.

The N and C abundances in WN stars are higher than in normal main-sequence stars. Nugis (1975) reports average abundance ratios $\langle$(N/He)$\rangle = 0.07$ (for O, B stars: 0.001), and $\langle$(C/He)$\rangle = 0.004$ (for O, B stars: 0.0015). For WC stars the C abundance is still higher: $\langle$(C/He)$\rangle = 0.8$.

TABLE XV

The H/He ratio in Wolf–Rayet stars

WN3	WN4	WN5	WN6	WN7	WN8	References
0.8	0.4	0	0.4	1.0	2.3	Smith (1973)
			0.13			Review: Rublev (1975)
		<0.06	0.2			Nugis (1975)

				WC7	WC8	
				0.11	0.07	Review: Rublev (1975)
				<0.03		Nugis (1975)

The relative abundances (C/N), (C/He), (N/He) were determined by Willis and Wilson (1978, 1979) who found, assuming a Zanstra temperature of 40 000 K (taking 30 000 K makes only little difference):

HD 50 896   (WN5, T_{el}=50 000 K): (C/N)=5.4×10^{-3}; (C/He)=9.0×10^{-5};
HD 192 163 (WN6;   50 000 K); 2.6×10^{-3} 9.0×10^{-5};
HD 191 765 (WN6;   50 000 K): 4.7×10^{-3} 9.0×10^{-5};
HD 192 103 (WC8;   30 000 K): 2.9×10^{-3} 9.0×10^{-3};

(N/He)=1.7×10^{-2};
          3.5×10^{-2};
          1.9×10^{-2};
          3.2×10^{-3}.

This result therefore suggests that the carbon abundance is the controlling factor in determining the difference between WC and WN stars. Furthermore, the low H abundance and the high abundances of C and N indicate that WR stars are evolved objects that have lost the greater part of their mantle, or in which nuclear reaction products are brought to the surface otherwise.

The *continuous spectra* of these stars have been observed over a very broad spectral range from the ultraviolet to the microwave range. A set of ultraviolet low dispersion spectra of nine stars was obtained by Willis and Wilson (1978) with the spectrograph S2/68 on the ESA satellite TD-1A. These cover three WC, three WN and three WC+O binaries. The ultraviolet fluxes, compared with model atmospheres of Kurucz *et al.* (1974), yielded colour temperatures of ~30 000 K. Similar results (25 000–35 000 K) were found by Nussbaumer *et al.* (1979) from IUE ultraviolet spectra of 15 WR stars. The largest survey of the continuous spectra of WR stars is by Van der Hucht *et al.* (1979b), who discussed observations of the energy distribution in the continuous spectra down to 1550 Å for 36 WR stars observed with the Astronomical Netherlands Satellite. Among these are 13 single WN stars and 8 single WC stars.

The continuous spectrum has in many cases an *infrared excess* and microwave emission (Hackwell *et al.*, 1974; Wright and Barlow, 1975; Cohen *et al.*, 1975; Florkowski and Gottesman, 1977; Cohen and Vogel, 1978). This emission is related to a stellar wind. For the WN star HD 50 896 Hartmann and Cassinelli (1977) found indications for the existence of an electron-scattering outer region, expanding with practically constant velocity.

Microwave emission has been observed for γ^2 Vel (Seaquist, 1976); HD 193 793 (Florkowski and Gottesman, 1977); HD 192 163 (Wendker *et al.*, 1975). This radiation must be emitted by a stellar wind with a temperature higher than that of the stellar photosphere.

A strong *stellar wind*, related with considerable *mass-loss* is a characteristic property of the WR stars. The P Cygni profiles of UV resonance lines have 'edge or: terminal velocities' v_∞ (=velocities defined by the wavelength of the short-λ edge of the line) between 1000 and 4000 km s^{-1}, dependent on the ionization potential IP (Conti and Van der Hucht, 1979; Nussbaumer *et al.*, 1979), with, for WC stars, roughly $v_\infty =$ $=-3000+120\times$IP km s^{-1}, with IP in eV. In Section 7.4 we discuss the mass-loss of WR stars, and show that on the average $-\dot{\mathfrak{M}}\approx10^{-4}\,\mathfrak{M}_\odot$ yr^{-1}. Thus, WR stars rank among stars with the highest known rate of mass-loss. The envelope, due to the stellar wind, must be highly ionized and flattened, as follows from the polarization which changes from about 1% in the continuous spectrum to smaller values in the cores of the emission lines (McLean *et al.*, 1979a).

In some cases WR stars are observed to be related to *gaseous shells*; apparently matter that has been emitted by the star. A pronounced example is the WN6 star HD 192 163, and the associated nebula NGC 6888 (Wendker *et al.*, 1975). The mass of the nebula is estimated from radiodata as $5\mathfrak{M}_\odot$. It seems likely that this nebula has been ejected by the Wolf-Rayet star but in which phase of its evolution? If the ejection took place during the WR phase, then this phase must have lasted at least 5×10^4 yr (assuming a constant mass loss of $10^{-4}\,\mathfrak{M}_\odot$ yr^{-1}, see Section 7.4). This is not unreasonable. The chemical composition of the nebula was determined by Parker (1978) who found that N and He are overabundant relative to 'cosmic' abundances, and who suggests that strong internal mixing must have occurred in the WR star before ejection.

An alternative suggestion is that the nebula has been ejected *before* the WR phase. But how then to explain the deviating abundances? Moreover, since WR stars are losing mass anyhow, at least part of the nebula *must* contain gas of WR origin.

Another way to study those parts of the gaseous shells that are closely around WR stars is by investigating the emission lines at eclipse in WR eclipsing binaries. Such an investigation, by Khaliullin and Cherepashchuk (1976), of several ionic lines shows for instance that the distance-dependence of the 3D level population of the He$^+$ ion is described by $n_{3D}(r)\approx3900\times(r_0/r)^3$ cm^{-3} with $r_0=2.6R_\odot$, and $n_e/n_{3D}\approx$ $\approx4\times10^9$. The polarisation of the continuum and (to a lesser degree) the lines are also suggestive of an extended envelope (McLean *et al.*, 1979a), and may for a part be due to a dust envelope.

The latest-type WC stars, spectral type WC7–9 have *circumstellar dust* envelopes as appears from the reddening and for a part also from the polarization of the observed continuous spectra. The infrared excess near 3 to 4 μm has a colour temperature of about 800 to 1800 K, which must be caused by dust particles heated by the star

(Cohen and Vogel, 1978). Particularly interesting is the eclipsing binary HD 168 206 (CV Ser, WC8+B0). The dust temperature is $T_c=1400$ K, as found from the shape of the infrared excess curve. The excess shows slight eclipse dependence, which indicates that the dust may be located partly *between* the two stars. More detailed photometry of this system would certainly help to better localize the dust cloud. The WC9 star Tr 27–28 has a dust shell (?) with $T_c=1400$ K (Thé *et al.*, 1980b). Interesting was the (sudden?) formation of dust around HD 193 793 (WC7+O5) which occurred between June 1976 and September 1977 (Williams *et al.*, 1978), after an initial nearly continuous decrease of the infrared excess between 1970 and 1976 (Cohen *et al.*, 1975). The dust colour temperature, which was 900 K in May 1978 decreased to 780 K in August 1978 (Williams and Antonopoulou, 1979). For three WC9 stars investigated by Cohen and Kuhi (1977) the total extinction in the visual spectral range is 4.0 ± 0.5 mag (for the stars AS320 and MR82) and 7.7 ± 1.1 mag (for Ve 2–45). The degree of polarization increases rapidly towards short wavelengths, thus indicating Rayleigh-type scattering, probably at small condensation nuclei.

Visual absolute magnitudes and T_e-values have been determined by various authors. Table XVI is a compilation drawn from the reviews of Smith (1973), Barlow *et al.* (1980), and from Van der Hucht *et al.* (1979b). Since only a few WC stars were used for Table XVI, the results are rather uncertain for these stars.

TABLE XVI

Luminosities and T_e-values of Wolf–Rayet stars. The M_b-values for the WN-stars were calculated assuming average M_v- and average BC-values from the various authors.

Type	WN3	WN4	WN5	WN6	WN7	WN8	References
M_v		−4.3	−3.9	−4.8	− 6.8	− 6.2	Smith (1973)
	−3.7	−3.7	−3.7	−4.4	− 6.3	− 5.0	Crampton and Georgelin (1979)
		−3.9	−4.3	−4.8	− 6.8	− 6.2	Barlow *et al.* (1980)
				−3.9	− 6.9		Van der Hucht *et al.* (1979b)
	−5.4						Breysacher and Azzopardi (1979a)
$T_e/1000$ K		53	46	33		23	Morton (1969b)
		42	55	60	36	38	Barlow *et al.* (1980)
BC		−4.6	−4.2	−3.3	−2.3	−2.3	Morton (1969b)
		−3.1	−3.6	−4.1	−2.6	−3.2	Barlow *et al.* (1980)
M_b	−8.0	−8.0	−9.0	−8.5	−9.0	−8.5	Smith (1973)
		−7.0	−7.9	−8.9	−9.4	−9.4	

Type		WC5	WC6	WC7	WC8	WC9	References
M_v		−3.2	−3.2	−4.4	−4.8	−4.8	Crampton and Georgelin (1979)
		−4.4	−4.4	−4.4	−4.8		Barlow *et al.* (1980)
						−3.9	Cohen and Kuhi (1977)
			−3.7		−4.7		Van der Hucht *et al.* (1979b)
						−6.4	Thé *et al.* (1980b)
$T_e/1000$ K			60	54	46	38	Barlow *et al.* (1980)
BC			−4.4	−3.8	−3.8	−3.0	,, ,,
M_b			−8.8	−8.2	−8.2	−7.8	,, ,,

The results show a tendency for $-M_v$ to increase with increasing spectral subclass number. While there is a reasonable agreement on M_v-values among authors, there is strong disagreement on T_e-values, hence on bolometric corrections, hence on L-values and on the position of the WR stars in the HR-diagram. The stellar temperatures are determined with the hydrogen-Zanstra method, and are essentially based on a comparison of Balmer line emission in the envelope with the optical continuous spectrum of the star; the helium-Zanstra method refers to the He II λ1640 line (Willis and Wilson, 1978).

Morton (1969a, 1973) has attempted to apply Zanstra's method to the radio emission of the envelopes and found fairly low T_e-values, in contrast to values based on the conventional hydrogen-Zanstra method (Rublev, 1964), which gives temperatures of the order of some 8×10^4 K, in agreement with ionization temperatures determined from the He II line emission of the nucleus, but not with the colour temperatures of Willis and Wilson (1978, 1979) and of Nussbaumer *et al.* (1979), and neither with their He-Zanstra values, which yield temperatures around 30000 K. Rublev (1975) ascribes the lower values derived by Morton (25000–55000) to the fact that the *envelope* would have low temperatures, approx. $T_{env} \approx 35000$ K $\pm$ 10000 K. This low value would be reflected in the radio-emission from the shell, so that the Zanstra-temperature, which is in actual fact a colour temperature would be too low. This argument is not very strong.

A support for the high temperatures of WR stars is the determination of brightness temperature (80000 K) and bolometric magnitude ($M_b = -10.2$) from a comparison of the WR and O7 components of CQ Cep (Kartasheva, 1974).

In conclusion: the *effective temperatures* and *absolute luminosities* of WR stars offer still considerable problems. For the time being it is perhaps best to realize that for the well studied object γ^2 Vel (see later this section) for which most authors find a colour temperature of $\sim$30000 K, a comparison with model computations, and spectral line studies yields $T \sim 25000$ to 50000 K. I therefore assume for the WR stars $T_e \approx 40000$ K $\pm$ 15000 K. The corresponding values of BC, L, and the position in the HR diagram are similarly uncertain: T_e-values of 30000 K and 80000 K would correspond with $-BC = 3.2$ and 5.2 mag respectively, this corresponds with a factor 6 in luminosity. With a view to their average mass ($\langle \mathfrak{M} \rangle \approx 10 \, \mathfrak{M}_\odot$) the WR stars are, hence, stars that are close to or perhaps *above* the corresponding Eddington limit of $\log(L/L_\odot)_{lim} = 5.5$.

In connection with the high He abundance it is interesting that – in case the temperature is high – also the position of the WR stars in the HR-diagram (Figure 2), at the high temperature side of the H–ZAMS may indicate that He is abundant in the cores of these stars (see also Section 4.2). So WR stars may be in the He-burning, or in the He-H burning phase. When the temperature would be around 30000 K the stars would be close to, slightly to the right-hand side of the ZAMS.

Several attempts have been made to explain the position of the WR stars in the HR diagram by some *evolutionary scenario* – see Section 4.11 – but the main difficulty lies in their fairly small mass of approx. 10$\mathfrak{M}_\odot$ (Table XIV). It is hard to explain how such undermassive stars can have the large observed luminosities: the curves of Figure 2 show that evolved stars that have lost a considerable part of their mass have about the same luminosity as stars of the same $\mathfrak{M}$ and T_e that have evolved *without* mass-loss.

For an understanding of the evolutionary status of WR stars, also in relation to the Of stars, the 'WN7/WN8 transition stars', (see earlier this section) may be crucial. Do the transition stars descend from Of stars by mass-loss of the latter? With their high rate of mass-loss, 5 to 20×10^{-6} $\mathfrak{M}_{\odot}$ yr^{-1}; see Section 7.4) the Of stars would produce during their evolution a more and more impoitant envelope from the hydrogen-rich outer layers. In the Wolf–Rayet phase the envelope would be strongly developed, giving rise to intense broad emission lines, while the star's photosphere eventually would show the helium-rich core, or even a core in a further stage of evolution. Hence, during this evolution the photospheric H/He ratio should continuously decrease. Some support for this evolutionary scenario may be found in the finding by Moffat (1978) that the single stars HD 93 129 Å (O3, f*), HD 93 162 (WN7), and HD 93 131 (WN7) seem to follow one sequence, as judged from their spectra. Conti (1978b, c), found in the atmospheres of these latter two stars a very strong outward motion.

It should further be stressed that in this picture it is not of first importance whether the Of or the resulting WR is member of a binary or not: fundamental is that the rate of mass-loss should be – even in single stars like ζ Pup – sufficient to produce a developed envelope in a time comparable to or shorter than the evolutionary time scale of the star. The 'transition WR-stars' should then be those objects in which some hydrogen has still remained.

There are several objects that illustrate this scenario. An example is HD 193077 (WN5) which has absorption lines in its spectrum, an attribute that occurs normally in spectra of WR *binaries,* and is then assumed to be due to the companion. But HD 193077 is practically certainly single, and may be a 'new' WR star still showing some Of characteristics (Massey, 1979). A different case is the binary BD+40° 4220 (Bohannan and Conti, 1976; see pertinent data in Tables VIII and IX, and see Section 3.1) consisting of two Of stars of which the secondary, with $\mathfrak{M} \sin^3 i = 11 \mathfrak{M}_{\odot}$ is overluminous and shows all the characteristics of a Wolf–Rayet star; also this observation indicates a genetic relation between Of-stars and WR-stars. Support for this hypothesis is also provided by the binary HDE 228 766 (07.5I+05.5f); the secundary has $\mathfrak{M}/\mathfrak{M}_{\odot} \approx 24$ but is losing mass at a rate of 10^{-5} $\mathfrak{M}_{\odot}$ yr^{-1}. It shows some spectral characteristics, like He II $\lambda 10124$ in emission, that are similar to features of WR stars. The time needed to reach a typical WR mass (~ 10 $\mathfrak{M}_{\odot}$) would be $\sim 10^6$ yr (Massey and Conti, 1977). The evolution of WR stars is further discussed in Section 4.11.

A further question is what is the ultimate fate of a WR star. Sanduleak (1971) suggests that WC stars with strong O VI emission (the 'O VI-class') may develop into O VI planetary nebulae. This is still rather hypothetic. What becomes out of WN stars is not known.

Prototypes: The binary γ^2 Velorum ($M_v \approx 2.2$) contains, as its brightest member, the brightest Wolf–Rayet star. The star is a double-lined spectroscopic binary classified as WC8+O8 (Baschek and Scholz, 1971b) or WC8+O9 I (Conti and Smith, 1972). It is separated by 45″ from a fainter component, γ^1 Vel ($m_v \simeq 4$; B1 IV). The *orbit* was determined by Ganesh and Bappu (1967), who found $e = 0.17$ and $(\mathfrak{M}/\mathfrak{M}_\odot)$ $\sin^3 i = 13.0$ and 46.3, respectively, for the WC8 and O9 I star. From the absence of an eclipse in the continuous spectrum (although an eclipse is clearly visible in the stellar wind – see later) it is concluded (Moffat, 1977) that $i_{max} = 73°$ leading to $\mathfrak{M}$ (WC8) $\geq 15\mathfrak{M}_\odot$. For later mass-determinations see Table XIV.

Observations of the *continuous spectrum* as obtained with various resolutions from observations made from rockets (Stecher, 1970; $\Delta\lambda = 10$ Å; Burton *et al.*, 1975; $\Delta\lambda = 0.3$ Å; Brune *et al.*, 1979; $\Delta\lambda = 12$–15 Å) and satellite (Boksenberg *et al.*, 1974; Wilson and Willis, 1978; $\Delta\lambda = 35$Å) are partly summarized by Burton *et al.* (1976). These data are shown as Figure 31, which includes also observations in the visual range. Burton *et al.* (1976) conclude that in the ultraviolet part of the spectrum the t st agreement with model computations of the continuous spectrum is obtained with a model from Cassinelli (1971a) of an extended spherical atmosphere, with $T_e = 48\,900$ K at $\tau = 0.67$, and a so-called 'extension parameter' $R(\tau = 10^{-3})/R(\tau = 0.67) = 1.89$. Such conclusions are, however, weakened by the unknown contribution of the O-star component, which is in the visual part of the spectrum 1.5 mag brighter than the WC8 component: the continuous slope in the visual corresponds to a temperature of 32 000 K.

Contrasting to the above result are those of Willis and Wilson (1978, 1979) who

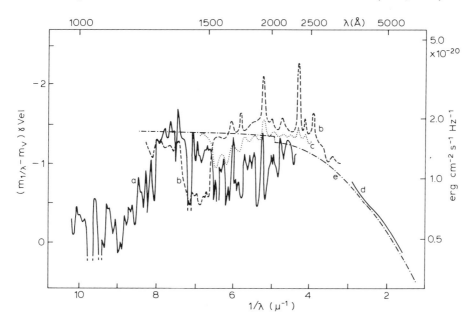

Fig. 31.   Measured absolute energy flux distributions for γ^2 Vel. The data from Burton *et al.* (1975) are shown as a curve '*a*' (————); those by Stecher (1970) in curve '*b*' (– – –); by Boksenberg *et al.*. (1973) in curve '*c*' (.........), and by Aller and Faulkner (1964) in curve '*d*' (————). Curve '*e*' (—·—·—) shows a model atmosphere calculation by Cassinelli (1971a). (From Burton *et al.*, 1976).

determined the effective temperature by using the measured fluxes, the angular diameters of the WC8 and O9 I components measured by Hanbury Brown *et al.* (1970) and the observation that the O9 component is brighter than the other by 1.4 to 1.8 mag in the ultraviolet. They thus found through an iterative procedure for the WC8 component: $T_e = 30\,200$ K. Similarly, Brune *et al.* (1979) from UV rocket spectrophotometry, found $T_e = 30\,600 \pm 2000$ K. These values are close to the colour temperatures and the Zanstra temperatures for other WR stars derived by Willis and Wilson (1978, 1979).

These results again show the uncomfortable situation with regard to the effective temperatures of the WR stars.

The *line spectrum* was described by Morton (1975); Johnsson (1978) published the high-resolution *Copernicus* UV-spectrum. It shows a number of P Cygni type profiles in the ultraviolet, including lines of C II, C III, C IV, and Si IV. The intersystem C III transition at 1908.7 Å has a weak absorption line at 1900.7 Å, at the shortwavelength edge of a strong emission feature. A suspected short-term variation in the emission line strengths was not confirmed (Haefner *et al.*, 1977). An analysis of the UV C III lines by Castor and Nussbaumer (1972) gives for the envelope an electron (kinetic) temperature $T_{el} \approx 20\,000\text{--}25\,000$ K, hence lower than the effective temperature of the photosphere.

As mentioned, the ultraviolet spectra show clear eclipse effects. This was initially studied by Moffat (1977) and in more detail by Willis *et al.* (1979). Particularly the emission lines C IV $\lambda 1550$, C III $\lambda 2297$, C III] $\lambda 1909$, and Si IV $\lambda 1400$ reduce greatly in emission intensity near phase 0.5. This is interpreted as an eclipse of part of the stellar wind of the WR star by the O8 component and this, in turn, supports the existence of a strong stellar wind.

The existence of an envelope with a fairly intense mass flow is also apparent from the microwave emission of the star; Seaquist (1976) communicates emission at 5.0, 6.27, and 8.87 GHz. The emission is due to an extended envelope consisting of ionized gas, radiating through free-free emission, and with a density gradient comparable to that expected from an outward flowing gas. A summary of the values for the rate of mass-loss is given in Section 7.4; the best value is between 3 and $10 \times 10^{-5}\,\mathfrak{M}_\odot\,\text{yr}^{-1}$. The envelope has an effective emitting radius of $\gtrsim 10^{10}$ km, much larger than the calculated Roche lobe of the WC star being 6×10^7 km, and also larger than the orbital radius of 2.4×10^8 km. The envelope, hence, extends far beyond the extent of the binary system.

The *second brightest Wolf–Rayet star* is θ Muscae (HR 4952, HD 113904 $v = 5.69$ mag., WC6 + O9.5 I/B0 Iab. From a study of the spectrum in the visual region Moffat and Seggewiss (1977) found that the velocity amplitudes of the C IV and C III emissions at 5801/12 Å and 5696 Å are 173 km s^{-1} and 493 km s^{-1} respectively, with the v_R-variation of the C III emission phase-shifted by P/4. This is interpreted by a gradient in the outstreaming motions, with large outward velocities (≈ 1500 km s^{-1}) in the outer (C III emitting) layers, and smaller velocities (≈ 1200 km s^{-1}) in the inner (C IV) layers.

The O companion hardly shows any orbital v_R-motion which would lead to an extremely large mass ratio, unless one assumes that the system consists of three stars:

(a) the observed WC6 star ($\mathfrak{M}_1 \approx 11 \mathfrak{M}_\odot$);

(b) a nearly invisible companion of mass $\mathfrak{M}_3$ which orbits $\mathfrak{M}_1$ in the observed period of 18.34 days;

(c) the observed O9.5/B0 Iab star of mass $\mathfrak{M}_2$, located relatively far from the pair $\mathfrak{M}_1$–$\mathfrak{M}_3$.

(The masses given in Table XIV are derived on the basis of this hypothesis, and are therefore uncertain.)

3.5. The P Cygni Stars and P Cygni Profiles

These stars show in the visual spectral region one or more lines with the *P Cyg characteristic*, which consists of a nearly undisplaced or slightly longward displaced emission component accompanied by one or more shortward displaced absorption components. The feature is explained as due to one or more expanding envelopes around the star. Most P Cygni stars resemble the spectral type Be, but there are also Ae types. There are transitions to Of and WR stars. It should be doubted whether the P Cygni stars form a separate group, and it seems advisable rather to speak of *P Cygni characteristics* or the *P Cygni phenomenon* (as described in Section 3.2; cf. Figure 27; and Section 6.2) which may occur for stars of different spectral types. Particularly many early-type *supergiants*, actually all those near the tip of the main sequence, have (ultraviolet) resonance lines with P Cygni shapes, indicating violent outstreaming motions (see for a discussion of outstreaming motions Chapter 7). A summarizing review of such motions was given by Morton (1975) who describes some 80 far-ultraviolet scans (resolution 50 km s^{-1}) of hot stars obtained with the Copernicus satellite. Snow and Morton (1976) published a catalogue of P Cygni lines in many hot stars. See also the detailed description of P Cyg profiles in ζ Oph (O9 Ve) and ζ Pup (O4ef) by Morton (1976, and see Morton and Wright (1979) for a description of O VI resonance line profiles in early-type stars. P Cygni profiles are observed for resonance lines of various states of ionization in all O-type stars, and the hottest B supergiants. The absorption components are strong; their shortward displacements correspond with outward velocities of between 1000 and 2000 km s^{-1}, with maximum velocities up to 3000 km s^{-1} at the edges. In some stars also the low lying exited levels of C III and N IV have P Cygni profiles. It *seems* remarkable that the intensity of the P Cygni characteristics varies inversely as the mass-flow rate (Hutchings, 1976c) but this is understandable qualitatively: the stronger the mass-flow, the denser the stellar wind and the closer to LTE is the situation in the line-forming region. For *main sequence* B-type stars like τ Sco there is no shift at the line centres, but the short-wavelength wings of the N V and Si IV doublets are depressed, showing outward motions; also in that case there is no P Cygni profile. Actually, Henize *et al.* (1975) find, from ultraviolet stellar spectra obtained with Skylab, that marked P Cyg profiles in C IV and Si IV, indicator of significant mass flow, occur in all stars with M_{bol}

brighter than −8.4 mag. So, P Cygni profiles are a general feature of hot B-type supergiants, and of O-type stars. If, however, one would wish to follow the classical line and would *define* the group of P Cygni stars as "hot stars, irregularly variable, with occasional fairly sudden brightness jumps, after which the star *may* stay fairly constant at the new brightness level, and with broad and generally diffuse absorption lines of H and He II, over which are superimposed narrower and variable emission lines with a central absorption", then still a fairly ill-defined group emerges, of which the brightest members are given in Table XVII. The apparent magnitude can range between wide limits (several magnitudes) with time scales of years; the spectrum is mostly early B; the stars have *not* all supergiant characteristics; there are often various shells of expanding matter with greatly varying velocities – related to the sequence of observed outbursts? The increasing narrowness of lines emitted from apparently higher layers may indicate a rotational velocity decreasing with height, or a decrease of the turbulence. An example of the confusion around this 'group' of stars is that the Be star γ Cas is sometimes classified as a P Cyg star.

TABLE XVII

Some of the brightest P Cygni stars

Star	m	Spectral type
P Cyg	3 ... 6	B1 eq; B0–B1 Ia; B1p
χ Oph	4.4 ... 5.0	B3 V pe
AG Car	7.1 ... 9.0	B5e ... Ae*
HR Car	8.2 ... 9.6	B2 eq
Z CMa	8.8 ... 11.2	Beq
TX CVn	9.5 ... 11.8	B1–2 eq**

  * Thackeray (1977) gives B0I ... A1 Ieq.
 ** Detailed description by Mammano and Taffara (1978).

The P Cyg star AG Car has $M_v = -8.5$ (Viotti, 1971) to −7 (Thackeray, 1977). It is surrounded by a ring-shaped nebula (Thackeray, 1950, 1977). The nebula is a radiosource: at $f = 5$ GHz the flux is $S_\nu = (285 \pm 40) \times 10^{-29}$ W m^{-2} Hz^{-1} (Milne and Aller, 1975). On the basis of these data and spectral line observations one finds that the nebula has $N_e = 1.7 \times 10^3$ cm^{-3}, and a mass of $(0.18 \pm 0.04)\mathfrak{M}_\odot$ (Johnson, 1976) or 0.25 $\mathfrak{M}_\odot$ (Thackeray, 1977), with $T_e = 7.5$ to 10×10^3 K. If AG Car loses mass at the rate of $10^{-4} \dots 10^{-5}$ $\mathfrak{M}_\odot$ yr^{-1} as other P Cyg stars do (see Section 7.4) then the nebula could have been ejected in 2×10^3 to 2×10^4 yr, and could have reached the present size with average speeds of 10 to 100 km s^{-1}. These quantities are all reasonable so that most likely the nebula has been ejected by the star. The (occasional?) occurrence of forbidden lines, such as the [Fe II] lines in the spectrum of AG Car and HR Car (reviewed by Viotti, 1971) is another indication for the presence of expanding shells presumably emitted at irregular intervals. This irregular emission is also manifest in the reported variations in the violet-displaced components of the P Cygni profiles of P Cyg (Viotti and Nesci, 1974). The P Cygni star TX CVn has $M_V \approx -3.7$, and systems of line components with $v_{exp} = -325, -140, -90$, and -20 km s^{-1}; the latter velocity represents the photospheric value (Mammano and Taffara, 1978).

Yet the problem remains: do the P Cyg stars form a physically homogeneous group? And: why is it that one near-unstable star becomes a WR star and another obtains P Cyg characteristics? What are the evolutionary relations of P Cyg stars to WR and Of stars and to an object such as η Car (Section 6.17)?

The *prototype* is P Cygni (HD 193237; B0–B1 Ia or B1 eq). The star was discovered by W. J. Blaeu on August 18, 1600, as a novalike object. It increased in brightness till the third magnitude, and remained at that brightness level for six years. There were other increases in brightness between 1654 and 1659 but since that time the star stayed more or less at the fifth magnitude, with irregular brightness variations.

The absolute visual brightness was determined by De Groot (1969, p. 270) and Hutchings (1976a) who found $M_v = -8.0$ and -8.4 respectively. However, Ambartsumian *et al.* (1979) determined the UV colour excess and found the star to be less reddened than assumed previously. This was confirmed by Hutchings (1979d), and yields a distance of 1.8 ± 0.7 kpc, not in disagreement with a distance of 1.2 kpc determined by Beals (1950). The effective temperature, as determined by Ambartsumian *et al.* (1979) is $(1.7 \pm 0.3) \times 10^4$ K, but Underhill (1979) who determined T_e and θ with the method of Blackwell and Shallis found $T_e = (1.2 \pm 0.2) \times 10^4$ K. Let us assume $T_e = 1.4 \times 10^4$ K. Ambartsumian *et al.*, and Hutchings derive $M_v = -7.6 \pm 1.0$. With the T_e-value the BC can be determined, hence $M_b = -8.5$ and $L/L_\odot = 2.0 \times 10^5$. Hence, with Equation (1.2;2): $R = 75 R_\odot$. If we assume $\mathfrak{M} = 40 \mathfrak{M}_\odot$ (on the basis of the star's position in the HR-diagram) then $g = 200$. (De Groot (1969) and Barlow and Cohen (1977) give the radius as $50 R_\odot$ and $90 R_\odot$, respectively.)

For the interpretation of the spectrum it is of importance to disentangle the absorption and emission components of the profiles. An example is given in Figure 32 (De Groot, 1969). This latter investigation also gives a detailed analysis of the spectrum, line identifications, intensities, radial velocities. Analyses of spectral lines shows that there are three shells, each producing differently shifted absorption lines. The two closest can not always be separated, and may also be close together in space. The sizes, given by De Groot are:
- radius of the star: $R \approx 50 R_\odot$;
- radii of first and second shell: $62 R_\odot$; and
- radius of third shell: $160 R_\odot$.

Slightly different values would be found if we had taken the stellar radius assumed above ($75 R_\odot$).

The velocities of the absorbing particles in the shells are determined from the absorption lines components, and particularly from the UV spectrum, which shows mainly blue-shifted absorption lines of once and twice ionized metals. The singly ionized ones have expansion velocities like those observed in the Balmer lines, of -210 to -250 km s^{-1}, and show in addition a second absorption at about -177 km s^{-1}; doubly ionized ones have rather -80 km s^{-1}, showing the outward decrease of the

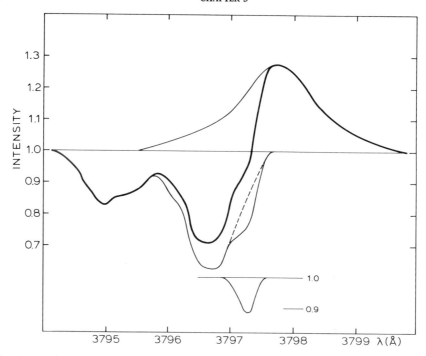

Fig. 32.   Profile of the Balmer line H10 λ3797 in P Cygni, derived by De Groot (1969). *Thick line:* observed line profile. *Thin lines:* assumed contributions from absorption and emission components

ionization temperature. The excitation temperature of the outer shell is $12\,000\pm900$ K (Cassatella *et al.*, 1979b).

It is remarkable that in the UV spectrum P Cygni profiles show only in the fairly strong resonance lines of low excitation; this points to a rather small value of the line source function in the shell.

There is a variability with a period of 114 days in the outer shell. This periodicity is only observable in the third component of the Balmer lines. It seems, though, that the observed variations in the radial velocity can be explained by assuming the *same* periodicity for all three shells, but in the inner two with a small, and not always observable amplitude. It may be that sound waves, traveling through the extended atmosphere play some part in the observed variation of radial velocity and of absorption line intensity. The mass-loss is discussed in Section 7.4, where we conclude to a 'best' value of $2\times10^{-5}\,\mathfrak{M}_\odot$ yr^{-1}.

The short-term light variation of P Cyg was investigated by Kharadze and Magalashvili (1967) who suggested that the star would have W UMa characteristics: a quasi-sinusoidal light variation, due to a close binary pair in a common atmosphere. The period should be $0\overset{d}{.}5006$. This is not confirmed by De Groot (1969, p. 268f). – see also Chapter 8.

The star emits microwave radiation and has an infrared emission excess, measured by various authors. A summarizing curve of flux-density versus frequency, covering the range from 3 μm up to 10 cm was compiled by Barlow and Cohen (1977) from data from five different sources and is reproduced as our Figure 33. Barlow and Cohen (1977) determined an empirical velocity-distance law, hence a density-distance law, from which a theoretical emission spectrum could be deduced (dashed line in Figure 33), that could be reconciled with this spectrum. It should be mentioned here, that Kunasz and Van Blerkom (1978) found that the Balmer line spectrum can be explained best by a linear law: $v(r)(:)r$, the velocity increases slowly with distance to the star, unlike O and B-type stars. This was confirmed by Ambartsumian *et al.* (1979),

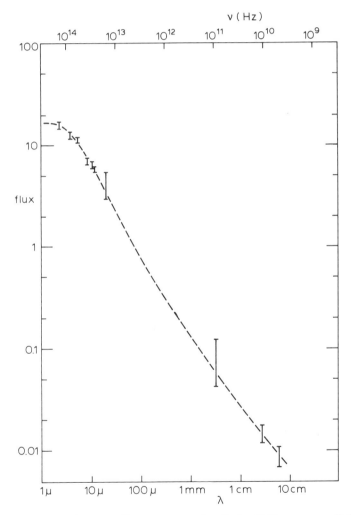

Fig. 33.   Infrared and microwave flux versus wavelength for P Cyg, as compiled from various sources by Barlow and Cohen (1977). The dashed line represents the emission calculated for an envelope obeying a certain assumed velocity-distance relation and a mass loss rate $-\mathfrak{M}/v_\infty = 5 \times 10^{-8}$ $\mathfrak{M}_\odot \, \text{yr}^{-1} \, (\text{km s}^{-1})^{-1}$.

who also stressed the low ionization in the wind. There is no doubt that the relation between velocity and distance for P Cyg differs greatly from that of giants or super-giants at the same location in the HR diagram: the terminal velocity v_∞ is about 300 km s^{-1}, about seven times smaller than in early-type supergiants, but the mass-loss is larger by a factor of $\sim$3 to 10.

3.6. The Oe, Be and Shell-Type Stars

These three groups are all related objects as has been shown for the Oe and Be stars by Frost and Conti (1976); for the relation between Be and shell stars we refer to Hutchings (1976d). An excellent review paper on Be stars was written by Slettebak (1979).

Oe stars show "emission in the hydrogen lines, but without emission in N III λ6434, 40, 41 or λ4686 He II. Some Oe stars also show emission in λ5876 He I. The emission profile is roughly centered with respect to the velocity of the stellar photosphere and often has a central reversal absorption feature giving the *appearance* of double emission lines". (Conti and Leep, 1974). Some O stars have emission only in Hα, without any in the other Balmer lines. Examples are HD 46056 (Conti, 1974), and ζ Oph (Niemelä and Mendez, 1974). These are classified as O(e).

Be stars: The first was discovered by Secchi in 1866 who, visually, saw Hβ in emission in γ Cas. In 1911, Curtiss was the first to classify emission-line stars. The present designation was introduced in 1922 by Commission 29 of the International Astro-nomical Union (chairman: Walter S. Adams) during its Rome meeting. It refered initially to B stars with emission lines. Nowadays they are, similarly as the Oe stars, defined as such stars with only the Balmer lines in emission. Often only Hα is in emission; such stars could therefore be called B(e) stars, but this notation is only rarely used. No emission occurs in lines of other elements. Most of these stars have luminosity classes V, IV (and III) (Henize *et al.*, 1976b). No supergiants occur among them.

Ae stars are defined similarly as Oe and Be stars. They are fairly rare and the notation is seldom used; a well-known example is 14 Com (Domini and Smith, 1977). Attention should be paid to the possible confusion with Herbig's (1960) class of Ae/Be stars: pre-main-sequence stars surrounded by shells, and characterized by strong mass fluxes $(-\mathfrak{M} \approx 10^{-6} \dots 10^{-7} \, \mathfrak{M}_\odot \, \text{yr}^{-1})$ (Garrison, 1978), several orders of ten more intense than the mass flux of Be stars. Examples are LkHα198, V380 Ori, R Mon.

The *shell stars* are closely related to the Be stars and a clear distinction between the two is not always made; the two types are even often confused. The term was intro-duced by Otto Struve to denote a star with an extended atmosphere. Nowadays the name refers to stars with detached and relatively speaking dense envelopes. "Spectro-scopically a shell is seen by the existence of sharp and unusually deep absorption lines of H, He I and of metals such as Si II, Fe II, Ti II. The usual emission lines are often enhanced. The absorptions may be asymmetrical, and show differential radial veloci-ties, particularly as a Balmer progression" (Hutchings, 1976d). Others (e.g. Bidelman and Weitenbeck, 1976) call shell stars those with an Hα line showing double emission and central absorption. This definition is not followed by all authors, since e.g. Plavec (1976) describes Hα spectra of Be stars with the same characteristic. Often also the idea is advocated that shell stars are Be stars seen edge-on so that the flat disk seen in the equatorial plane is observed at great optical depth, which then accounts for the above-mentioned 'dense envelopes'.

Heap (1976a) describes the difference between B-shell stars and Be stars as: shell stars show "absorption lines in the visual regions of the spectrum which regular Be stars do not". This apparently contradicts some above definitions.

Wackering (1970) published a catalogue containing more than 5000 early-type emission line stars. A list of early-type shell stars, including Of stars and O, B, A stars with P Cygni profiles was published by Bidelman (1976). The percentage of emission line stars decreases with increasing spectral type in the O and B classes; it is 14% for O stars (Frost and Conti, 1976), 11% for B0–B3 stars (Merrill, 1933), 9% for B5 stars, 3% for B7–B8; 1% for B9 and 0.2% for A2 stars (Swings, 1976). The ratio of numbers Be (:) B stars varies greatly from cluster to cluster (Meadows, 1960; Schild and Romanishin, 1976), an unexplained phenomenon.

The spectral differences between O and Oe, and B and Be stars refer only to the hydrogen emission lines, the effect being restricted to the relatively strong lines, while the continua are in broad outline the same (Briot, 1978). Hence the envelope, responsible for the emission lines must be optically thin for continuous light, and opaque in the lines, similarly as in the solar chromosphere. The Oe and Be stars must be stars with extended but not spherically symmetric 'chromospheres', contrary to normal O and B-type stars.

So far, high-dispersion ultraviolet spectra of Oe and Be stars have been observed for about a dozen of Be stars, among which ζ Tau (B4 IIIp; Heap, 1976b), HD 102 567 (B1 Vne), X Per (O9.5 III–Vep) and γ Cas (Marlborough et al., 1978, Hammerschlag-Hensberge et al., 1979a). In the latter star Mg II shows (optically thin) emission peaks, and the strong Si IV absorption lines near 1400 Å have emission peaks in their cores. This, again, indicates the 'chromospheric' character of the emission features. No prominent N V lines could be detected near 1240 Å. However strong Fe III absorption lines characterize the UV spectra between 1120 and 1140 Å of 17 Be and shell stars, described by Snow et al. (1979); this is again a clear shell feature. Interesting is that γ Cas does not show the Fe III absorption.

Schild (1966, 1978) and Briot (1978) have determined the *continuous energy distribution* of Be stars. Photometric observations (Feinstein and Marraco, 1979) show that nearly all Be stars have an ultraviolet excess. The early Be stars have filled-in Balmer jumps. In the spectra of some of them the Balmer continuum may even appear in emission, but this does not show in Briot's (1978) results. More clarity is brought into this matter by realizing that Be stars have often *two* Balmer jumps (Barbier and Chalonge, 1939), one of which is from the star, the other from the shell. The latter can occur in emission *or* absorption. Differentiation between the two is only possible by very accurate spectrophotometry. Among the middle- and late-type Be stars several are found with larger than normal Balmer jumps. The Balmer jump discrepancy seems related to the Hα emission line equivalent and is apparently caused by the shell. Some Be stars also show emission in the Paschen lines. Briot (1977) noted that the continuous radiation of such stars shows an infrared excess as compared to those which have the Paschen lines in absorption (see also Schild's (1976) review). Such stars have also stronger emission in the first few members of the Balmer series than the others. This indicates more developed shells in stars with the Paschen lines in emission. In such shells the free-free emission of electrons in the fields of ions produces an excess infrared flux with relative intensity increasing with wavelength. For several members of this latter group of stars the first members of the neutral helium-series (singlets and triplets) are also in emission while later members give the appearance of filled-in profiles. About 20% of the early-type emission line stars, and 15% of the Be-stars have infrared Ca II emission (Polidan and Peters, 1976).

Not for all stars the emission features are permanent. Frost and Conti (1976) found that out of a sample of 28 sufficiently bright main sequence O-type stars, of spectral type later than O7.5, four (i.e. 14%) have shown the Oe features at some time. Two of them have shown it intermittently in Hα and should therefore be classified O(e). The two others, X Per and HD 45314 are real Oe stars.

In the range of B stars, 20% of the B1–B5 stars have shown emission at some time (Massa, 1975).

The *photospheric structure* is best studied from the continuous spectrum and the weak absorption lines, while log$g$-values are derived from the wing profiles of the Balmer lines. Peters (1976a) finds for Be stars $\langle \log g \rangle = 3.7$ and for shell stars $\langle \log g \rangle = 3.35$, while it seems that the T_e-values do not deviate significantly from the values for main sequence stars of the same spectral types. For the B4 III p star ζ Tau Heap (1976b) found reasonable agreement with observations for $T_e = 27\,500$ K and log $g = 4.0$. For 27 CMa (B3 IVe) Danks and Houziaux (1978) found $T_e = 20\,000$ K and log $g = 3.8$ from spectral observations in the range 0.14 to 4.7 μm.

In this connection it makes a difference when the star is observed pole-on or equator-on. Rapid rotators, as Be stars presumably are, should be flattened at the poles. The theory of rotating stars shows that the radiation flux varies on the equipotential surfaces proportional to g. Hence their surface temperature is higher at the poles than at the equator. Because the ultraviolet flux depends stronger on T than the visual flux, an early-type rapid rotator seen pole-on may therefore have a marked UV-excess as compared with the same star seen equator-on. Kodaira and Hoekstra (1979) found such indications for the stars ι Her and ζ Cas, which are apparently Be stars seen pole-on. For the pole-on stars υ Cyg (B2 Ve) and μ Cen (B2 IVe) Peters (1979) finds T_e and log g: 24000 K; 3.8, and 20500; 3.7 respectively.

Various models have been proposed for the *structure of the shell*. All observational evidence shows that the shell should be mainly equatorial and either be disc shaped, as originally proposed by Struve (Figure 34a) or have the shape of an excentric elliptical ring, as proposed by Mc Laughlin and by Huang (Figure 34b). A third alternative is provided by the binary hypothesis. It is not easy to distinguish *obser-*

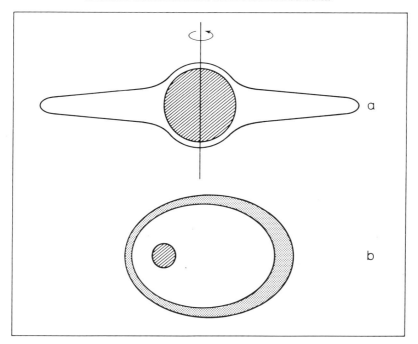

Fig. 34.   (a) The classical model of a Be star: a rapidly rotating star (which should be greatly flatten-ed) with an equatorial shell (Struve). (b) The elliptic ring model proposed by McLaughlin and by Huang.

vationally between the disk or ring alternatives. Křiž (1979a, b) has developed the elliptical ring concept in detail and found – as could be expected – that it is able to explain the observed spectral features, such as the shape of the line profiles, the V-R changes (due to the motion of the periastron of the ring), the origin of absorption cores and 'shell lines' (if the ring's inclination is close to 90°). A minimum ring density of $n_e \approx 10^{11}$ cm^{-3} is needed to have emission lines. The same consideration would, however, also apply to an equatorial disc (apart from the V-R changes). Problems of the stability of the elliptical ring may be a reason to prefer the more classical disc model. The infrared continuous spectra of these stars give additional information; at short infrared wavelengths the spectrum does not fall off rapidly with increasing wavelength, but for $\lambda > \lambda_c$, with $\lambda_c \approx 10$ μm a sharper decrease is often noticed. Woolf *et al.* (1970), Gehrz *et al.* (1974), Scargle *et al.* (1978) found the near IR excess ($\lambda \lesssim 10$ μm) due to hydrogenic bound-free and particularly free-free emission from a shell of hot H$^+$ gas, with $n_e \approx 3 \dots 6 \times 10^{11}$ cm^{-3}.

Cassinelli and Hartmann (1977) and later Hartmann (1978) found that the steeper decrease beyond λ_c can be explained by assuming that the disc around the star has a fairly rapid density decrease with distance; typically $n(:)r^{-\alpha}$, with $2 \lesssim \alpha \lesssim 3$.

The observed polarization, of up to $\sim 2\%$, of the continuum intensity, discovered by Behr (1959) (review by Coyne, 1976; see also Jones, 1979), and the increase of polarization p with the value of $v \sin i$ (McLean and Brown, 1978; McLean, 1979b) support the axially symmetric flattened disc model; an example is p Car B4 Ve. The wavelength dependence of the polarization and the infrared truncation described above are both linked to the electron density in the shell, and to the inclination of the plane of the disk with respect to the field of view (Hartmann, 1978). (It is obvious that a disc seen edge-on has maximum polarization; seen from the pole the polarization would be zero). Linear polarization is observed so far for about half of the Be stars. An overview is given by Poeckert *et al.* (1979). The intrinsic polarization decreases abruptly across the Balmer limit, a smaller decrease occurs also across the Paschen limit. This is in accordance with the idea that the polarization is related to scattering of radiation in a flattened envelope or shell.

A quantitative discussion shows n_e of the shell to increase towards the star. To have a coherent picture, describing at the same time the infrared excess radiation (due to the shell) and the λ-dependence of polarization (related to shell thickness), Jones (1979) found it necessary to adopt $T_{\text{shell}} \approx 11\,000$ K, and $2 \times 10^{11} < n_e < 4 \times 10^{12}$ cm^{-3}. For comparison, McLean (1979b) found from the continuum polarization: $n_e \approx (3$ to $5) \times 10^{11}$ cm^{-3}. It is gratifying that similar results are found from the *emission lines*: see e.g. Houziaux and Andrillat (1976), Heap (1976b), Coyne (1976), Peters (1976b). They find $T_{\text{shell}} \approx 10^4$ K and $(n_e)_{\text{shell}}$ ranging between 10^{11} and 10^{12} cm^{-3}, but there are large deviations from this average. Brosch *et al.* (1978) describe the B3e III object MWC 342, for the shell of which $n_e \approx 5 \times 10^{10}$ cm^{-3}. From a study of the UV Fe II and Fe III lines near 1130 Å shell column densities for H of approx. 10^{20} cm^{-2} are derived. The lines show negative wavelength displacement, indicating outstreaming motions in the shell. The degree of ionization increases with the wind velocity (Snow *et al.*, 1979). The radius of the shells is estimated to range between $2R_*$ and $10R_*$ with an average value of $5R_*$, although Kitchin (1976) found $R_{\text{shell}}/R_* \approx 25$.

Another source of information with important diagnostic possibilities is the polarization observed in *spectral lines* like Hα (Poeckert and Marlborough, 1977) or Hβ (McLean *et al.*, 1979b). Typically, the degree of polarization decreases towards the line center, but has a local maximum in the line center, showing the influence of the shell. The variation of the position angle of polarization with distance to the line center may give information on the sense of rotation. The detailed theory has yet to be elaborated.

Extending beyond the (Balmer emitting) shell, several Be and shell stars have a *circumstellar envelope*, detectable in the cases of the least developed envelopes by an infrared excess and by the polarization of the star's light. An example is 27 CMa (B3 IVe) which has a shell with $T_{\text{el}} = 14\,400$ K, and $n_e \leq 6 \times 10^{11}$ cm^{-3} (Danks and Houziaux, 1978). If the shell is more strongly developed its existence is shown by the infrared excess *and* the radio emission. Purton (1976) describes the microwave emission of the objects MWC 349 (Bep), MWC 957 (Be) and a few others with radiospectra apparently emitted by an envelope, optically thick in radio waves. Sometimes the envelope is even visible around the star. Not all authors would call such objects Be stars; they are sometimes classified as planetary nebulae.

About 10% of the Be stars and most Ae stars are surrounded by *dust shells*; see the discussion in Section 6.15. The variable Ae star HR 5999 owes its variation to changes in the dust envelope (Tjin a Djie and Thé, 1978).

About $\frac{3}{4}$ of all Be stars show *photometric variability* (Feinstein and Marraco, 1979) on various time-scales.

Long-term variations of Be stars are perhaps caused by the expulsion of a shell or of changes therein. The variations in the shell properties are partly revealed by variations in the *continuous spectrum*, as was the case for π Aqr (B1 Ve) for which the colours U, B, and V increased steadily from 1957 to 1975 (Nordh and Olofson, 1977), while $(B-V)$ increased and $(U-B)$ decreased, which indicates mainly brightness increases in the near ultraviolet and the red spectral regions. Also *the Hα line* is very sensitive to changes in the shell. Slettebak and Reynolds (1978) observed the Hα profiles of 35 Be stars during a two-years period, and found definite and often drastic changes in Hα over periods of 2 months to a year.

We describe a few interesting and well studied objects. During the period 1968–1977 the colour and spectrum of the shell star 88 Her showed long-term variations; the colour, initially (1968) resembling that of a B8 V star changed to B6 V (1974) and thereafter returned to B7 V in 1977 (Harmanec *et al.*, 1978). A comparable case is o And. The star emitted a new shell in July 1975 (Bossi *et al.*, 1976). In the few years thereafter the star showed brightness fluctuations which gradually reduced in amplitude while also the brightness reduced to its earlier value (Guerrero and Mantegazza, 1979).

An interesting and well-known example of a star with long-term variations in the line spectrum is *Pleione* (HD 23862; B8 IV–Ve) of which the shell spectrum has gradually changed after 1972. The observation of this star suggests that at that time a new shell phase started, characterized by increasing strength of the emission equivalent widths of the first Balmer lines – clearest visible in Hα and metallic lines (Figure 35a). The observations are interpreted (Hirata and Kogure, 1977, 1978) by the formation of an equatorial shell with highest electron density ($\approx 10^{11}$ cm^{-3}) in the equatorial plane with a shell of smaller electron density ($\approx 5 \times 10^{10}$ cm^{-3}) at either side at higher latitudes. The local mass of the shell increased from 10^{-11} $\mathfrak{M}_{\odot}$ in 1973–1974 to 1.4×10^{-10} $\mathfrak{M}_{\odot}$ in 1976 (Figure 35b). A broad absorption feature in the K line of Ca II that gradually formed is interpreted as due to a more outside rapidly rotating equatorial region ($v_{rot} \approx 430$ km s^{-1}).

Shorter-term variations, from night to night, have also been observed but are less pronounced than the long term variations: The variability of the linear polarization of λ Cep (Oe) on a time scale of a day (Hayes, 1978) indicated the presence of an extended asymmetric atmosphere or shell, the variability of which may be due to changes in velocity and density of the outflowing material. The star HD 88661 (B2 pe) shows an irregular variability in V of 0.1 mag with a time scale of the order of a few days (Cousins and Stoy, 1963; Stift, 1978). The UV emission cores of γ Cas are variable in intensity and position with a 'period' of months to a few years (Marlborough *et al.*, 1978). The B4e star HD 174237 (CX Dra) shows variations of the radial velocity of the helium lines with a total amplitude of 70 km s^{-1} and a period of 6.698 days (Koubský, 1978).

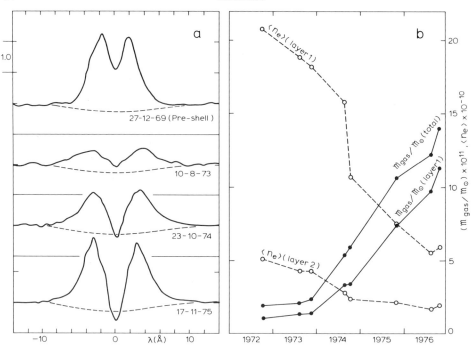

Fig. 35.   (a) Profiles of the Hα line in Pleione, 1969–1975. The dashed curve is the profile assumed for the star's photosphere, rotationally broadened with $v_{eq} \sin i = 340$ km s^{-1}. (b) The variation of n_e and $\mathfrak{M}_{shell}$ for the shell of Pleione. 'Layer 1' is the densest equatorial part of the shell; 'layer 2' the more tenuous part at somewhat higher latitude. (After Hirata and Kogure, 1977, 1978).

Rapid variations in line profiles have indeed been claimed as observed, although caution is needed in the interpretation. There is no doubt that part of the 'irregular variations' mentioned in the literature is merely an overoptimistic interpretation of photon or plate noise! Hutchings *et al.* (1971) and Hutchings (1976d) describe excellent observations of the Hα profile, with high spectral resolution of γ Cas which give evidence of small irregular variations on time scales between approx. one hour down to a minute. The variations range from more than 25% to the lower observational limit of 2–3%. Bahng (1971, 1976) reports variations in the Hα equivalent width in ζ Tau, α Col, PP Car and δ Cen. The variations are of a few percent, barely exceeding the limit of observational accuracy, and time scales of 1 to 3 min were found. Doazan (1976) described rapid, minutes-scale, variations of Hβ, Hγ and Hδ in o And, γ Cas and χ Oph. No photometric accuracy is indicated, and the significance of the observations is less clear than for the previously described ones. Variations (∼0.1–0.15%) in the continuum polarization of γ Cas on time scales of hours are attributed to inhomogeneities in the rotating circumsteller envelope (Piirola, 1979).

The next important question refers to the *origin* of *the shell*. Is it due to rapid rotation or are the stars binaries? The variability of many Oe or Be stars may support the binary hypothesis, but sofar the binary character has only been established for a few such stars. Polidan (1976) describes the binaries 17 Lep (B9 Ve+M2 III), AX

Mon (B0.5e+K2 II), HD 218393 (B3e+K1 III) and HR 894 (=HD 18552: B8eV+ gG9:), and Hutchings (1976d), and Harmanec and Kříž (1976) mention nine other eclipsing Be binaries. The latter authors also advance the hypothesis that *all* Be's are binaries! This, however, is far from being proven, and both photometric detection of the binary character as well as spectroscopic detection seems most difficult. Most probable seems that the origin of the shell is caused by *rotation* and thus associated with *mass-loss*. A study of the 46 Be stars by Slettebak (1976) shows a correlation between the Balmer emission-line widths and the rotational velocity, which supports this hypothesis. The observed rotational velocities are close to the computed equatorial critical velocities for main sequence stars (Figure 36), but they are smaller on the average. The difference may be removed by assuming that a typical Be star is not a main sequence star but slightly evolved into the subgiant or giant region – this suggestion is also supported by the low surface gravity found for shell stars (see also Figure 17). It seems therefore allowed to assume the rotational hypothesis to be

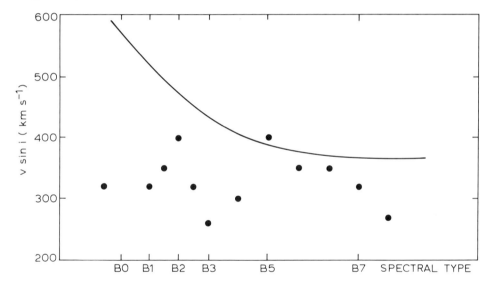

Fig. 36. Comparison of the largest observed rotational velocities for Be stars of spectral types O9.5 to B9 (dots) with computed equatorial critical velocities (at which the equatorial centrifugal force balances the gravitational force) for main sequence models of the same spectral type range (full-drawn line). (After Slettebak, 1976).

correct. The hypothesis that the shell is due to outstreaming motions is further supported by the observation of the P Cygni profiles in the ultraviolet spectra of Be stars (review by Heap, 1976a) where velocities of 120 to 900 km s^{-1} are reported. The differences, found by Heap (1977) for ζ Tau, between $v \sin i$ determined from the visual and ultraviolet spectra regions (300 and 100 to 150 km s^{-1} respectively) is certainly due to differential limb darkening effects (see our Section 2.10).

From ultraviolet observations with *Copernicus* of Be *and* B main sequence stars Snow and Marlborough (1976) found a correlation between mass-loss (indicator:

asymmetry of spectral lines) the projected rotational velocity, and the occurrence
of emission lines, suggesting that an outer shell, emitting emission lines is produced
by stellar winds in those stars where the equatorial effective acceleration of gravity
is greatly reduced by rotation. Lamers and Snow (1977) found that for luminosities
below $M_{bol}=-6.0$ nearly every class IV or V star of spectral type B which has a
chromospheric ('warm') wind containing Si IV is a Be or shell-star. Since Be characteristics may be linked to rotation the presence of the chromosphere should be
related to rotation too.

Attempts to set a quantitative basis to the theory of the origin of Be shells were
initiated by Huang (1976), who assumed – in view of the variability of many shell
features – that mass-ejection in Be stars is transient and sporadic. In the equatorial
disc, thus formed, relaxation of angular momentum and of the velocity distribution
('thermalization') takes place, as well as dissipation of kinetic energy. Particles with
low angular momentum and velocity will fall back, those with low angular momentum
and high velocity will escape, while the others form a rotating ring around the star.
Observed rapid variations in emission line profiles would indicate non-uniformity of
the ring. Disappearance of the emission characteristics would mean that the gaseous
ring is destroyed temporarily. This picture is more or less supported and partly
modified by Marlborough et al. (1978). Their observation of emission peaks in ultraviolet lines of γ Cas forced them to assume that the turbulence induced in the outstreaming stellar wind by the rotation leads to the formation of a corona by dissipation. This corona would mainly be formed in the less dense regions of the wind, at
high stellar latitudes. In the equatorial region the terminal velocity of the wind should
be low leading to higher densities, and thus to the 'equatorial shell'. It seems, however,
that outstreaming motion occurs at all stellar latitudes.

Finally we mention Morgan's (1975) attempt to summarize 'characteristic data'
for the central star and disc of an average Be star. The data may be useful for a semi-
quantitative discussion of these stars; see Table XVIII. From the data in this table we
derive a bolometric stellar luminosity $M_b=-6.5$.

TABLE XVIII

Characteristic data for central star and disc of a
typical Be-star (Morgan, 1975)

Stellar mass	$10\,\mathfrak{M}_\odot$
Stellar equatorial radius	$10R_\odot$
Surface temperature	25 000 K
Equatorial rotational velocity	340 km s^{-1}
Disc temperature	10000 K
Average disc density	10^{-12} g cm^{-3}

As a prototype we present the Be star γ Cassiopeia (HD 5394=HR 264, presently
at $m_v=2.6$, but variable both spectroscopically and photometrically (m_v ranges between 1.4 and 3.0). Interesting was the period in the mid-1930s when the star brightened to 1.4 mag, accompanied by important variations of the spectral shell features.
A weak and variable ($P<1^d$) X-ray source was detected (Jernigan, 1976; Mason et al.,

1976) only $10''$ from the star (error circle $1'$). Observations of the visual radial velocity show this to be constant with time, without any clear periodicities ($2\overset{d}{.}5 \leqslant P < 4000^d$), but the UV emission peaks of Mg II and Si IV show a V/R variability which forced Marlborough et al. (1978) to assume the existence of a neutron star companion ($\mathfrak{M} \approx 1 \mathfrak{M}_\odot$) revolving in about four years. So, the star is most probably a binary. It has, furthermore, a visual companion ADS 782B, separated by $2\overset{.}{.}2$ and approx. 8 mag fainter. The distance of γ Cas is 250 pc, hence $M_v = -4.1$; $M_b = -6.85$, and the separation of the visual companion should be approx. $10^5 R_\odot$ (Marlborough, 1977).

The spectral type is usually given as B0.5 (or B0) IVe; sometimes as B0p. The parameters of the star and its shell were investigated by Poeckert and Marlborough (1977, 1978a, b) who found $\mathfrak{M} = 17 \mathfrak{M}_\odot$, $R = 10 R_\odot$; $T_e = 25\,000$ K and log $g = 3.5$.

A description of the long-term spectral and brightness variations until 1942 as given by Struve and McLaughlin was summarized by Merrill and Burwell (1943); later reviews are from Edwards (1944, 1956), Kitchin (1970), and Baliunas and Guinan (1976). For short-periodic spectral variations (months to minutes) of the Balmer lines, reference is made to Hutchings et al. (1971), Hutchings (1976d; Figures 1 and 5), Bahng (1971, 1976), Doazan (1976), Marlborough et al. (1978), but these were not confirmed by Slettebak and Snow (1978); see also earlier this section.

The spectrum is characterized by a nearly complete absence of sharp features. The early Balmer lines are double peaked; they are in emission as is the Balmer discontinuity. A review of UV spectral observations is given by Snow et al. (1979). The absorption lines are rotationally broadened. The photospheric rotational velocity is determined as $v_{eq} \sin i = 230$ km s^{-1} (Slettebak, 1976), and is given as 300 km s^{-1} by Marlborough and Snow (1976), and 570 km s^{-1} by Poeckert and Marlborough (1978a, b). From a comparison of UV and visual line-width Hutchings and Stoeckley (1977) derive $v_{eq} = 420$ km s^{-1}, and $i = 47°$.

Outstreaming motions are shown by the P Cygni profile of the C IV doublet, of which the absorption component of each line is displaced shortward by -450 km s^{-1} (Marlborough and Snow, 1976). A study of these and other UV resonance lines yielded a mass-loss of $-\dot{\mathfrak{M}} = 7 \times 10^{-9} \mathfrak{M}_\odot$ yr^{-1} (Hammerschlag-Hensberge et al., 1979a). A sudden erratic flaring-up of Hα and the ultraviolet lines Si IV and Mg II, lasting about one hour, was noticed by Slettebak and Snow (1978). A model for the shell was first proposed by Struve (1931) and refined and quantized by McLaughlin (1961), Limber and Marlborough (1968) and Hutchings (1970b), Marlborough et al. (1978), Scargle et al. (1978) to mention the most important papers. The model assumes that gas ejected from a star rotating near the critical velocity and accelerated radiatively forms a gaseous envelope or ring, densest in the equatorial regions, with coronal characteristics at higher stellar latitudes. Assuming that the observed X-ray emission is due to a shell around the star Marlborough (1977) finds for the shell $n_e \leqslant 10^{11}$ cm^{-3}, and $T \gtrsim 2 \times 10^7$ K. From near-infrared spectrophotometry (1 to 4 μm with about 2% spectral resolution) Scargle et al. find the infrared excess mainly due to hydrogenic bound-free and free-free emission in a disc for which the following quantitative data were derived: $T_{el} \approx 18\,000$; τ (1 μm) ≈ 0.5; $n_e \approx 10^{12}$ cm^{-3} and radius of $\sim 2 \times 10^7$ km. Poeckert and Marlborough (1978a) find some what larger values: $T_{e1} = 20000$ K, $n_H = 3 \times 10^{13}$ cm^{-3}, while the envelope extends till $50R_*$ to $250R_*$ ($= 3.5 \times 10^8$ to 2×10^9 km).

Information on the *shape* of the equatorial disc can be derived from the fact that the continuous spectrum and the Hα line shows linear polarization, which decreases towards the center of Hα, while the direction of polarization varies in an asymmetric way over the line profile (Figure 37). Poeckert and Marlborough (1977) can explain these features by a *disc-like* (explaining the polarization) envelope, (inclined at an angle of 47° to the line of sight) which rotates in a *clockwise* direction as seen from the Earth

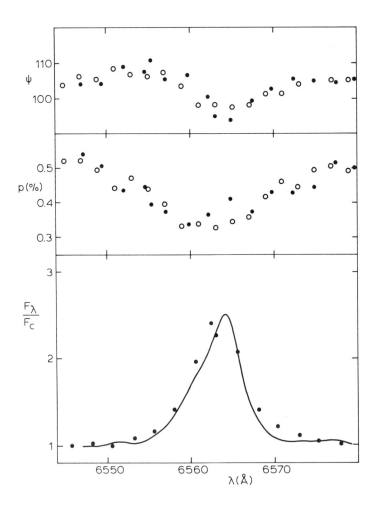

Fig. 37. Position angle ψ, percentage polarisation p, and line profile F/F_c of Hα for γ Cas. Filled circles represent one scan on 1976, October 1. Open circles are mean values from seven scans taken between 1974, October 21 and 1976, October 1. The solid line is the line profile obtained 1976, October 1. (From Poeckert and Marlborough, 1977).

(deduced from the sense of rotation of the position angle). Short-term variation in the continuum polarization are perhaps caused by inhomogeneities in the rotating shell (Piirola, 1979). See also Section 2.11.

3.7. The CNO-Stars (or: OBCN Stars)

Not forming a special group in the strict sense of the word it seems useful to mention these stars because their existence may have important evolutionary implications. The group was discovered in classification work on O- and B-type stars by Jaschek and Jaschek (1967) and a compilation giving data on about 30 stars was later given by Jaschek and Jaschek (1974). Walborn (1974) gives another list. A short review was also given by Baschek (1975), and a subdivision in OBN and OBC stars was proposed by Walborn(1976a). The stars all belong to the stellar population I, and show *only* anomalous strengths of C, N and/or O, not of other elements. Walborn (1971a, 1972a) notes that a strengthening of the N lines (N-type anomaly) usually correlates with a weakening of C and vice versa (C-type anomaly) thus showing similarity to the WC, WN dichotomy, but *this* is questioned by the Jascheks. The C-type anomaly seems mainly confined to the main sequence. At least half but perhaps all OB–N stars seem spectroscopic binaries while none of the OB–C stars are (Bolton and Rogers, 1978). Many CNO stars show variations in spectrum and radial velocity; they have a slow rotation. The abundance anomalies are fairly modest: quantitative spectrochemical analyses do not give larger deviations than factors 5 to 10, with a rather large scatter.

Relative abundances have been determined for the supergiants HD 96 248 (B1 Iab; Dufton, 1972, 1979), o^2 CMa (B3 Ia; Van Helden, 1972), HD 96 159 and HD 96 214 (Dufton, 1979). The three supergiants studied by Dufton are all nitrogen-deficient with a deficiency factor of about a factor 6. In contrast the hypergiant HD 269 896 (O9.7 Ia$^+$; $M_v = -8.1$) is N-rich and C-deficient, as follows from all relevant spectral features (Hyland and Bessell, 1975). However, no abundances have been determined for this star. Walborn (1977) classifies it as ON.

Walborn (1976a) notes that all O9.5 ... B0.7 supergiants in the Orion Belt and in NGC 6231 (the nucleus of the Scorpius OB1 association) are systematically N-deficient, relative to supergiants of the same spectral types.

The *explanation* of the deviating abundances in these stars must be related to the fact that in advanced evolutionary phases C, N and/or O from the star's interior may be mixed toward outer layers. If at the same time the outer mantle is lost by intense mass-loss, such regions with changed abundances arrive at smaller distances to the surface. It seems unlikely that these anomalies refer to the primordial matter out of which the stars were formed since there are several OBCN stars that are members of a group, whereby the anomalies do not occur in other group members. Auer and Demarque (1977) have noted that relatively high C and/or N abundances can also occur in population II stars, where in late B horizontal-branch stars of a globular cluster (M92) the observed values of the Balmer discontinuity can only be explained by assuming a (C, N)/H-ratio $\sim 10^3$ times the cluster average. The suggested explanation is that of surface enrichment through mixing with a helium-burning core in the course of stellar evolution.

An *example*, not so much a prototype of a type I CNO star is the B3 Ia star o² CMa, studied by Van Helden (1972). For this star $M_b = -8.9$. From a model analysis Van Helden derived $T_e \approx 20000$ K, log $g \approx 2.5$. The stellar mass as determined from the model analysis is 22 $\mathfrak{M}_\odot$ (Van Helden, 1972; Sterken, 1976a) see our Table IX. The logarithmic C, N, O abundance deviations – as compared with those in solar-type stars – are: [C] ≈ -1.1 (Baschek, 1975)

[N] ≈ -0.3

[O] ≈ -0.8.

Note that for this star the abundance deviations are *negative*; no explanation has so far been advanced to explain such negative deviations.

3.8. The Super-Supergiants (Hypergiants)

This class, which we will sometimes abreviate as S–SG, and generally call 'hypergiants' (following a suggestion by A. M. van Genderen), was introduced by Feast and Thackeray (1956) – see Section 2.1. It should be noted (see also Figure 2) – that the hypergiants are not the most luminous stars: Of stars are brighter! Initial discoveries of stars of this type were made in the Magellanic Clouds, and in their fundamental paper Feast *et al.* (1960) list a dozen of hypergiants (their Table II), most of spectral types B and A, and four later-type stars of extreme brightness. In addition the early type LMC star HD 269896 O9.7 Ia⁺ is a hypergiant, $M_v = -8.1$ (Hyland and Bessell, 1975; Walborn, 1977). The brightest star of the Small Magellanic Cloud (HD 7583) is also a hypergiant, A0 Ia⁺, $M_v = -9.3$ (Wolf, 1973).

Furthermore, at least seven galactic hypergiants are known: ϱ Cas–F8 Ia⁺, (Sargent, 161); ζ¹ Sco–B1.5 Ia⁺ (Stothers, 1972a); HR 4337–G0 Ia⁺ (Humphreys *et al.*, 1971); HD 217476=HR 8752–G0–G5 Ia⁺ (Sargent, 1965, Smolinsky, 1971; Luck, 1975; Lambert and Luck, 1978; Harmer *et al.*, 1978), HD 119796 = HR 5171–G8 Ia⁺, with $M_b = -9.75$ (Humphreys *et al.*, 1971; Warren, 1973a); HD 160529–A2 Ia⁺ (Wolf *et al.*, 1974) and HD 80077–B2 Ia⁺e, with $M_b \approx -11.5$ (Moffat and FitzGerald, 1977). Obviously these latter stars can be better studied spectroscopically than those in the Magellanic Clouds because of their relative proximity. All hypergiants show in their spectra indications of the instability associated with their high luminosity. So, ϱ Cas varies in brightness and spectral type, has strong outflow of matter (Sargent, 1961), and has a circumstellar shell (Beardsley, 1961). Others, like HD 217476 and HD 119796 show spectral line variability (Smolinski, 1971; Warren, 1973a). The star HD 217476 has nebular emission of [N II], which suggests a shell of higher kinetic temperature than the stellar photosphere. Moreover, this shell appears to be variable. The Hα line shows a P Cyg profile. For more details on the variability of these stars we refer to Chapter 8 of this volume.

In the following we describe some of the better studied hypergiants.

A nearly classical example is *the galactic star* ϱ Cas=HD 224014 (F8 Ia⁺) for which the proposed membership of the IV Cas association is still uncertain (Lambert and Luck, 1978). Its distance was determined along various ways, resulting in absolute magnitudes: $M_v = -8.4$ and $M_b = -8.5$ mag. (Sargent, 1961). With $T_e = 5000$ K the radius would be $600 R_\odot$. On the basis of evolutionary tracks of supergiants, Sargent estimated the mass at $\sim 25 \mathfrak{M}_\odot$, yielding $g = 1.4$ cm s^{-2}. The spectrum varies greatly: before 1930 it had a K-type appearance, but Morgan et al. (1943) classified it as F8 Ia. However, during the years 1946–1947 the visual magnitude decreased temporarily by about 1.5 mag and during several months the spectrum showed M-characteristics, including TiO bands. The brightness decrease and late-type spectrum were most probably caused by an ejected circumstellar shell. Greenstein (1948) found some evidence for matter falling back on the star after mass ejection. There are small irregular variations in brightness ($\Delta m \approx 0.2$ mag), while variations of the Balmer emission with time scales of the order of one or several days have been observed (Joshi and Rautela, 1978). Many mainly low exitation (<2.9 eV) spectral lines are double with one component displaced by -40 km s^{-1} indicating a mass-flux with that outstreaming velocity. Furthermore there are emission lines of neutral atoms (Fe, Ni, Ca), apparently originating from a circumstellar shell, with excitation and ionization temperatures of the order of 4000 K, and a microturbulent velocity component of 7 km s^{-1}. In the star's photosphere $v_M = 20$ km s^{-1} for ionized and 14 km s^{-1} for neutral atoms, while v_μ decreases from 10 km s^{-1} for resonance lines to 6 km s^{-1} for ions of increasing degree of excitation.

The continuous spectrum was compared by Joshi and Rautela (1978) with that of the F8 Ia star δ CMa. It appears that the two stars have very similar spectra, but ϱ Cas has in addition circumstellar features such as an enhanced continuous spectrum shortward 4000 Å; perhaps a Balmer continuum emission. Also in the Paschen continuum there are differences: the colour temperature for ϱ Cas was higher than that of δ CMa by about 650 K in 1970 and 100 K in 1974.

In 1955–58 the mass of the envelope was 2×10^{28} g, $n_H = 10^{10}$ cm^{-3} and $n_e = 10^6$ cm^{-3}; the rate of mass-loss was estimated at 2.5×10^{-5} $\mathfrak{M}_\odot$ yr^{-1}. In 1959 all these values decreased to about one-tenth of their initial values in less than 260 days, which probably indicated the ejection of the envelope, which started in 1957 (Bidelman and McKellar, 1957). In 1960 the parameters values increased again to the previous value in less than 300 days. So mass-loss appears to be a sporadic and in any case a variable aspect of the star. The radius of the shell is estimated to be about twice the stellar radius; hence approx. $1200 R_\odot \approx 10^9$ km.

The spectrum of the hypergiant HR 8752 = HD 217476 (G0–G5 Ia$^+$) is similar to ϱ Cas. In its spectrum velocity gradients occur; we list the relative velocities of some lines: (Sargent, 1965; Harmer et al., 1978).

	1964	1975	1976
Hα, abs.		-84	-88 km s^{-1}
Hα, em.	-40	-48	-52
H and K	$+40$		
Fe I		-57	-66
[N II]	-17	-71	-71

The [N II] shell should have $T_{el} \approx 10^4$ K and $n_e \lesssim 10^6$ cm^{-3}. Both the [N II] line as well as Hβ show occasionally a splitting in two components separated by $\sim$20 km s^{-1} (Smolinski, 1971). The star has a BIV-type companion, assumed to be responsible for the [N II]-ionisation (Stickland and Harmer, 1978).

Sargent (1965) derived a kinematic distance of about 5 kpc, and this would yield $M_b \approx -9$ mag, similar to the absolute magnitude found for ϱ Cas, but the star varied in absolute brightness in the last 30 yr. Osmer (1972) redetermined this value and found $M_V = -6.6 \pm 0.3$. With a bolometric correction of -0.7 we may thus assume that M_b varies: $-7.5 \leqslant M_b \leqslant -9.0$.

The spectrum has been described by Harmer *et al.* (1978), and was analyzed by Luck (1975) who subjected it to a differential curve-of growth analysis, with the Sun and ε Vir; the material was also compared to model photosphere computations by Johnson (1974). A (Newtonian) acceleration of gravity log $g = -2$ seems the best value. In that case the physical extent of the atmosphere is so large that the plane-parellel assumption breaks down, and so does then the basis of the analysis. Hence, the analysis should be redone for a model based on spherical symmetry. The derived photospheric data ($T_e \approx 4000$ K; log $g \approx -2$; $v_\mu = 7$ km s^{-1}; $v_M = 11$ km s^{-1}) should therefore be used with caution.

Interesting and similar to that of ϱ Cas, is the life history of the star (Lambert and Luck, 1978). Around 1950 the spectrum was G0 Ia; since, it went slowly to the cool region of the HR diagram, reaching in 1973 a spectral type of K2–K5 Ia. After 1973 the spectrum returned to an earlier type and in 1978 the star was described as G0 to G5. Shorter-term ($\sim$one year) variations have also been observed (Harmer *et al.*, 1978). In line with the photometric variations it seems that one may assume that the star lost a fairly dense shell of matter in the years of its redward migration. With the subsequent thinning out or falling back of the shell (the latter assumption is more likely: the 1977 observations showed the line-forming region to descend relative to the photosphere with $\sim$30 km s^{-1}) the star assumed its initial conditions around 1977.

Indications for an excited shell are not only given by the [N II] observations and by many emission lines in the infrared, but also by the radio-emission at 10.5 GHz, having a flux of 25 mJy (Smolinski *et al.*, 1977).

The spectrum of *the galactic* S–SG HD 119796=HR 5171 (G8 Ia$^+$) was investigated over the spectral range 0.4–18 μm by Humphreys *et al.* (1971). It is a visual binary, with a B0 Ibp companion at 9″.7 distance. From a comparison of the apparent binary magnitudes $M_v = -8.9$ was derived. The distance is estimated to be $\sim$3.6 kpc. From intrinsic colours of the B- and G-type stars a total visual interstellar absorption of 3.18 mag is found for the B-star while the G-star yields 3.66 mag, nearly 0.5 mag more! This finding may be related to the infrared spectral observations – Figure 38 – which indicate the presence of an emission peak near 10 μm, usually attributed to the emission by a silicate dust shell. The hatched area in Figure 38 is suggestive for the radiation absorbed by the shell if it is assumed that the shell-absorption has the same wavelength dependence as the interstellar absorption. This energy would be sufficient to explain the silicate emission at 10 μm.

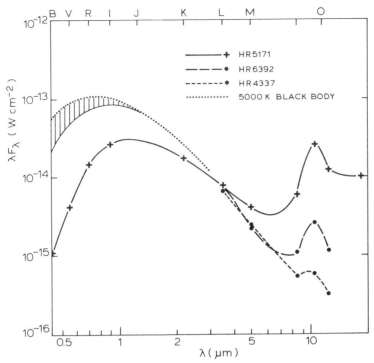

Fig. 38. Plot of λF_λ in energy per unit area versus wavelength for the G supergiants HR 5171 (G8 Ia⁺) (pronounced emission feature), HR 6392 (G5 Ia) (intermediate emission), and HR 4337 (G0 Ia⁺) (small excess). For HR 5171 the + refers to the observational points, the dotted line represents a 5000 K blackbody normalized at 3.6 μm, and the thin solid line represents the observations of HR 5171 corrected for interstellar extinction as determined from its B component. (From Humphreys *et al.*, 1971).

Apart from this cold outer shell there must also be a hotter shell, presumably nearer to the star, as indicated by a weak but persistent [N II] emission in the spectra (Warren, 1973a), similar as in HR 8752. Curious is that emissions of [S II], [O II] and [O III] that often accompany the [N II] emission are not observed.

The photospheric spectrum was the subject of an analysis by Warren (1973a) who derived an effective temperature of 4900±200 K. The photospheric abundances do not deviate from the solar values.

Some of the physical characteristics of the star and its cold shell are given in Table XIX.

The galactic hypergiant HD 160529 (A2 Ia⁺) $V=6.67$ mag was studied by Wolf *et al.* (1974). From a curve-of growth analysis of lines of singly ionized metals an excitation temperature of 7900 K and $v_\mu \approx 10$ km s⁻¹ was derived. There are large variations of the radial velocity, with amplitudes of 40 km s⁻¹, while variable line splitting up to a separation of 42 km s⁻¹ indicates inhomogeneous atmospheric motions. Also, variations in brightness and colour show the instability of the star's outer envelope. Wolf *et al.* (1974) suggest a connection between the variable velocity

TABLE XIX

Physical characteristics of the galactic hypergiant HD 119796=HR 5171
(G8 Ia$^+$), and its shell. (After Humphreys *et al.*, 1971; and Warren, 1973a)

Parameter	Star	Shell
Distance (kpc)	3.6	3.6
Angular diameter (arc sec)	2.8 10^{-3}	0.08
Radius ($R_\odot$)	10^3	3×10^4
Temperature (K)	4900	500
Power radiated (W)	2.4×10^{32}	2.4×10^{31}
M_b	-9.75 (incl. shell)	-7.25

field and the variations in line intensities, brightness and colour. The emission features – such as a variable emission component in Hβ, at one time showing a P Cygni-profile, and at other times just being of the inversion type – point to a highly variable stellar chromosphere.

The hypergiant HD 80077 (B2 Ia$^+$–e) was described by Moffat and Fitzgerald (1977). The bolometric magnitude, although uncertain, is estimated at -11.5. Its varying radial velocity suggests it to be a binary. The Hβ-profile has a P Cygni-profile, indicating mass loss. The mass is estimated at 60 $\mathfrak{M}_\odot$.

The most early-type *galactic hypergiant* is ζ^1 Sco (HD 152236; B1 Ia$^+$). It is a member of the Sco OB1 association, at a distance of $\sim$2 to 2.4 kpc. Its spectral and photometric properties were studied by many authors including Beals (1950), Rosendhal (1973b), Sterken (1977), Sterken and Wolf (1978a, b), Humphreys (1978), Hutchings (1979b, d), Appenzeller and Wolf (1979), Wolf and Appenzeller (1979). The work of Humphreys, Appenzeller and Wolf and of Hutchings leads to the following parameters: $T_c \approx 19\,000$ K in the ultraviolet range, $M_v = -8.7$, $M_b \approx -10.5$, $\mathfrak{M} \approx 60$ to $65\mathfrak{M}_\odot$, $R \approx 100 R_\odot$. The pronounced P Cygni profiles of the ultraviolet resonance lines show that strong mass-flow occurs with $v_\infty \gtrsim 400$ km s^{-1}, and $\dot{\mathfrak{M}} \approx -10^{-5} \mathfrak{M}_\odot$ yr^{-1} (Table XL in Chapter 7). The star is photometrically variable with in the visual range an average brightness amplitude of 0.1 mag on a time-scale of 15 to 20 days (Sterken, 1977). It is sometimes classified as a P Cygni star, but there are no really strong reasons for doing that.

3.9. The Most Luminous Extragalactic Star

The most luminous star known outside the Galaxy is the hypergiant HD 33579 (A3 Ia$^+$(e)) in the Large Magellanic Cloud; the spectrum was determined by Feast *et al.* (1960). With the distance modulus $m - M = 18.5$ of the LMC of Tifft and Snell (1971) one obtains $M_v = -9.53$. Humphreys gives -9.8 (1979). With a bolometric correction of $+0.06$: $M_b \approx -9.6$ (Wolf, 1972) to -9.9. The spectrum and the resulting physical parameters were studied by Przybylski (1965, 1968), Wares *et al.* (1968), and Wolf (1972). The main photospheric parameters as determined from various model atmospheric analyses are in reasonable agreement, as shown by Table XX. The photospheric chemical abundances were determined by Wares *et al.* (1968) and by Wolf (1972); both investigations agree that they are solar-like. The photometric data were studied and compared with results of model computations by the Walravens (1971), and by Van Genderen (1979a, b). The results, also given in Table XX, are in fair mutual agreement. The Balmer discontinuity is 0.26 (the Walravens) and 0.30 according to Przybylski (1968), but the Walravens note it to be variable: Van Genderen (1979b) found that it had decreased to 0.245 in the period 1973–1978. Some dilution effects are observed in Si II λ4128–30 and Mg II λ4481, which suggest that these lines are formed in a shell. The presence of an outer shell moving outward with velocities up to 170 km s^{-1} is shown by the asymmetry of all absorption lines, red-shifted Ca II emissions, and a strong red-shifted emission in Hα. Wolf (1972) estimates the age of the star from its positions in the HR diagram and finds it between 3×10^5 and 5×10^5 yr.

TABLE XX

Main parameters of the A3 Ia$^+$ hypergiant HD 33579 in the Large Magellanic Cloud. After Wolf (1972); Wares *et al.* (W, 1968); Przybylski (P, 1968); the Walravens (WR, 1971); and Van Genderen (G, 1979b)

T_e $\quad$ =8130 K $\pm$ 300	8000 (W)	8060 (P)	8700 (WR)	8200–9000 (G)
$\log g$ $\quad$ =0.7 $\pm$ 0.2	0.2 (W)		0 (WR; $\log g_{\mathrm{eff}}$)	0.7 (G)
$\log P_e$ =0.25 at $\bar{\tau}$=0.7		0.0 (P)		
v_t $\quad$ =6 km s^{-1} at $\bar{\tau}$=0.2				
$\quad$ 26 km s^{-1} in outer shell				
$L/L_\odot$ =530000				
$R/R_\odot$ =360				
$\mathfrak{M}/\mathfrak{M}_\odot$=25				

It is of interest to compare HD 33579 with the brightest star of the *Small* Magellanic Cloud, which was studied by Wolf (1973). The spectral type is A0 Ia$^+$. It is variable by 0.04 mag in V, 0.04 mag in B-V and 0.10 in U-B. With a distance module m-M= 19.2, and V=9.86 one obtains M_v=-9.34. A bolometric correction of 0.19 yields M_b=-9.53. Hence, this star is only slightly less bright than the brightest star of the LMC, HD 33579.

The photospheric parameters, determined from a spectral fine-analysis are T_e= 8960 ± 300 K; $\log g$=0.87 ± 0.10. The photospheric chemical abundances are of the solar type; there is in this star no indication of a difference in chemical abundance between the SMC and the Galaxy, contradictory to some other results. Chromospheric phenomena are shown by a strong red-shifted emission in Hα, and emission in the core of the Ca II H and K lines.

With M_b=-9.53 and T_e=8969 K one finds $R/R_\odot$=295 ± 50. With $\log g$=0.87 one has $\mathfrak{M}/\mathfrak{M}_\odot$=$24 \pm 11$. From a comparison with evolutionary tracks for stars of this mass Wolf (1973) finds that the star is in the He-burning phase, and must have an age of approx. 5×10^6 yr.

In connection with these stars we should also consider *the brightest known strongly**

* We have to add the adverb ‚strongly' because all supergiants are variable!

variable supergiant: S Doradus (HD 35343, A2–5 eq) also a member of the Large Magellanic Cloud. The star is of spectral type A0 Ia (eq); the variability has been claimed to be of the Algol-type (Kukarkin *et al.*, 1958), but later investigators show it to be an irregular variable (Thackeray, 1973), with detectable variations of the order of 0.01 mag on time scales of a few hours. Important are the long-period variations, summarized for the period 1948–1978 by Van Genderen (1979a); fluctuations of 2 mag occur at a time scale of ten years.

The blue magnitude ranges between 9.2 and 11.5, which should be compared with the blue magnitude of HD 33579 which is 9.4 according to Przybylski (1968) and 9.0 according to the Walravens (1971). Hence, S Doradus *can* even be as bright as HD 33579.

More stars of this type have been discovered in other nearby galaxies. Sharov (1975) gave a brief summary.

3.10. G- and K-type Supergiants; Zeta Aurigae Stars

Equivalent widths of spectral lines between 5000 and 6650 Å in seven G- and K-type Ib supergiants (Van Paradijs, 1973a) were used by Van Paradijs (1973b) and by Van Paradijs and De Ruiter (1972) to derive the atmospheric parameters of these stars, and to deduce chemical abundances. *Effective temperatures* were derived from the curves-of-growth of Fe I and Ti I lines by requiring that the scatter in these curves should be minimum. By a differential procedure, comparing the equivalent widths with solar data, values for $\Delta\theta_{ex} = \theta_{ex}^{star} - \theta_{ex}^{sun}$ were found (with $\theta = 5040/T$). Assuming then $\Delta\theta_{ex} = \Delta\theta_e$, yields T_e, which appears to range between 5100 K (for μ Per, G0 Ib) to 3650 K (for ξ Cyg, K5 Ib). The temperature scale as derived from these results agrees best with previous determinations by Johnson (1966) and Böhm-Vitense (1972). Luck (1977a) found values that are systematically 100 K lower, which is insignificant, and caused by the v_μ-value assumed for the Sun in these comparative studies.

Values for the *acceleration of gravity g* were found by Van Paradijs and De Ruiter (1972) by assuming that chemical abundances derived from neutral elements and ions should be the same. Use was made of photospheric models of Carbon and Gingerich (1969), extrapolated to smaller g-values. Although the strong dependence of g on the assumed value of θ_e badly influences the result (actually $\Delta \log g/\Delta\theta_e \approx -13$!), apparently reliable values for log g could be found; they range between 0.75 and 1.75, average value 1.2. The average value for the G0–G5 Ib stars is 1.5, that for the G8–K5 Ib stars is 1.0. Since these values were derived from a comparison with model computations they should refer to the geometrical g-values, not to g_{eff}. It is, therefore, gratifying that the log g-values found do not exceed significantly ($+0.16 \pm 0.19$) the values found from the mass-luminosity law. These results were in an overall way confirmed in the spectroscopic study by Luck (1977a).

The *chemical composition* of the photospheres were determined by Luck (1977b) for fourteen Ib supergiants, of spectral types G0–K5. Investigated were lines of some 24 metals ($Z > 10$). The metal abundances appear to exceed the solar values by an average value of 0.3 dex (factor 2), but with a mean error of ± 0.2 dex; the excess is therefore hardly significant. Insofar as the metals are concerned, there are no deviating abundances of such kinds that would indicate nuclear processing in the interiors and subsequent transport of processed material to photospheric regions. There is neither correlation between C^{12}/C^{13} abundances and most metal abundances. Only Li is, with maximum logarithmic abundance $\log(\text{Li}) \equiv \log[\varepsilon(\text{Li})/\varepsilon(\text{H})] = -10.5$, approx. 30 to 100 times or 10 times (Warren, 1973b) less abundant than the 'cosmic' value (Suess and Urey, 1956; Luck, 1977b). These abundances correlate stepwise with the $^{12}C/^{13}C$-ratio; the Li abundances smaller than -11.5 (logarithmically) correspond to a $^{12}C/^{13}C$-ratio smaller than 20. No satisfactory explanation other than a tentative suggestion of meridional circulation has sofar been given.

The *microturbulent velocity* components v_μ, derived by Van Paradijs (1972) from curve-of-growth analyses ranges from 2.8 to 3.9 km s^{-1}, average 3.3 km s^{-1}. Luck (1977a) found $v_\mu = 2.5 \pm 0.5$ km s^{-1}, and $v_M = 5 \pm 3$ km s^{-1}.

The *ultra-violet spectra* of these stars are interesting. In main-sequence stars and giants (large g-values) chromospheric features are shown, in contrast to the coolest high-luminosity objects. A review was given by Dupree (1980). See also Chapter 6.

The Zeta Aurigae stars. A small group of binaries, consisting of a K- or M-type supergiant with an early-type main sequence companion. Apart from these criteria, the members of the group have in common that the orbital period is long, as well as the eclipse period. For instance ε Aur has a period of 27.1 yr, a full eclipse period of 714 d and a period of totality of 330 d.

For all group members the companion is small enough as compared to the extension of the primary's atmosphere that ingress and egress can be used for sounding the supergiant's atmospheric structure and its variation with distance to the primary's center. The properties of the five best studied members of the group are listed in Table XXI, deduced basically from the excellent reviews by Wright (1970, 1973) with a number of later additions.

Other stars, that may belong to this group but that have not yet been studied in sufficient detail are GG Cas, AZ Cas KU Cyg, CH Cyg, RZ Oph; AR Pav, V777 Sag, V381 and V383 Sco, BL Tel, Boss 1970, and Boss 1985. The latter object studied by Cowley (1965) consists of a primary star of type M2 ep Iab (Bidelman, 1954) or M1 Iab (Jaschek and Jaschek, 1963) and an early B-type secundary – as judged from the far UV spectrum (Cassatella and Viotti, 1979); it performs one revolution in 27 yr in a highly eccentric orbit. The system is surrounded by an envelope, as is seen from the [Fe II] emission lines, which do not share the v_R-variation of the H and Fe II absorption lines. Masses are estimated at 40 to 100$\mathfrak{M}_\odot$ (M-star) and 20 to 60$\mathfrak{M}_\odot$ (B-star). For AZ Cas (M0 Ib+B0–1) the component masses are 18 and 13$\mathfrak{M}_\odot$, and $M_b = -6.5$ and -5.3 (Cowley *et al.*, 1978).

The system ε Aur is enigmatic since there is no secondary eclipse while also spectral data on the secondary star are lacking. So, there has been much speculation on the properties of the secondary member. While most authors advocate it to be a B type star, at least 3 mag fainter than the primary, there are also indications that it might

TABLE XXI

Physical and orbital parameters of the best studied Zeta Aurigae stars

Parameter		ζ (=8) Aurigae		32 (=o²) Cygni[d]		31 Cygni		VV Cephei[c]		ε (=7) Aurigae	
		Primary	Secondary	Primary	Secondary	Primary	Secondary	Primary	Secondary	Primary	Secondary
Spectrum		K4 Ib	B6+V	K5 Iab	B4 IV-V	K4 Ib	B4	M2 Iabep	B-A	F0 Ia	B: (?)[b]
Luminosity, M_v											
Spectroscopic data		−3.7	−1.5	−4	−1	−4.6	−2.0	−4.0	−2	−8.7[f]	−7:
Eclipse data		−2.7	−2.0			−2.7	−1.6			−5.9[a]	
Temperature T_e (K)		3550	15400	3500	17600	3550	17600	3000		7800[f]	15000
Bolometric correction		−1.1	−1.1	−1.1	−1.6	−1.1	−1.6	−1.2:		0.1	−3:
Diameter ($R/R_\odot$)											
Spectroscopic		200	4.1	250	5	320	4.9	1600	13	277[f]	2
Eclipse data		130	5.1			135	4.0	1620		295	
Mass; $\mathfrak{M}/\mathfrak{M}_\odot$		8.0±1.2	5.8±1.1	15	7	9.2±2.2	6.2±2.5	18.3	19.8[e]	28[f]	13.7
Period, P (days)		972.162		1147.8		3748.3		7430.5		9890	
Eccentricity, e		0.406± 0.011		0.301± 0.013		0.222± 0.008		0.346± 0.01		0.200± 0.034	
Longitude of perioastron ω^0		336.0 ± 2.2		218.2 ± 3.0		201.1 ± 2.4		59.2 ± 2.8		346.4 ±11.0	
Periastron passage, τ											
J.D. 2430000+	calculated	4585.74 ± 4.46		3141.80 ± 7.74		7169.73 ±22.29		8461.0 ±48.2		3346 ±278	
Mid eclipse		5478.4		6675.4		7660.4		5914.4		5543	
J.D. 2430000+	observed	5470.0		6668.1		7685.7		5931.4		5638	
Semi-amplitude, km s⁻¹ K_1		24.58 ± 0.40		16.95 ± 0.28		13.98 ± 0.13		19.43 ± 0.33		15.00 ± 0.58	
K_2		31.4 ± 2.8		34:		20.8 ± 3.8		19.14 ± 0.68		17.0:	
Semi-major axis, km $a_1 \sin i$		3.00×10^8		2.55×10^8		7.09×10^8		19.14×10^8		20.0×10^8	
$a_2 \sin i$		3.38×10^8		5.10×10^8		10.55×10^8		18.86×10^8		22.6×10^8	

[a] From a detailed astrometric study of the apparent orbit of ε Aur by Van de Kamp (1974) and improved in a later study (Van de Kamp, 1978).

[b] Other models are also proposed; – the practically invisible – companion might be a dust and gas disc (Handbury and Williams, 1976). From far UV observations Hack (1979) finds the communicated values for M_v, T_e, and R. Hack and Selvelli (1979) find that there is a late B companion *and* a gaseous ring, and postulate that the B dwarf is an evolved object and that the ring may be gas ejected by the B star during its evolution.

[c] Data mainly from Wright (1977) and Van de Kamp (1977), and for the secondary from Hack (1979).

[d] Physical data from Saijo and Sato (1977), and Stencel et al. (1979).

[e] Hutchings and Wright (1971).

[f] Castelli (1978).

be an opaque dust and gas cloud with a particle temperature of about 500 K (Hand-bury and Williams, 1976).

For the two best studied systems ζ Aur and 31 Cyg the mass ratio $\mathfrak{M}_1/\mathfrak{M}_2$ is less than 1.5. The average masses of the (late type) primary and (early type) secondary are respectively $8\mathfrak{M}_\odot$ and $6\mathfrak{M}_\odot$; the average diameters are $135R_\odot$ and $5R_\odot$ respectively, with a large scatter in the former values.

Many studies of strong resonance lines, particularly of the K line of Ca II during ingress and egress show that the behaviour of the line can be very different during these two periods. The line is often split in various components, which suggests that the chromospheres of the supergiants consist of a number of individually moving

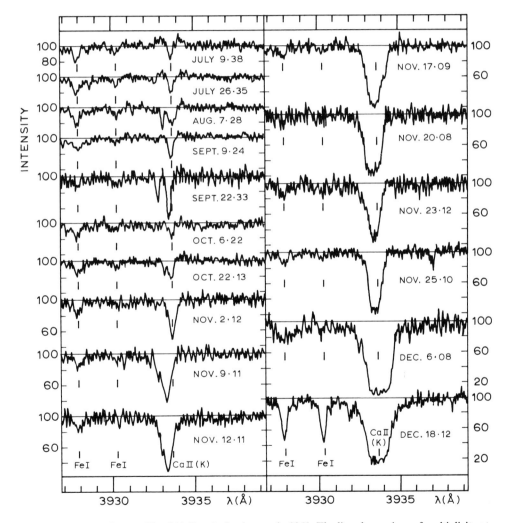

Fig. 39a.   The K line profile of 31 Cyg during ingress in 1961. The line shows signs of multiplicity at various occasions, while also the large width of the line close to ingress indicates multiplicity. (From Wright, 1972).

elements without much coherence in the respective motions. See e.g. Wright's (1970) description of the spectrum variations of 32 Cyg during ingress, and our Figure 39a, after Wright (1972), showing these data for 31 Cyg. The increasing width of K (Ca II) must be due to multiplicity.

Some of the stars, in any case ζ Aur, 31 and 32 Cyg are surrounded by circumstellar matter as is evidenced for the former two objects by observations of the Ca I $\lambda6572$ line, which does not participate in the radial velocity variation of either of the two members of the binary (Saito and Kawabata, 1976). The line density of Ca atoms

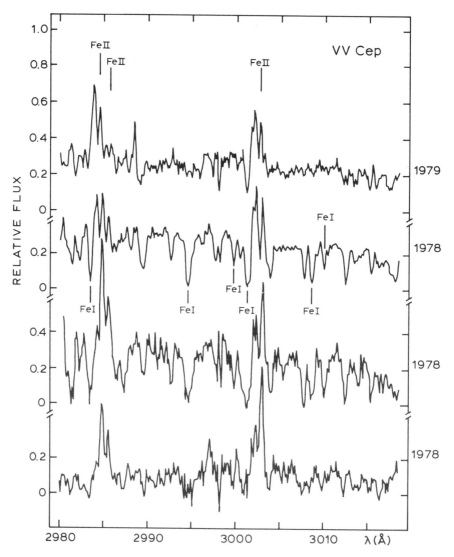

Fig. 39b. Part of the near-UV spectrum of VV Cep during the egress of the secondary from eclipse. Note the gradual weakening or disappearance of Fe I absorption features: neutral iron extends less far in the atmosphere than Fe^+-ions. (From Hagen *et al.*, 1980).

contributing to this line varies from one eclipse to the other, which suggests that mass-loss of these stars is a non-continuous phenomenon. This latter remark can be related to the observations that the extent and density of the *atmospheres* of these stars as studied during eclipses can vary fairly strongly from one eclipse to the other. Ultra-violet spectral observations of 32 Cyg show P Cygni resonance lines with multiple absorption components; the B star is within the K-chromosphere (Stencel *et al.*, 1979).

In the system VV Cep the B component is surrounded by an emission shell, while gas streams move from the cool supergiant to the shell (Hutchings and Wright, 1971). For conservation of angular momentum the shell will assume a disk shape around the B-component, as was concluded from spectroscopic observations during the ingress phase 1976/77 of the first Balmer lines, Ca II, H and K, and a few other lines including [Fe II] $\lambda 4287.4$ (Möllenhoff and Schaifers, 1978). Asymmetry of the Balmer lines points to the existence of cool gas outside the main part of the ring. The rotational velocity of the ring should be ~ 100 km s^{-1}, as can be concluded from the displacement of the Balmer lines at ingress. Assuming this to be the Kepler velocity, and with the mass $20\mathfrak{M}_\odot$ for the B-star the ring should have a radius of $\sim 380 R_\odot$. From the time difference between the disappearance of the blue and red emission components of Hβ it is concluded that the inner radius of the Hβ emitting ring is $96 R_\odot$. Several observers have noted a small degree of polarization ($\approx 1\%$) of the total light (Koch and Pfeiffer, 1978). It seems to show some periodicity. No doubt the polarisation must be due to asymmetric gas clouds or shells in the system but a quantitative explanation has not yet been given and would be fairly difficult.

3.11. The Coolest of the Bright stars

Various groups of red stars in the upper right part of the Hertzsprung–Russell diagram would have more right than any other type to carry the name 'supergiant' because of their large radii: for a star with $T_e = 3000$ K and the same bolometric magnitude as a B-type star of 30000 K the diameter must be a hundred times larger.

An essential difficulty with the interpretation of stellar observations in this part of the Hertzsprung–Russell diagram is that of ambiguity: young contracting stars as well as evolved stars with He, C or heavier-metal cores occur in this region; the first are very often embedded in thick clouds of gas and dust; the latter are too, but the clouds are thinner. Quantitative analyses of the atmospheres of cool stars became well possible after the developments of the techniques of (high-resolution) infrared and (sub)-mm spectroscopy, and of spatial interferometry allowing the measurements of angular diameters. Reviews on infrared stellar spectroscopy are by Spinrad and Wing (1969) and by Merrill and Ridgway (1979). An atlas of infrared stellar spectra was published by Johnson and Mendez (1970). We mention further the IRC *Two Micron Sky Survey* (Neugebauer and Leighton, 1969), and the AFCRL/AFGL *Infrared Sky Survey* (Walker and Price, 1975; Price and Walker, 1976).

Let us describe the various spectral groups of cool stars!

The *M-type stars* are characterized principally by TiO-bands, relatively weak for early M's but with increasing strength for later types. Bands of H_2O at 1.8 and 2.7 μm are prominent in later M-giants. Well-known examples: β And (M0), α Ori (M1.5), π Aur (M3), ϱ Per (M4). The additional letter e signifies that at least one of the Balmer lines is in emission.

The list of standard stars of Morgan and Keenan (1973) ends with M8 and for supergiants even with M5. Extensions up to M10 have been proposed by several authors; the classifications are based on the strength of the infrared molecular bands of TiO and VO. Solf (1978) describes classification criteria for M5 to M10, and gives also luminosity criteria, enabling one to distinguish between classes I, II, and III. A list of more than 30 standard stars is added.

In addition to the M-classification there are stars of other spectral types in the region of low temperature stars:

R-type stars are distinct from M-stars by the strong bands of CN and C_2 instead of TiO-stars. The subdivision is R0 ... R9, according to the intensity of the C-bands.

N-type stars are like R-stars but show large depressions in the violet and UV. The subclasses are N0 ... N7. The distinction between late R-type stars and N-type stars cannot always be made, so there is some confusion.

The *Carbon stars* (types C0 ... 9) were defined by Keenan and Morgan (1941). They are characterized by strong absorption bands of C_2, CH and CN. Late C-stars have bands of HCN and C_2H_2 in the 3.1 μm region (Ridgway *et al.*, 1978) and have emission around 11 μm (Treffers and Cohen, 1974). The C stars belong also to the spectral types G5–M4; R(R0=C1, R5=C4), N, in a sequence of decreasing temperature. In actual fact the C-class may be considered as partially replacing the – older – R and N classifications, and various authors, particularly those using high-dispersion spectra, prefer the C-class, as a better temperature sequence than the R and N-classes (Sanford, 1950; Fujita, 1966; Utsumi, 1970). In addition to the subclasses C0–C9, we should mention the peculiar Carbon stars (cf. Mikami, 1975): the metal-poor CH stars and the Hd (H-deficient) stars.

An interesting example of the latter class is U Aqr, a C-rich, H-deficient R CrB star ($M_v = -4$ mag) which shows strong C_2 and CN bands and strong lines of the *s*-process elements Sr and Y (Bond *et al.*, 1979). Bouigue (1954) and Yamashita (1972) have introduced the designation J-type for those Carbon stars that have very strong ^{13}C features. Examples are RX Peg (C3 ... 4), Y CVn (C5) and WZ Cas (C9). Reviews on C stars were published by Fujita (1970, 1980) and Wallerstein (1973). In his 1980 review Fujita dealt in great detail with spectroscopic classification. A *catalogue* containing 3219 Carbon stars was published by Stephenson (1973).

The – rare – *S-type stars* (characteristic objects: π^1 Gru, $M_b = -4.9$, R Gem) have, apart from strong lines of Ca II, lines of Ca I 4227, Ba II 4554. The relatively few 'pure' S-stars show no TiO bands. The stars also show lines of the unstable element Tc I, and are particularly characterized by oxyde bands of *s*-process elements: ZrO, YO and LaO. A classification scheme based on the red spectra was established by Ake

(1979). A list of the S-type stars with the largest apparent brightnesses was published by Wing and Yorka (1977). The number of known S-type stars is $\sim$700.

Hence, in the domain of cool stars we distinguish between *three spectral types*: M, C ($\approx$R+N), and S. These three types are primarily *abundance classes* and, to a lesser extent, they indicate different temperatures and luminosities. Because these three types differentiate between abundances, it is only natural that transition types between them do occur – indeed there are cool stars that cannot easily be fitted in any of the three classes M, S, or C. This has forced one to introduce the class MS, and intermediate between the S and the C stars the classes CS and SC (see Catchpole and Feast, 1971; Catchpole, 1975). It appears that the sequence M–MS–S–SC–C is one of increasing carbon-to-oxygen ratio (Table XXVI).

The *SC-stars*, of which a twenty are known, are characterized by weakness of molecular features in the visible spectral range – though ZrO and CN are both present – and by considerable strength of lines of elements such as La, Y, Ba, Zr, Sr, which are *s*-process elements. The suggestion has been made that these stars have a C/O ratio very close to unity so that nearly all C and O is locked up in CO. Examples are VX Aql (Wyckoff and Wehinger, 1976), UY Cen and FU Mon.

In the spectra of *CS stars* the ZrO lines are absent or very weak, and the CN lines are stronger than in SC stars. Prototypes are R CMi and R Ori.

Many of the stars of the classes M, S, C and interim classes belong to the long-period variables of the Mira-type.

Let us next discuss the *luminosities and T_e-values* for the various groups of stars described in this section. The brightest are the early M-type supergiants (M0–M4). For these: $-8 \lesssim M_v \lesssim -5$ as could be judged from the homogeneous sets of such objects in clusters and nearby galaxies, collected by Stothers and Leung (1971) and by Humphreys (1978, 1979). For such stars M_b would then be ≈ -7 to -10 (see Table VII). For the Large Magellanic Cloud Cowley and Hutchings (1979) find $M_b \gtrsim -9$. Among the later-type objects the N-type stars (or late C-type stars) and the late-type Mira stars are the intrinsically brightest, with M_b-values up to -9 for late M sugergiants. The *average M_v-values* are -6 for M-type supergiants and for giants $\langle M_v \rangle$ is around 0. Extremely bright is the object VYCMa (Section 6.16), with $M_b \approx -9.5$.

The determination of *effective temperatures* is obviously difficult. For K0–M3 giants Van Paradijs (1976) found a relation between the *R-I* colour index and T_e. Similarly, Helt and Gyldenkerne (1975) found correlation between spectral and luminosity classes and stellar colours. In the same track Rautela and Joshi (1979) found a well-defined relation between $m_{0.65\ \mu m} - m_{0.54\ \mu m}$, T_e and luminosity class. Late C-(N-type) stars have T_e-values between 2500 and 3000 K as could be determined from diameter measurements and infrared photometry (cf. Table XXII). The T_e-values decrease from 3500 K (K5e-type Mira stars) over 3700 (M1 … 2 supergiants) to $\sim$2700 K (N5e) and hence the radius R increases from $\sim 80 R_\odot$ to $1000 R_\odot$. Consequences of these large radii are the small values of g_{grav} and *a fortiori* of g_{eff}, and hence very large atmospheric extensions (see Section 5.2) and the breakdown of the planar approximation for the atmospheres. The radii of Mira Ceti and R Leo, measured in the TiO bands are about twice those measured in the continuous spectrum (Labeyrie *et al.*, 1977), yielding an atmospheric extension comparable to the stellar radius.

In addition, *rotation* can produce consideral deviations from spherical symmetry. Linear polarization has been detected in the TiO-bands in the spectra of several cool giants and supergiants (cf. the review by Coyne and McLean; 1979). For the semi-regular variable V CVn (M4e–M6e) the polarization of 4 to 7% changes with time, suggesting a *pulsating* molecular scattering atmosphere of a *rotating* star with an equatorial rotational velocity of ≈ 11 km s^{-1} (rotational period ≈ 7 yr). This would produce a pole-to-equator temperature difference of 200 to 300 K, sufficient to explain the observed polarization (Coyne and Magalhães, 1979). Another consequence of the extension of these atmospheres is that the g-value varies greatly through the atmos-phere. It is relevant in this connection that all M-giants, all K-supergiants and G Ia supergiants have circumstellar spectral lines and have appreciable mass-loss (e.g. Reimers, 1977); see further Chapter 7.

With regard to the *position* of these stars in *the HR-diagram* (Figure 40) it is fundamental that the reddest giants occupy a region that runs more or less parallel to the Hayashi-lines (Section 4.1) and to the late evolutionary tracks for stellar masses of the order 1–10$\mathfrak{M}_\odot$, and that the stars with the lowest temperatures take the position of the Hayashi-line for approx. 1$\mathfrak{M}_\odot$ or even smaller. Apparently the stars, to be discussed in this section, are fully convective stars of about one to several solar masses.

Roughly, the following generalities may apply:
– The stars discussed here that belong to the giant luminosities (III) occupy a fairly broad band ($\Delta \log T_e \approx 0.1$) between the Hayashi-lines of 3 to approx. 0.5$\mathfrak{M}_\odot$.
– The non- or slightly variable K- and M-type supergiants occupy a region to the upper-left of the hatched area corresponding to larger masses, up to $\mathfrak{M} \approx 20\mathfrak{M}_\odot$. Actually, Stothers and Leung (1971) determined masses of 20$\mathfrak{M}_\odot$ for such objects.
– The Mira variables occupy a narrow strip, approximately between the Hayashi-lines of 1$\mathfrak{M}_\odot$ and a few $\mathfrak{M}_\odot$.
– The C-stars occupy a parallel band, partly overlapping with that of the Mira's, but shifted towards higher masses, very roughly between 1.5$\mathfrak{M}_\odot$ to 3 $\mathfrak{M}_\odot$.
– The late-type R and N-type stars are found in the upper part of the C-area ($M_{bol} = -4$ to -7).
– The early-type R stars (R0–R4) form a group that is distinctly separated from the later types; they lie at slightly higher temperatures below the C-area.
– The S-type stars are more or less evenly distributed over the region of the Mira's, and in the low-T part of the C-stars area.

At this stage we also draw attention to the thick line in Figure 40; it gives the locus of the first He–shell–flash. The deviating abundances of many of the stars found to the low-T side of this line may be related with transport of nuclear fusion products to the surface in stars after the He flash. The *massive early M-type* supergiants like α Sco, α Ori, μ Cep, S Per are all to the left of this line. We refer further to Section 4.6.

In the next part of this Section we pay some further attention to two of the three

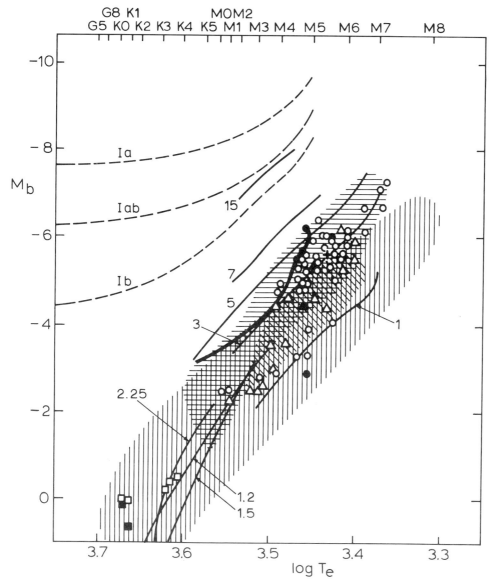

Fig. 40.   Composite HR diagram for late-type stars partly after Scalo (1976). *Vertically hatched area:* mean locus of stars of luminosity class III. *Filled circles:* Carbon stars with individually determined distances. *Open circles:* Carbon stars with absolute visual magnitudes, assigned according to statistical studies. *Filled and open squares:* same, but for R0–R4 stars. *Filled and open triangles:* same, but for S stars. *Horizontally hatched area:* area of C stars according to Mikami (1975). *Diagonally hatched area:* Mira stars, according to Smak (1966), and Eggen (1975a). The thick line is the locus of the first helium shell flash, partly drawn after Sweigert and Gross (1978). Thin lines are Hayashi tracks labeled with the mass in solar units.

main spectral groups described above: the C-, (R-, and N-)stars and the S-stars, and we describe the Mira-variables. Thereafter we describe two prototypes of M-type supergiants and a prototype of a Mira star.

The sub-classification of *the Carbon stars* as based on the intensities of several spectral features is described by Yamashita (1967, 1972, 1975). An interesting attempt to arrive at a quantitative spectral classification system, on the basis of narrow-band photometry of selected spectral features is by Yamashita *et al.* (1977). See also Fujita's review (1980).

Many of the C-stars are variable; there is a strong overlap with the Mira-stars. Quantitative information on the physical parameters of these stars is scarcely available: only for one star is the parallactic distance known, but badly so: 90 to 330 pc for χ Cnc. Bergeat *et al.* (1978) give distances for 38 C-stars, based on a statistical study of proper motions. This enabled them to determine M_v, but effective temperatures are not well known, since reliable data on the absolute fluxes are absent for these stars. The latter are hard to determine because of the large infrared contribution to the stellar radiation, and the consequent large bolometric corrections. Colour temperatures, which range from approx. 5000 K for C0 to 2200 K for C9 (Baumert, 1972) cannot easily be transformed into effective values since model atmospheres for very cool stars are not available. Yet it seems from some infrared observations that bolometric corrections for C-stars range from 0.2 mag for stars with $T_e=5000$ K through approx. 5 mag for stars with $T_e=2200$ K (see Wallerstein, 1973, Figure 2). There is a large scatter in plots of infrared colour temperatures versus C-type, which is due to the blocking effects of the CN absorption bands and to interstellar reddening (Yamashita, 1975). Such variations may also be related to circumstellar dust clouds, which are known to occur also around many of the latest type supergiants (see Section 6.15). The most heavily obscured C-star known sofar is the infrared C-rich object GL 3068, studied by Jones *et al.* (1978). The colour temperature of the dust shell is 325 K; its angular diameter (0".4, – Lebofsky and Rieke, 1977) corresponds at an assumed distance of 2 kpc to a size of 10^{11} km (10^3 AU).

For four C stars for which angular diameters have been measured T_e-values could be determined. These values are reproduced in Table XXII together with some values based on infrared photometry. The small range and the scatter in the data is obvious. There is also no clear temperature progression with increasing C subclass. Colour temperatures (Yamashita, 1965) are comparable to these T_e-values.

TABLE XXII

Effective temperatures of some Carbon-stars (from a summary by Fujita and Tsuji (1977), based chiefly on Ridgway *et al.* (1977b), and Dearborn *et al.* (1976); added: value for AQ Sgr (Walker *et al.* (1979)).

Star	Spectral type	T_e (K)	Source
χ Cnc	C5	2490 ± 140	angular diameter
DS Peg	C5	2840	infrared photometry
U Hya	C6	2860	,, ,,
γ Tau	C7	2700 ± 100	angular diameter
19 Psc	C7	3060 ± 150*	,, ,,
AQ Sgr	C7	2680 ± 110	,, ,,
WZ Cas	C9	2560	infrared photometry

* For 19 Psc infrared photometry yields $T_e=2890$ K, in fair agreement.

The *absolute visual magnitudes* were determined by Mikami (1975) on the basis of a statistical analysis of radial velocities and proper motions. They were transformed into M_b- and T_e-values by assuming effective temperatures of Mendoza and Johnson (1965), and resulted in the data given in Table XXIII. A similar track was followed by Gordon (1968), Cohen (1979), and by Bergeat *et al.* (1978) for galactic C-stars, while C-stars in the Large Magellanic Cloud were studied by Crabtree *et al.* (1976) using stars for which VRI photometric data were available. As a matter of fact only bright late-type C-stars could be observed. By fitting a blackbody curve to the $(R-I)$ colours the value of T_c is derived, which is thereupon identified with the effective temperature T_e. For the average of these bright late-type C-stars: $T_e \approx 2500$–3000 K; $M_b \approx -5$ to -7.

TABLE XXIII

Effective and colour temperatures, absolute magnitudes and radii of C-stars, after Yamashita (1965), Gordon (1968), Mikami (1965) and Cohen (1979).

Type	M_v		M_b		T_e	T_c	$R/R_\odot$
	(M)	(G)	(M)	(G)	(M)	(Y)	(C)
C0–3	−1.34	−2.0	−2.20	−2.8	3890	3150	110
C4–5	−2.64	−2.4	−5.28	−4.3	2800	2830	270
C6–7	−2.59	−2.7	−5.66	−3.8	2700	2560	550
C8–9		−3.0		−6.7		2300	1250
R0–3	−1.67		−2.58				
R4–9	−1.69		−3.92				
N	−2.39		−5.23				

The *brightest known galactic C-star* is W CMa. Its distance and absolute visual brightness could be estimated on the basis of its likely membership of the association CMa OB1 and is at maximum: $M_v = -4.7$ (Gordon, 1968; Herbst *et al.*, 1977). With bolometric corrections by Olson and Richer (1975) one finds $M_b = -7.2$ a value comparable to the maximum brightness found for C-stars in the LMC.

The average absolute bolometric magnitudes and effective temperatures for the various groups of red giants, as summarized above, and partly also derived from other sources are given in Table XXIV.

The Mira stars. After the fundamental review by Smak (1966) important work has been done in this field by Eggen (1975a, with references to earlier work); see also Cahn and Wyatt (1978).

The stars have periods P of pulsation ranging from $\sim$100 to 650 days. Distances, and hence absolute visual magnitudes could be determined by statistical investigations based on stellar motions. Whether these results are correct or have to be modified depends on the degree of certainty of the statement (Reid, 1976; Reid and Dickinson, 1976) that the maser emission line groups are displaced symmetrically around the stellar radial velocity.

TABLE XXIV

Average M_v- and M_b-values and effective temperatures T_e for red giants and supergiants (partly after Scalo, 1976; and some other authors). Most data are only crude averages.

Group	$\langle M_v \rangle$	$\langle M_b \rangle$	$\langle T_e \rangle$	Ref.
C stars	-2.1	-6	5000(C0)–2200(C9)	(2)
M0, 1 giants (III)	-1.0 ± 0.2	-2.7		(3)
M2–4 giants (III)	-1.0 ± 0.2		3700	(4)
M5, 6 giants (III)	-0.4 ± 0.3			(3)
M7, 8 giants (III)	$+0.5\pm0.6$			
M supergiants (I)	-5.8 ± 0.5	$\begin{cases} -7 \text{ (early M)} \\ -9 \text{ (late M)} \end{cases}$		(3)
R0–R3	$+0.5$	-0.2	4800–4000	
R5–R8	-1	-5	3600–2700	
N	-2.3	$-5 \ldots -6$	3200–2300	(2)
S (invar)	-1	-5.5	2800	(5)
S (Mira)	-1.5 to -2 (max light)	-3 to -3.5		(5)
SC	-2	-5	3500–2000	
MS	-0.5	-2.5	2700	
Miras	see Table XXV		3500(K5)–2000(N5)	(1)

References:
(1) Loyd Evans (1978a).
(2) Lloyd Evans (1978b).
(3) Mikami (1978).
(4) Tsuji (1979b).
(5) Yorka and Wing (1979).

Bolometric corrections could be determined with some accuracy after the observation of a considerable part of the infrared spectrum during the Statoscope II flight. Eggen (1971) found that the bolometric corrections correlate well with the $(R–I)$ colour: $BC = 0.60 - 2.5\,(R–I)$ mag, yielding $\langle M_b \rangle = -0.65 - 2.5\,\langle (R–I) \rangle$, where $\langle R–I \rangle$ is taken at phase 0.25.

There is a $M_v(P)$ relation such that $M_v \approx -2.5$ for $P \approx 150d$ to $M_v \approx -0.5$ for $P \approx 500d$; on the average $\langle M_v \rangle \approx -1.25$ mag. After applying the bolometric correction the relation becomes $\langle M_b \rangle = +0.50 - 2.25 \log P$.

Table XXV gives the relation between M_v, BC and M_b with spectral type. Eggen (1975a) using the relation between $(R–I)$ and T_e, as derived by Johnson (1966) – but this relation is very uncertain for cool stars! – derives a relation between T_e and M_b for these stars. Pairs of $(\log T_e; M_b)$ values are: (3.50; -3.4); (3.48; -4.2); (3.46; -4.6); (3.44; -5.2).

TABLE XXV

Bolometric corrections and absolute magnitudes of Mira-stars. (After Smak, 1966)

Spectral type	M1.9	M2.7	M3.7	M4.2	M5.3	M6.0	M6.2	M6.5
$\langle M_v \rangle$	-1.67	-2.74	-2.10	-2.03	-0.93	-1.17	-1.05	-0.31
BC	-2.3	-2.55	-2.8	-3.0	-3.6	-4.25	-4.45	-4.8
$\langle M_b \rangle$	-4.0	-5.3	-4.9	-5.0	-4.5	-5.4	-5.5	-5.1

The *mass* is only determined directly for the binary o Ceti (Mira) for which it is between 0.7 and $1.8\mathfrak{M}_{\odot}$ (Fernie and Brooker, 1961).

The effective temperatures, of around 3000 K, are apparently inconsistent with the occurrence of H_2O in the atmospheres, detectable through strong bands near 1.5 and 1.9 μm: water dissociates above about 2500 K. The solution is that the H_2O bands are formed in outer layers. In the well-studied case of R Leo (Hinkle and Barnes, 1979) there appear to be *two* outer shells, one with $T_{ex} \approx 1000$ K and a less important warmer one with $T_{ex} \approx 1700$ K. The latter shell is closer to the photosphere, as can be inferred from the wavelength displacement of the lines as functions of the phase of the pulsation. It may be that the warm shell is formed behind the shock front due to the gas motions in the atmospheric pulsations.

Additional information on the occurrence of various 'shells' or 'elements' in the extended atmospheres of Mira stars are the multiple absorption components in spectral lines; these vary in wavelength with the same period as the brightness and yield excitation temperatures for these 'elements' ranging between 1500 and 3500 K (Hinkle, 1978).

All Mira stars have circumstellar envelopes, detectable by thermal and maser microwave emission (Section 6.14) and dust (Section 6.15). Mass-loss is a common property and it may be that the Mira stars are an important source of interstellar molecular dust (Gehrz and Wolf, 1971), mainly detectable by infrared emission features (see Section 6.15).

Spectra of *pure S stars* (showing no TiO-bands) were described by Wyckoff and Clegg (1978). Comparison with dissociation calculations of the relevant molecules shows that pure S stars have C/O abundance ratios between 0.98 and 1.02. For $C/O \approx 1$ the balance between the various molecules containing C or O is very sensitive to small deviations of this ratio, while also the calculated abundances of molecules such as SiO, H_2O, CS, ... etc., are sensitive to the assumed molecular constants. Therefore, the small number of pure S-type stars, as well as of intermediate CS-stars (each containing approx. 2% of all known S-type stars) reflects the narrow range of C/O ratios to which the S classification applies: S-stars *are rare because only few stars have* $(C/O) \approx 1$!

The reddest supergiants: infrared spectrophotometry of stars has opened the way to detect very red stars. There appears to exist a large number of fairly bright infrared sources that have no clear visual counterpart. Many of these are carbon-rich stars embedded in their own ejecta, or in the dust out of which they originated. There are also less extreme cases and it seems that some late type M-stars, particularly those showing simultaneously fairly strong VO and CN absorption bands in the spectrum, are indeed Ia-type supergiants. The best known examples are NML Cyg (spectral type M6?); VX Sgr (M4p; TiO spectrum: M5, VO spectrum: M8.5, but the spectral type may even move to M9.5 – Humphreys and Lockwood, 1972); VY CMa M3–5e I; VO spectrum: M7.0), S Per (M4p; TiO spectrum: M5.5; VO spectrum: M7.5), and V 1804 Sgr (M9 Ia) – see Wing (1974), and Fawley (1977).

In this connection we should also mention the 'infrared objects'. Many of these are bright M-type stars, embedded in dust. Gürtler *et al.* (1979) list 60 of them. The spectra can be best described by the assumption that there is a central star embedded in a cooler dust cloud, emitting mainly in the infrared ($T \approx 300$ K) and producing silicate absorption at $\sim$10 μm. Warner and Wing (1977) mention the star IRC 30308 (M7 Ia). It has an apparent colour temperature of 1000 K but it is heavily reddened; the colour temperature of a M7 Iab star may rather be $\sim$1800 K, but it is still very uncertain. It seems (Cohen, 1979) that a fair fraction of the central objects are C-stars, and the hypothesis may be worth investigating that these objects are the endproducts of the evolution of C-stars which are gradually disappearing in the accumulating dust, condensing in the stellar wind.

Gas and dust shells around cool stars, maser effects. Particularly the long period variables are surrounded by large envelopes of gas and dust, typically of $\sim$10^{11} km diameter ($\approx 10^3$ AU), and characterized by circumstellar absorption and (often forbidden) emission lines, microwave radiation, both thermal and nonthermal. The lines of the latter kind are emitted by pointlike nuclei and have brightness temperatures $\gtrsim 10^8$ K. Their emission mechanism is ascribed to maser action. The dust particles *per se* are detected by broad near-infrared emission and absorption features.

For a detailed description of these features we refer to Sections 6.14 and 6.15. An interesting and well-studied object, that could be considered a *prototype* is VY CMa (M3–4.5e I), an irregulai variable. It is described in some detail in Section 6.16.

Crucial for the *evolutionary interpretation of the red giants* and supergiants are the *abundance anomalies* observed in their photospheres. These abundances are mostly hard to determine because of the dense packing of blended and satuiated lines of so many molecular bands. Table XXVI gives a compilation. There are several indications that nuclear reaction products may reach the photospheres: the presence of the unstable element Tc (half life time $\approx 2 \times 10^5$ yr), high abundances of *s*-type elements;

TABLE XXVI

Some abundances in cool giants, mainly after Motteran (1972), Pagel (1977) with some later additions and modifications. [ ] denote logarithms of relative ratios; see in the list '*Symbols and Notations*'.

Type	M, MS	S	SC, CS	C
[Fe/H]	$\sim$0			
(H/(C+N+O))	1	1		
(C/O)	$\leqslant$0.5	0.7 to 0.1	0.98 to 1.02	$\geqslant$1[b]
(^{12}C/^{13}C)	5 to 20	6 to 25		2 to 100[a,c]
(N/C)	0.2 to 2	$\sim$1		
[N/Fe]				$\sim$0
[O/Fe]				0:
[Zr/Ti]	-3 to 0	$+0.4$ to $+1.2$		$+0.9$ to 1.9
Tc?	trace (Miras)	enhanced	RZ Peg	some
[Li/Ca]	-2.1 to $+0.4$	$\leqslant$3.3	1.0 (SC)	$\begin{cases} -1 \text{ to } +2 \\ (\text{WZ Cas} \sim +5) \end{cases}$

[a] Fujita and Tsuji, (1977),
[b] Gow (1977),
[c] Dominy (1978): V 460 Cyg: 32 ± 9.

for certain stars a deviating $^{12}C/^{13}C$ ratio, and anomalous C/O ratios. The importance of a precise determination of the C/O abundance ratio is that it gives information on the physical conditions in the stellar core, as well as on interaction between the core and the atmosphere. The 11 μm band of SiC appears to be an excellent discriminator between stars with (C/O)<1 and those with (C/O)>1, the 'carbon-rich' stars; cf. Merrill and Stein (1976).

The C/O-ratio of cool giants was investigated by Thompson (1977) who attempted to match observed spectra of the long period variables RS Cyg and RR Her, and the irregular variable Y CVn with computed spectra, and who found the best agreement to occur for C/O ratios somewhat above unity. This result was refined by Gow (1977) who investigated the red part of the spectra of 75 carbon stars and found that half of them had C/O ratios between 1.0 and 1.8; only 20% have C/O ratios above 3.0. Johnson (1978) found a relation between the C/O ratio and luminosity for C-stars; typically: $C/O \approx 2$ for $L \approx 5 \times 10^3$, and 1 for $L \approx 2 \times 10^4$.

These observations should perhaps be related to that of the *isotope ratio* $^{12}C/^{13}C$ in cool stars: deviating ratios must be due to nuclear reactions. A list of references is given by Gow (1977). Values between 2 and 50, and 2 and 30 were published by Scalo (1977), and by Olson and Richter (1979) respectively (Table XXVII). Values between 3 (for WZ Cas) and 100 (V 460 Cyg ≡ DS Peg) were summarized by Fujita and Tsuji (1976, 1977). The latter also attempted to systematize the values for the isotope ratio. A summary of their results, also compared with those of others, is given in Table

TABLE XXVII

$^{12}C/^{13}C$ abundance ratios in cool stars

(a) Average values

Spectrall class	Type or name	Subtype	$^{12}C/^{13}C$ ratio	References
K	giants and subgiants		3–50	Dearborn *et al.* (1975), Deming (1978)
K	supergiant		5	Lambert and Tomkin (1974)
G8–K	giants		10–15	Tomkin and Lambert (1974) Ridgway (1974)
M	α Her	M5 II-III	5	Maillard (1974)
	R	C1 . . . 3	4–16	Fujita and Tsuji (1977)
C stars	Hd	C1 . . . 2	>100 to <500	,,
	J	C3 . . . 9	2–8	,,
	N	C4 . . . 7	10–100	,,

(b) Values for some specific C-stars

Star	Scalo (1977)	Olson and Richter (1979)
19 Psc	15–25	21
Z Psc	50	18
X Cnc	22	31
UU Aur	20–25	25
Y CVn	2–5	2.4

XXVII. Fujita and Tsuji's (1977) C-isotope ratios for R- and N-type carbon stars are larger than those of other authors by factors ranging from 3 to 8, in some cases even more. This aspect, ascribed by them to their incorporation of the influence of saturation on the line profiles needs further examination. That extreme care is indeed needed and is shown by a detailed analysis of DS Peg ($\equiv$V 460 Cyg) by Dominy et al. (1978) who find $^{12}C/^{13}C = 32 \pm 9$, a factor 3 less than Fujita and Tsuji! Climenhaga et al. (1977) found even the value 7.9 for this star, which seems too low. Tsuji (1979b) showed the influence of the assumed T_e-value on the results. Yet, a correction factor $\frac{1}{3}$ to Fujita and Tsuji seems perhaps appropriate.

Another interesting aspect noted by Fujita and Tsuji (1977) is the (rough) dependence of the $^{12}C/^{13}C$ ratio on the luminosity in G-, K-, and M-type stars, as reviewed in Table XXVIII: the relative abundance of ^{13}C seems to increase with increasing luminosity. Deming (1978) mentioned that population I K giants with very low $^{12}C/^{13}C$ ratios (<15) have weak-CN bands.

TABLE XXVIII

$^{12}C/^{13}C$ ratio in G; K- and M-type stars, in dependence on luminosity (Fujita and Tsuji, 1977)

$\log(L/L_\odot)$	$^{12}C/^{13}C$ ratio		Number of stars
	Range	Average	
<1.1	30, 50, 51	–	6
$1.1 < \ldots < 1.6$	12 to 34	22	6
$1.6 < \ldots < 3$	6.5 to 25	14	
$3 \ \ < \ldots < 4$	5.1, 17	–	2
$4 \ \ < \ldots < 5$	7, 12	–	2

How to explain these abundance-anomalies. The observed $^{12}C/^{13}C$-ratios in R and N-type stars cannot be explained by the CNO-cycle: it would produce a $^{12}C/^{13}C$ ratio of approx. 4, while most ratios found are higher.

The C/O ratio, of about unity, would be produced by the CNO-bi-cycle at temperatures of about 30 MK. It is often assumed that the ^{12}C over-abundance could be due to triple-α process produced ^{12}C-particles brought to the surface. Such particles would then be produced in a He-burning shell and brought upward by large-scale convection initiated by thermal instability flashes. Alternatively, one could think that the outer mantle of the stars may have been lost by strong stellar mass-loss. Apparently, essential research has still to be made in this field; cf. also Section 4.6.

Important are the low relative $(C)/(C_\odot)$-values and the relatively low $^{12}C/^{13}C$ abundance ratio for three M-giants and supergiants investigated by Geballe et al. (1977), Table XXIX. Although the authors suggested production by the CN cycle and subsequent mixing, the $^{12}C/^{13}C$ ratio seems too high to justify this assumption. An escape may be provided by the hypothesis that the envelope has undergone only partial CN processing. The low relative C/H-ratio in α Sco and α Ori is assumed to be due to the CN process, which reduces the C-abundance, until new C is formed by

TABLE XXIX

Abundances of 'strategic' elements in some well studied cool giants or supergiants (after Geballe *et al.*, 1977; and others)

Star	Spectrum	$(C)/(C_\odot)$	$^{12}C/^{13}C$	$^{16}O/^{17}O$	$^{17}O/^{18}O$
α Her	M5 II-III	0.2	30; 17[c]	450	>1
α Ori[a]	M1.5 Iab	0.08; 0.05[b]	10; 7[c]; >20[d]	500	
α Sco	M1.5 Iab	0.06	15	500	

[a] For this star Spinrad and Vardya (1966) find $(O/C)=1.05$; $(O/H)=10^{-3}$; $(N/C)=2$.
[b] From Tsuji (1979b), who finds in addition for α Ori: $[N]=+0.24$; $[O]=-0.81$.
[c] Lambert *et al.* (1974); Hinkle *et al.* (1976).
[d] Gautier *et al.* (1976).

the triple-α process after the onset of shell He burning when the (C/H)-value should start to increase again.

For a further discussion of abundance-anomalies in relation to the late phases of stellar evolution we refer to Section 4.6.

Prototypes: α Sco, α Ori, and o Cet. The foregoing description of the many different types of cool stars should have made clear that there are differences between on one hand the coolest types of giant stars, with masses not exceeding $\sim 5\mathfrak{M}_\odot$, most of which show clear abundance anomalies, and the supergiants of slightly earlier spectral type, early M and K with larger masses (typically up to 20 or $30\mathfrak{M}_\odot$) which have still abundances closer to the 'cosmic' values, but only in broad outline: while G and K giants have still solar C-abundances, the C abundance (and that of O) are clearly reduced in α Sco and α Ori, and this may be generally so for M supergiants; see Table XXIX. The division line between the two groups of stars seems to coincide remarkably well with the locus of the first He shell flash, (Figure 40) suggesting an evolutionary relation. Two stars of the second group, hence objects that have presumably not yet undergone the first He shell flash, may be α Scorpii (Antares, M1.5 Iab), and α Ori (Betelgeuze, M1–2, Iab), the third and second brightest stars in the sky after Sirius. The first, α Sco is also interesting because it is a binary. The main data are given in Table XXX. They are for α Sco based on an adopted distance *d* of 180 pc and for Betelgeuze on $d=140$ pc (Eggen, 1973).

The primary of *the α Sco system*, α Sco A is variable with *V* ranging between 0.8 and 1.8 mag. The apparent separation of the two components changed between 1847 and 1957 from 3".3 to 2".9 (Jeffers *et al.*, 1963); the latter value would correspond to a projected linear separation of $\Delta=522$ AU.

Physically, the system is interesting because of the possibility offered by the B-type companion to probe the circumstellar envelope of the M supergiant at the fairly large distance of separation of the companion, while at the same time the ionization in part of the envelope of α^1 Sco is locally still changed by the ultraviolet radiation of the B star.

The ultraviolet spectra of the stars was described by Van der Hucht *et al.* (1979c). A detailed analysis of the photospheric spectrum and comparison with model studies is still lacking for α Sco. The spectrum of α Ori was compared with model results based on plane-parallel models, by Tsuji (1976b, 1978a, 1979a), who assumed $\mathfrak{M}=15\mathfrak{M}_\odot$, $R=600R_\odot$, hence log $g=0$. He found $T_e=3900\pm150$ K. This latter value is also found (Tsuji, 1978a) from a simultaneous determination of T_e and θ according to the method of Blackwell and Shallis (1977; see Section 2.3). However, there are two sources of error: the existence of a dust halo may influence the result (Section 2.5), and also – see Section 5.2 – a plane-parallel model is not acceptable for this star. Watanabe and Kodaira (1978) found the best agreement with spherical model calculations for $T_e=3900$ K, but log $g\leqslant-0.5$, which would then mean that $\mathfrak{M}\lesssim5\mathfrak{M}_\odot$ instead of 15$\mathfrak{M}_\odot$! This is an *important* discrepancy, which needs thorough further investigation because of the fundamental implications of this result for considerations on the late stages of stellar evolution. From infrared continuum photometry Scargle and Strecker (1979) found $T_e=3580$ K. We *assume* this value, but note the important difference with Tsuji!

TABLE XXX

Main data of the α Sco binary system (based on compilations by Kudritzki and Reimers, 1978; and Van der Hucht *et al.*, 1979c), and of the supergiant α Ori (Tsuji, 1976b; Watanabe and Kodaira, 1978). T_e values are from Scargle and Strecker (1979). The upscripts a, b ... refer to the footnotes below.

	α Sco A	α Sco B	α Ori
Spectral type	M1.5 Iab	B2.5 V	M1–2 Iab
m_v	1.2 (0.9 ... 1.8)[a]	5.17	
A_v	0.6	0.6	
M_v	−5.7	−1.71	
M_b	−7.1	−3.61	−7.5[d]
$\log L/L_\odot$	4.7	3.34	4.85[c]; 4.90[d]
T_e (K)	3650	18500	3580
θ (in arc sec)	41		52 (47[e])
$R/R_\odot$	810	4.6	600
	640[b]	5.2[b]	
$\mathfrak{M}/\mathfrak{M}_\odot$	18 ... 24	7 ... 8.5	20 ... $\lesssim$5
log g	+0.2 ... −1.4	4.0	0 (for plane-par atm.)
			$\leq$ −0.5 (for spherical atm.)

Remarks:
[a] Irregularly variable, characteristic time one year (?).
[b] From L and T_e.
[c] Dyck *et al.* (1974).
[d] Eggen (1973).
[e] Scargle and Strecker (1979).

We next turn to a discussion of the outer layers of α Sco A, mainly the envelope, which expands. The average velocity of expansion, already discovered by Adams and MacCormack (1935) and measured from the asymmetry of the stellar photospheric lines is

for Ti II	-17 to -18 km s^{-1}	(Swings, 1973; Kudritzki and Reimers, 1978)
Ca II (H and K)	-18 km s^{-1}	(Deutsch, 1960, p. 109)
Other circumstellar lines	-11 km s^{-1}	(Deutsch, 1960, p. 109)
Ultraviolet circumstellar lines	-18 ± 4 km s^{-1}	(Van der Hucht et al., 1979c)

So, virtually all data converge to about -18 km s^{-1}, a value that we will adopt. The star is apparently losing mass; different values are given. The most detailed determination gives $\sim 7 \times 10^{-6}\,\mathfrak{M}_\odot$ yr^{-1} (Van der Hucht et al., 1979c), see also our Table XLIII.

In contrast to this is the situation with α Ori, where the emission lines of Fe II, Ca II H and K are redshifted with respect to the photosphere lines by $+5$ and $+6$ km s^{-1}, and where at larger distances to the star ($3''$ to $5''$) gas appears to leave the star with ~ 10 km s^{-1} (Bernat and Lambert, 1976b). This suggests an inhomogeneous, large-scale motion pattern where mass-loss at large distances concurs with matter that is *infalling* near to the star! In the infalling matter the lines have a large width and this may be caused by a velocity gradient or may indicate considerable micro- and macro-turbulent velocity components (Boesgaard and Magnan, 1975). Boesgaard (1979) presents a model in which the shell is at $1.8R_*$ whereby matter is accelerating inward with $v \approx 15$ km s^{-1} at $1.8R_*$ to 60 km s^{-1} at R_*. This picture may be related to the hypothesis of very large moving volumes ('giant convection elements') as suggested by Stothers and Leung (1971) and Schwarzschild (1975) – cf. Section 5.10. This is supported by measurements of the polarization of α Ori by Tinbergen et al. (1980) indicating the existence of $\sim$ one element at a time with life-time of approximately one year.

Like other similar stars, and like α Ori, the spectrum of α^1 Sco has displaced emission lines, apparently of circumstellar origin. As compared to those in α Ori they are weaker (Deutsch, 1960) an effect presumably due to the B-companion, who may ionize (part of) the envelope. Bernat and Lambert (1976a) calculated for the Ca II shell a radius of $4R_*$. For comparison, the Ca II shell of α Ori has a radius of at least $30R_*$, while observations of Fe II emission lines suggest that they originate in a region of material with a radius of approx. $2R_*$.

A comparative study of the infrared fluxes of α Ori and α Sco in 5 wavelengths bands between 1 and 3.5 μm by Nordh et al. (1978) support the hypothesis that the main difference between the two stars would be the existence of a cool chromosphere ($T \approx 5000$ K) at α Ori (Lambert and Snell, 1975) while α Sco would not have one. This surprising result certainly needs further investigation.

At a few frequencies microwave radiation has been detected from α Ori. We summarize:

10.7 GHz	7 ± 1 mJy	Altenhof and Wendker (1973)
	9 ± 4	Altenhof and Wendker (1973)
14.9	8.7 ± 1.1	Bowers and Kundu (1979)
90	90 ± 32	Schwartz and Spencer (1977).

The most likely explanation seems that this radiation is caused by the stellar wind.

The position of α Sco B (α^2 Sco) along the line of sight was determined by Bernat et al. (1978a) and Van der Hucht et al. (1979c): the photospheric light from α^2 Sco is partly and selectively absorbed by the outstreaming matter. Thus the spectrum shows lines of Ti II. Hence the observed displacement of the absorption lines with respect to the wavelength of rest gives information on the relative positions of the stars: for example if α^2 Sco would be situated far behind α^1 Sco the Ti II lines would not be displaced. The expected size of the Strömgren sphere around the B-star (where the Ti II lines would not occur) should be included in the calculations. The observed displacement coupled to the Doppler velocity of α^2 Sco (8 km s^{-1}), and a Strömgren sphere of 300 AU, yields, on the basis of model computations, that α^2 Sco is 420 AU closer to Earth than α^1 Sco. This brings the lower limit of the separation between the stars at 670 AU; the period of revolution should be $\sim$3000 yr.

Herzberg (1948) reported the detection of fairly strong Fe II emission lines in the near-ultraviolet (3150–3300 Å) and mid-ultraviolet (2000–3000 Å) respectively. These lines are also superimposed on the spectrum of the B-companion, which means that the shell extends to beyond the separation of the companion.

The primary star is surrounded by a dust shell, as indicated by a weak spectral emission feature accounting for about 30% of the flux at 11 μm. The size of the shell was measured by Sutton et al. (1977) by spatial heterodyne techniques who found that the dust particles are contained in an optically thin shell outside about 0''4 (corresponding to the region outside $12R_*$).

Betelgeuze has a similar dust shell, with at 10 μm an optical thickness of 0.04–0.1 (Tsuji, 1979b). The inner radius is at about $12R_*$ (Sutton et al., 1977) but should be between 1 and $3R_*$ according to Tsuji (1979b). Half of the dust emission at 11 μm is confined in an area with radius $30R_*$ (McCarthy et al., 1977); weak extensions are noted by the sky polarization around the star and extend to $\sim$30'', which is $\sim600R_*$ (see also Section 6.15). Tsuji (1979b) finds that the dust mass density decreases with distance as r^{-2}, so that the dust grains may be formed close to the star and driven out in a constant-velocity flow. The existence of this dust shell is even noticeable in angular diameter determinations in the visible spectral regions, yielding too large values (see the discussion in Section 2.5). The star is also surrounded by a gas halo of the same size, as noted by the circumstellar emission of the K-resonance line at $\lambda7699$ which is observed till that distance (Bernat et al., 1978b).

The B-companion of α Sco has strong [Fe II] emission lines, which originate from a region with a diameter of 6'' centrered on α Sco B (Struve and Swings, 1940); later, other emission lines were also found, as well as radio-emission (Wade and Hjellming, 1970; Hjellming and Wade, 1971), also centered on the B-star. This shows that there is interaction of the gas that streams away from the primary, with that emitted by the hot secondary but the details of the emission mechanism, the interaction, and the structure of the cloud are not yet understood.

Mira (o) Ceti (M5e–9e) is another prototype of the stars described in this section, and has all the characteristics of the very red, late type variables at the extreme right of the HR diagram.

The photospheric parameters and a few other properties of Mira were assembled by Cahn and Wyatt (1978), partly directly from observations but partly also from an interpretation of observational data on the basis of a diagnostic diagram (Figure 60, to be discussed in Section 4.7). This yields the following data, the accuracy of which depends partly on the reliability of Figure 60:

	distance	115	± 8 pc
	period	330	days
	$\log \langle L/L_\odot \rangle$	3.9	
hence:	$\langle M_b \rangle$	-5.0	
	ΔM_b	1.1	mag
	$\mathfrak{M}/\mathfrak{M}_\odot$	0.8	
	$\mathfrak{M}$	-6×10^{-7}	$\mathfrak{M}_\odot$ yr^{-1}.

Preliminary calculations of the absorption-line spectrum show agreement with the observations for $T_e = 3000$ K, $\log g = -1$ and $v_\mu = 5$ km s^{-1} (Panchuk, 1978), but from infrared fluxes Scargle and Strecker (1979) find the more precise value $T_e = 2470$ K. From these data we derive an average photospheric radius of $\sim 500 R_\odot$.

An interesting and perhaps fundamental aspect of Mira Ceti is that it is a visual binary. The companion is a B-type white dwarf with an orbital period of ~ 750 yr. From observations of the P Cygni profiles of Balmer and Fe II emission lines in the spectrum of the B companion Yamashita and Machara (1977, 1978) derive that the companion must be surrounded by a rotating and expanding gaseous disc. The expansion velocity varies between 100 and 300 km s^{-1} with a period of ~ 14 yr; a similar period applies to other Be-activity. This expansion is confirmed by ultraviolet observations (Cassatella et al., 1979) yielding a velocity of -350 km s^{-1}. It is assumed that the disc is an accretion disc so that Mira must lose matter to the dwarf. The wealth of ultraviolet emission lines found by Cassatella et al. (1979) is most probably related to this accretion disc.

That mass-loss occurs in Mira is supported by the observation that Mira is surrounded by extended gas and dust shells. While spectral interferometric observations show that Mira itself has an apparent diameter $\lesssim 0.1$ arc sec (Foy et al., 1979), interferometric observations of ths dust shell at 11 μm (the wavelength of the 'silicate emission') (Sutton et al., 1978) showed this to have an angular size of $\sim 0\rlap{.}''7$. The temperature of the particles is ~ 500 to 700 K. The size and brightness of the dust shell are largest ($>1''$) at maximum phase (McCarthy et al., 1978).

3.12. The Central Stars of Planetary Nebulae

These stars are the faintest of the 'brightest stars', but are so closely related – evolutionary and physically – to various of the types of stars mentioned in this chapter, that they should be mentioned, at least with a few words. The spectra of the central stars are of early type (O, Of, WR, and 'cont.') Evolutionary it seems that the planetary nebulae, which have kinematic properties that rank them among old population I stars must have arisen from more massive stars after having lost part of their mass. Figure 41 composed from work by O'Dell (1968), Pottasch *et al.* (1977), and Weidemann (1977) shows their positions in the Hertzsprung–Russell diagram: O'Dell (1968) located the stars in a large area of the HR-diagram between $\log(L/L_\odot)=0$ and $+5$, and $\log T_e$ between 4.5 and 5.3. This location was revised by Pottasch *et al.* (1977) on the basis of UV photometry with the Astronomical Netherlands Satellite. The effective temperatures were redetermined with the method described in Section 2.3, while M_b was determined from distances, apparent magnitudes and spectral photometry. Later, Weidemann (1977) in a rediscussion of these data could make acceptable that the distance scale used by Pottasch *et al.* had to be so corrected, that their luminosities had to increase by about $\Delta \log L \approx 0.3$. There is no *essential* difference between these distances and the distance scale derived by Acker (1978).

It is Pottasch's data as changed by Weidemann that are given in Figure 41. The diagram suggests that at least part of the central stars of Planetary Nebulae are fairly bright. The suggestion that there may be two groups of Planetary Nebulae (Grieg, 1971) seems to make sense. Furthermore, (Figure 41) the link with the white dwarfs (which occupy approximately a band between $(\log T_e, \log(L/L_\odot))=(4.7; 0)$ and $(3.8; -4)$) is suggestive. Indeed, Pottasch *et al.* (1977) estimate $\log g = 8$ for some central stars of Planetary Nebulae. It has been suggested by many authors, for the first time by Vorontsov-Vel'yaminov (1948), that Planetary Nebulae evolve to white dwarfs. This is supported suggestively by Cahn and Wyatt (1976) who find: (a) for planetary nebulae in the solar environment a local rate of birth and death of 4 to $6 \times 10^{-3}\,\mathrm{pc}^{-3}\,\mathrm{yr}^{-1}$; which values, with Weidemann's (1977) improved distance scale becomes 2 to $3.5 \times 10^{-3}\,\mathrm{pc}^{-3}\,\mathrm{yr}^{-1}$; (b) for stars with main sequence masses between 1 and $5\mathfrak{M}_\odot$ a local death rate of 2 to $3 \times 10^{-3}\,\mathrm{pc}^{-3}\,\mathrm{yr}^{-1}$, values that are of the same order as (c) Weidemann's (1977) birth rate of white dwarfs of $2 \times 10^{-3}\,\mathrm{pc}^{-3}\,\mathrm{yr}^{-1}$.

After having thus touched the ultimate lot of the nuclei of planetary nebulae we have to discuss their origin. Information on the chemical composition of the photospheres may help to this end. The He/H ratio is often enhanced and ranges from 0.08 to above 0.2; also the N/O ratio is enhanced as compared to solar values; the two ratios show a clear correlation (Kaler, 1978a, b). Furthermore, the relative abundances of C and N with regard to H are enhanced as compared to solar values; N by a factor 3 to 5, and C by perhaps an order of magnitude (Torres-Peimbert and Peimbert, 1976). All this information suggests that nuclei of planetaries are evolved stars, whereby the relative abundances of C, N and O help to ascertain in stars of what initial masses the nuclear processes occurred (for the ultimate relative abundances of C, N and O depend on the internal temperature, hence on the mass of the star – see Section 4.6, and Figures 56 and 57). It appears that the reactions must have operated at fairly high temperatures; the stars must have lost a considerable part of their mass in order that reaction products have reached the surface (Finzi and Yahel, 1978). Indeed Kaler *et al.* (1978) found that the observed He/H and N/O abundance ratios point to initial main-sequence masses *in excess* of 3 to $5\mathfrak{M}_\odot$. These stars must have gone in their later evolution through the red giant phase, and the *immediate* progenitor

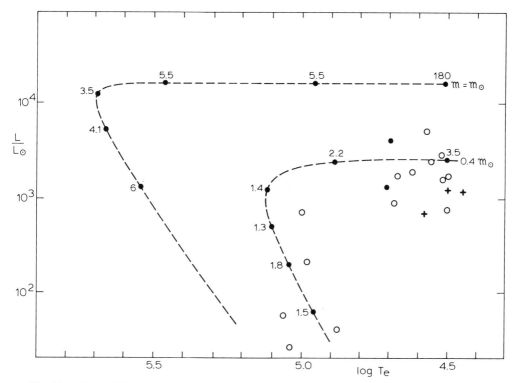

Fig. 41. HR positions and evolution of the central stars of planetary nebulae. Positions of the stars in the HR diagram according to Pottasch *et al.* (1977), with luminosities corrected according to Weidemann (1977). *Open circles:* O-type spectra; *crosses:* WR spectra; *dots:* continuous spectra. The dashed lines give theoretical evolutionary tracks, after Yahel (1977), for stars of two masses (1 and $0.4\mathfrak{M}_\odot$). The numbers labeled to the tracks are the masses of the non-degenerate envel- opes in units of $10^{-6}\mathfrak{M}_\odot$ in the $1\mathfrak{M}_\odot$ case, and in units of $10^{-4}\mathfrak{M}_\odot$ in the $0.4\mathfrak{M}_\odot$ case.

of a planetary is assumed to be an inhomogeneous red giant (a Mira star??) with a core mass of approx. $0.6 \ldots 0.8\mathfrak{M}_\odot$.

A dramatic example of a sudden transformation of a red star into a planetary nebula may be V 1016 Cyg, which has been observed as a red star before 1964 and thereafter brightened by 5 mag and went into an emission-line stage (Boyarchuk, 1968; Fitzgerald and Pilavaki, 1974; Ciatti *et al.*, 1975, 1978). Optical observations show a cool gas shell (typically M6) expanding at 34 km s^{-1}, and a stellar wind with $v=105$ km s^{-1} (Kwok, 1975; Ahern *et al.*, 1977). The wind had a fairly high degree of ionization and excitation which increased continuously with time. In the mid-seventies lines of Fe II, [Fe VII] appeared with increasing intensity (Ciatti *et al.*, 1979).

Another star of this kind may be HM Sagittae (Ciatti *et al.*, 1977; Kwok and Purton, 1979) which increased in brightness by 6 mag during 5 months in 1975 and thereafter showed spectral indications for the existence of a new shell which, by inter-action with the M-star wind, yielded the brightness increase. At minimum it had a red colour but at maximum the spectral type was BQ (nova-like) with a stellar photospheric temperature of 30000–50000 K and a nebular kinetic temperature of 10^4 K (Arkhipova et al., 1979). In the course of 1977 Wolf–Rayet features appeared, but one year later the broad emission features were hardly visible so that the WR-phase was obviously transitory (Ciatti et al., 1979). It is not excluded that the star is a binary (Puetter et al., 1978). The star flares occasionally ($\sim$0.4 mag) with a characteristic time of a few days (Belyakina et al., 1978). It also shows radio emission and a strong infrared excess and high excitation lines as [N II], [O III], [Ar III], [Ar IV] ..., all characteristics for an expanding envelope of enhanced temperature. It is surrounded by a dust shell (Puetter et al., 1978). The nebulae of V1016 Cyg and HM Sge are fairly dense, $n_e > 10^6$ cm^{-3} for V1016 Cyg (Flower et al., 1979) and $\sim 5 \times 10^7$ cm^{-3} for HM Sge (Kwok and Purton, 1979), which suggests that these two objects are *young* planetary nebulae, although we should also note that there is also some analogy in the spectral behaviour with a slow nova like RR Tel (see Chapter 9).

It is not certain whether the peculiar variable FG Sagittae belongs to the same class. It consists of a faint circular envelope (radius 18″: 0.2 pc) around a relatively bright central star; the spectrum shows sharp emission lines of Hα, β, and γ, with Hγ showing a P Cygni profile (Henize, 1961). The brightness increased steadily (Richter, 1960) from $m_{pg} = 13.6$ in 1894 to B=9.6 in 1965 (Herbig and Boyarchuk, 1968). The spectrum changed from B4 I in 1960 via A5 Ia in 1965, F2 Ia in 1969, F7 Ia in 1971 (Chalonge et al., 1977), to G2 Ib in 1975 (Smolinsky et al., 1976). In the same period the P Cygni-like emission fringes at the Balmer lines faded out. Rapid changes in spectral class, luminosity and colour occurred between June and September 1970. In 1977 the star had a pure absorption line spectrum (Kipper, 1978). Most interesting and undoubtedly *unique* is the fact that beginning in 1967 the abundances of elements synthesized by s-type processes seemed to increase (Langer et al., 1974): FG Sge is the only star known so far in which the enrichment of the atmosphere with nucleo-synthesis products was observable in a time span shorter than a human life! The gas shell visible in 1960 may have been caused by an ejection that started a few thousands of years ago (Herbig and Boyarchuk, 1968), while the recent phenomena can certainly be interpreted as related to a second ejection of an – initially opaque – shell of gas.

The infrared spectra of these stars are interesting. The existence of the 'silicate feature' around 10 μm in the spectra of HM Sge and V 1016 Cyg indicates the presence of dust in the shell, while the 'background' radiation of HM Sge yields an infrared brightness temperature of 950 K. There is in both stars evidence for CO absorption at 2 to 3 μm (Puetter et al., 1978).

Other possible proto-planetary nebulae are Vy 2–2 and M1–92, described in some detail in Section 6.14.

The way in which stars transform into nuclei of planetary nebulae must – most probably – go through ejection of an outer mantle, but the mechanism has not yet been settled: Härm and Schwarzschild (1975) assume that loss of the outer mantle and the consequent transformation of the red giant into a blue nucleus is caused by the transformation of nuclear and gravitational energy into heat and calculate a time scale for the transition of 5000 yr. Finzi *et al.* (1974) assume the ejection due to radiation pressure. Including turbulent pressure Yahel (1977) calculated a time scale of 3000 yr for the ejection of $0.1\mathfrak{M}_{\odot}$ of hydrogen-rich shell material, during which time T_e increases from 6500 to 8000 K, while L remains constant. Trimble and Sackmann (1978) assume the mantle to be ejected by nuclear runaway shell flashes and thus explain the occurrence of planetary nebulae with more than one shell. Katz *et al.* (1974) found that the parameters of nuclei of planetary nebulae are well explained by $\mathfrak{M}=0.65\mathfrak{M}_{\odot}$, and a hydrogen-rich envelope with a mass $\mathfrak{M}_H=10^{-3}\ \mathfrak{M}_{\odot}$. Figure 41 shows evolutionary tracks for the nuclei of planetary nebulae with $\mathfrak{M}=0.4\mathfrak{M}_{\odot}$ and $1.0\mathfrak{M}_{\odot}$, where the time-parameter is mimiqued by the variation of the mass of the hydrogen-rich envelope. Characteristic time scales for the transition to a white dwarf are found to be of the order to 10^3 yr ($1\mathfrak{M}_{\odot}$ case) and 10^4 yr ($0.4\mathfrak{M}_{\odot}$ case).

However, the discovery (Bond *et al.*, 1978) of the fact that UU Sge is an eclipsing binary consisting of a sd O primary of $\sim 0.9\mathfrak{M}_{\odot}$ and $0.4R_{\odot}$ with a dK companion of $\sim 0.7\mathfrak{M}_{\odot}$ and $0.7R_{\odot}$ separated by $3R_{\odot}$ may lend support to the assumption that nuclei of planetary nebulae may descend from binary M stars. Bond *et al.* (1978) suggest that with time it may transform into a cataclysmic variable ('subnova').

The physical parameter of the shell (the 'nebula') are tentatively explained by Kwok *et al.* (1978) as caused by interaction of the fast stellar wind from the – newly formed – nucleus, being the core of the red giant progenitor, with the old, slower, stellar wind of the red giant, a suggestion that is supported by the observations of the new planetary nebula V 1016 Cyg. At the interface between the two winds a denser but thin shell is formed. Numerical calculations applied to the available observational data on the stellar wind and shell velocity of V 1016 Cyg yield $\dot{\mathfrak{M}}\approx 3\times 10^{-5}\ \mathfrak{M}_{\odot}\ yr^{-1}$, and a present shell mass of $7\times 10^{-5}\ \mathfrak{M}_{\odot}$, which should increase to $\sim 0.04\mathfrak{M}_{\odot}$ in 5000 yr.

The various hypotheses seem to agree that the transition from a red giant to a planetary nebula takes about 10^3–10^4 yr while the planetary nebula would take 10^3 to 10^4 yr to transform into a white dwarf. Hence the planetary nebula phase may be a very brief *intermezzo* in the life of a star with an original mass of a few solar masses.

CHAPTER 4

THE EVOLUTION OF MASSIVE STARS

4.1. The First Phases of the Evolution of Massive Stars

The pre-main sequence of stars goes through the following phases (ref.: Symposium-proceedings edited by De Jong and Maeder, 1977):

(a) During a period of about one to ten million years after interstellar matter has been met by a galactic density wave (a spiral arm) new stars appear, according to the 'birth-law' $d\varrho_{star}/dt(:)\ (\varrho_{gas})^{1.8}$ (Schmidt, 1959; Guibert, 1979). Conglomerations of matter originate; very few of these are gravitationally stable and these act as contracting centres for their surroundings (see the review by Mezger and Smith, 1977). During a period when the interstellar gas is still sufficiently tenuous, the contraction is best described by the conditions of 'free fall', for which a characteristic time is defined by $\tau_{ff} = (d\ln\varrho/dt)^{-1}$. For a homogeneous sphere with density ϱ(g cm^{-3}), hence mass $\mathfrak{M} = 4\pi R^3\varrho/3$ the time for a contraction such that the density is increased by a factor e is then derived assuming that at $t=0$ at the surface (radius R) the velocity of fall is equal to that of escape: $v(0) = (2G\mathfrak{M}/R)^{1/2}$. Then, integration of the velocity yields that a decrease of R to $R/e^{1/3}$ needs an approximate time:

$$\tau_{ff} \approx 0.2(R^3/G\mathfrak{M})^{1/2} = 0.2\,(\tfrac{3}{4}\pi G\varrho)^{1/2} \approx 40\varrho^{-1/2}\ \text{s} \qquad (4.1;1)$$

(note that this expression gives a ten times smaller value than those customarily found in literature, where the velocity is integrated from starting value zero or over the full radius R).

(b) During the subsequent phase of contraction, when the object has already stellar dimensions, the contracting cloud appears to be fully in convective equilibrium; it moves quickly through the upper right-hand part of the HR-diagram.

(c) In the next phase of the contraction the energy transfer regime in the cloud changes eventually to radiative equilibrium; in the HR-diagram the object approaches the main sequence; it moves much slower than in the previous phase.

(d) When T and ϱ in the star's centre increase to sufficiently high values, nuclear reactions become the star's principal energy source. The star arrives at the main sequence; its subsequent evolution is again slower than in phase (c).

During the contraction phases (b) and (c) and perhaps even during the initial part of phase (d) the stars are still surrounded by large masses of dust and gas that practically fully hide the star ('coccoon stars') and are hence characterized by a stellar-like radiation flux and low surface brightness. These envelopes will be shed off at the end of this phase, and massive stars may then appear as Herbig's Ae/Be-stars (see Section 3.6.).

In this section we give a brief elaboration of some aspects of the last part of phase (b) and phase (c), and we estimate the time necessary to reach phase (d), while we give a description of the observational aspects relevant to the time during which the stars are still embedded in clouds.

A. CIRCUMSTELLAR CLOUDS AROUND CONTRACTING STARS

There are several observations that at least the massive contracting stars are surrounded by large clouds – shells or discs – of gas and dust, out of which they tend to emerge only when the star has already arrived near or is at the main sequence. Such clouds are often detected by their infrared or microwave emission. The largest of such clouds have $\mathfrak{M} \gtrsim 2 \times 10^5 \, \mathfrak{M}_\odot$ and sizes $\lesssim 100$ pc (Stark and Blitz, 1978). In several such clouds the strong microwave H_2O lines at 22 GHz, and OH-lines at 1665 and 1667 MHz (Habing *et al.*, 1972) amplified by maser action are characteristic. Very long baseline interferometry shows that the H_2O clouds have sizes of the order of 10^{11} km $\approx 10^3$ AU (see Section 6.14 for a more detailed discussion of such clouds). Such molecular clouds lose their energy quite easily (Suchkov and Shchekinov, 1977), as follows also from a consideration of their contraction time (see Equation (4.1; 9)). The large integrated infrared flux of the brightest – far infrared emitting – clouds makes it clear that their emission must in many cases arise from a very luminous star, presumably of the O- or B-type; as an example we mention the infrared complex AFGL333, which has the luminosity of a B0.5 main-sequence star (Thornton and Harvey, 1979). Seven H_2O masers are known to be related to Herbig–Haro objects (Rodriquez *et al.*, 1978), objects thought to be stars in the process of formation. Genzel and Downes (1977b) who listed positions and spectra of 82 H_2O sources at 22 GHz, have attempted to relate a number of these sources to the various possible phases of the early evolution of a O- or B-type star. Their challenging scheme is reproduced in our Table XXXI. More specific details have been worked out by Walker *et al.* (1978).

B. CONVECTIVE CONTRACTION, THE HAYASHI-LINE

When, during the contraction phase, the temperature becomes high enough for H_2 to dissociate, the specific heat $\gamma = c_p/c_v$ decreases to below the value $\frac{5}{3}$ and thus the star can get into convective equilibrium (see also discussion in Section 4.2). The Russell–Vogt theorem (Russell, 1933; Vogt, 1926) states that in a *restricted* area of the Hertzsprung–Russell-diagram (see for a discussion of the degree of restriction: Kähler, 1972) the internal structure of a star is fully determined by its total mass and chemical composition (because the fundamental differential equations of stellar structure depend only on mass $\mathfrak{M}$ and composition (X, Y, Z)). We use this theorem to show that a convective star of given mass is located at a well-defined track in the HR-diagram. At the photospheric level of any star, the pressure P takes the value

$$P_{\text{phot}} = \frac{G\mathfrak{M}}{R^2} \int_R^\infty \varrho \, dr \approx \frac{2}{3} \frac{G\mathfrak{M}}{\kappa_{\text{phot}} R^2} . \qquad (4.1; 2)$$

TABLE XXXI

Possible phases of H_2O emission in an evolutionary scheme for OB stars (From Genzel and Downes, 1977b)

Evolutionary stage	Duration	Dynamical effects	Observational appearance	Possible examples
1) Pure accretion phase	—	Infall of protostellar cloud at 2 to 5 km s^{-1} (scale 10^{18} cm)	Dense molecular cloud with low excitation molecules; CO self-absorption.	?
2) Stellar core forms and Star turns on	3×10^5 yr	Radiation pressure stops accretion flow at $\sim 10^{13}$ to 10^{14} cm; inner opaque cocoon forms; kinetic energy of accreting material heats the dust.	Far-IR source; OH maser, pumped by IR; H_2O emission with a narrow velocity range (small Δv) appears toward the end of this phase.	W3 IRS5, H_2O source near W3OH/IRS8
3) Cocoon begins to expand	2×10^4 yr	Radiation pressure increases further, due to nuclear burning of stellar core. Dust front/inner opaque cocoon expands from 10^{14} to 10^{16} cm at a velocity of 3 to 10 km s^{-1}. Possible outer cocoon at 10^{17} cm. The material outside the cocoon is still accreting.	Strong Far-IR source; H_2O 'shell' type spectra ('triples') become well defined with a range of 3 to 10 km s^{-1}; OH maser lines occur at the same velocities at H_2O.	NGC7538S, W75N–OH
4) Development of a super-compact H^+ region	—	Strong UV from star creates supercompact, dust-bounded H^+ region which expands at 1–10 km s^{-1}. A radiation-driven stellar wind begins to blow some fragments out of the dust shell. The outermost parts of the envelope may still be accreting.	Far-IR and near IR source; low velocity H_2O 'shell' emission dominant, but some weak, high velocity features appear over a limited velocity range ($v^H \sim 30$ km s^{-1}). Supercompact H^+ region may be detectable in the radio continuum. Most of the UV photons are absorbed by the dust, however, so radio flux is not proportional to IR flux.	NGC7538-IRS1, ON1

Stage	Time	Dynamics	Observations	Sources
5) Shell begins to fragment; stellar wind phase	10^4 yr	Strong stellar wind at 2×10^3 km s^{-1} blows out parts of shell from 10^{16} to 10^{17} cm; no significant accretion any more; compact H$^+$ region expands to 10^{16} cm but is still dust bounded.	Strong near and Far-IR emission; H$_2$O has strong low velocity emission as well as high velocity features over $\geq \pm 100$ km s^{-1}, possible SiO emission from low velocity shell (?); 'hot' high velocity flow in CO and high excitation molecules, the compact H$^+$ region may be detectable in the radio continuum and the ratio of 20 μm to 6 cm flux is relatively small.	W49N/H$_2$O W51Ma in Orion KL
6) Shell breaks up	–	The inner cocoon is unstable and breaks up; the expanding compact H$^+$ region is ionization bounded.	Strong near-IR emission; low-velocity H$_2$O emission looses its symmetrical pattern and its intensity decreases. The H$_2$O is highly variable, and the high velocity features are strong. The compact H$^+$ region is strong in the radio continuum.	W51N G0.55-0.85
7) Compact H$^+$ region expands	10^4 yr	H$^+$ region reaches 10^{17} cm (location of a possible outer cocoon?); H$_2$O molecules are broken up by UV, OH masers are pumped by UV.	Near IR radiation from the hot dust in the H$^+$ region. No H$_2$O maser is present. OH masers may be located at interface (10^{17} cm) with ionization front. The optically-thin radio continuum flux density is constant in time, while the turnover frequency is slowly decreasing. The radio flux is proportional to the near infrared flux.	Continuum and OH maser in W3OH/IRS8 (earlier pase); W3A/IRS1 (later phase)
8) H$^+$ region reaches edge of protostellar cloud (10^{18} cm)	3×10^4 yr	H$^+$ region becomes density bounded.	No H$_2$O or OH masers are present. The H$^+$ region is a near-IR, visible and UV source.	Trapezium Region in Orion

Hence

$$P_{\text{phot}} = P_{\text{phot}} (\mathfrak{M}, R, X, Y, T_{\text{phot}}).$$

On the other hand, in the interior of a star in fully convective equilibrium:

$$P = KT^{5/2}; \quad \text{with} \quad K = K(\mathfrak{M}, L, R, X, Y), \tag{4.1; 3}$$

while in the ionization zone just below the photosphere:

$$P_i = P(\varrho, X, Y, T). \tag{4.1; 4}$$

Since there must be continuity between the functions P (Equation (4.1; 3)) and P_i (4.1; 4), with P_{phot} as the photospheric boundary value, there must be a one-to-one relation between the stellar radius R and the photospheric temperature T_{phot} $(\approx T_e)$ for a fully convective star of a given mass $\mathfrak{M}$ and composition X, Y, Z: $R = R(\mathfrak{M}, X, Y, T_e)$. Since also $L = 4\pi R^2 \sigma T^4$, the elimination of R from the last two equations yields a one-to-one relation between L and T_e. This relation defines in the HR-diagram the position of fully convective stars, of given mass $\mathfrak{M}$ and composition. It is called the *Hayashi-line* for that mass.

Computations of these curves in the HR-diagram have been made by Hayashi (1961, see also Hayashi, 1966a, b), and later by others, and are shown in Figures 2 6 and 40. For *massive stars* the Hayashi tracks have not yet been computed. The curve is nearly vertical so that fully convective stars of a given mass have all nearly the same surface temperature. It is of importance that stars which are only partly convective, must be found to the left of the Hayashi line, because for such stars the same $\mathfrak{M}, L$-values would correspond to a higher T_e-value. Reversely, since stars cannot be more than completely convective, no stable stars can exist in the region to the *right* of the Hayashi-line for a given mass. By consequence, a contracting star will find its first hydrostatic equilibrium position at the Hayashi-line, and, when further contracting, will slide downward till there is no longer full convective equilibrium. At that phase it moves through a transition configuration till it will eventually shift onto the line for contracting *radiative* stars.

C. Radiative contraction

This track can easily be computed by simple homology formulas. Let $L_0, R_0, (T_e)_0 \dots$ etc. be the stellar parameters at a certain moment during the radiative contraction, then during that contraction

$$\pi F / \pi F_0 = (L/L_0) (R/R_0)^{-2},$$

while

$$\pi F / \pi F_0 = [T_e/(T_e)_0]^4.$$

Hence

$$T_e/(T_e)_0 = (L/L_0)^{1/4} (R/R_0)^{-1/2}. \tag{4.1; 5}$$

Eddington's mass-luminosity equation for stars in radiative equilibrium with constant values for $\mathfrak{M}$, μ, P_{rad}/P and the chemical composition reduces to the L, R-relation:

$$L/L_0 = (R/R_0)^{-1/2}. \qquad (4.1;6)$$

From Equations (4.1; 5) and (4.1; 6) one finds

$$L/L_0 = [(T_e/T_e)_0]^{0.8}, \qquad (4.1;7)$$

which equation describes the track of a fully radiative contracting star in the HR-diagram.

The time τ_c needed for contraction is easily estimated. Since with Equation (4.2; 3);

$$L = -dE/dt = -\frac{3\gamma - 4}{3(\gamma - 1)} \, d\Omega/dt, \qquad (4.1;8)$$

in which E is the total internal energy of the star, $\gamma = c_p/c_v$, and Ω is the star's potential energy, one obtains:

$$\tau_c = \frac{3\gamma - 4}{3(\gamma - 1)} \int_{\Omega}^{\Omega_0} \frac{d\Omega}{L}, \qquad (4.1;9)$$

where the subscript 0 refers to the initial value. Integrating Equation (4.1; 9) with solar values for contraction out of infinity and assuming a mono-atomic gas (for which $\gamma = \frac{5}{3}$) yields: $(\tau_c)_\odot = 6 \times 10^7$ yr, while for other stars:

$$\tau_c \approx 6 \times 10^7 \, \frac{\Omega}{\Omega_\odot} \, \frac{L_\odot}{L} \text{ yr.} \qquad (4.1;10)$$

This contraction time is usually called the *Kelvin–Helmholtz time-scale*. For massive stars the influence or radiation pressure should still be included in this (approximate) expression. We would then find that the hot star ζ Pup (O4ef) should have contracted to the main sequence in 2×10^6 yr. The expression (4.1; 10) can also be used to calculate the contraction time of *part* of a star, e.g. the stellar core.

4.2. The Main Sequence, Life Times, Stability

For stars of which the radiation is due to nuclear reactions, the *life time at the main sequence* is the time needed for depleting the nuclear (hydrogen) fuel in the core, hence roughly

$$\tau_{nf} = \mathfrak{M}_{core} A_f \, \varepsilon/L \, (:) \, \mathfrak{M}/L \, (:) \, \mathfrak{M}^{1-n}, \qquad (4.2;1)$$

where $\mathfrak{M}_{core}$ is the mass of that part of the star where nuclear reactions can occur, A_f the abundance of nuclear fuel, ε the nuclear energy generation (erg g^{-1} s^{-1}), and n the exponent in the mass-luminosity relation (Equation (1.4; 1)). Since for stars with masses up to $\sim 50\mathfrak{M}_\odot$ n is of the order of 3 to 4, τ_{nf} is a rapidly decreasing function of stellar masses, even if we take into account that for larger masses, where radiation pressure would dominate gas pressure through the whole of the stellar body, n tends to decrease to values of the order of unity. The approximate values of τ_{nf} plotted against stellar bolometric magnitude are given in Figure 42, in which we also indicate the existence of a vibrational instability for stars brighter than $M_b \approx -10$ mag (cf. Table XXXII). Later in this section we shall return to that instability. It is clear from Figure 42 that the average lifetime of very bright stars on the main sequence is short, cosmologically speaking. The post-main sequence evolution of stars will be treated in Sections 4.3 and following.

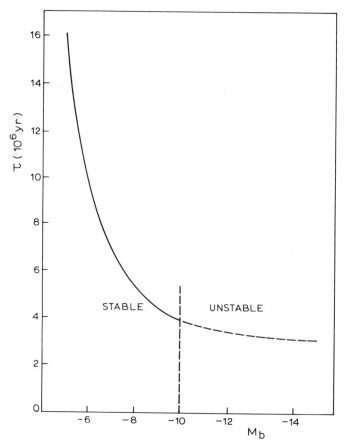

Fig. 42.   Life times, in 10^6 yr, of stars on the main sequence, as a function of the stellar bolometric magnitude. (After Kippenhahn, 1971).

Next we discuss the *position of the main sequence* with respect to the helium abundance. For a first orientation it is useful to consider sequences of stars with various masses ($10\mathfrak{M}_\odot$ to $0.5\mathfrak{M}_\odot$ – helium stars with masses exceeding $\sim 10\mathfrak{M}_\odot$ are not stable) with helium cores of different extent. Model computations were performed by Kippenhahn and Weigert (1967a) for stars with a core with mass $\mathfrak{M}_r$, consisting solely of helium, while the extent of the core was defined by the parameter

$q_0 = \mathfrak{M}_r/\mathfrak{M}$. The various sequences are shown in Figure 43. It is important to compare the main sequence for pure He-stars (the He-zero age main sequence: He–ZAMS) with computations by others, like those by Cox and Guili (1961) for $Y=1$ and $Z=0$; the He–ZAMS of Cimino *et al.* (1963) for $Y=0.98$ and $Z=0.02$; the one by Nariai (1970) for H : He : C : Fe $= 10 : 1000 : 10 : 1$; and the one by Paczynski (1971c) for $Y=1$. The differences between these curves – not all are drawn in Figure 43 – is small. Larger are the differences with the He–ZAMS derived by Stothers and Chin (1978) using a set of opacities derived by Carson (1976).

It is often supposed that the He–ZAMS should be the locus of Wolf–Rayet stars, but this is not the case. On the contrary, the WR-stars are located closer to the H–

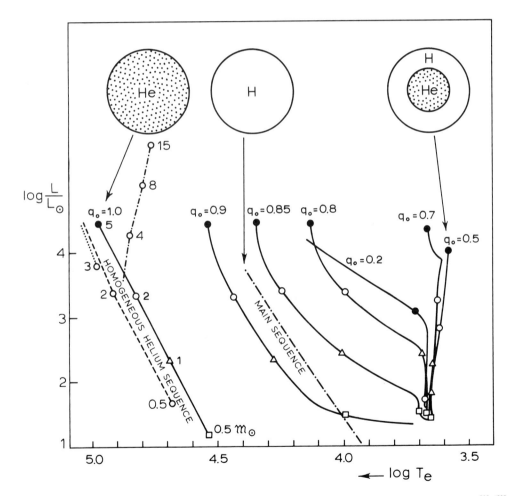

Fig. 43.   Positions of the main sequence for different assumed values of the parameter $q_0 = \mathfrak{M}_r/\mathfrak{M}$ where $\mathfrak{M}_r$ is the mass of a core consisting of pure He. The dashed and dotted He–ZAMS's are for $Y=1$, $Z=0$, and $Y=0.98$, $Z=0.02$, respectively (see text). The dashed-dotted curve is Stothers and Chin's (1978) ZAMS for pure He, using Carson's (1976) opacities. Partly after Kippenhahn and Weigert (1967a).

ZAMS; nevertheless they are believed to have a high He-abundance. In this connection it is illustrative to learn from Figure 43 that *nearly*-pure He-stars ($q_0=0.9$ or 0.85) occupy positions close to the H–ZAMS.

A next item of interest is the *width of the core-hydrogen-burning strip* in the HR-diagram. It is often – intuitively – assumed that this strip should coincide with the ZAMS, but that is not true for the massive stars. The introduction of appreciable convective overshooting from the inner convective region during this phase may considerably enlarge the star, thus widening the core-H-burning strip, – under extreme, but perhaps not very likely, conditions even up to the Hayashi-border (Massevitch *et al.*, 1979).

A similar result could be obtained with increased values for the opacity as was demonstrated by Stothers and Chin (1977). A change from the Cox–Stewart opacities, based on 'hydrogenic' atomic models to new – and perhaps unreasonably large – opacities by Carson (1976), based on the more refined 'Thomas–Fermi' atomic models produced a fundamental change in the position of that region. The larger opacities increase the star's sizes, which is particularly valid for stars of masses $\gtrsim 30\mathfrak{M}_\odot$. Hence with these opacities the region occupied by stars in the core-H-burning phase then extends over most of the upper part of the HR-diagram – see Figure 44. The dependence on the Z-value is striking from the two curves giving the positions of stars for $Z=0.02$ and 0.04. The essential aspect of these computations is clearly that with larger opacities core-hydrogen burning occurs over a much broader area of the upper part of the Hertzsprung–Russell diagram, than has been assumed before.

At the time of writing (1980) no decissive argument has yet been advanced for one or the other set of opacities. Undoubtedly, the Thomas–Fermi atomic models better represent the atomic electron cloud than the hydrogenic model, but offhand one would not expect the differences between the two opacities to exceed a few tens of percents; therefore the large differences found are surprising.

The *stability of very massive main sequence stars* has several aspects. With increasing stellar masses an instability sets in which is essentially due to the fact that the radiation pressure increases strongly with mass, because of the increasing temperature and luminosity. Consequently the value of $\gamma=c_p/c_v$, which is $\frac{5}{3}$ for a monoatomic gas tends to approach $\frac{4}{3}$, which is the value for isotropic blackbody radiation.

In an elementary way it can be shown that stars with $\gamma=\frac{4}{3}$ are close to instability: If $U=$ the integrated internal energy of the stellar gas, E_{kin} the star's kinetic energy, Ω its potential energy, and $E=U+\Omega$ the total energy, then thermodynamics shows that with $U=c_vT$,

$$E_{kin} = \tfrac{3}{2}\mathfrak{R}T = \tfrac{3}{2}\left(c_p - c_v\right)T = \tfrac{3}{2}\left(\gamma - 1\right)U, \qquad (4.2;\,2)$$

with $\gamma=c_p/c_v$.

Since according to the virial theorem, in a closed system neglecting rotational energy:

$$2E_{kin} + \Omega = 0, \qquad (4.2;\,3)$$

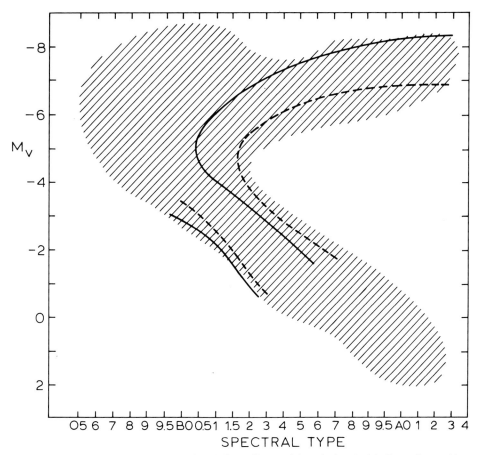

Fig. 44. Theoretical main-sequence bands for stellar models, calculated with Carson's opacities and with hydrogen burning core, are superposed on an observational HR diagram for bright early-type stars. The assumed initial metal abundances are $Z=0.02$ (continuous lines) and $Z=0.04$ (dashed lines). The two boundary lines are the ZAMS (chemically homogeneous star), and the TAMS (the line at which T_e reaches a minimum during core-H-burning). (After Stothers and Chin, 1977).

the combination of Equations (4.2; 2) and (4.2; 3) yields $\Omega=3\,(1-\gamma)\,U$, and hence

$$E = U + \Omega = (4 - 3\gamma)\,U = \frac{3\gamma - 4}{3(\gamma - 1)}\,\Omega.$$

Hence, a star with $\gamma < \frac{4}{3}$ would have a positive total energy and would hence be unstable, and one with $\gamma = \frac{4}{3}$ would just be at the fringe of instability. Quantitative computations specify the character of this instability.

Ledoux (1941) was the first to give these considerations a more physical basis. He could show that in chemically homogeneous stars and for γ-values approaching $\frac{4}{3}$ the star would pulsate nearly in a homologous way, in other words its pulsational amplitude would be finite all over the star, also in the core. The consequent change of the nuclear reaction rate during the pulsations has a destabilizing effect which is stronger than the stabilizing effect of radiative transfer in the envelope. Hence, stars with very large masses may become pulsationally unstable. On the basis of more accurate models Schwarzschild and Härm (1959) assuming homogeneous stars with $\mathfrak{M}=28\mathfrak{M}_\odot$ to $200\mathfrak{M}_\odot$ could show that the imaginary part of the phase coefficient of the stellar pulsation amplitude changes from negative to positive for values of about $60\mathfrak{M}_\odot$. Ziebarth (1970), Simon and Stothers (1970) – see also Hansen's (1978) review – have dealt with this question, using linear pulsation theory and they found limiting mass values depending on the mass fraction of He (Y) and of heavier elements (Z), see Table XXXII. (Similar calculations by Aizenmann *et al.* (1975) yielding $\mathfrak{M}_{lim}=$ 60 for the abundances $Y=0.25$, $Z=0.03$ are not fully comparable because they simply assumed Kramer's opacity law with electron scattering.) Since all existing supergiants must be young in the cosmological sence, their composition should be

TABLE XXXII

Stellar lower mass values for which vibrational instability sets in. The computations were made for homogeneous stars (main sequence objects), assuming linear pulsation theory with various values of the mass abundances of He (Y) and of heavier elements (Z). After Stothers and Simon (1970), and Ziebarth (1970, denoted by a + sign).

$Z=$	0.05	0.04	0.03	0.02	0.01	0.002
$Y=0$	165		134		120	106
0.18				91+		
0.2	140		110		90	86
0.28				87+		
0.38		84+		76+	72+	
0.4	114		88		70	64
0.48				67+		
0.9–0.95	60		41		26	21

close to that of population I-type stars, e.g. $Z\approx0.03$ and $Y\approx0.3$. This would lead to the conclusion that the upper mass for vibrationally stable main sequence stars is somewhere around 90 solar masses, depending on the composition. For more massive stars the limit of instability would be defined by a line fanning out from the main sequence towards lower effective temperatures. For stellar models based on Carson's opacities the limit is found at smaller masses, roughly $L/\mathfrak{M}\approx10^3$ where L and $\mathfrak{M}$ are expressed in solar values (Stothers, 1976) – see Figure 45. For rotating stars $\mathfrak{M}_{lim}$ increases greatly (Stothers, 1974). Parallel to the above research Appenzeller (1970a, b), Ziebarth (1970), Talbot (1971a, b), and Papaloizou (1973) have improved the discussion of the problem, by extending it to *non-linear effects*. They found, as could be expected, that in that case the pulsations do not increase infinitely but reach a limiting amplitude, and, therefore, do not have the disrupting effect as would result from the linear analyses. For instance, rates of mass-loss are estimated still as small as $3\times10^{-5}\,\mathfrak{M}_\odot\,yr^{-1}$ for a $100\mathfrak{M}_\odot$ star (Talbot, 1971b), and $4\times10^{-5}\,\mathfrak{M}_\odot\,yr^{-1}$ for a $130\mathfrak{M}_\odot$ star (Appenzeller, 1970b). Since, furthermore, massive stars rapidly move out to the right-hand side part of the HR-diagram, where stars are more

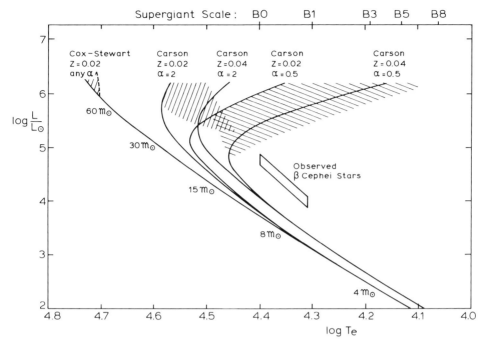

Fig. 45. Vibrational instability for stars of large mass. Vibrational instability exists in the hatched areas. The notions 'Cox-Stewart' and 'Carson' refer to the two different set of opacities used; α is the ratio between convective mixing length and pressure scale height. (After Stothers, 1976).

stable, there is little reason to expect slightly evolved stars of such masses to be vibrationally unstable. Another aspect of the problem is that of the possible occurrence of *non-radial pulsations* of massive stars, and was studied in a first orientation by Simon (1957) and later by Wan and Van der Borght (1966), Aizenman *et al.* (1975), and Vemury and Stothers (1977). It seems that resonance interaction may occur between the first radial harmonic and the non-radial p_r mode of oscillation for stars with $\mathfrak{M} \gtrsim 60\mathfrak{M}_\odot$. Hence, although stars with masses larger than $60\mathfrak{M}_\odot$ should not necessarily be instable, small amplitude variations may occur.

The problem of the *upper limiting mass* of main sequence stars is obviously not answered by these studies. A more promising approach is therefore, to follow Eddington (1926, pp. 15–18), as Larson and Starrfield (1971) did, and to consider the problem from another point of view: whether stars of the large masses around $100\mathfrak{M}_\odot$ *can originate* at all: In brief, their point is that during the formation of a *protostar* the radiation flux is so large that *radiation forces* would just prevent a collapsing gas-cloud, more massive than approx. $50\mathfrak{M}_\odot$ (uncertainty: factor 2) to be formed, in the time span available for stellar contraction out of a primordial cloud.

Let us at this stage summarize the situation with regard to the upper limit of stellar masses:

(a) Linear instability analyses yield a limit at $90\mathfrak{M}_\odot$. Above that limit stars would show some form of vibrational instability, but the non-linear approach yields higher limiting values and predicts $-\mathfrak{M}=4\times10^{-5}\,\mathfrak{M}_\odot\,\mathrm{yr}^{-1}$ for a $130\mathfrak{M}_\odot$-star. Such values for the rate of mass-loss are indeed observed for the early-type Of stars (see Table XXXIX in Chapter 7), while evolutionary calculations predict these masses for the earliest type stars.

(b) Considerations on stellar origin yield an upper limit of $\sim50\mathfrak{M}_\odot$ ($+50$; -25).

(c) The Eddington criterium (Sections 1.3 and 1.4) yields an upper limit of $\sim50\mathfrak{M}_\odot$.

(d) The most massive binary-components (Table VIII in Chapter 2) have a mass of 60 to $80\mathfrak{M}_\odot$ (Plaskett's star, and BD+40.4220).

We conclude therefore that the largest stellar mass is $\sim100\mathfrak{M}_\odot\pm40\mathfrak{M}_\odot$. Further stability analyses of main sequence and slightly evolved stars, as well as a more detailed discussion of the stability of a contracting protostar would certainly be very useful.

4.3. Post-Main-Sequence Evolution of Massive Stars

The evolution of massive stars at and off the main sequence (De Loore, 1980; – review paper) is determined by a few essential factors.

The first aspect is that the energy generation in the stellar core, after arrival at the main sequence is defined by *nuclear energy generation*. Information on the nuclear energy generation in very hot stellar interiors is roughly summarized in Figure 46. The density of the fuel is always assumed 100%, while the material density is 10 g cm^{-3} for He-burning and 10^5 g cm^{-3} for the other reactions. For neutrino processes $\varrho=10^5$ g cm^{-3} has been taken and $\mu_e=2$; matter was assumed not to be degenerated.

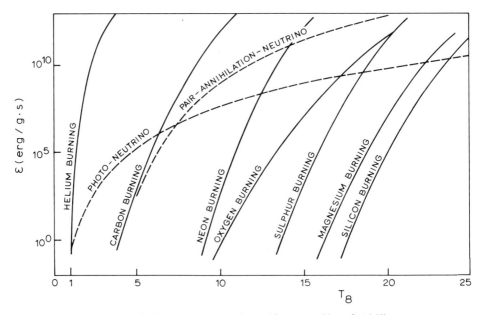

Fig. 46   Nuclear energy generation. (After Hayashi *et al.*, 1962).

Secondly, the *life-time at the main sequence* τ_{ms} is defined by the time for exhaustion of nuclear (hydrogen) fuel, Equation (4.2; 1).

Thirdly, when nuclear fuel of a certain species is depleted in the stellar core, and when the core temperature is too low for nuclear reactions of other core material to start; when in addition there is no 'shell source', the stellar radiation flux is maintained by core contraction. For this process the *characteristic time of core-contraction* τ_c is defined by Equation (4.1; 10) in which one should read for Ω the potential energy of the contracting core, L remains the star's luminosity. From Equation (4.1; 8) it follows that for a star with $\gamma = \frac{5}{3}$ (mono-atomic gas) half of the decrease of potential energy goes into stellar radiation, while the other half is used for heating the core material: because Equation (4.2; 3) and the virial theorem yield $dE_{kin}/dt = -\frac{1}{2} d\Omega/dt = L$: the increase in the core temperature is simply a consequence of the virial theorem.

Thus the scenario of stellar evolution at and off the main sequence is simple in principle: stellar radiation is provided by nuclear fuel (time scale: τ_{nf}); when the fuel in the core is exhausted, stellar core contraction is the source for stellar luminosity (time scale τ_c); at the same time the internal temperature increases, till another nuclear fuel can take over.

It is further of importance that all main-sequence stars with $\mathfrak{M} \gtrsim 3\mathfrak{M}_\odot$ have well-developed convective cores, so that by convective mixing the chemical composition is at any time equal in the whole core, and depletion occurs 'at the same time' in the whole core. This is, however, no longer valid during subsequent evolution, when the size of the core can vary greatly.

An outline of *the evolution of an average massive star* – say of approx. $5\mathfrak{M}_\odot$ – is drafted in Figure 47. During the process of H-depletion in the stellar core the star moves only little in the HR-diagram till H is depleted over so large a fraction of the core that nuclear burning comes to a temporary end. Since during this phase the star obviously continues to radiate (the temperature difference between core and photosphere remains, so there is a radiation flux) this radiative loss must be compensated internally, which takes place by *core contraction* (characteristic time scale τ_c). During this contraction of the core the increase of the central temperature causes the outer mantle of the star to expand, thus increasing the radius. The net effect is a decrease of T_e. In the HR-diagram the star moves rightward, and slightly upward if mass loss is neglected. This comes to an end when the temperature in the central part rises high enough for *hydrogen shell burning* to start in a shell around the H-depleted convective core.

The consequent contraction of the outer mantle causes the star to move a small distance leftward, till another contraction phase starts, that of the He-core, accompanied by a rightward motion over a larger distance in the HR-diagram. Schmalberger (1960), and others among whom Lesh and Aizenman (1973) and Eggen (1975b) have drawn attention to the positions of the short-periodic variables of the β Cep ($= \beta$ CMa) – type in the HR-diagram – Figure 48. Accurate photometry shows that the β Cep stars occupy a narrow strip (Shobbrook, 1978a; Jerzykiewicz and Sterken,

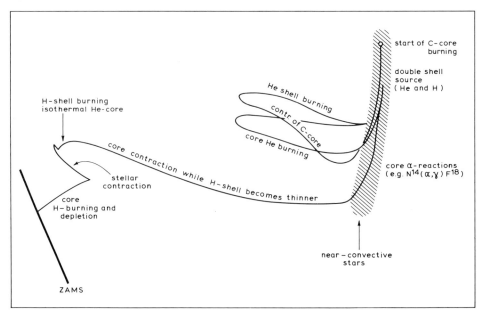

Fig. 47.   Phases in the evolution of a fairly massive star.

1979). About three quarters of the stars in this strip are β Cep stars (Shobbrook, 1978b). It appears that this strip coincides with a region that is traversed three times by a star during its post-main-sequence evolution: first during the core-H-burning phase, then in the second contraction phase, and finally in the shell-H-burning phase. Initially, Schmalberger suggested that this strip may be identical with the brief phase between the start of H-shell burning (after the first core-contraction), and the beginning of the secondary contraction phase, at the end of the H-shell burning phase, but Shobbrook (1978b), who compared the density of variable stars in the strip with the expected lifetimes of stars in the three parts of the evolutionary track within the strip, concluded that β Cep instability should just occur *before* the first bend in the evolutionary track, so that the β Cep stars are very near to the end of core hydrogen burning. Various proposals have been advanced to explain the instability. The suggestion by Stothers and Simon (1969b) to relate the instability to accumulation of H and He in the atmosphere of the least massive of an evolved binary does not seem correct since most of these stars are not binaries. The origin of the instability may be related to resonance of the eigenmode of radial oscillations of the whole star with overstable convection in the rapidly spinning core of a star (Osaki, 1974). It is significant that the degeneracy of these two frequencies is only possible in the late stages of core hydrogen burning of massive stars! In addition, Stellingwerf (1978) proposes to relate the instability to a destabilizing effect of the He$^+$ ionization edge: at that level the local work done, $\oint P\,\mathrm{d}V$ is just non-zero. But it seems unlikely that these mechanisms explain the observed fierce expanding accelerations of these atmospheres (Burger *et al.*, 1980).

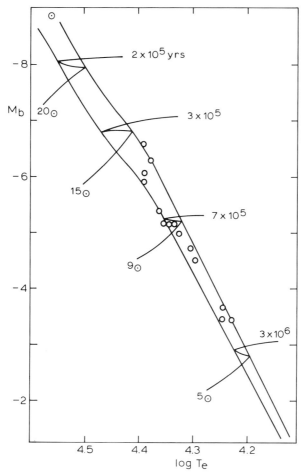

Fig. 48.   The double bend in the evolutionary tracks corresponding to the contraction phase before
the onset of the H-shell burning may be an instability strip characterized by β Cephei-variability.
Open circles represent β Cep stars. The numbers alongside give the 'contraction time' for the four
evolutionary tracks. (From Eggen, 1975b).

At the end of the He-core contraction the increase of the internal temperature and
of the temperature gradient make the star move into the region of near-completely
convective stars and the star *ascends* in the HR-diagram along a line very similar
to the Hayashi tracks along which the near-completely convective stars *descended*
in their pre-main-sequence contraction. Detailed computations of the evolution of
stars of $0.7\mathfrak{M}_\odot$ to $2.2\mathfrak{M}_\odot$ up to the moment of start of He-core burning were made
by Sweigart and Gross (1978). Figure 49 shows a few examples. It is important that
for stars of the moderate masses considered in that Figure the He-core is degenerate
in nearly all cases. This is relevant to the subsequent evolution because: (a) a
degenerate core is nearly isothermal because of the large heat conductivity of degenerate matter; and (b) the weak temperature dependence of P in degenerate material

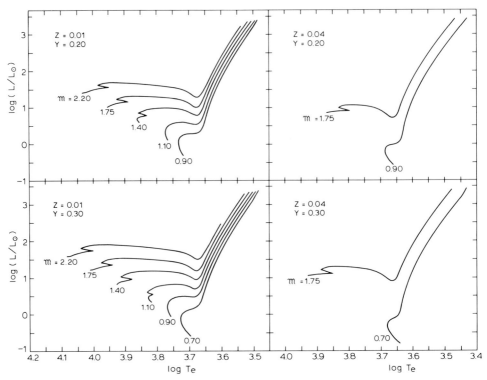

Fig. 49.   Evolutionary tracks for red giants of various masses and *Y*- and *Z*-values. (From Sweigart
and Gross, 1978).

will cause a thermal runaway at the time the triple He reactions would start. This
is called the He-flash (see also Section 4.6 and later this section).

In the subsequent phase of *core helium burning* the star moves again leftward in
the HR diagram, and performs a loop till at the end of that phase the contraction
of the consequently originated C-core starts. The precize shape of this loop and
particularly its extent towards the high-temperature region is still rather uncertain.
This is a consequence of the feature that the fairly massive stars considered here develop important convective regions – in the core, the intermediate stellar region and
the stellar mantle as well – during at least part of their evolution. As will be shown
in Section 4.4, there is still some uncertainty as to what criterion should be applied
in defining the stellar regions where convection occurs. Stothers and Chin (1975,
1976) calculated evolutionary tracks for stars of $15\mathfrak{M}_\odot$ to $64\mathfrak{M}_\odot$ for various assumptions about convection (the Ledoux and Schwarzschild criteria, cf. Section
4.4) as well as for different assumed chemical abundances. For stars of solar or
Population I-type chemical abundances there appear to be differences in the shapes
of the evolutionary tracks for stars of different masses. Stars with $\mathfrak{M} \gtrsim 20\mathfrak{M}_\odot$ do
not develop a blue-ward loop. Such stars go more or less directly to to red part
of the HR diagram and reach the vicinity of the Hayashi-track towards the end

of their phase of core-helium-burning. Stars with $\mathfrak{M} \lesssim 17 \mathfrak{M}_{\odot}$ develop a blue loop, with a blue-ward extension depending on the convection criterion.

Similar but less detailed conclusions were reached earlier by others: Chiosi and Summa (1970), Simpson (1971), and Kippenhahn (1971) found that core-helium burning in massive stars occurs only in the low-temperature region of the HR diagram. This was confirmed by Massevitch and Tutukov (1974) who showed that for masses below ∼$40 \mathfrak{M}_{\odot}$ core-helium burning occurs in the blue supergiant region; for more massive stars nearly all helium seems to be burnt in the red supergiant phase. The medium-temperature part of the first loop is identified with the region in the HR diagram occupied by the Cepheid variables: these would hence be stars in which core He burning is nearly finishing – quantitative data of a few such Cepheids is given by Cox *et al.* (1978) for the short periodics AC And and TU Cas, for which $\mathfrak{M} \approx 3 \ldots 4 \mathfrak{M}_{\odot}$ is found from the various pulsational modes.

When the contraction of the carbon-core has caused a further increase of the temperature in the central parts *shell-helium burning* sets in. This may cause a second loop, as shown in Figure 50 (Höppner *et al.*, 1978). While the *existence* of the two loops seems beyond doubt for fairly massive stars such as the $7 \mathfrak{M}_{\odot}$ and $9 \mathfrak{M}_{\odot}$ stars shown in Figure 50 their precise position is yet very much debated and appears even to depend too much on the computational program used (cf. Paczynski, 1970b).

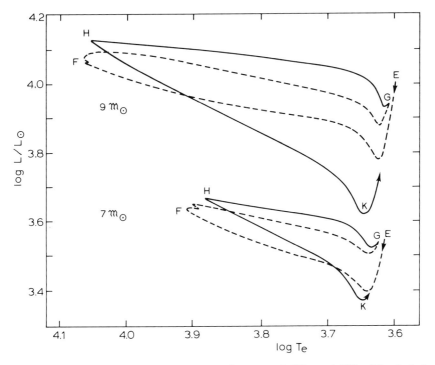

Fig. 50. Evolutionary tracks in the HR diagram for stars of $7 \mathfrak{M}_{\odot}$ and $9 \mathfrak{M}_{\odot}$. The dashed part (*E-F-G*) shows the first loop (central He burning); the solid part (*G-H-K*) the second loop (beginning of the He shell source burning phase). (From Höppner *et al.*, 1978).

At the end of the carbon-core contraction phase the star has moved up fairly high along the Hayashi-track at the low temperature part of the HR-diagram; this continues until core-carbon burning can start.

Before closing this Section a few of the existing uncertainties should be recalled:
– We mentioned the uncertainty in the stellar opacities, which may greatly change the positions and tracks of stars in the HR-diagram.
– Other difficulties in the computation of the evolution of massive stars are not only the imperfect knowledge of the relevant physical data (apart from the opacities also the nuclear burning rates) but also fairly fundamental matters regarding the convective stability in chemically inhomogeneous stellar regions; see for this aspect the excellent review by Dallaporte (1972), and our Section 4.4 on the problems of convection and semi-convection.
– One should not forget that convective overshooting in the inner parts of a star greatly influences its structure: it enlarges the radius and reduces T_e.
– In the last phases of stellar evolution the core density in stellar cores of medium mass becomes so high that the core is degenerated. It is a property of degenerated matter to cool down at compression (contrary to non-degenerate matter) and to heat up at expansion. If, for some reason nuclear reactions start in a degenerated stellar core (for example when the temperature would exceed a certain limit) this would lead to expansion, hence further temperature increase and increased nuclear heating, etc. This phase is called the *flash-phase*. The temperature would continue to increase till degeneracy is removed.

The *helium flash* occurs in a stellar core of degenerate helium. The *helium-shell-flash* occurs in a double shell-burning star and may sensibly influence the surface temperature (cf. e.g. Paczyńsky, 1977). The *carbon flash* would occur in a degenerate carbon core, etc. It is thought (Ivanova *et al.*, 1977) that the thermonuclear burning of the developing carbon flash may lead to gravitational collapse, with the production of a neutron star. For stars of intermediate masses, $\sim$7 to $9\mathfrak{M}_\odot$, the neutrino losses cause cooling of the core, and hence a temperature inversion, causing carbon to ignite at some distance from the core (Paczyński, 1971c; Ergma and Vilhu, 1978). This is called the *carbon shell flash*.

4.4. Convection and Semi-Convection

In an evolved star the central parts have another chemical composition than the outer parts. The central parts are chemically homogeneous when this core is in convective equilibrium. When in the course of the evolution the convective core would contract, it will be surrounded by layers with decreasing $\mu(r)$, or even with a discontinuity in $\mu(r)$. This may cause a discontinuity in the opacity κ. If electron scattering is the main contributor to κ – which is the case in massive stars, with approximately $\mathfrak{M} \geq 10\mathfrak{M}_\odot$ – then $\kappa \approx 0.2\ (1+X)$ will show a clear discontinuity at the $\mu(r)$-discontinuity, because H produces more electrons per unit mass than He. The situation that then arises may have consequences for the energy transport regime. The question is what condition would define the occurrence of convection in such circumstances. The classical condition for radiative equilibrium is Karl Schwarzschild's, who assumed that a density gradient would be required smaller than its adiabatic value:

$$\left|\frac{d\varrho}{dr}\right| \lesssim \left|\frac{d\varrho}{dr}\right|_{ad}. \qquad (4.4;\,1)$$

In a chemically *homogeneous medium* this is equivalent to:

$$\left(\frac{dT}{dP}\right)_{rad} \lesssim \left(\frac{dT}{dP}\right)_{ad}$$

or, as is customary written

$$\nabla_{rad} \lesssim \nabla_{ad} \qquad (4.4;\,2)$$

with $\nabla = d \log T / d \log P$.

This criterium is often called the *Schwarzschild* or the *temperature criterion*.

A condition valid for an *inhomogeneous medium* was derived by Ledoux (1947), elaborated by Sakashita *et al.* (1959, 1961), and Stothers and Chin (1968). The condition (4.4; 1) leads in that case, after some algebra, to the *Ledoux* or *density criterium*:

$$\nabla_{rad} \leq \nabla_{ad} + \frac{\beta}{4 - 3\beta} \frac{d \log \mu}{d \log P}. \qquad (4.4;\,3)$$

Apparently, the application of condition (4.4; 3) tends to stabilize the motions in the model of the star's interior, as compared to Schwarzschild's condition. Note that the difference between the criteria (4.4; 2) and 4.4; 3) is only significant in those stellar layers where the molecular weight μ varies with height; elsewhere they are equal. Hence, a comparison of these criteria makes only sense for *evolved stars*.

Before comparing the two criteria and deciding on their use, we describe another effect, which is also related to the chemical inhomogeneity in evolved stars: the existence of *sem-convective motions*. It looks intruiging to examine whether near the layer of the discontinuity in X, μ, and κ a small unstable region originates, where slow motions would tend to equalize the abundances or at least smooth the $\mu(r)$ discontinuity. Following Ledoux (1947), Schwarzschild and Härm (1958) we therefore define semi-convection as the process by which in (part of) the inhomogeneous region (generally situated above the inner convection zone) chemical homogeneity is – at least partly – re-established through – often slow – convective motions. Simpson (1971) defines semi-convection as 'the partial convective mixing that takes place in a convectively unstable region, where stability can be obtained by the results of the mixing before the region is completely homogenized'. Semi-convection appears to occur for stellar masses above 10 to $15\mathfrak{M}_\odot$ (Robertson, 1972), as can be readily understood: these stars start their life on the main sequence with a sufficiently large convective core, which later reduces in size.

The way how semi-convective mixing has to be treated in a numerical procedure has been developed by Stothers (1970) and by Simpson (1971) following a suggestion by Henyey, and was refined later by Sreenivasan and Ziebarth (1974) and discussed again by Sreenivasan and Wilson (1978a). Although there are differences in the equations and the numerical techniques applied by the various authors, their methods are virtually equal and so are the results. Semi-convection is treated as a diffusion of the abundance parameters, where a flux φ^i of the abundance χ^i is defined with

$$\varphi^i = -\lambda \partial \chi^i / \partial r,$$

while for each of the constituents a continuity equation is written:

$$\frac{\partial \chi^i}{\partial t} + \nabla \Phi = \left(\frac{\partial \chi^i}{\partial t}\right)_{\text{nucl.}}$$

where the right-hand term describes the extra change in the abundances due to nuclear reactions.

In addition to these two equations the run of the composition must be such that the temperature gradient remains at its prescibed value according to condition (4.4; 2) or (4.4; 3). The method is thus to vary the abundances till stability is reached. This, at the same time, defines the parameter λ. If $\lambda > 0$ the region is semi-convective; if $\lambda < 0$ it is not.

The influence of semi-convection on model calculations is far from negligible. As a rule models calculated without semi-convection are systematically bluer than those calculated with the same parameters but with the inclusion of semi-convection (Chiosi and Nasi, 1978). As an example we show in Figure 51a evolutionary tracks for a $15\mathfrak{M}_\odot$ star, without mass loss, where convection is treated with the temperature criterion (Schwarzschild, Equation (4.4; 2)). One of the tracks includes semi-convection; the other not. Illustrative is also the internal structure of the model with semi-convection, shown in Figures 51b and 51c: the evolution of semi-convective regions on top of convective ones is interesting. The calculations were made for $X=0.70$, $Y=0.27$, $Z=0.03$.

In addition, Figure 52 is used to show the influence of *two other effects*: the diagram shows evolutionary tracks for stars of various masses calculated for:
– the two convection criteria (Schwarzschild and Ledoux),
– two metal abundances.
Semi-convection was included in the calculations.

A comparison of the dashed line for $15\mathfrak{M}_\odot$ from Figure 52 with the dotted line of Figure 51 shows the important influence of the smaller Z-values. The effect appears comparable to the influence of including semi-convection. This can be understood: the effect of the inclusion of semi-convection is to locally reduce the temperature gradient, and is hence comparable to the effect of introducing smaller opacities.

Of secondary, but certainly not negligible importance, is the choise of the convection condition, while (thirdly) the heavy metals abundance influences the tracks also appreciably.

Another aspect that is of great importance for the structure of massive stars, but

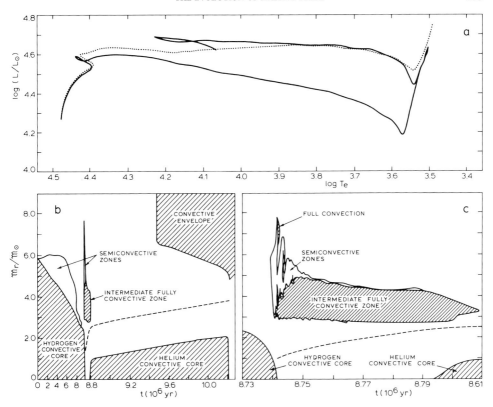

Fig. 51.   The evolution of a 15$\mathfrak{M}_\odot$ star. (After Sreenivasan and Wilson, 1978a). (a) Evolutionary
tracks. Full-drawn: without semi-convection. Dotted: with semi-convection. (b, c) The boundaries of
the convective regions (full-drawn) and of the helium core (dashed) versus time.

that has so far only been dealt with in a rather approximative way is *convective over-shooting*. This phenomenon tends to increase the star's size, and is therefore of importance for its stability and T_e-value, but it is difficult to quantize (Varshaskyj and Tutukov, 1972).

We now return to the *convection criteria* (4.4; 2) and (4.4; 3). It has been a matter of considerable debate which of the two convection criteria should be used in the region of the μ-discontinuity. In that connection Kato (1966) rightly remarks (see also Sreenivasan and Ziebarth, 1974): in a region where condition (4.4; 2) holds, the condition (4.4; 3) holds too, because μ and P both decrease with r so that the extra term on the right-hand side of condition (4.4; 3) is positive. Since the effect of semi-convection is to produce mixing and smoothing of the abundance gradients in a region where mixing would not occur, the simultaneous application of condition (4.4; 3) and semi-convection seems contradictory. In line with this Simpson (1971) found that numerically stable solutions of the equations for the internal structure of stars can only be obtained when the temperature (Schwarzschild) criterion is applied in the semi-convective region.

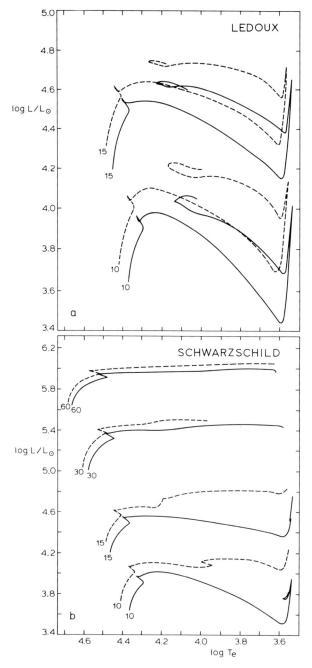

Fig. 52.   Evolutionary tracks in the HR diagram, calculated for various stellar masses, and using the Ledoux and Schwarzschild criteria for convection (Figures 52a and 52b respectively). Solid lines are for $X=0.739$, $Z=0.044$; dashed ones are for $X=0.739$, $Z=0.021$. (After Stothers and Chin, 1975, 1976).

Hence it seems that the Schwarzschild criterion (4.4; 2) should normally be used together *with semi-convection*.

Since the above reasoning seems convincing, but needs quantitative verification several authors have attempted to search in other ways for arguments enabling them to decide between the two criteria.

There are three ways to attack the problem of the selection of the convection criteria:

(a) Kato (1966) in a preliminary investigation, went the fundamental way by examining the stability of a region calculated according to the density (Ledoux) criterion, and found this region to be unstable for vibrations, which set in and produce sufficiently rapid mixing to, finally, attain chemical homogenity, so that the temperature criterion applies again. This work was followed by more research along the same line. Gabriel and Noels (1976, 1977) examined the evolution of a $30\mathfrak{M}_\odot$ population I star during its main-sequence phase, and later (1977) also some stars with masses around $1.1\mathfrak{M}_\odot$. Stephenson (1979) investigated the stability of the semi-convective region, and made numerical calculations for five stellar models, with masses between $0.66\mathfrak{M}_\odot$ and $121.1\mathfrak{M}_\odot$. In the inhomogeneous region, where semi-convection should occur the models have a temperature gradient larger than the adiabatic one, as derived with condition (4.4; 3). In this region, and for massive stars ($\gtrsim 15\mathfrak{M}_\odot$) growing overstable modes may provoke non-radial oscillations in the chemically inhomogeneous zone corresponding to spherical harmonics of high degrees. Most of these models are unstable, with time scales of the order of 10^3 to 10^4 year. In the semi-convective region the temperature gradient has to be just slightly overadiabatic, i.e. <1 to 2% over the gradient calculated for the Schwarzschild condition (4.4; 2) in order to make the layer maintaining a state close to convective neutrality. This may be considered as a support for the Schwarzschild condition. So far no quantitative non-linear computations have been made.

(b) The analyses described above are based on a *stability* treatment of the semi-convective region. Another aspect is to study the *physics of convection*. On that basis Schlesinger (1975) showed that, when during the course of stellar evolution an intermediate convective zone should originate between the stellar boundary, and a receding inner convective core, with a semi-convective zone in between, the two convective regions will merge due to *overshoot of convective elements* out of the intermediate convective region. Thus, sufficient mixing will occur, homogenizing the region and justifying the use of the temperature criterium.

(c) Another way to investigate the matter is the 'empirical' one, by considering the evolutionary tracks of massive stars and comparing these with the number-densities of stars in various regions of the HR-diagram. A drawback is that the observations are far from decisive in this matter; the values found for the ratio $N_{\rm BSG}/V_{\rm RSG}$ (Blue, Red, Super-Giants) ranges between 1.6 (Schild, 1970) and 10 (Stothers, 1972b).

Evolutionary calculations show that the time spent *between* H-exhaustion and He-ignition, when the stellar core contracts rapidly till a temperature is reached sufficiently high for He ignition, is relatively short. Most of the time away from the main sequence is therefore spent in the core He-burning phase. Therefore it is of importance to know the region in the Hertzsprung–Russell-diagram where the star is in the He-burning phase: then a comparison with colour-magnitude diagrams of star clusters would allow one to decide on the stellar primary emission mechanism. Usually one therefore computes the ratio t_{br} between the times spent by the star in the core hydrogen burning stage to the period spent in the evolution from core helium ignition to the end of the shell-helium-burning stage. It appears that for the case of the Schwarzschild criterion t_{br} increases with increasing mass: while in models based on the Ledoux criterion core-He burning commences only in the red-supergiant phase. In the first case the more massive supergiants spent a longer part of their life in the blue part of the HR-diagram. We refer, for an example, to Figure 52.

The problem has particularly been considered by Stothers and Chin (1975, 1976, 1977) who computed stellar evolutionary tracks for massive stars (10–$60\mathfrak{M}_{\odot}$) of different chemical abundances with subsequently the conditions (4.4; 2) and (4.4; 3) for convection. 'Zones of occupation' of blue supergiants in the HR diagram are calculated for both cases. The differences between stellar evolutionary models and evolutionary paths in the HR diagram (Figure 52) are small, and not of fundamental importance.

Although these latter investigations leave the matter undecided, we *conclude* that most arguments, seem to favor the application of the temperature (Schwarzschild's) criterion.

4.5. The Influence of Mass-Loss on the Evolution of Massive Stars

It has been a matter of much debate since the early fifties whether mass-loss influences the evolution of stars. The situation is clearer now, since we know that appreciable mass-loss occurs for the brightest early-type stars, with $M_b \lesssim -6$ to -7.5 (see Chapter 7), while also the rate of mass-loss for stars in the various parts of the HR-diagram is approximately known.

Fessenkov (1949) was the first to suggest the importance of mass-loss for the evolution of early-type stars. He assumed the mass-loss-law

$$-\dot{\mathfrak{M}} = KL. \tag{4.5; 1a}$$

The expression (4.5; 1a) assumes that the mass-loss is due to radiation pressure effects. Evolutionary calculations based on this relation were thereupon performed by Massevitch (1949) who extended these calculations in subsequent years (see for further references: Massevitch *et al.*, 1979). Much later the matter was also taken up by Tanaka (1966a, b) who too assumed a mass-loss-law, and in comparing his results with the conservative case found – in line with Massevitch *et al.* – that at the same stage of central hydrogen depletion the mass losing star has smaller L- and T_e-values a larger relative mass of the convective core and a longer hydrogen-burning life time. These early results were essentially – but not in all cases! – confirmed by the various

authors who, in later years, dealt with the same matter, in several degrees of refinement and computational perfection.

De Loore *et al.* (1977) similarly choose to describe the rate of mass-loss by

$$-\dot{\mathfrak{M}} = NL/c, \qquad (4.5; 1b)$$

where L is the stellar luminosity and c the velocity of light; N is a free parameter.
It is to be noted that for $N \gtrsim 1000$ the evolutionary paths coincide practically with
the main sequence. But this value of the parameter N is far too large. Note that the
track for $N \to \infty$ should coincide exactly with the ZAMS! A comparison of computed
evolutionary tracks for different N-values with observations – see Figure 110 in
Chapter 7 and Figure 53 – shows N to be about 100 for stars of spectral type B (cf.
also Lamers *et al.*, 1980). For stars of larger masses, practically Of and O-type stars,
Vanbeveren and De Loore (1980) find $300 \lesssim N < 500$. Computations for such stars,
with $\mathfrak{M} \gtrsim 50\mathfrak{M}_\odot$ were made (De Loore *et al.*, 1978a, b; De Loore, 1980) assuming
$N=300$. The main results of De Loore *et al.* are shown in Figure 54 for stars up to

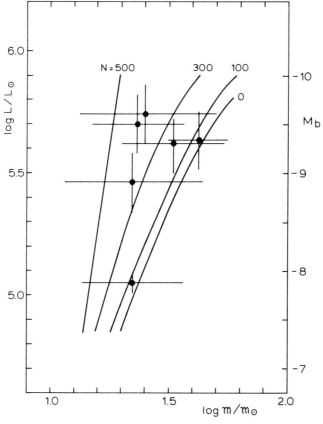

Fig. 53.   The masses and luminosities of the supergiants from Table IX compared with the predicted
mass-luminosity relation during the hydrogen shell burning phase, for different mass loss rates of
$0 \leqslant N \leqslant 500$. The length of the error bars corresponds to 1σ. (From De Loore *et al.*, 1977).

$50\mathfrak{M}_{\odot}$; for heavier stars they are given in our Figure 1. Figure 54 is based on $N=300$. Isochrones for time-steps of 10^6 yr are given (dashed); dotted are lines of equal mass. In later work De Loore *et al.* (1979a) extended these calculations to stars in the range $20-120\mathfrak{M}_{\odot}$, using the empirical $-\dot{\mathfrak{M}}$ values found for massive stars. Along similar

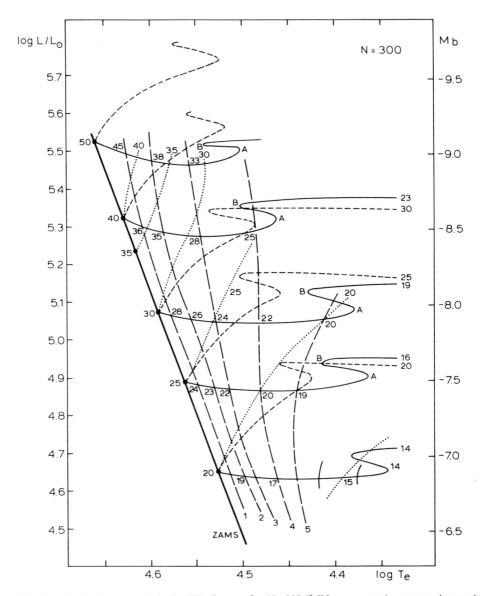

Fig. 54. Evolutionary tracks in the HR-diagram for $N=300$ (full heavy curves), conservative tracks (dashed curves) and equal mass curves (dotted curves) labelled with their masses (calculated by De Loore *et al.*, 1977). The numbers on the mass losing tracks indicate the changing mass during the evolution. The ZAMS is indicated, and also isochrones with time interval of 10^6 yr. The positions *A* and *B* mark the end of core hydrogen burning and start of hydrogen shell burning.

lines Stothers and Chin (1978, 1979) calculated evolutionary tracks using Carson's (1976) and Cox-Stewart's opacities, and assuming

$$-\dot{\mathfrak{M}} = kLR/\mathfrak{M},$$ (4.5; 1c)

where k is a (time-dependent) parameter. Their calculations were made for initial masses from 15 to $120\mathfrak{M}_\odot$. Figure 55 gives part of their results for the Cox–Stewart opacities. The importance of these calculations is that they are extended to the end of core He-burning, and thus enable one to discuss fundamental problems such as the absence of very massive red supergiants, the occurrence of WR-stars, etc.

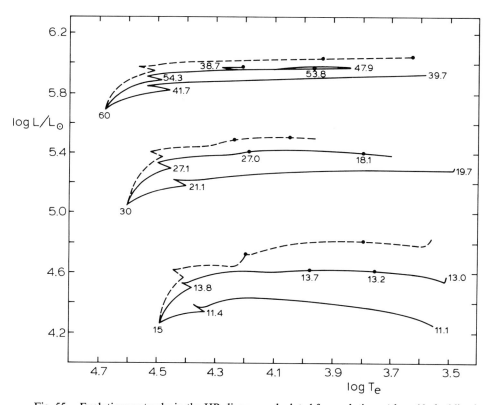

Fig. 55. Evolutionary tracks in the HR diagram calculated for evolution *without* (dashed lines) and *with* mass loss (solid lines). The Schwarzschild criterium was used throughout. The tracks refer to $k=0$, 1×10^{-11} and 3×10^{-11} (see Equation 4.5; 1c) in decreasing order of luminosity. Heavy dots mark the beginning and end of the stages of core helium burning. The tracks are labeled with the stellars masses in solar units. (After Stothers and Chin, 1979).

In parallel to these various investigations Massevitch *et al.* (1979) made calculations assuming

$$-\dot{\mathfrak{M}} = \alpha L/v_{esc}c,$$ (4.5; 1d)

an expression introduced by Lamers *et al.* (1976). The calculations were made for massive ($\geq 10\mathfrak{M}_\odot$) stars and different α-values. A comparison with observed rates of mass loss yield best agreement for $\alpha=0.3$.

Similar, but less detailed computations were made by Dearborn and Eggleton (1977) who used the tentative formula

$$-\mathfrak{M} \sim LR, \tag{4.5; 1e}$$

hence somewhat similar to the expression (4.5; 1c). The results for stars with $\mathfrak{M}=32\mathfrak{M}_\odot$ and $16\mathfrak{M}_\odot$ are in good accord with those of De Loore *et al.* for suitably chosen values of the proportionality constant.

Other computations of evolution with mass loss were made by Sreenivasan and Wilson (1978). Used expressions for the mass-loss were those by Lamers *et al.* (1976), Equation (4.5; 1c) where several expressions were used for the efficiency parameter α; for larger rates of mass-loss an expression was used given by Castor *et al.* (1975). The mass-loss chosen in that treatment is more based on theoretical considerations than in the semi-empirical discussion of De Loore *et al.* (1978a, b). The obvious disadvantage is a slightly worse agreement with the observed masses and luminosities.

The assumption that the efficiency factors N, k, or α would be constant during the whole evolution of one and the same star is – of course – not correct; see e.g. Section 7.4. However, it appears that during the hydrogen core burning phase the efficiency factor can be assumed to be constant. During the subsequent hydrogen shell burning phase the evolution is so rapid that no appreciable mass fraction is lost so that the evolutionary tracks are very similar to those of 'conservative case' stars.

De Loore (1980) gave a complete review of evolutionary calculations with mass-loss published up to 1978.

The question then arises whether computations such as those described can help to solve a few of the most pregnant problems of stellar evolution, such as those of:

– the WR stars (luminous He stars of relatively small mass);
– the CNO stars (massive stars with abundance anomalies in C and N);
– the absence of very luminous and massive ($>25\mathfrak{M}_\odot$) red supergiants.

The first problem, whether WR stars can be explained as a consequence of mass-loss from *single* star, and the associated problem whether WR stars may have a *binary* origin will be discussed in Section 4.12. It will be explained there, that WR-stars can result from binaries but also from single stars. The CNO stars can be explained as resulting from the evolution of single stars with mass-loss. This will be described here.

For the explanation of the *CNO stars* Dearborn and Eggleton (1977) used the fact that at the end of the core hydrogen burning phase of a $32\mathfrak{M}_\odot$ star the C abundance is increased in the inner 70% of the stars' mass and is not changed in the outer parts. If the total mass-loss would be $>0.3\mathfrak{M}_*$ then such abundance changes could come to the surface. These results were confirmed by De Loore *et al.* (1979a): assuming the large-$\dot{\mathfrak{M}}$-values found for the fairly massive stars of 60–$100\mathfrak{M}_\odot$ it appears – see also Figure 1 – that these can indeed lose a large fraction of their mantle (up to $0.5\mathfrak{M}_*$). Thus for a star with $\mathfrak{M}_{init}=100\mathfrak{M}_\odot$, and $-\dot{\mathfrak{M}}=2\times10^{-5}\ \mathfrak{M}_\odot\ yr^{-1}$ (typically: an Of star) the atmospheric composition changes from $X=0.7$; $Y=0.27$ to $X=0.36$, $Y=0.61$, while also the products of the CNO bi-cycle reach the atmosphere, at the end of core-hydrogen-burning. At that time the star has still a fairly high T_e-value; hence at least some of the OBCN stars can be explained by mass-loss of massive stars.

Another problem arises in this connection by the observation that stars with $M_b < -10$ and $\log T_e < 4.25$ are nearly absent in the HR diagram. It looks reasonable to assume that this could be due to strong mass-loss during the H *shell* burning phase since in that case the evolutionary tracks would move down- and right-ward in the HR-diagram. But the large rates of mass-loss needed to explain this phenomenon that way have not been observed. However, if there would be a very large rate of mass-loss during the H *core* burning phase, the stars would not finish their evolution toward the red giant phase but would eventually move leftward toward the Helium-ZAMS (De Loore *et al.*, 1977; Stothers and Chin, 1979). This however, would demand mass-loss ratios $> 10^{-5} \mathfrak{M}_\odot \, \mathrm{yr}^{-1}$ which has so far been observed for a very few stars only.

For the *M-type supergiants* the rate of mass-loss (Table XLII) is of the order of $10^{-6} \mathfrak{M}_\odot \, \mathrm{yr}^{-1}$; hence mass-loss is an important feature of such stars. With these values (Section 7.4) one finds appreciable fractional loss in the time estimated for the star to stay in that part of the HR-diagram. Consequently, during the evolution of red supergiants, such stars would slip onto the evolutionary tracks for stars of smaller masses. This effect may explain why very massive red supergiants with $\mathfrak{M} \gtrsim 25\mathfrak{M}_\odot$ are lacking. This explanation is related to the one discussed in our Section 1.6, where the possibility was considered of introducing an 'Eddington limit' due to turbulent energy dissipation. A quantitative comparative study of these two effects may prove to be important.

For completeness sake it is useful to mention a third possible cause for the absence of very luminous red supergiants: dust clouds around such stars will lower the T_e-value (not the L-value) and shift the star to the very low T part of the HR diagram. Such stars may then appear as infrared sources (Massevitch *et al.*, 1979).

4.6. Stellar Abundances and Nuclear Reaction; the Influence of Thermal Pulses

In order to understand the 'anomalous' abundances, as they are observed in the photospheres of – mainly – evolved stars, particularly in red giants and supergiants, one has to realize that:
– normally, nuclear reactions occur in stellar *interiors*, and their products will not reach the surface, unless:
– brought there by violent convective motions;
– or by intense large scale motions: meridional circulation;
– or they may appear when the outer mantle is shed off by strong mass-loss;
while it has further to be realized that:
– the CNO-bi-cycle predicts a certain equilibrium distribution of elements, which can be changed by reactions taking place at higher temperatures.

With regard to the latter aspect the *basic* CN cycle only involves the isotopes of C and N, and ^{15}O. It does not produce the other O isotopes. However, if O is initially present, as is presumably the case with second- and later-generation stars, ^{16}O and ^{17}O become also involved; because the reactions $^{16}O(p, \gamma)^{17}F(e^+\nu)$ $^{17}O(p, \alpha)^{14}N$ feed additional ^{14}N nuclei into the cycle. Furthermore, there is a small probability that ^{16}O is produced by $^{15}N(p, \gamma)^{16}O$. See further Table XXXIII.

TABLE XXXIII

The CNO-bi-cycle

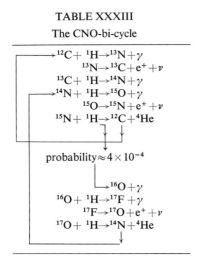

The equilibrium distribution of the most important isotopes involved in the H-burning reactions is shown in Figure 56. For a detailed discussion of the equilibrium problem reference is made to Clayton (1968), pp. 390ff.

There is, however, one aspect of the equilibrium problem that has to be touched here. The equilibrium distribution is a function of the various reaction rates. For a very stable element it takes a long time to arrive at equilibrium. It may even happen that the time needed for establishing the equilibrium distribution of an element may exceed the time during which the reaction concerned is principally operative. The clearest example is ^{16}O in the CNO cycle: at a temperature of 20 MK the equilibrium time τ_{eq} is approx. 10^8 yr so that in massive stars (which need much shorter time for their evolution) the equilibrium distribution as given by Figure 56 is not reached.

To illustrate this we show in Figure 57a the variation with temperature of τ_{eq} for a number of the intermediate elements in the H-burning reactions. Only for high temperatures, $T > 30$ MK, would equilibrium be reached in virtually all cases. However, even then the influence of convective mixing has to be carefully considered. As soon as large convective regions are developed, and when there is a chance for the elements involved to stay outside the nuclear burning region for a considerable fraction of the time the *effective* value of τ_{eq} would increase considerably. Quantitative treatment of that case is difficult since it demands a reliable discussion of convective mixing times in the central parts of stars.

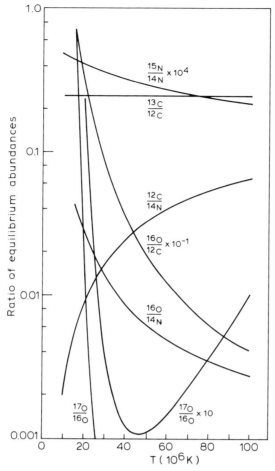

Fig. 56.   Ratios of abundances of CNO nuclei for a cycle operating in equilibrium. (After Caughlan and Fowler, 1962).

With respect to the variation of the element abundances in stars during the evolution we refer to Figure 57b which – albeit for a simplified case – shows the variation of mass-abundances with time during CNO H-burning, for an assumed constant temperature of 30 MK. Interesting is, for example, the considerable decrease of the C-abundance with time. Would this explain why certain M supergiants are C-deficient?

The crucial and still essentially unsolved problem is how gas with deviating abundances can reach the photospheres of evolved stars. In our previous section we showed that in certain cases mass-loss can shed off part of the outer mantle, but in most cases the region where nuclear reactions take or took place is thus not brought to the surface. In a search for other effects, initial attempts were based on the finding by Schwarzschild and Härm (1967) that after some time a He-shell may become thermally adiabatically unstable. Qualitatively, this is easily understood: when the shell pro-

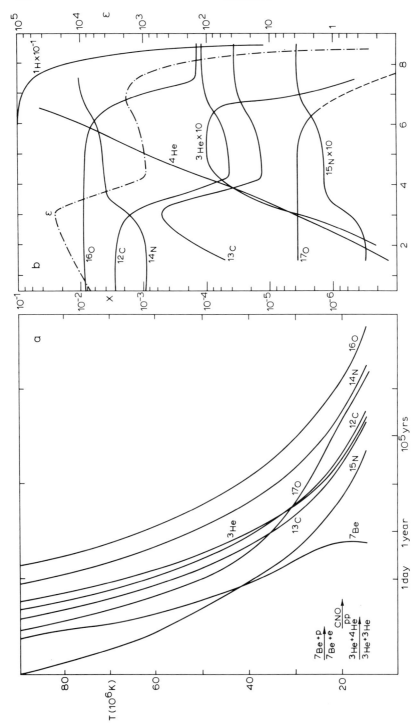

Fig. 57. (a) The variation with temperature of τ_{eq} (abcissa, logarithmic scale) of intermediate elements in H-burning reactions. (From Castellani and Sacchetti, 1978). (b) The typical variation with time of the abundance by mass of intermediate elements during H-burning. Constant temperature ($T = 30 \times 10^6$ K) and density ($\varrho = 1$ g cm^{-3}) are assumed. The dash-dotted line shows the corresponding variation of the expected output of energy (ε in erg g^{-1} s^{-1}). The abcissa gives the log of the time in years. (From Castellani and Sacchetti, 1978).

gresses outward the temperature gradient tends to increase. As a consequence a fairly extended outer convection zone originates which, finally, extends into those layers that contain hydrogen. Through convective mixing the H interacts with the He which leads to a radiation flash followed by an expansion of the outer layers, and a return to radiative equilibrium. The phenomenon is called a *thermal pulse* or a *thermal flash*. The first name is to be preferred because of the possible confusion with the He- and C-flashes. The pulses occur particularly when the shells are so thin that even a large expansion (due to the increased nuclear energy yield) does not decrease the pressure of the overlying layers, so that the flash does not immediately extinguish. After some time the process repeats itself; hence these 'thermal pulses' are a quasi-periodic phenomenon, that takes place in late phases of stellar evolution. Iben (1976) made numerical calculations for a $7\mathfrak{M}_{\odot}$ star with a carbon-oxygen core of $\sim 0.95\mathfrak{M}_{\odot}$, and found that the amount of matter containing fresh ^{12}C depends on the intensity of the pulse and ranges between $3 \times 10^{-5}\,\mathfrak{M}_{\odot}$ and $4 \times 10^{-4}\,\mathfrak{M}_{\odot}$. Whether such pulses have real influence on the chemical composition of the star's photosphere depends obviously on the total mass in the regions where the pulses occur. Sweigart (1974) and Fujimoto (1977a) made numerical calculations and found that the helium-shell-pulses generally do not cause sufficient mixing, the convection zone being hardly changed. The mechanism is only effective if the fraction of stellar mass contained in the outer hydrogen-rich envelope is as small as 10^{-5} to 10^{-6}. Hence, this mechanism operates only in very special cases.

It also does not seem that the *plume mixing* introduced by Scalo and Ulrich (1973) helps in this matter. Plume mixing would occur in the two-shell phase, when convection driven by thermal pulses from the innermost shell would bring C-enriched matter from that shell into the hydrogen shell. This would cause increased CNO-cycle burning, since that cycle is controlled by the ^{12}C-constant. It was suggested that this would cause enhanced meridional circulation, which would bring reaction products to the surface. See also Kippenhahn (1974).

But there is another possibility: cool giant stars have very deeply extended *outer convection zones*, and the ^{12}C produced by He-shell burning could be brought to the surface if the convection zone extends deep enough, thus causing an increase of the photospheric ^{12}C/^{13}C ratio.

Another way to transport nuclear reaction products upward would be by *turbulent transport* (Schatzman, 1977). A turbulent diffusion coefficient $D = \mathrm{Re}^* \nu$ is defined, where Re^* is a number similar to the Reynolds number, and ν the kinematic viscosity. Semi-empirically $D \approx 200$.

The first mechanism is perhaps only operative in red supergiants, and perhaps in the brightest N-type stars, but turbulent transport may apply to most giants (Genova and Schatzman, 1979), and thus one may assume that the fairly high ^{12}C/^{13}C ratio in N-type stars (see Table XXVII) is caused by ^{12}C addition from He-shell burning through turbulent transport in the stellar outer layers. The question is in any case worth to be investigated in more detail.

The production of heavier elements, by *s*-type nucleosynthesis, seems to be explainable by a mechanism developed by Iben (1975, 1976). At temperatures exceeding 3×10^8 K neutrons are produced in the convective zone encompassing the helium-burning shell through the reactions $^{14}N(\alpha, \gamma)^{18}F(\beta^+ \nu)^{18}O(\alpha, \gamma)^{22}Ne(\alpha, n)^{25}Mg$. The neutrons are absorbed to produce the s-type elements. In the declining phase of a thermal pulse the convective part of the (outer) hydrogen-rich envelope may penetrate downward till deep into the helium region, and thus may carry the s elements and carbon to the stellar surface. Fujimoto et al. (1976) and Fujimoto (1977b) showed that the mechanism works only for stars with a fairly large carbon core ($\mathfrak{M}_{core} \gtrsim \mathfrak{M}_\odot$). They suggest that this mechanism may explain the N-type (hence the most luminous) carbon stars, and the most luminous S-type stars.

In addition, Sweigart and Mengel (1979) have suggested to consider the influence of large-scale meridional circulation to bring CNO-processed material from the vicinity of the H-shell into the envelope of a red giant star. Semi-quantitative research shows that this suggestion is promising but it has still to be further worked out.

In *conclusion*, and with reference to the data summarized in Tables XXVI through XXIX we may state that the abundances in N-type carbon stars can be understood reasonably well, at least in a qualitative way. The R-type stars offer no special problem as far as their $^{12}C/^{13}C$ ratio is concerned – it agrees fairly well with the CNO-stability abundances – but the apparent general overabundance of C in Carbon stars remains to be explained. The $^{12}C/^{13}C$ ratio in H-deficient carbon stars, its relation to the CN-strength in K-giants, and the general problem of the deficiency of hydrogen in certain stars are other open problems.

4.7. The Mira Stars

These long-period variables take a key-function in the HR-diagram because their variability may help to better fix their position in the diagram, *via* theories on pulsational instability of giant stars, and also because they may be the progenitors of planetary nebulae (and – hence? – of white dwarfs). Whether these stars are related to supernovae is not very likely because of their small masses.

The Mira's have periods of variability P between ~100 and 650d. The distribution function of periods, $dn/d(\log P)$ is bell-shaped with a flat maximum near $P = 380d$, and a full width at half maximum of 200d. This information, and the reasonable assumption that Miras have finished He-core burning and are in the double-shell-source phase of evolution and are, hence, ascending the giant branch for the second time, formed the basic data for Wood and Cahn (1977), and Cahn and Wyatt (1978a) to determine the *parameters of Mira stars* through a *diagnostic diagram* for the analysis of Mira's in the HR diagram.

Essentially, the Mira star parameters are found as follows. In a diagram of stellar masses $\mathfrak{M}$ against luminosity $\log L$, the stellar evolutionary tracks are in first approximation horizontal lines, which bend slightly downward at the large L-side, because of mass-loss (Figure 58). In this diagram the lower limit of possible masses is a curve giving the core-mass *versus* luminosity for degenerate cores as calculated by Paczynski

(1971c); see also Section 4.8 and Figure 59. The mass-loss function assumed in calculating the tracks in Figure 58 was Reimers's

$$-\dot{\mathfrak{M}} = 10^{-13}\,(L/L_\odot)\,(R/R_\odot)\,(\mathfrak{M}_\odot/\mathfrak{M}),$$

which is our Equation (4.5; 1c) but with a constant, five times smaller than the value proposed by Reimers. The value of the constant used here was found by trial and error: with Reimers's value the Miras with long periods ($P \approx 600d$) should have masses around 2.5 or $3\mathfrak{M}_\odot$ which was considered too much.

Next, the periods of variability P are discussed. It was shown by Keeley (1970),

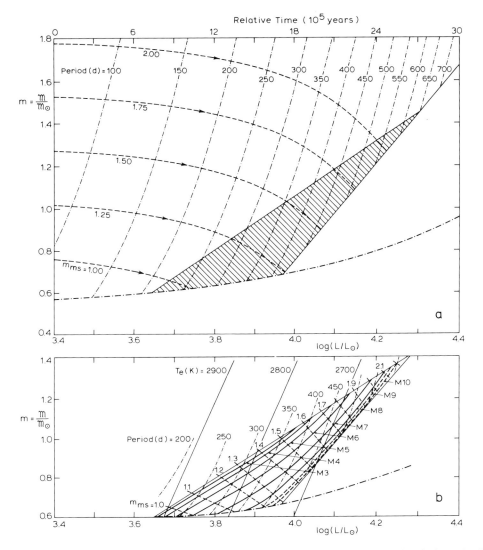

Fig. 58.   (a) Semi-theoretically derived position of Mira-instability region (hatched area). (b) Diagnostic diagram for the analysis of Mira-stars. (From Cahn and Wyatt, 1978a).

Langer (1971), and Wood (1975) that the observed periods are those from first-over-tone pulsations. Linear, non-adiabatic eigenvalue calculations for stars in the extreme red giant branch, made by Wood and Cahn (1977) show that for these stars the pulsa-tional constant

$$Q = P[(\mathfrak{M}/\mathfrak{M}_\odot)/(R/R_\odot)^3]^{1/2} = 0.04d. \qquad (4.7; 1)$$

The value of the constant Q is essential for a description of the state of pulsation of Mira stars. It is therefore gratifying that Stothers and Leung (1971) found a similar value from a discussion of Mira stars with reasonably well known masses and lu-minosities; they derived $Q=0.06d$. For Mira stars we will therefore assume $Q=0.04$ to 0.06d.

The expression (4.7; 1), with the assumed value $Q=0.04d$ defines the lines of con-stant P in Figure 58a. If we next *assume* that Paczynski's lower limit represents one boundary of the Mira instability region, and – just for simplicity – that the other boundaries can be represented by straight lines in the ($\mathfrak{M}$, log$L$) diagram, a com-parison with the distribution function of periods yields such a straight line and de-termines the hatched region in Figure 58a.

The way to obtain this result is undoubtedly beset with some assumptions that need further testing, and the result of Figure 58a is surely not unique; indeed Tuchman *et al.* (1979) find slightly different results along a similar way. But the approach is promising: the possibility to use Figure 58 as a *diagnostic diagram* for the analysis of observations of Mira stars is obvious, particularly if following Cahn and Wyatt (1978a) the relation between period P and spectral type at maximum is introduced in the diagram (Figure 58b); knowledge of a star's period and spectral type at maxi-mum define a star's mass and L-value, while also the time may be derived, spent as a Mira-variable.

What happens after the 'Mira-phase' is still an open problem. There are observa-tional indications – see Section 3.12 – that nuclei of planetary nebulae originate from Miras. Tuchman *et al.* (1979) who investigated the pulsational stability of red giants found that these pulsations – which resemble those of Miras – may eventually di-verge. Thereafter the star may lose a considerable amount of matter which would make it possible to develop into a planetary nebula.

4.8. The Carbon Burning and Later Phases

When He is fully exhausted in the star's core, and when during the subsequent phases of gravitational core contraction the central temperature has increased up to ~0.3 GK (GK=Giga-Kelvin=10^9 K) carbon reactions may start. The star has at that moment arrived in the red giant-supergiant part of the HR-diagram, depending on its mass. For stars with masses of 3 to 15$\mathfrak{M}_\odot$ C-burning starts at about $L\approx 3\times 10^4 L_\odot$; log$T_e\approx$ 3.5 (computed without mass-loss).

A very important fact is that the physical conditions in the core depend strongly on the stellar mass. As Paczynski (1970c) and Alcock and Paczynski (1978) showed by evolutionary computations without mass-loss, stars with $\mathfrak{M} \gtrsim 10 \mathfrak{M}_\odot$ have non-degenerate cores at carbon ignition, while less massive stars have degenerate cores, Figure 59. It is also important to notice that carbon-ignition starts *approximately* at the moment at which the nuclear burning rate equals the neutrino energy losses (dotted line). A consequence of the core degeneracy is that for stars with masses $\gtrsim 10 \mathfrak{M}_\odot$ no 'carbon flash' will occur in the centre, but the flash may be important for less massive stars.

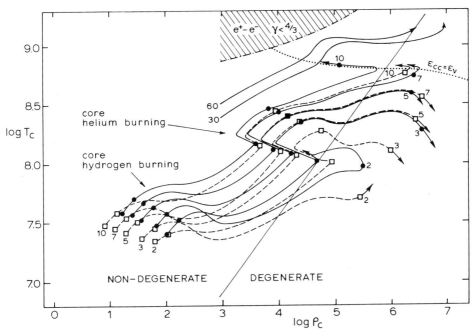

Fig. 59.   The evolution of the central temperature and central density for stars with $Z = 0.03$ (dashed lines and open squares) and stars with $Z = 0.0004$ (solid lines and filled circles). The masses of the stars are written beside the initial and final models. The locus along which carbon burning supplies as much energy as neutrinos carry away is marked with a dotted line ($\varepsilon_{cc} > \varepsilon_\nu$ above the line). A straight line marks the points where gas pressure $\approx$ electron degeneracy pressure. In the hatched upper left area the star is dynamically unstable. The arrows mark the expected direction of future evolution. (After Alcock and Paczynski, 1978; and Sugimoto and Nomoto's review, 1980).

In stars in which carbon reactions occur the number of other possible nuclear reactions involved becomes large, while many of the relevant physical parameters are still uncertain: two reasons for leaving the results of the computations unreliable. Arnett (1972) has given arguments that the main nuclear reactions during the C-burning phase could be restricted to $^{12}\text{C}(^{12}\text{C}, \alpha)^{20}\text{Ne}$; $^{16}\text{O}(\alpha, \gamma)^{20}\text{Ne}$, and $^{20}\text{Ne}(\alpha, \gamma)^{24}\text{Mg}$, hence neglecting reactions such as $^{12}\text{C}(\alpha, \gamma)^{16}\text{O}$ and $^{24}\text{Mg}(\alpha, \gamma)^{28}\text{Si}$.

With still further increasing temperatures more and more nuclear species enter into the picture and the problem finally becomes of an overwhelming complexity, the results accordingly more unreliable. Another aspect is that of the contribution of the neutrino reactions: for high temperatures the neutrino-anti neutrino pair producing process will act as an energy sink in the central part of the star, and thus will be important or even crucial for the star's life history. Indeed, Weaver *et al.* (1978) who were the first to study the evolution of massive stars (15 and $25\mathfrak{M}_\odot$) till iron-core collapse find – confirming Paczynski, Figure 59 – that starting with the carbon burning phase neutrino losses ($\sim 10^{39}$ erg s^{-1}) exceed the optical luminosity. In the silicon burning phase the neutrino energy flux is even $\sim 10^6$ times the radiation flux!

The carbon burning phase of stars with the smaller masses (typically $3 < \mathfrak{M}/\mathfrak{M}_\odot < 10$) was first studied by Hayashi and Cameron (1962), Stothers and Chin (1968), and Paczynski (1970c). Carbon burning in stars with *relativistic* electron degeneracy was dealt with by Ergma *et al.* (1976). For these stars, for which the degenerate core may have a mass of $\sim 1.4\mathfrak{M}_\odot$ the evolution has been examined by Ivanova *et al.* (1977); although partly based on now obsolete results of Paczynski (1970c), the main conclusions remain valid. For the range of central densities $2 \times 10^9 < \varrho_c < 3 \times 10^{10}$ g cm^{-3} carbon burning starts with a carbon flash. The above lower limit represents the core density limit for the evolution of a single star in that mass-range, the upper limit is the Chandrasekhar density limit for carbon stars. For stars in the range of these core densities the study of the hydrodynamics of the flash process suggest that the main consequences will be the collapse of the core that will be surrounded by a tenuous H, He-mantle; the star will appear as a red giant. Whether the collapse will be accompanied by mass-ejection at a relatively large rate is still uncertain.

More advanced stages of stellar evolution are characterized by the burning of neon (Arnett, 1974a), oxygen (Arnett, 1974b), and silicon (Arnett, 1977). In the last case the central temperature and densities have values of approx. $T_e = 3$ to 5×10^9 K and $\varrho_c = 3 \times 10^7$ to 3×10^8 g cm^{-3}. Weaver *et al.* (1978) have started computations on the evolution of massive stars from the ZAMS through iron core collapse.

Salpeter (1971) has computed theoretical evolutionary tracks for stars of ~ 1 solar mass, consisting mainly of carbon and oxygen, for various masses and mantle helium abundances – see Figure 60. The computation suggests that during part of their evolution certain very evolved stars – presumably the central stars of planetary nebulae – may attain considerable bolometric brightness and high effective temperatures before evolving into white dwarfs. Such stars would be best observable in the far UV region of the spectrum.

The problem which stars should be identified with theoretical stellar models in the carbon burning phase has only been touched so far. The paucity of very massive red supergiants (Stothers, 1972b) shows that the carbon burning and later evolutionary phases most be more rapid than calculated: the only possible mechanism to accelerate the nuclear processes seems neutrino emission. The interesting fact that there are no C- and S-type massive supergiants may also be due to neutrino emission. An attempt to determine the position of the Mira variables of spectral type S in the HR diagram is due to Motteran (1972). The observation that the *brightest* Mira-type S-stars have L-values of $\sim 10^5$ to 10^6 $L_\odot$ suggests that these stars occur in the upward tracks leading to C-core burning fairly massive stars. In view of the fact that S-type variables

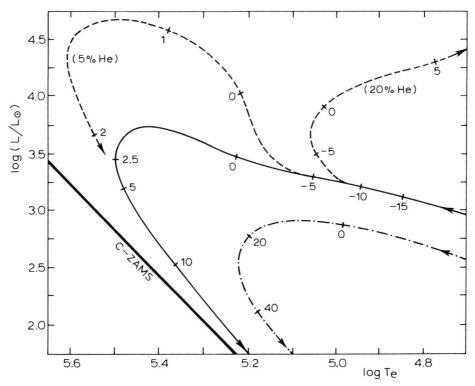

Fig. 60.   The theoretical L-T$_e$ diagram for stars consisting mainly of carbon and oxygen. The upper curves are for total mass 1.02$\mathfrak{M}_\odot$: the solid curve with no helium envelope, the two dashed curves for stars with a helium envelope containing 5% and 20% of the mass, respectively. The dash-dot curve is for a 0.75$\mathfrak{M}_\odot$ star without helium envelope. The evolutionary age (from an arbitrary zero point) in units of 10^4 yr is shown at a few places along each track. (After Salpeter, 1971).

are characterized by anomalous abundances in their atmospheres, this identification does not seem unreasonable. The interpretation of the deviating abundances may be related to the observation (see Figure 40) that in the HR-diagram the peculiar red giants of types R5–R8, N, S, and MS lie to the right of the line marking the onset of the helium shell flash, which suggests a relation between the helium flash and the abundance peculiarities.

4.9. Supernovae

Theories on stellar interiors show that stars with final masses exceeding a value of approx. 1.4 to 2.5$\mathfrak{M}_\odot$ cannot stabilize after the exhaust of nuclear energy sources and must collapse to a compact object (Oppenheimer and Snyder, 1939). It is assumed that this collapse gives rise to the supernova phenomenon. However, the rate of supernova occurrence seems to be considerably smaller than that of the formation of stars with $\mathfrak{M} > 1.4\mathfrak{M}_\odot$ so that supernova progenitors must be massive stars that have lost mass (see Chapter 7 for stellar mass-loss and Chapter 9 for a review of supernova properties).

The question what stars really become supernovae will be dealt with in our Section 9.11, where we consider the various pieces of information available on masses and stellar population types of pre-supernovae, on the basis of data such as statistics of stellar birth-rates, supernova, and of the end-products: pulsars, and galactic supernovae remnants. Anticipating: type II supernovae arise from massive stars with approximate masses between 5 and $10\mathfrak{M}_\odot$, and type I pre-supernovae have masses between 1 and $3\mathfrak{M}_\odot$.

The *classical scenario for the origin of a supernova* is derived by considering the processes that may take place in very evolved stars, and which can be schematized, from Figure 46, by the following crude statements:
- at temperatures of 0.5 GK: C, O and perhaps Ne start to react;
- at $T \approx 1.2$ GK: reactions occur like $2\,^{16}O \rightarrow {}^{28}Si + {}^4He + energy$;
- at $T \approx 2$–3 GK: reactions occur like $2\,^{28}Si \rightarrow {}^{56}Fe + energy$.

All above reactions are exoergic reactions: the synthesis of heavy nuclei from lighter yields energy. This is no longer the case for nuclear synthesis involving nuclei of and above $\sim^{56}Fe$, since nuclear binding energy reaches a maximum value for ^{56}Fe. So it seems that in an end phase of evolution stellar cores should consist of Fe. However, during the subsequent contraction of that core the temperature will rise, and for temperatures ≈ 5 GK the flux of γ-radiation in the core of the star becomes so large, that the endoergic reaction $^{56}Fe + \gamma \rightarrow 13\,^4He + 4n$ would set in. This would lead to cooling and consequent collapse of the core; the consequent heating of the more outer layers causes that Si$\rightarrow$Fe reactions could set in, etc. It is believed that this sequence of reactions could give rise to the supernova phenomenon.

The above-described process is illustrated in the classical picture of Fowler and Hoyle (1964) – Figure 61 – and gives the evolution of a $30\mathfrak{M}_\odot$ star, which is thought to yield a type II supernova. The center of the diagram shows the presupernova; the type II supernova effects are shown on the right. It is mostly assumed that, immediately after the endoergic nuclear reaction, by some kind of braking action perhaps caused by rotation, the stellar envelope explodes; the explosive burning of oxygen is assumed to be the main source of energy in the supernova phenomenon. Unprocessed material, as well as the results of the various nuclear reactions would thereupon be shed out; a compact object should remain after the events. The scenario is crude but gives the basic elements.

This classical picture has been elaborated and greatly refined by many authors. In particular it is now clear that the final fate of a star depends crucially upon its initial mass.

Therefore, we first review what happens to stars of various initial masses; a review by Sugimoto and Nomoto (1980) contains specific details and many references.

(1) $\mathfrak{M}_{init} < 4\mathfrak{M}_\odot$. As shown in the preceding section (cf. Figure 60) such stars eventual transform into white dwarfs.

(2) $4\mathfrak{M}_\odot \lesssim \mathfrak{M}_{init} \lesssim 8\mathfrak{M}_\odot$. The mass of the C–O-core is smaller than the Chandrasek-

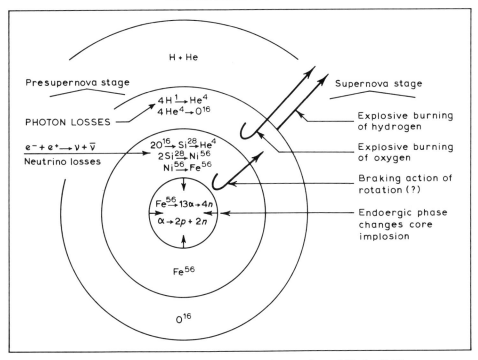

Fig. 61.   The supernova phenomenon. (After Fowler and Hoyle, 1964).

har limit ($\approx 1.4 \mathfrak{M}_\odot$) so that the electron-gas of the core is degenerated. When the core increases in size and contracts till $\varrho_c \approx 2 \times 10^9$ g cm^{-3} a carbon-flash starts the disruption of the star. During the subsequent T-increase also heavier nuclei are processed, up to iron-peak elements.

(3) $8 \mathfrak{M}_\odot \lesssim \mathfrak{M}_{\text{init}} \lesssim 12 \mathfrak{M}_\odot$. The electron-gas of the C–O-core is initially not degenerated, but degeneracy sets in when finally an O–Ne–Mg-core is formed. By electron capture heavier nuclei are formed. At $\varrho \approx 2.5 \times 10^{10}$ g cm^{-3} an oxygen flash starts. The flash is fairly weak, however, and does not seem to result in core-disruption. Rather, oxygen-burning starts in each shell, that is subsequently compressed to $\varrho \gtrsim 2.5 \times 10^{10}$ g cm^{-3}. It is not yet clear whether this process leads to supernovae as we observe.

(4) $12 \mathfrak{M}_\odot \lesssim \mathfrak{M}_{\text{init}} \lesssim 30$ to $40 \mathfrak{M}_\odot$. A central iron-core is formed. Collapse occurs by photodissociation of Fe-nuclei. When the star explodes the mantle is ejected; it contains elements up to Si. The upper limit of this range of masses is uncertain; the quoted values are from Barkat *et al.* (1967) and Fraley (1968). Sugimoto and Nomoto (1980) place the upper limit at $\sim 100 \mathfrak{M}_\odot$.

(5) Higher masses. Such stars become eventually unstable ($\gamma < \frac{4}{3}$) because of the creation of electron-positron-pairs (cf. Figure 59). The question is, however, whether such heavy stars ever reach the state in which dense cores can originate. It is most probable that stars $\gtrsim 120 \mathfrak{M}_\odot$ do not originate at all, while stars with $\mathfrak{M} \gtrsim 30$ to $40 \mathfrak{M}_\odot$ will lose a large part of their mass during their main-sequence evolution and shortly thereafter.

With this information it is now clear how the classical scenario of Fowler and Hoyle has to be modified. Since it is fairly certain that supernova progenitors have $\mathfrak{M} \lesssim$ $10\mathfrak{M}_\odot$ it is improbable that a supernova is caused by the formation and subsequent collapse of an Fe-core. In these less massive stars it is rather a C-flash in a degenerate C–O core that starts the explosion (Arnett, 1969; Bruener, 1971, 1972). The details of the process are difficult to handle quantitatively but it seems, after work by Chechetkin *et al.* (1980) that a C-flash in a C–O core with $2 \times 10^9 \lesssim \varrho \lesssim 9 \times 10^9$ g cm^{-3} would lead to disruption of the whole star and a kinetic energy loss of $W \approx 10^{50}$ to 10^{51} erg (type II supernova?), while a flash in cores with $9 \times 10^9 \lesssim \varrho \lesssim 3 \times 10^{10}$ g cm^{-3} would only lead to partial envelope ejection, with $W \approx 10^{49}$ to 10^{50} erg (type I?). More research is certainly needed.

Some authors have considered additional effects that might seem of secondary importance but that could be crucial. The possibility that angular momentum of a rapidly rotating neutron core is transferred to the envelope was suggested and elaborated by Bisnovatyi–Kogan *et al.* (1970, 1976). That the magnetic field of the star should be considered during the collapse phase has been argued by Leblanc and Wilson (1976) and in more detail by Meier *et al.* (1976).

It should be clear that the complicated situation in the interior part of the star where so many different nuclear reactions could occur will make it very hard to test the semi-qualitative picture by more detailed quantitative model-computations. However, the picture is supported by computations such as those of which the results are given in Figure 62, which shows the distribution of element abundances for a silicon gas that has been assumed to have burnt for 10 s at a temperature of 4.2 GK. About 35% of the original silicon appears to remain. The open dots give abundances of the various natural elements; isotopes are connected by dashed lines. The filled dots represent the natural solar abundances; their isotopes are connected by drawn lines. There is a remarkable similarity between the abundances of the most abundant nuclei which strongly suggests that silicon burning during short intervals of time in extremely hot stellar interiors can have been important, at least in those supernovae that were responsible for the production of the material out of which our primordial solar nebula has been formed.

'Direct' observation of the chemical composition of supernova ejecta would be another way to test the above described scenario. But supernova remnants are contaminated with swept-up interstellar gas. The best objects are therefor young remnants. It is interesting that the X-ray spectrum of Tycho's supernova-remnant does show overabundances by factors $\gtrsim 6$ for Si, S, and Ar, while Mg and Fe have solar-type abundances (Becker *et al.*, 1979a). This suggests significant Si-formation and ejection, and less or no Fe-ejection, an interesting result that confirms the suggestion that Fe-production is not preponderant in supernovae.

More attempts to 'observationally' determine the physical conditions inside a supernova have naturally been made several times, in various degrees of refinement. A detailed review of the implications of the 'r-process' abundances, by Hillebrandt (1978) shows the poor agreement between various authors. For the time being it seems that

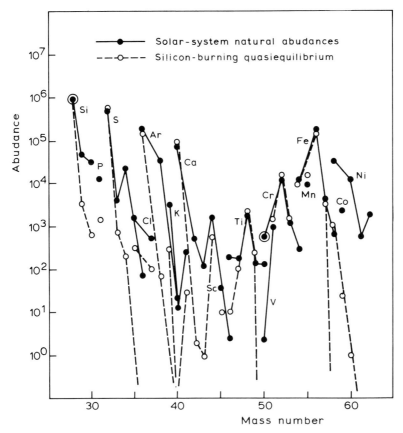

Fig. 62.   Comparison of abundances obtained after ^{28}Si-burning during 10 s in a 4.2 GK temperature environment. (After Bodansky *et al.*, 1968).

temperatures of a few times 10^9 K and neutron particle densities of $\sim 10^{28}$ cm^{-3} must have governed the origin of these elements.

4.10. Evolution of Close Massive Binaries

Binaries are common objects. Up to a distance of 20 pc from the Sun 690 single stars are known, and 475 stars that are objects of binaries or multiple systems. Hence, at least 41% of all stars are members of pairs or multiple systems. Any possible incompleteness of this statistics can only increase this percentage (Herczeg, 1963). Reviews on the evolution of binaries were given by several authors, including particularly Plavec (1968), Paczynski (1971a), Rees (1974), Kraft (1975), Van den Heuvel (1976). The consequences of mass-transfer were reviewed by Thomas (1977).

Since during several phases of their evolution stars increase in size, it may happen that a member of a binary system becomes large enough for matter to stream from that component to the other. This form of mass-loss can be much larger than the mass-loss of individual stars and this effect will result in a change of the orbital parameters of the binary, and – most important – in a loss of the outer stellar mantle so that eventually the photosphere of the primary star or even of the companion may contain material with a chemical composition that shows the results of nuclear processes that have taken place in the core. Thus the existence of stars with higher photospheric abundances of elements like C and N may be explained, as well as photospheres with $^{12}C/^{13}C$ ratios deviating from the 'normal' values, and in the extreme case one may thus even be able to explain the origin of objects such as pure helium stars or even pure carbon stars. In this process of mass exchange the notion 'Roche surface' is important (Section 4.10.1).

Close binaries often rotate synchronously with their revolution, because of tidal interactions between the components. For not too extreme conditions the rotational angular momentum of the stars can be neglected as compared with the orbital angular momentum. During the phase of mass-exchange the angular momentum of the individual stars may change. So the variation of the angular momentum during the evolution has to be considered (Section 4.10.2).

4.10.1. THE ROCHE SURFACE

The equipotential surfaces in a co-rotating coordinate system in a binary containing the first Lagrangian point L_1 is called the Roche surface; the two lobes around the members of the binary are the Roche lobes. The importance of the Roche surface for binary evolution was already mentioned by Kuiper (1941): if in the course of its evolution one of the two stars expands so far that it finally completely fills its Roche lobe, matter may stream through L_1 into the partly 'empty' Roche lobe surrounding the other star. Thus, mass-exchange between members of a binary may take place at a fast rate.

Kopal (1959) has tabulated values of the mean radius r_1 of the Roche lobe. Since the lobes are not spherical these values are found from the lobe's volume V_L with $V_L = \frac{4}{3}\pi r_1^3$. Paczynski (1966, 1971a), by introducing $q = m_1/m_2$ and $a =$ semi-major axis of the binary system, gave approximate expressions for r_1, with a 2% accuracy:

$$r_1/a = 0.46\,(q/(q+1))^{1/3} \qquad q < 0.8;$$
$$= 0.38 + 0.2\log q \qquad 0.3 < q < 20.$$

Very useful for numerical calculations are the expressions by Horn et al. (1969):

$$q > 0.1: r_1/a = 0.37771 - 0.20247\log q + 0.01838\,(\log q)^2 + \\ + 0.02275\,(\log q)^3,$$

$$q < 0.1: r_1/a = 0.3771 - 0.2131\log q - 0.0080\,(\log q)^2 + \\ + 0.0066\,(\log q)^3.$$

It has not always been fully appreciated that the notion of a closed Roche surface, common to the two stars, with its classical double-lobed 8-like shape, loses sense when one or both of the binary members produce a stellar wind with accelerating forces that are comparable to the stellar acceleration at the position of the Roche surface (Schuerman, 1972). McCluskey and Kondo (1976) have calculated the equipotential surfaces for such a case (the binary UW CMa, with an O7f star as the primary) and found – as may be expected – that the lobe opens behind the component with the smallest wind acceleration – see Figure 63. In these calculations the shadowing effect of the secondary star has been neglected. This was rectified in similar but more detailed calculations presented by Vanbeveren (1977b, 1978) who also considered asynchronous rotation. More important is however that in none of the calculations presented sofar the influence of the *line radiation force* (see Section 5.6) has been included. This force may be about one order of magnitude stronger than the force

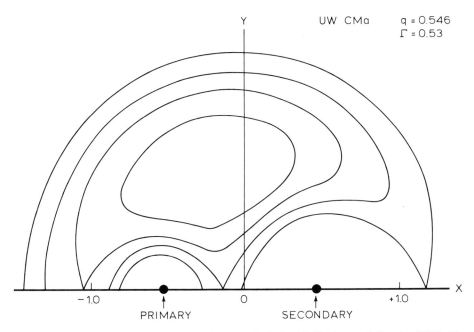

Fig. 63.   Equipotential surfaces computed for UW CMa by McCluskey and Kondo (1976). The primary is the O7f-component and is assumed to produce appreciable stellar wind acceleration. Note the opening of the Roche lobe around the secondary, as compared to the classical case with no stellar winds. Here q=mass ratio primary:secondary; and Γ=ratio of radiation pressure force to gravitation force for the primary star.

arising from the continuous radiation. But even the present obviously very preliminary calculations show that mass that would leave the star may enter into a circumbinary orbit instead of entering the region around the cooler component. This is one way to explain why several early-type or luminous binaries are surrounded by shells of gas (e.g. β Lyrae, and others). In addition, Drobyshevski and Resnikov (1974) found that in a system in which the stellar rotation and their orbital revolution is synchronous, about 40% of the mass expelled by the primary is lost through the *external* Lagrangian point. Furthermore, Amnuel and Guseivnov (1977) have shown that even in a 'closed' Roche lobe system more than 80% of the matter passing through the *internal* Lagrangian point may escape from the system, when the rotation of the primary is not synchronous with the revolution of the system as a whole.

Hence, it is *concluded* that in close binaries mass-exchange between the components can yield much larger $-\dot{\mathfrak{M}}$-values than those for single stars; part of this mass may enter into a circumbinary shell, while a considerable part of the exchanged matter will be totally lost from the system, by various possible effects. This *non-conservative* case is an important aspect of binary evolution. Its main aspects are described in our sub-Section 4.10.3.

4.10.2. EVOLUTION OF CLOSE BINARY STARS; THE CONSERVATIVE CASE

In an analysis of this topic distinction should be made between two cases: close-binary evolution *without* and *with* loss of matter and/or momentum to the system. The first case is the least common among *massive* binaries (the topic of this book). But it is important to discuss first the conservative case before one treats the more difficult topic of binary evolution with mass-loss.

In *the conservative case* one assumes conservation of mass:

$$\mathfrak{M}_1 + \mathfrak{M}_2 = \text{constant}, \tag{4.10; 1}$$

and conservation of angular momentum:

$$J^2 = Ga\frac{(\mathfrak{M}_1\mathfrak{M}_2)^2}{(\mathfrak{M}_1 + \mathfrak{M}_2)} = \text{constant}. \tag{4.10; 2}$$

From Equations (4.10; 1) and (4.10; 2) it follows that

$$a/a^0 = (\mathfrak{M}_1^0\mathfrak{M}_2^0/\mathfrak{M}_1\mathfrak{M}_2)^2, \tag{4.10; 3}$$

where superscripts ^{0} denote the initial situation. Equation (4.10; 3) shows that mass-loss from the (most massive) primary results in a decrease of a till mass-equality is reached. Thereafter a increases again.

The kind of evolution of a binary is greatly defined by the phase of the primary's evolution during which the Roche lobe gets completely filled. Following Kippenhahn and Weigert (1966, 1967a, b) and Lauterborn (1970) it is customary to define three evolutionary phases:

phase I: hydrogen core burning; the star stays virtually on the main sequence and expands only little;

phase II: hydrogen is exhausted in the core; the core of the star contracts but the star as a whole expands, becomes redder in colour, and moves rightward in the HR-diagram;

phase III: helium core burning and later phases.

In binary evolution three cases are considered, called cases A, B, and C and defined according to whether the primary star fills its Roche lobe in phase I, II, or III respectively. Case B is the most common; about 80% of the O5–F binaries evolve according to this case. This is not surprising: phase II lasts longer than phase III; in phase I the star expands hardly or not at all!

The values of masses and radii in the three cases are described in an illustrative diagram after Thomas (1977), Figure 64. The diagram is based on theoretical computations of binary evolution, and gives radii versus masses. Shaded is the main-sequence band, which contains the case A objects. The zig-zag line gives the boundary between cases B and C: it is at this line that He ignition starts. The dashed lines indicate orbital periods at the beginning of mass transfer for an assumed mass-ratio of 2:1.

The *significance* of the subdivision is partly also illustrated by Figure 64: the various evolutionary types result in different kinds of binaries. The evolutionary *case*

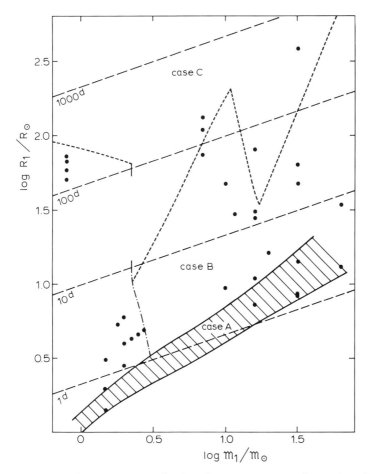

Fig. 64.   Distribution of radius and mass for the primary components of a number of binaries. The main-sequence band is shaded. Ignition of helium takes place at the zigzag line; to the left of the dash-dotted line mass transfer in case B will produce a helium white dwarf. Dashed lines indicate periods at the beginning of mass transfer for a mass-ratio of 2:1. (After Thomas, 1977).

A yields Algol-type close binaries. What is obtained in case **B**, depends on the mass of the primary star. The *massive case B* (in which we are mainly interested in this book) is defined by Plavec (1968) as the case with a primary mass $> 9\mathfrak{M}_\odot$. Kippenhahn and Weigert (1967a) (see the dash-dotted line in Figure 64) locate the boundary at smaller masses: $\approx 3\mathfrak{M}_\odot$. For smaller masses than this limit the mass transfer in case **B** will eventually produce a helium white dwarf. For the larger masses the end of the mass transfer comes with the start of the helium burning process, when the star's envelope starts to contract again. At that moment the greater part of the envelope may have already been transferred to the secondary, and the remainder of the primary may be just a pure helium star. Such a case was first studied by Kippenhahn (1969), who investigated the evolution of a pair consisting of a $25\mathfrak{M}_\odot$ star and a $15\mathfrak{M}_\odot$ star. It appeared that after 4.7×10^6 yr the primary would fill its Roche lobe and thereafter would transfer $16.5\mathfrak{M}_\odot$ to the secondary in only 7000 yr. The period of orbital revolution, initially $7^d.8$, increases to $20^d.3$. In these and later calculations of mass exchange made sofar it is generally assumed that all matter that would move out of the Roche surface is immediately transferred to the other object; no matter would be lost from the system, or brought into a circumbinary envelope. We know that this assumption is generally not correct.

A summary of theoretical models of the evolution of massive binaries in case **B** is given in Table XXXIV. See also the reviews in Webbink (1975), De Grève (1976), De Grève and De Loore (1976b, 1977). As an example we give in Figure 65a evolu-

TABLE XXXIV

Theoretical model computations of the evolution of massive binaries for case B of mass transfer

Initial masses (in $\mathfrak{M}_\odot$)		Reference
7	6	Plavec *et al.* (1973)
	4.5	
9	3.13	Kippenhahn and Weigert (1966, 1967b)
10	9.4	Tutukov *et al.* (1973), Massevitch *et al.* (1976)
	8	De Greve and De Loore (1976b), De Greve *et al.* (1978)
	3.7	Ziolkowski (1976)
10, 12, 13, 15	8	De Greve and De Loore (1977)
15	8	De Greve (1976), De Greve *et al.* (1978)
16	15	Tutukov *et al.* (1973), Kraitcheva (1974, 1978), Massevitch *et al.* (1976)
	10.67	Paczynski (1967)
20	14	De Greve (1976), De Loore and De Greve (1975)
	10	De Greve (1976)
	8	De Greve (1976), De Greve *et al.* (1978)
	6	De Greve (1976)
	4	De Greve (1976)
25	15	Kippenhahn (1969), De Greve *et al.* (1978)
30	24	De Greve (1976), De Greve *et al.* (1978)
	16	De Greve (1976)
	8	De Greve (1976)
32	30	Tutukov *et al.* (1973), Kraitcheva (1974, 1978), Massevitch *et al.* (1976)
	2	Massevitch *et al.* (1976)
64	60	Massevitch *et al.* (1976)

tionary tracks of the system $(20\mathfrak{M}_{\odot}+14\mathfrak{M}_{\odot})$ in the HR-diagram. In the figure the letters A through F refer to: A: initial model; B: start of mass-loss; C: minimum luminosity; D: He-ignition; E: end of mass-loss; F: C-ignition. An important feature

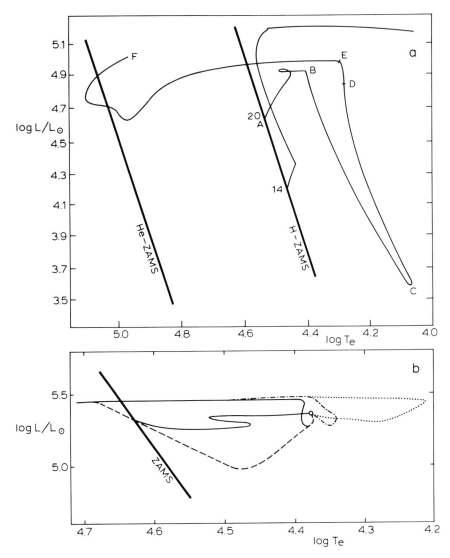

Fig. 65. (a) Evolutionary tracks of a binary $20\mathfrak{M}_{\odot}+14\mathfrak{M}_{\odot}$ in the theoretical HR diagram. (After De Loore and De Greve, 1975). (b) Evolutionary tracks of the primary of a system initially consisting of $40\mathfrak{M}_{\odot}$ and $20\mathfrak{M}_{\odot}$. The end of the stellar wind phase is marked by $\bigcirc$; thereafter different forms of loss of mass and angular momentum are considered:

$$\text{———}: \beta=1, \alpha=0,$$
$$\text{--------}: \beta=0, \alpha=3,$$
$$\text{········}: \beta=0, \alpha=0,$$
$$\text{—·—·—·—}: \beta=0, \alpha=1.$$

(From Vanbeveren *et al.*, 1979).

is that after the formation of a He core and loss of the primary's mantle a nearly pure He star remains, so that the star moves to the Zero Age Main Sequence for He stars (He ZAMS). This may be important for the explanation of Wolf–Rayet stars (Section 4.11).

In case C the mass-exchange occurs during the period of rapid contraction of the core and expansion of the outer layers, which takes place after the core He-burning phase and before the onset of core C-burning. Computations of case C were carried out by Lauterborn (1969, 1970), Barbaro *et al.* (1969), and by Tutukov and Yungelson (1973).

4.10.3. EVOLUTION OF CLOSE BINARIES IN THE NON-CONSERVATIVE CASE

This subject was first treated in the seventies, but a large literature has already originated. Reference is made to early studies by Paczynski and Ziolkowski (1967), Tutukov and Yungelson (1971), to a review paper by Thomas (1977), and a paper by Vanbeveren *et al.* (1979), to mention just a few.

It makes sense to follow Vanbeveren *et al.* in making distinction between two phases. The first coincides with core hydrogen burning. The stars expand hardly or not at all, and both components undergo stellar-wind loss. During the second phase the primary star grows so much that Roche lobe overflow occurs, with various possible consequences such as mass-exchange and loss, and the formation of a common envelope to the system.

During *the first phase* the mass ratio between the components and the orbital period changes. To calculate this, one has to relate the mass-loss to *loss of angular momentum*, while also the *mass-exchange* should be included in the mathematical formalism. This was done – albeit still in a rather formal way – by Vanbeveren *et al.* (1979). We write ΔJ for the loss of total orbital angular momentum J, with $C \equiv \Delta J/J$. The quantity C is assumed to be related to the mass-loss by

$$C = 1 - \left(1 - \frac{\Delta \mathfrak{M}}{\mathfrak{M}_{1i} + \mathfrak{M}_{2i}}\right)^{\alpha}.$$

Here $\Delta \mathfrak{M}$ is the total amount of mass lost by the system; the suffix i refers to the initial state; suffixes 1 and 2 denote the components; α is an arbitrary parameter ≥ 0. The mass-exchange between the components is described by a second parameter:

$$\beta = (\mathfrak{M}_2 - \mathfrak{M}_{2i})/\Delta \mathfrak{M}.$$

Apparently the parameter $(1-\beta)$ defines the relative mass-fraction that leaves the first component. In that case the component distance and period can be calculated as functions of the parameters α and β. The parameters are arbitrarily chosen in the apparent absence of a physical theory that should define these quantities. The two parameters α and β were called f and g by Paczynski and Ziolkowski (1967) and are comparable to parameters α and ψ of Massevitch and Yungelson (1975).

As an example we show in Figure 65b the track in the HR diagram of the primary component of a system with initial masses $40\mathfrak{M}_\odot + 20\mathfrak{M}_\odot$. In the first (stellar wind-)phase simple mass-loss was assumed, while only in the second phase various combinations of α and β were tried. Such calculations are evidently helpful to understand the possible evolution of close binaries undergoing mass-loss, but it is clear that this aspect of research is still in the 'heuristic' phase. Further developments have to be based on a clear understanding of the principal processes defining mass-exchange, and by relating the phenomena of mass-loss and of angular momentum.

Another important aspect of close-binary evolution is the formation of common envelopes to binaries. Preliminary calculations performed so far (Yungelson, 1973; Ulrich and Burger, 1976; Kippenhahn and Meyer-Hoffmeister, 1977; Neo *et al.*, 1977; Popova *et al.*, 1978) indicate that if the timescale of accretion is shorter than the thermal timescale of the envelope of the accreting star, the latter rapidly expands. The binary then becomes a contact system, and a common envelope seems to form.

A common envelope may also originate when the accreting component is a white dwarf or a collapsed star and the accretion rate exceeds the value corresponding to the critical Eddington luminosity.

4.10.4. THE LAST PHASES OF BINARY STAR EVOLUTION

The last phases of binary star evolution are, like those of single stars, for the most part still hidden in the realm of speculation. The *massive case B* seems the most interesting to study. Since in that case the end phase of the primary is nearly a pure helium star one has to consider the further evolution of such stars. In order to understand what happens then, it is important that those stars for which $\mathfrak{M} \lesssim 4\mathfrak{M}_\odot$ eventually increase in size; so they lose mass through a second process of mass-exchange and may end as white dwarfs (Biermann and Kippenhahn, 1972). Some detailed calculations were made by De Grève and De Loore (1977) for binaries with initial primary masses of 10 to $15\mathfrak{M}_\odot$. A second stage of mass-exchange appears to occur for primaries of $\sim 14\mathfrak{M}_\odot$. The stars with larger masses develop into neutron stars, the others into white dwarfs. Detailed evolutionary calculations were made for a system of $10\mathfrak{M}_\odot + 8\mathfrak{M}_\odot$. This confirms that after the end of core-helium burning a second stage of mass transfer from the primary occurs. In a degenerated core carbon ignition occurs. The remnant of the primary is a white dwarf of $1.1\mathfrak{M}_\odot$ (De Grève and De Loore, 1976b).

The more *massive Helium stars* expand hardly during helium- and carbon burning, and thereafter they shrink again. It is believed – see Section 4.9 – that such stars finally undergo a supernova explosion.

4.11. The Origin of Wolf–Rayet Stars

The position of the *Wolf–Rayet stars* in the HR-diagram was already mentioned in Section 4.2 in connection with the positions of the He–ZAMS and the ZAMS. Effective temperatures of WR stars are badly known, so their position in the HR-diagram is uncertain. They seem to occur, though, only slightly to the low-temperature-side of the ZAMS, but this does not necessarily mean that their composition is 'cosmic': Figure 43 shows that models with a He-core with q_0 values between 0.9 and 0.85 may explain the observed position of the WR stars in the HR-diagram. Let us assume that model. The high Y/X-ratio in the atmosphere must then be caused by processes like convective mixing in the outer mantle, and *suggests* the existence of a He-core. Hence, this model is chiefly based on photospheric data and not in conflict with the position in the HR-diagram.

If Wolf–Rayet stars consist for the greater part of He they must have lost the hydrogen mantle. In Section 3.4 we described a possible scenario for the evolution of the outer layers, given by Conti (1976), and suggesting that WR stars originate from Of stars. In Of stars the rate of mass-loss is high, and particularly if the star is member of a binary, dramatic mass-loss can occur. The position of WR stars in the HR-diagram (Figure 2) does not conflict with this assumption. Let us therefor examine the scenario of the origin of WR stars from massive stars (single *or* binaries) which, in the course of their evolution, have lost a great deal of their outermost layers. We recall (Section 3.4) that ~40% of the WR stars have been observed to be members of a binary, but Vanbeveren and De Grève (1979) remark that WR binaries with very small periods may be observed as single stars, so the true percentage may be larger.

First, we have to investigate whether mass loss of early-type stars can indeed lead to the formation of stars with the physical characteristics of WR stars. In order to arrive through mass-loss at nearly pure He stars of $\sim 10 \mathfrak{M}_\odot$ *considerable* mass-loss seems necessary!

A mechanism to arrive in a fairly short time at a large mass-loss is by *mass-exchange in a binary*. In this thought Paczynski (1967) and Kippenhahn (1969) suggested that the WR-ancestors should have been massive members of a binary, in which mass-exchange has occurred. The remaining object should consist nearly of He and be in the He-burning phase, with central temperatures exceeding 10^8 K. Following Paczynski (1973) we may suggest that WN stars are those in which outstreaming of mass has gone down to the CNO-shell, while the WC stars should have lost the mantle down to the 'triple-alpha' core, where C is produced. The WC stars should therefore be the farthest evolved objects, and in a rapid state of evolution.

Later computations of binary evolution with a view to the WR star problem were made by Tutukov *et al.* (1973), De Grève (1976), De Grève *et al.* (1978), Vanbeveren *et al.* (1979), Vanbeveren and De Grève (1979), Vanbeveren and Packet (1979), Vanbeveren and Conti (1980). The first concluded that the original mass of the primary should have been $\sim 15 \mathfrak{M}_\odot$, while the others found that it may be $\gtrsim 25 \mathfrak{M}_\odot$, a value also assumed by Kippenhahn (1969). Vanbeveren *et al.* (1979, 1890) presented calculations of binary evolution which show that Of members of a binary can indeed transform, by mass-exchange, into stars with WR characteristics. Vanbeveren and De Grève (1979) found that the surface H/He-ratio decreases during the evolution and can assume the observed values. Thereafter Vanbeveren and Conti (1980) could show that late WN stars evolve into early-type WN stars by intense mass-loss (for single stars) or mass-transfer (binaries). Early-type WN stars would then be He-burning stars with a low atmospheric H-content. Further mass-loss would thereupon bring C to the surface, and this would then yield WC stars. Hence, on the average WC stars should be less massive than WN's, but that is not obvious from the scanty data of Table XIV. Vanbeveren and Packet (1979) who studied the above scenario found that the observed number of WC stars can only be consistent with the calculated life-times of the WC-phase for rather restricted values of the rate of mass-loss, e.g. $500 < N <$ 1000 for a $32\mathfrak{M}_\odot$ star, and $2000 < N < 3000$ for a $10\mathfrak{M}_\odot$ star [N is defined in Equation (4.5; 1b)].

We therefore *conclude* that mass-exchange in a binary with a massive (Of?) component *can* lead to the formation of Wolf–Rayet stars, but this explanation in no case excludes a possible origin from single stars.

The alternative possibility is therefore to assume that WR stars originate through *mass-loss from single stars*. In that case the immediate problem is whether the large rate of mass-loss that is demanded can indeed transform a massive main-sequence star into a near-He star at the position occupied by WR stars in the HR diagram. As we showed in Section 4.5 a *large* rate of mass-loss brings a massive main sequence star vitually down the main sequence ($N \approx 1000$). Thus, Dearborn *et al.* (1978) and Dearborn and Blake (1979) showed that with a mass-loss-parameter defined as $\zeta = -\tau_{\mathrm{ms}}\dot{\mathfrak{M}}/\mathfrak{M}$ an evolutionary track nearly down the main sequence or slightly to the right is obtained for ζ_{initial} (:) $\zeta_{\mathrm{max}} \approx 1$, and found that WR stars result if $-\dot{\mathfrak{M}} \approx 7.5 \times 10^{-10}$ $\mathfrak{M}^{2.35}$ $\mathfrak{M}_\odot$ yr^{-1}. Hence, a large rate of mass-loss during the main sequence phase of stars would be necessary to yield Wolf–Rayet stars. Similarly, Chiosi *et al.* (1978) and Massevitch *et al.* (1979) require large rates of mass-loss to yield WR stars from red supergiants. Along a similar line of thought De Loore *et al.* (1978a, b) examined whether the Wolf–Rayet stars could be explained by loss of the outer stellar mantle, with $\dot{\mathfrak{M}}$ according to Equation (4.5; 1b) during the first evolutionary phase.

Let us compare this result with the observations. From Table XXXIX and Section 7.4 we derive rates of mass-loss for Of stars of $(1 \text{ to } 2) \times 10^{-5}$ $\mathfrak{M}_\odot$ yr^{-1} and for WR stars of $(3 \text{ to } 10) \times 10^{-5}$ $\mathfrak{M}_\odot$ yr^{-1}. The above expression by Dearborn *et al.* requires rates of 10^{-5} and 4×10^{-5} $\mathfrak{M}_\odot$ yr^{-1} for stars of 60 and $100\mathfrak{M}_\odot$. The other authors mentioned come to similar requirements. This is compatible with the observations.

We *conclude* that WR stars *can also* originate from *single* stars through mass-loss. Yet there remain some difficulties: these refer to the chemical composition of the envelope: the high He-content and the WN/WC-dichotomy. It is hard to see how sufficient He can be brought to the stellar surface, and it is still harder to explain the WC stars: if they have a C-core, they must be *very* evolved stars, and have perhaps gone through the red-giant phase, contrary to the WN stars.

4.12. The Origin of X-Ray Binaries; Mass-Transfer Onto a Compact Companion

The explanation of *the origin of massive X-ray binaries* started with the suggestion of Van den Heuvel and Heise (1972) that a binary with an original period of about

$3d$ and masses of about $16\mathfrak{M}_\odot$ and $3\mathfrak{M}_\odot$ may evolve into a system like Cen X-3 via case B mass-transfer, and with a Wolf–Rayet star in an intermediate phase. In a further elaboration Van den Heuvel (1973) and Van den Heuvel and De Loore (1973) postulated the attractive picture that in a binary where the most evolved star is an almost pure He star, the secondary, enlarged in mass with the hydrogen from the outer layers of the primary becomes a massive OB-main sequence star and thus begins to evolve again practically as if it were at zero age. The He star evolves much more rapidly than this rejuvenated secondary. If the mass of the He star exceeds four solar masses (the initial primary mass should then have exceeded 15 solar masses) it will finish its life as a type II supernova within 2×10^6 yr after the first stage of mass exchange, leaving a neutron star or black hole (Van den Heuvel and Heise, 1972). The supernova explosion need not to disrupt the binary system, because it is the less massive component which explodes. Some 4 to 6×10^6 yr after the explosion the OB secondary leaves the main-sequence, fills its Roche lobe and begins to transfer matter to the collapsed object; the impact of gas on the compact object causes the X-ray emission – Figure 66.

If the above picture is correct all massive O and B-type stars, with masses exceeding 15 to $25\mathfrak{M}_\odot$ and which are members of a binary system should at one time transfer most of their outer envelope to the binary, become a Wolf–Rayet, nearly pure He-star; this star would then eventually become a supernova. What remains is a neutron star or black hole – generally: a compact object, – which, in a later phase, may capture matter from its growing companion and thus becomes an X-ray source. Thus, WR stars would be *progenitors* of X-ray binaries, and can be the *companion* of a collapsed object in the second phase. An example of this latter case may be HD 50896 (WN5+?) for which $P=3.76d$ (Firmani *et al.*, 1979).

We consider the problem of X-ray emission by *mass-transfer in evolved binaries*, consisting of a compact object and a non-degenerate massive star. There are two ways in which mass can flow from the massive star towards the compact object. One case is the '*Roche lobe overflow*' in which the radius exceeds (partly or wholly) its Roche lobe, so that mass-transfer occurs. The other case is called the '*stellar wind*': the radius of the massive star is smaller than that of its Roche lobe, and the symmetrically outflowing stellar wind of the massive star just hits the compact object. In either case X-ray emission may be expected by the impact of the exchanged matter on the compact star or on a circumstellar accretion disk: the matter will impact with velocities of the same order of that of escape from the impacted object – the particles come virtually from 'infinity'. Such velocities correspond with energies of the order of keV's.

We first deal with the intricate problem of the mechanism of X-ray emission by an object onto which gas is accreted. Thereafter we will compare the effects of the two possible flow mechanisms.

The relation between the rate of accretion of matter by a star and the consequent X-ray emission was considered by Davidson and Ostriker (1973), Van den Heuvel (1975a), and Lamers *et al.* (1976). See also the summary in a paper by Den Boggende

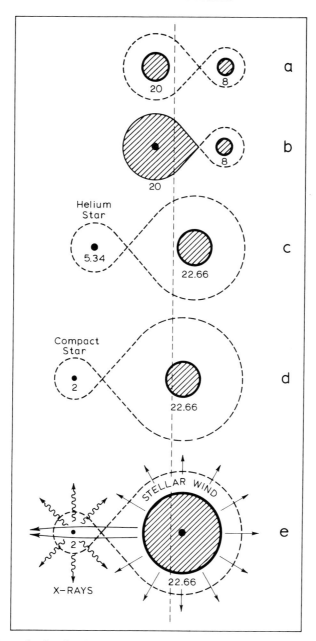

Fig. 66.   Possible mechanism for the origin of a massive X-ray binary. (After Van den Heuvel and Heise, 1972; computed numbers by De Loore *et al.*, 1975). (a) $t=0$; $P=4.54d$. (b) $t=6.17\times10^6$ yr; $P=4.56d$. Onset first stage mass exchange. (c) $t=6.20\times10^6$ yr; $P=10.86d$. End first stage mass exchange, begin WR stage. (d) $t=6.76\times10^6$ yr; $P=11.70d$. He star (=WR star) has exploded as supernova, becomes compact object. (e) $t=11.18\times10^6$ yr; $P=11.70d$. Primary becomes supergiant. Mass transfer to compact star produces X-radiation.

et al. (1979). An *upper limit* for $|\dot{\mathfrak{M}}|$ can be deduced from the consideration that the star can not accrete more than would correspond to a limiting luminosity derived according to the Eddington criterion (see our Sections 1.3 and 1.4). Thus it is found (Davidson and Ostriker, 1973; Shakura and Sunyaev, 1973), that

$$- \dot{\mathfrak{M}}_{\text{crit}} = (1.5/\zeta)\ 10^{-9}\ \mathfrak{M}_X\ \text{yr}^{-1}, \tag{4.12; 1}$$

where $\mathfrak{M}_X$ is the mass of the compact object, and ζ an efficiency factor, of the order of 0.1.

This upper limit may, however, be reduced because the heating of the infalling gas by X-rays emitted by the compact object may raise the gas temperature at a given distance to the star to such high values that the mean molecular velocity may exceed the escape velocity at that distance. In that case the gas will not fall onto the compact object, and computations (Ostriker *et al.*, 1976) indicate that the critical mass-flow can even be a few orders of ten smaller than the one given by Equation (4.12; 1). This result then leads to the assumption that the accretion necessary for a star to become an observable X-ray source should be between 10^{-11} and $10^{-8}\ \mathfrak{M}_{\odot}\ \text{yr}^{-1}$.

In case that much matter is accreted, a *disk* would form as was inferred for SMC X-1 by Van Paradijs and Zuiderwijk (1977) – see also Table XXXV. For very high accretion rates the disk would *absorp* the X-radiation, so that the disk would behave as an optical star. This is reached (Shakura and Sunyaev, 1973) for

$$\frac{\dot{\mathfrak{M}}}{\dot{\mathfrak{M}}_{\text{crit}}} \geq 3 \times 10^3\ (\alpha \mathfrak{M}_{\odot}/\mathfrak{M}_X)^{2/3}, \tag{4.12; 2}$$

where α is the so-called viscosity parameter of the disk. Equation (4.12; 2) then defines an upper limit for the accretion for the stars still to emit X radiation.

Next, the mechanisms for mass transfer in a binary with one compact object are discussed. For the case of Roche lobe overflow Paczynski (1970a) has shown that the massive component loses its envelope in a time scale:

$$\tau = 3 \times 10^7\ \mathfrak{M}^2/RL\ (\text{yr}),$$

yielding an accretion rate on the compact object of

$$- \dot{\mathfrak{M}}_{\text{accr}} = 0.8\ RL/3 \times 10^7\ \mathfrak{M},$$

with $\mathfrak{M}$, L, and R in solar units.

Since R increases, and L does not change notably during the evolution (at least for not too violent mass-loss) a lower limit to $|\dot{\mathfrak{M}}_{\text{accr}}|$ is found by inserting the main-sequence values for $\mathfrak{M}$, R, and L, with, for $\mathfrak{M} > 2\mathfrak{M}_{\odot}$: $R \sim \mathfrak{M}^{0.75}$ and $L \sim \mathfrak{M}^{3.25}$. Hence: $\mathfrak{M}_{\text{accr}} \sim \mathfrak{M}^3$, and $\tau \sim \mathfrak{M}^{-2}$. Values, thus computed for $\dot{\mathfrak{M}}_{\text{accr}}$ and τ were derived by Van den Heuvel (1975a, b).

In the *stellar wind case* we assume that the stellar wind, with mass flux $\dot{\mathfrak{M}}_W$ (g s^{-1}), is spherically symmetric: $|\dot{\mathfrak{M}}_W| = 4\pi r^2 \varrho_W(r)\ v_W(r)$. That fraction of the wind flux will be accreted that passes the centre of the degenerate companion within the impact distance (Davidson and Ostriker, 1973; following Bondi and Hoyle, 1944):

$$r_a = 2G\mathfrak{M}_X/v_{rel}^2 \approx 1.3 \times 10^{10} \, \mathfrak{M}_X v_{rel}^{-2} \, cm,$$

where r_a is roughly the distance at which a particle moving past the compact object with velocity v_{rel} will be deflected by $90°$; $\mathfrak{M}_X$ is the mass of the compact object, v_{rel} the relative wind velocity at 'infinity'. Then, the accretion rate is (Davidson and Ostriker, 1973):

$$\dot{\mathfrak{M}}_X \approx \pi r_a^2 \varrho_w v_{rel} = -\dot{\mathfrak{M}}_w r_a^2/4a^2 = -\dot{\mathfrak{M}}_w (G\mathfrak{M}_X/a)^2/v_{rel}^4,$$

where $-\dot{\mathfrak{M}}_w$ is the total stellar mass flux (g s^{-1}) and a the orbital radius. This yields (Lamers *et al.*, 1976b):

$$\dot{\mathfrak{M}}_X = -3.65 \times 10^{10} \dot{\mathfrak{M}}_w (\mathfrak{M}_X/a)^2 \, v_{rel}^{-4} \mathfrak{M}_\odot \, yr^{-1},$$

and an X-Ray luminosity (Blumenthal and Tucker, 1974):

$$L_X = 1.2 \times 10^{41} \, (\mathfrak{M}_X R_X) \dot{\mathfrak{M}}_X \, ergs^{-1}$$

We refer to Tutukov and Yungelson (1973), and Van den Heuvel (1975a).

Refinement of the basic theory given above, shows that in actual practice Roche-lobe overflow is more common than would be expected, particularly when massive stars with small g_{eff}-values are involved. The main reason why this is so, is that Roche lobe overflow starts already when the stellar radius is smaller than the Roche lobe radius by a few scale-heights – after all, even then a fraction of the atmosphere extends already beyond the Roche lobe. Hence Roche lobe overflow takes place much more gradually than was initially assumed, and blocking of the X-radiation by too much accumulated gas would not occur.

Detailed investigations were made by Savonije (1978a, b), who found, that Roche lobe overflow can be the basic mechanism of mass transfer, and the main source of X-radiation in X-ray binaries of small as well as in those of large masses. In massive X-ray binaries with primaries of the order of 15 to 20$\mathfrak{M}_\odot$ the beginning of the mass transfer is in case A. For the binaries Her X-1 and Cen X-3 it takes some 10^5 and 10^4 yr before $|\mathfrak{M}|$ exceeds the critical accretion rate, which would mark the end of the X-ray emission phase; these periods can therefore be considered as the lifetimes of these binaries.

A review of the main properties and parameters of the best studied *massive X-ray binaries* is given in Table XXXV. Typically, the systems consist of a hot early-type supergiant and a compact companion. The X-rays are produced by accretion of matter onto the companion. The luminous primary can best be studied in the ultra-violet spectral region. An interesting feature is that most primaries seem undermassive for their luminosity, which is another indication for past mass-loss.

TABLE XXXV

Some well-studied X-ray binaries, and their principal physical parameters

Source	Spectral type primary	$\mathfrak{M}_p/\mathfrak{M}_\odot$	$\mathfrak{M}_s/\mathfrak{M}_\odot$	$P(d)$	$R_p/R_\odot$	$i(°)$	log $(L_x/L_\odot)$ [a]	log (L_x/L_{opt}) [d]	$-\dot{\mathfrak{M}}_i/\mathfrak{M}_\odot$ (yr^{-1}) [d]	M_b primary	Pulsar?	References
4U 1700−37 [a] =HD 153919	O6.5f	30	2	3.41180	15	90	2.5	−3.3	1.5×10^{-5}	<−8.9		Vanbeveren (1977a)
		27.1	1.3		20					−10.2	97 min[e]	Fahlham et al. (1977); Van Paradijs et al. (1978)[c]; Mantilsky et al. (1978); Conti (1978a)
4U 0900−40 =Vela X-1	B0.5 Ib	21.7	1.5 to 3	8.96	27		2.5	−3.1			283 s	Avni (1976); Charles et al. (1978), McClintock et al. (1975)
=HD 77581		45	2.6									Hutchings (1974); Van Paradijs et al. (1976, 1977)
		21.2	1.6							−8.4		Branduardi et al. (1978)
		24			32				$>5 \times 10^{-7}$			Hammerschlag Hensberge et al. (1979); Rappaport and Joss (1980)
Cyg X-1 =4U 1956+35 =HDE 226868	O9.7 Iab	>25	9	5.607	22	27	3.6 to 4.5	−2	2.6×10^{-6}			Mason et al. (1974), Avni and Bahcall (1975b)
		2.39	1.48							−8.7		Conti (1978a)
Cen X-3 =4U 1119−60 =Krzeminski's star	O6.5 IIIe	18	1	2.087	13	90	4.1	−7?			4.84 s	Tananbaum and Tucker (1974); Hatchett and McCray (1977)
		30	2		17							Conti (1978)
		17.2	1.4							−9.0		Rappaport and Joss (1980)
		18			12							
4U 1538−52		20±8	≲2	3.73	16 to 20		2.7	−2.7			529 s	Davidson et al. (1977); Cowley et al. (1977)
		20±4	2±0.4			70						Hutchings (1978)
		19			16							Rappaport and Joss (1980)

Source	Sp. type								Period	References
SMC X-1 = 4U 0115−37 = Sk 160	B0.5 Iab	15	0.9	3.893	70	5.2	−0.6	10^{-5}	0.71 s	Van Paradijs and Zuiderwijk (1977)
		15 to 24	0.8 to 4.5	14 to 22						Avni and Milgrom (1977)
		16.2±1.4	0.8 to 1.8	1.02±0.2	17		−8.8			Primini et al. (1977)
		12.5	0.8				−8.1			Hutchings et al. (1977)
		18			15					Conti (1978a)
										Rappaport and Joss (1980)
LMC X-4	O8 V-III	20	1.3	1.4083			−0.5			Chevalier and Ilovaisky (1977)
						3.4^b				White and Davidson(1977), White(1978)
	O7	>6	>1.1							Li et al. (1978)
		25	2 to 4							Hutchings et al. (1978b)
		22.5	2.5				−8.4			Conti (1978a)
4U 0352+30 = X Per	O9.5 (III−V)pe	~20	>2?	580?		0.4	−3.4		835 s	De Loore et al. (1979b)
V861 Sco = HD 152667 = OAO 1653−40?	B0 Ia	~25	~8	7.84825	33		−10.1	~$5×10^{-6}$	− 9.85	Hutchings (1979a)
		48±10	≥12.5±2^f			1.7^h	−3.7	~10^{-5}		Wolff and Beichman (1979)
										Tanzi et al. (1979), White and Pravdo (1979)

^{a} The system is remarkable because the secundary moves with supersonic velocity through the *dense* (Dupree et al., 1980) stellar wind of the Of primary. An extended wake is expected to trail the secundary (Hensberge et al., 1973), and is actually observed (Fahlham 1977). Dupree et al. (1977) detected outflowing [Fe xiv] emission, presumably from a volume around the X-ray source, but this was not confirmed by Lester (1979). A variable rate of mass-loss is suggested. Considerable flaring activity occurs, up to a factor 25 (Pietsch et al., 1980).

^{b} Flares occur with peak intensities of $9×10^{38}$ erg s^{-1} (White, 1978).

^{c} These authors find it impossible to represent the light curve by the T_{eff}, $\mathfrak{M}$, and M_b values.

^{d} In this column L_{opt} is the optical luminosity of the source; Conti (1978a) and Hutchings (1979).

^{e} The reality of this period is doubted by Hammerschlag-Hensberge et al. (1979b).

^{f} Companion may be a black hole (Wolff and Beichman, 1979).

^{g} Mainly from Hutchings (1978).

^{h} Variable X-ray emission (Polidan et al., 1979).

4.13. Stellar Rotation and Evolution

Figure 67 shows a composite HR-diagram after Van den Heuvel (1968). The diagram gives the main-sequence and also the theoretical Zero Age Main Sequence; and further the observed positions of stars of different luminosity classes, the evolutionary tracks for various masses, as derived by different authors. The average equatorial rotational velocity is indicated along the ZAMS. Using these data, and available information about stellar evolution, the dashed lines were computed in the diagram; they indicate lines of constant values of $\langle v_e \sin i \rangle$, derived assuming (1) that the star rotates as a solid body, and (2) conservation of angular momentum in shells. A comparison of the velocities given here with the 'observed' values as given in Figure 19 shows that the latter are considerably larger. Kraft (1966) concluded from such a result that the communicated values for the equatorial rotational velocities for stars brighter than luminosity class II would not be due to rotation but rather to macro-turbulence.

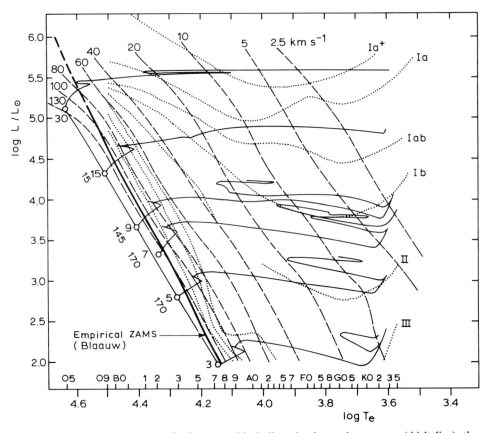

Fig. 67. The Hertzsprung–Russell diagram with indicated: the main-sequence (thick line), the theoretical Zero Age Main Sequence, observed positions of stars of different luminosity classes (shaded), evolutionary tracks of stars of various masses, and (dashed) lines of constant $\langle v_e \sin i \rangle$. (After Van den Heuvel, 1968).

Whether this conclusion is correct can only be determined from a very careful investigation of the line profiles of massive stars, using very high spectral resolution, a badly needed investigation!

The difference between the observed and calculated velocities is *enlarged* if one considers the loss of angular momentum associated with mass-loss: The matter lost to the star carries angular momentum away. If the momentum of inertia of the star is written as $\mathfrak{M}(k_s R)^2 \omega$, when $k_s R$ is the 'radius of gyration' of the stellar matter, one may write, following Williams (1969) and Sreenivasan and Wilson (1978b):

$$d(\mathfrak{M}(k_s R)^2 \, \omega) = (k_d R)^2 \, \omega \, d\mathfrak{M}, \qquad (4.12; 1)$$

where $k_d R$ is the radius of gyration of the ejected matter. Equation (4.12; 1) is transformed into:

$$(k_d^2 - k_s^2)R^2 \, \omega \, d\mathfrak{M} = \mathfrak{M}k_s^2 R^2 \, d\omega + 2\mathfrak{M}k_s^2 R\omega \, dR,$$

which yields the rate of spin-down:

$$\frac{\dot{\omega}}{\omega} = \frac{k_d^2 - k_s^2}{k_s^2} \frac{\dot{\mathfrak{M}}}{\mathfrak{M}} - 2\,\frac{\dot{R}}{R}. \qquad (4.12; 2)$$

Actually, if $\dot{\mathfrak{M}}=0$ this expression described the conservations of angular momentum for different shells. The influence of mass-loss on the spin-down rate as compared to the case in Figure 67, where mass loss in not included, can be estimated with Equation (4.12; 2). Assuming $\dot{\mathfrak{M}}/\mathfrak{M} \approx -10^{-6}$ yr^{-1}, $\dot{R}/R \approx 10^{-7}$ yr^{-1} (i.e. doubling the size during the H-burning time, $\sim 10^7$ yr), $k_s \approx 0.5$, $k_d \approx 1$, the first term in Equation (4.12; 2) appears to be much larger than the second.

First-order computations, essentially based on (4.12; 2), but further elaborated in order to include also loss of translational and rotational energy by radiation, were made by Sreenivasan and Wilson (1978b) for a $15\mathfrak{M}_\odot$ star. It appears that such a star loses the main part of its spin in a time ($\sim 10^7$ yr), similar to the H-burning lifetime. For a $60\mathfrak{M}_\odot$ star the $|\dot{\mathfrak{M}}/\mathfrak{M}|$-term is larger than for a $15\mathfrak{M}_\odot$ star by about two orders of magnitude, while also $\dot{R}/R$ is larger, because of the shorter H-burning time of the star. Hence a $60\mathfrak{M}_\odot$ star spins down much faster than a $15\mathfrak{M}_\odot$-star. Similar detailed computations on the evolutionary development of rotating stars we carried out by Packet *et al.* (1979) for initial masses of 15 to $100\mathfrak{M}_\odot$. These computations confirm again that mass-loss by a stellar wind leads to a decrease of the rotational velocity. Hence the problem of the large observed rotational velocities for developed stars, as obtained in the first part of this section, remains.

CHAPTER 5

THE STRUCTURE OF VERY TENUOUS
STELLAR ATMOSPHERES

5.1. Atmospheric Models; Radiative or Convective Equilibrium

The structure of a stellar atmosphere is mostly described by a *model*, which is a table giving the variation of various physical parameters as functions of the physical depth, mass column density, or a monochromatic optical depth τ_λ. Very often the optical depth is taken at one selected wavelength, e.g. 0.5 μm. Here $d\tau_{0.5} = \kappa_{0.5}\varrho\ dz$ where $\kappa_{0.5}$ is the continuous absorption coefficient per gram of matter at 0.5 μm, and z is the height ordinate. Input parameters used to specify a model are usually the effective temperature, T_e, the acceleration of gravity g, and the chemical composition.

In order to be able to compute a theoretical model, the physical parameters such as the absorption coefficient, the average molecular weight, the density, etc. must be known as functions of pressure and temperature. It is also necessary to specify the pressure balance in the atmosphere. Hydrostatic equilibrium is commonly assumed, so that $dP_g = -g_{eff}\varrho\ dr$ where g_{eff} is the effective acceleration of gravity,

$$g_{eff} = g_{grav}\ (1 - \Gamma_F). \tag{5.1; 1}$$

Furthermore the mechanism of energy transport in the atmosphere plays an important role. In the stellar atmospheres with which we are concerned only two mechanisms come into question: *radiative* and *convective* transport. The numerical techniques for computing an atmospheric model in radiative equilibrium are very advanced compared to the rather crude methods that are applied to atmospheric regions were convective transport dominates. It is a queer situation that many of the published model atmospheres are based on the assumption of radiative equilibrium even in those cases where one is nearly certain that it does not occur, as if mathematical exactness should prevail over physical correctness!

With the very extended atmospheres of the extremely tenuous supergiant atmospheres that are an important part of this book, other difficulties enter, partly of a computational but partly of a physical nature, related to the curvature and the instability of the atmosphere. Only few atmospheric models in *radiative* equilibrium have been computed so far with $\log g_{eff} \lesssim 1$. A few authors have computed grids of atmospheric models in *convective* equilibrium with $\log g$-values down to and below unity. For nearly all luminosities convection occurs only in those stars with effective temperatures below approximately 10 000 K. Stars with larger effective temperatures should have atmospheres essentially in radiative equilibrium.

5.2. The Assumption of Plane-Parallel Atmospheric Layers

In many investigations plane-parallel atmospheric layers are assumed i.e. the mean-free-path of a photon in the photosphere is thought to be small as compared to the stellar radius. We should now verify this assumption (see De Jager and Neven, 1975).

The mean-free-path of a photon is comparable to the optical scale height θ, defined by $dz=\theta$ d $\log\tau$, with z=geometrical depth, τ=optical depth. It was shown by Van Bueren (1973) that in most stars and at most wavelengths (including spectral lines) $\theta \approx H$, the density scale-height, to within a factor of 2 (although in some cases larger deviations *can* occur!).

Hence the assumption of plane-parallel layers reduces to the condition that

$$H < \delta R, \qquad (5.2;\ 1)$$

where δ is a small number, of the order 0.1.

With $H=\Re T/\mu g_{eff}$, $R=(G\mathfrak{M}/g_{grav})^{1/2}$ and $g_{eff}=\varepsilon g$, writing g for g_{grav}, condition (5.2; 1) reduces to:

$$g_{\lim} = \frac{T^2}{\varepsilon^2 \delta^2 \mathfrak{M}} \frac{\Re^2}{\mu^2 G}. \qquad (5.2;\ 2)$$

For $g > g_{\lim}$ the plane-parallel approximation is justified.

To illustrate the limits imposed by condition (5.2; 2) we take:

$$\mu = 1.4 \text{ for } T < 8000 \text{ K},$$
$$= 0.6 \text{ for } T > 8000 \text{ K},$$

values that follow from available data on stellar atmospheres. For average supergiant masses we assume

$$T > 8000 \text{ K}: \mathfrak{M} = 40\mathfrak{M}_\odot$$
$$< 8000 \text{ K}: \mathfrak{M} = 20\mathfrak{M}_\odot.$$

Further we assume $\delta = 0.1$ and $\varepsilon = 0.1$, both perhaps reasonable values, and thus find from Equation (5.2; 2) the following limiting g-values:

$T=$	30000	20000	10000	5000	2500 K,
$g_{\lim}=$	6	3	0.7	2	0.5 cm s^{-2}.

We have to consider this result together with the one from Section 1.5 (see Equation (1.5; 1) and Figure 3) that for the *hot* stars g_{grav} always exceeds the here calculated $g_{\lim}$-values. For such stars atmospheric curvature may only be important when ε takes much smaller values than the one assumed here (0.1). However, for cool stars this reasoning does not hold. For these the present calculation leads to the *conclusion* that only photospheres with $g \gtrsim 2$ can be treated as plane-parallel. These conclusions do not apply to the outstreaming stellar winds, because there $H \approx R$, since g_{eff} can even reach negative values! Such regions are in any case extended and curved.

5.3. Models of Very Tenuous Atmospheres; the Plane-Parallel Case

For very tenuous *hot* atmospheres, the theoretical computations are facilitated by the fact that the atmospheric opacity is caused practically entirely by scattering of radiation by free electrons, for which the scattering coefficient is frequency-independent. The atmosphere then approximates the 'gray' case. On the other hand, this same fact complicates the computations since in that case the source function $S_\nu(\tau)$ can not be defined by the local temperature but rather, by virtue of the scattering properties of the atmosphere, by the global behavior of the radiation field. The dependency of $S(\tau)$ on $T(\tau)$ is less the smaller the contribution of absorption of radiation is to the total opacity. For such models we assume that the atmosphere can still be considered plane-parallel (see the discussion in Section 5.2). Then the depth-variation of the monochromatic source functions $S_\lambda(\tau_\lambda)$ can be computed according to the iterative procedure

$$S_{n+1}(\tau_\lambda) = \frac{\sigma_\lambda}{\kappa_\lambda + \sigma_\lambda}(\tau_\lambda)\,[J(\tau_\lambda) - B(\tau_\lambda)] + B(\tau_\lambda);$$

$$J(\tau_\lambda) = \tfrac{1}{2}\int_0^\infty S_n(x)\,E_1(|\tau_\lambda - x|)\,\mathrm{d}x,$$

where the first approximation is chosen conveniently, in practice equal to the Planck function of an assumed $T(\tau_\lambda)$ relation. For stars with very small g_{eff}-values and high temperatures $\kappa_\lambda/\sigma_\lambda \approx 0$, and the iteration converges extremely slowly. Deviations from LTE occur in the models for $\tau < 1$ but their influence is small except in the far UV (De Jager and Neven, 1975).

An interesting aspect of the models obtained is that in the upper parts of the photospheres ($\tau_5 \lesssim 1$) the radiation field at the same geometrical depth can *not* be described by one unique temperature (Figure 68). This is caused by the fact that the scattering does not contribute to the 'thermalization' of the radiation field – each scattering electron just changes the direction of the photons but it does not adapt the energy-distribution-function of the photons to that of the particles, as would happen in an absorbing atmosphere. For such considerations the 'thermalization depth' is a useful quantity; it is the depth z_{th} at which $\tau_{\mathrm{abs}} = \int_{z_{\mathrm{th}}}^\infty \kappa\varrho\,\mathrm{d}z \approx 1$. Apparently, in the example of Figure 68 this thermalization depth is reached at $\tau_{0.5} \approx 2$ in the cool supergiant, but at $\tau_{0.5} > 3$ in the hot one.

Another aspect is that for the most extreme stars (high T_e- and low g-values) for virtually *any* initially assumed $T(\tau_\lambda)$-relation the *same* final $S(\tau_\lambda)$-relation is found, which again means that actually the photospheric source function is not coupled to the star's local kinetic temperature but rather is fixed to the boundary conditions imposed at the 'bottom' of the model, where $\tau_{\mathrm{abs}} \approx 1$.

For these *two* reasons the notion 'temperature' is not suitable for describing the radiation field in atmospheres of hot stars with very low effective g-values.

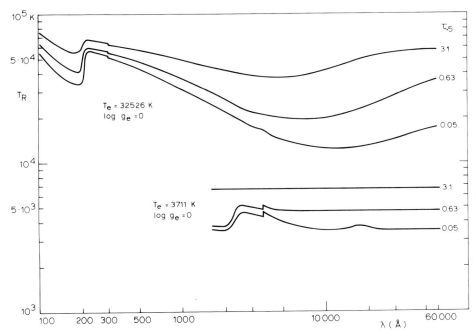

Fig. 68.   The radiation temperature varies strongly with wavelength in the outermost parts of super-giant photospheres. The diagram shows the radiation temperature $T_r(\lambda)$ for three $\tau_{0.5}$-values in photospheric models of a hot ($T_e=32526$ K) and a cool ($T_e=3711$ K) supergiant (De Jager and Neven, 1975).

For cooler atmospheres, even for the very tenuous ones, absorption is always important. The construction of models for the coolest atmospheres is complicated by the many molecular species that originate at low temperatures. As a rule the LTE-computation of the number density of the molecules no longer poses essential problems, but calculations of their absorption-coefficients do!

For atmospheres of cool supergiants reference is made to the reviews by Vardya (1972) and Carbon (1979). At very low temperatures the opacity is dominated by H_2O, TiO, and MgH in the visible and near-infrared regions. Opacities in molecular bands have been reviewed by Vardya (1972). For CN, C_2, and CO reference should be made to Querci *et al.* (1971). The models of such cool stars are characterized by convective instability, which starts already at fairly small optical depths because of the effects of H_2 dissociation. As an example we mention one of Auman's (1969) models with $T_e=3000$ K, log $g=1$, where convection sets on at $\tau=0.015$. In more refined models convective overshooting should be included. In addition, in such extreme conditions considerable heating of the outer 'radiative' regions will take place due to dissipation of mechanical energy (Vardya and Kandel, 1967). Furthermore, the influence of the blanketing due to TiO opacity produces a global photospheric heating of the order of $\sim$100 to 300 K for stars with T_e values in the range 2500–3400 K (Krupp *et al.*, 1978). The emergent flux shows local maxima at $\lambda=1.6\,\mu$m and 2.4 μm caused by gaps in the H_2O opacity. This, however, may be smoothened by the effect of CN absorption (not yet included in the computations).

A comprehensive bibliography of computed plane-parallel photospheric models published before 1976 was given by Burger (1976). Table XXXVI is essentially based on her publication, with some additions and deletions – the latter because we restrict ourselves to massive stars. The models are given in order of increasing T_e-values. In interpreting the g-values one has to consider that sometimes the parameter is g_{eff}. For hot stars, where radiation pressure is important or for fast rotating stars, or stars with appreciable dissipation of turbulence that value may be much smaller than $g_{grav} = G\mathfrak{M}/R^2$ (see Section 2.6). For the most luminous stars, for which g_{grav} is close to the limiting g-value it makes therefore sense to compute models with the parameter

TABLE XXXVI

Bibliography of computed plane-parallel photospheric models, based on a compilation by Burger (1976). Usually solar type abundances and LTE are assumed unless remarked otherwise.

T_e-range (K)		$\log g$-range		Rad. or incl. Conv.	non-LTE	Line blank-eting	Remarks; κ or chem. abundances	References and further remarks
2000	4000	−2	5	C		−	H_2O	Auman (1969)
2500	5000	−2	4	R		yes		Johnson (1974)
2500	3600	0	1	C		−	H_2O, CN, CO	Alexander and Johnson (1972)
2500	41600	1	5	C		−	Yamshita (1962)	De Loore (1970)
2800	3200	−0.9	+1	R		−	spherical models	Watanabe and Kodaira (1978, 1979)
3400	4500	−1	1	C		−	CN, CO, C_2 (C stars)	Querci et al. (1974)
3600	4000	−0.5	0	C		yes	v_t varied	Tsuji (1976a)
3700	33000	1	5	R ($T_e>10^4$)		−		De Jager and Neven (1975)
3750	6000	0.75	3	R		yes	$-3 \leqslant [A/H] \leqslant 0 \to 0$	Bell et al. (1976), Gustafsson et al. (1975)
5000	8500	2	4.5	R		yes	various [A/H]ratios	Peytremann (1974a,b, 1975), Kurucz et al. (1974, 1976)
5500	50000		≤4.5	R		yes	various [A/H]	Kurucz (1979)
7000	11000	2	4.44	C		−		Mihalas (1965)
7000	13000	2	4	R		yes		Mihalas (1966)
7500	10000	0.45	−1.1	R		−		Wolf (1972, 1973), variable v_t; g_t
8000	50000	2	5	R		yes		Kurucz et al. (1974)
8500	50000	1	5	R		yes	three [A/H] ratios	Kurucz (1979)
	9170		1.13	R		−		Groth (1961), g_t; one model
10000	15000	4		R	n	yes		Borsenberger, Gros (1978)
10000	40000	2.5	4.5	R		yes		Klinglesmith (1971)
11000	50000	2	5			yes		Kurucz (1969)
14400	25200	3	4.5	R		yes		van Citters and Morton (1970)
15000	55000	2.5	4	R	n	−		Mihalas (1972a,b)
30000	200000	3.4	7.5	R				Hummer and Mihalas (1970)
	39500		3.45	R	n	−		Mihalas and Hummer (1974a, b)

g_{eff}, since very small differences in g_{grav} may have large effects on g_{eff} and hence large consequences on the atmospheric structure. Such models have so far only rarely been calculated. Very extensive models for hot plane-parallel atmospheres were computed by Kurucz *et al.* (1974). These models, in which the blanketing influence of many absorption lines is included, represent the best effort, so far, in the field of constructing plane-parallel model atmospheres. The g-range varies with T_e; for instance, for T_e=16000 K the models cover the log g-range 2.0 (0.5) 4.5; for T_e= 25000 K: 3.5 (0.5) 5.0, and for 50000 K: 4.5 and 5.0. For $T_e < 16000$ K the log g-range is 2.0 (0.5) 4.5. The range of T_e-values is 8000 (500) 10000 (1000) 16000 (2000) 20000 (5000) 50000 K. Updated models were published later by the same authors (Kurucz *et al.*, 1976; Kurucz, 1979). The latter models cover the range 5500 to 50000 K for gravities from main sequence values down to the radiation pressure limit, and for various sets of chemical abundances.

Table XXXVII gives some of the characteristics of plane-parallel atmospheres with log g_{eff}=1, hence still within limit of applicability of the plane-parallel assumption.

Limb darkening coefficients for atmosphere models of late-type stars, including giants and supergiants, were calculated by Manduca *et al.* (1977) and Manduca (1979); for earlier-type models by Mutsahm (1979).

TABLE XXXVII

Characteristic parameters for plane-parallel model atmospheres (De Jager and Neven, 1967, 1975; De Loore, 1970) with $\log g_{eff}$=1; P_g and P_e are given for $\tau_{0.5}$=0.45; v_{max} is the predicted maximum convective velocity (km s^{-1}).

T_e (K) (approx.)	$\log P_g$	$\log P_e$	v_{max}
2500	2.3	–	1.0
3100	2.2	–	3.7
3700	2.0	–	6.
4760	3.5	−0.7	6.5
5940	2.8	0.2	–
8300	1.5	1.1	–
9500	1.1	0.8	0
16600	1.15	0.85	
30000	1.2	0.9	
42000	1.15	0.85	

5.4. Models of Extended and Spherical Stellar Atmospheres; the Static Case

It has been suggested by various authors that the apparent difficulties in explaining at the same time the continuous energy distribution and some spectral line features of supergiants, Of-, P Cygni-, Wolf–Rayet stars, nuclei of planetary nebulae and similar hot stars, on the basis of plane-parallel photospheric models, may result from the fact that for the photospheres of such stars the plane-parallel concept does not hold. It makes sense to examine whether this is true, because our discussion of Section 1.5 shows that photospheres of hot stars have as a rule log $g \geqslant 2.5$ to 3, while our Section 5.2 showed that the photospheres these types of hot stars will only deviate from the plane-parallel case if $g_{grav} \leqslant 5$ cm s^{-2}, while we assumed $g_{eff} = 0.1$ g_{grav}, so g_{eff} should be $\leqslant 0.5$ cm s^{-2}. That g_{eff} is fairly small indeed follows from our Figure 17, from which one reads that for the above-given groups of stars $g_{eff} \approx 0$. The precise value of g_{eff} is not known, however, hence research should still be done. An additional feature of the atmospheres of these groups of stars is that they have a fairly dense stellar wind, so dense that the level $\tau_\lambda \approx 1$ for different λ-values can correspond to very different geometrical heights in the star – in other words: the radius R is wavelength-dependent!

The extended stellar atmospheres to be considered in this section have pressure scale-heights H, or optical scale-heights θ, that are comparable to the radius of the star. Early developments in the theory of such models were made by Kosirev (1934) and Chandrasekhar (1934). An important fact brought forward by these first studies was that extended atmospheres yield spectra with a lower colour temperature than planar atmospheres. The colour temperature appears also to be wavelength-dependent. Only much later was further progress made; partly initiated by a note by Chapman (1966) that the radiation field in the outermost layers of an extended atmosphere is fairly sharply peaked outward. For increasing distance r from the center of the star the specific intensity $I(r, \mu)$, where $\mu = \cos\theta$, becomes more and more peaked around $\mu = 1$. This fact can be understood easily: for a detector at large distance from a star the source of radiation, i.e. the star, occupies a small fraction of the celestial sphere. Hence, the classical Eddington approximation

$$\frac{K}{J} \equiv \frac{\frac{1}{2}\int\limits_{-1}^{+1} I(r, \mu)\, \mu^2\, d\mu}{\frac{1}{2}\int\limits_{-1}^{+1} I(r, \mu)\, d\mu} = \frac{1}{3},$$

breaks down in the range of small optical depths. Curves showing the change of $I(\mu)$ as a function of the optical depths τ are given in Böhm's (1973) review, and also in papers by Peraiah (1973) and by Rybicki and Hummer (1975). In order to deal with this problem Auer and Mihalas (1972) introduced a variable 'Eddington factor' $f \equiv K/J$, with as boundary values $\lim_{r \to \infty} f = 1$ and $\lim_{r \to 0} f = \frac{1}{3}$.

In the years around and after 1970 several investigations were made to develop model atmospheres of extended stars. We mention the work by Hummer and Rybicki (1971) who dealt with the gray case, followed by Cassinelli (1971b) who discussed gray and non-gray cases, Castor (1974a), Mihalas and Hummer (1974a, 1974b). In this relation we refer also to the excellent review in Mihalas's (1978) book: *Stellar Atmospheres*, particularly Chapters 7, 14, and 15.

In the investigations described above the work of Mihalas and Hummer was based on non-LTE considerations; the other are LTE-treatments. Since extended stellar atmospheres, being at their limit of stability, must have appreciable outward mass-flow, it is only natural that later investigations incorporated the effects of stellar winds, see e.g. Castor *et al.* (1975), Mihalas *et al.* (1975b), and Cassinelli and Hartmann (1975). This series of investigations started with the basic research on stellar winds by Lucy and Solomon (1970). Since it will be found (Section 5.4) that curved atmospheres *cannot* be static and must lose mass, and that allowance for a gas-flow leads to quite different model atmospheres than the static treatment does, it is of importance to fully include the atmospheric expansion in the mathematical formalism. However, chiefly for didactical reasons it seems advisable to discuss in this section first the mathematics of non-expanding extended atmospheres. In that case the model atmospheres to be computed are forced to satisfy *hydrostatic* and *radiative* (or *convective*) equilibrium.

Difficulties that one meets in a complete theory are that the two basic equations, those of hydrostatic equilibrium and of transfer (assuming radiative equilibrium) have to be solved simultaneously: the function $I(r, \mu, \nu)$ occurs in both equations, and there is often an appreciable contribution of radiation pressure to the general pressure term. This implies that the model's pressure-structure is strongly affected by the radiation field, which in turn is determined by the model; the two solutions are thus strongly coupled. Additional difficulties are that the 'radius' of the star is frequency-dependent, so that the term 'effective temperature' loses meaning. Finally the acceleration of gravity depends on the distance from the star's center r, and on the optical depth, and thus varies over the depth of the atmosphere in contrast to the case of a 'thin' plane-parallel atmosphere.

Atmospheric models for extended stellar atmospheres can be defined by the following parameters: the stellar mass $\mathfrak{M}$, its luminosity L, and the 'radius' $R(\tau_\lambda)$ at a given monochromatic or average optical depth; an additional input should be the chemical composition of the atmosphere. Of course, other equally satisfactory selections of initial parameters are possible.

In the case of spherical geometry the transfer and hydrostatic equations change as compared to the planar case, as will be shown next.

The monochromatic *equation of transfer*, in which – for simplicity – frequency subscripts are omitted, and written in polar coordinates, reads:

$$\mu \frac{dI(r, \mu)}{dr} + \frac{(1 - \mu^2)}{r} \frac{dI(r, \mu)}{d\mu} = k(r)\varrho[S(r) - I(r, \mu)]. \qquad (5.4; 1)$$

Here, r is the radial distance, $\mu = \cos\theta$, I is the radiation intensity. Further $k_\nu = \kappa_\nu + \sigma_\nu$ is the total opacity per unit of mass, being the sum of absorption and scattering. For hot stars $k_\nu = \sigma_e$ the Thomson scattering, independent of frequency, but not isotropic. However Mihalas and Hummer (1974b) have shown that the deviations from isotropy in the electron scattering term have only minor influence on the average radiation intensity, so that the isotropy assumption for σ_e is acceptable. Furthermore the source function is

$$S_\nu = \frac{\kappa_\nu B_\nu + \sigma_\nu J_\nu}{\kappa_\nu + \sigma_\nu},$$

where J_ν is the average intensity of radiation:

$$J_\nu(r) = \tfrac{1}{2} \int_{-1}^{+1} d\mu \, I_\nu(r, \mu),$$

being the zeroth moment of the radiation field I. Following Eddington, we define the first and second moments:

$$\left.\begin{array}{c} H_\nu(r) \\ K_\nu(r) \end{array}\right\} = \tfrac{1}{2} \int_{-1}^{+1} d\mu\, \mu^n I_\nu(r, \mu), \quad \text{with} \quad \begin{cases} n = 1 \\ n = 2. \end{cases}$$

With $d\tau = -k\varrho\, dr$ and taking first and second moments against μ of Equation (5.4; 1) one obtains

$$\frac{1}{r^2} \frac{\partial}{\partial \tau} (r^2 H) = J - S, \tag{5.4; 2}$$

$$\frac{\partial K}{\partial \tau} - \frac{3K - J}{k\varrho r} = H. \tag{5.4; 3}$$

With the Eddington factor f, Equation (5.4; 3) is transformed into

$$\frac{\partial (fJ)}{\partial \tau} - \frac{(3f - 1)}{k\varrho r} J = H. \tag{5.4; 4}$$

Following Auer (1971) and Mihalas and Hummer (1974b) we introduce a *sphericity factor* q with

$$\ln(qr^2) = \int_{r_{min}}^{r} \frac{(3f - 1)}{f} \frac{dr'}{r'} + \ln(r_{min})^2. \tag{5.4; 5}$$

Then, from Equation (5.4; 5)

$$\frac{1}{k\varrho r^2 q} \frac{\partial}{\partial r} (fqr^2 J) = \frac{\partial (fJ)}{\partial \tau} - \frac{(3f - 1)}{k\varrho r} J. \tag{5.4; 6}$$

Hence, Equation (5.4; 4) can be rewritten as

$$\frac{1}{k\varrho r} \frac{\partial}{\partial r} (fqr^2 J) = r^2 H. \tag{5.4; 7}$$

We substitute Equation (5.4; 7) into Equation (5.4; 2) and thus obtain the *combined moment equation*

$$\frac{\partial}{k\varrho\, \partial r} \left(\frac{1}{q} \frac{\partial}{\partial \tau} (fqr^2 J) \right) = r^2 (J - S). \tag{5.4; 8}$$

Defining a new independent depth coordinate

$$dX = -qk\, dr = q\, d\tau$$

one obtains:

$$\frac{\partial^2 (fqr^2 J)}{\partial X^2} = \frac{r^2}{q} (J - S) \tag{5.4; 9}$$

which is the *moment equation of transfer* adapted to the case of an extended spherical stellar atmosphere.

As a next step outer and inner *boundary conditions* have to be derived. For $\tau=0$ there should be no incident radiation, hence $I(-\mu)=0$, and with

$$f_H \equiv \int_0^1 I(\mu)\mu \, d\mu \Big/ \int_0^1 I(\mu) \, d\mu = H(0)/J(0),$$

Equation (5.4; 7) becomes

$$\frac{\partial(fqr^2J)}{\partial X} = f_H r^2 J \quad \text{for} \quad \tau = 0. \tag{5.4; 10}$$

At the inner boundary we assume that the diffusion limit is reached, hence that the photon mean-free-path becomes small as compared to the density scale height H_d, so that

$$\frac{dJ_\nu}{d\tau_\nu} = \frac{dB_\nu}{d\tau_\nu}, \tag{5.4; 11}$$

hence

$$H_\nu(\infty) = \frac{-1}{3(\kappa_\nu + \sigma_\nu)\varrho} \frac{dB_\nu}{dr}\Big|_\infty = \frac{-1}{3k_\nu\varrho} \frac{dB_\nu}{dr}\Big|_\infty. \tag{5.4; 12}$$

If $(\kappa_\nu+\sigma_\nu)_R \equiv k_R$ is the Rosseland mean opacity, defined by:

$$\frac{1}{k_R} = \frac{\pi}{4\sigma T^3} \int \frac{1}{\kappa_\nu + \sigma_\nu} \frac{\partial B_\nu}{\partial T} \, d\nu,$$

then

$$H = \frac{-1}{3(k_\nu + \sigma_\nu)_R\varrho} \frac{dB}{dT} \frac{dT}{dr},$$

and

$$\frac{H_\nu}{H} = \frac{(\kappa_\nu + \sigma_\nu)_R}{(\kappa_\nu + \sigma_\nu)} \frac{dB_\nu/dT}{dB/dT} = \frac{(k_\nu)_R}{k_\nu} \frac{dB_\nu/dT}{dB/dT}$$

with

$$4\pi H = \frac{L}{4\pi R_{min}^2}. \tag{5.4; 13}$$

The equation of *hydrostatic equilibrium* reads

$$\frac{dP}{\varrho \, dr} + \frac{G\mathfrak{M}\varrho}{r^2} - \frac{dP_{rad}}{\varrho \, dr} - \frac{dP_t}{\varrho \, dr} = 0, \tag{5.4; 14}$$

where the last term is mostly neglected in literature, because the velocity field in stellar atmospheres, particularly in giants and supergiants, is so badly known. We shall do the same, realizing that we make most probably an intolerable but regretfully still unavoidable approximation.

The radiative acceleration is

$$\frac{\mathrm{d}P_{\mathrm{rad}}(r)}{\varrho\,\mathrm{d}r} = \frac{4\pi}{c}\int_0^\infty (\kappa_\nu + \sigma_\nu)\,H_\nu\,\mathrm{d}\nu. \tag{5.4; 15}$$

Already earlier we defined the Rosseland mean opacity k_R; next we define the *flux mean opacity* k_F

$$k_F = \frac{1}{H}\int (\kappa_\nu + \sigma_\nu)\,H_\nu\,\mathrm{d}\nu. \tag{5.4; 16}$$

Thus one obtains with Equation (5.4; 13):

$$\frac{\mathrm{d}P_{\mathrm{rad}}}{\varrho\,\mathrm{d}r} = \frac{k_F}{c}\frac{L}{4\pi r^2}, \tag{5.4; 17}$$

which, introduced in Equation (5.4; 14), gives:

$$\frac{1}{\varrho}\frac{\mathrm{d}P}{\mathrm{d}r} + \frac{G\mathfrak{M}}{r^2}(1 - \Gamma_F) = 0, \tag{5.4; 18}$$

where Γ_F is defined by Equation (1.3; 2). In a fully scattering atmosphere without lines k_F is equal to σ_e. Then Γ_F in Equation (5.4; 18) should be replaced by

$$\Gamma_s = \frac{\sigma_e L}{4\pi c\, G\mathfrak{M}}.$$

However, spectral lines also contribute to the radiation pressure, so their influence has to be calculated.

To that end it is sometimes convenient, and in any case in agreement with current practice, to split the radiative acceleration term into one contribution of the continuous radiation, and another of the lines. If k_{FC} would be the continuous flux-mean-opacity (hence neglecting the lines), and if we define

$$\Gamma_C = k_{FC}L/4\pi c G\mathfrak{M}, \tag{5.4; 19}$$

then Equation (5.4; 18) would read

$$\frac{1}{\varrho}\frac{\mathrm{d}P}{\mathrm{d}r} + \frac{G\mathfrak{M}}{r^2}(1 - \Gamma_C) + \Sigma g_L = 0, \tag{5.4; 20a}$$

where g_L is the acceleration due to the Lth spectral line. This latter term is further elaborated in Section 5.6 and is described by Equations (5.6; 1) and (5.6; 4). The *radiative force multiplier M*, defined there, is the ratio between the acceleration exerted by the lines and the continuous radiation. The introduction of this term would then yield

$$\frac{1}{\varrho}\frac{\mathrm{d}P}{\mathrm{d}r} + \frac{G\mathfrak{M}}{r^2}\{1 - \Gamma_C(1 + M)\} = 0. \tag{5.4; 20b}$$

We have still to add that M is small in the photosphere – lines do not contribute greatly to photospheric instability – but can take values of between 10^1 and 10^2 in the stellar winds.

As a next step the expression (5.4; 20) is elaborated (Cassinelli, 1971b) by using the equation of state

$$P = \varrho kT / \mu m_H$$

and defining

$$\zeta \equiv \frac{GMm_H}{kT_1 r} \equiv \frac{A_\zeta}{r},$$

where A_ζ is a constant and T_1 a reference temperature. Then Equation (5.4; 20) can be written in *logarithmic* form

$$\frac{\mathrm{d} \ln P}{\mathrm{d}\zeta} = \frac{\mu(\zeta)}{T(\zeta)/T_1} \left\{ 1 - \Gamma_c(1 + M) \right\}, \qquad (5.4; 21)$$

which is useful because the pressure varies over many powers of ten in a stellar atmosphere.

The interpretation of Equation (5.4; 21) is particularly simple if T, μ, and k_F are constant. The latter condition is fulfilled in very hot, tenuous atmospheres, where $k_F \approx \sigma_e$ and M is small. Then

$$P = P_0 \exp(\xi - \xi_0), \qquad (5.4; 22)$$

with

$$\xi = \frac{\mu}{T/T_1} (1 - \Gamma)\zeta = A_\zeta / r,$$

and

$$\mathrm{d}\xi / \mathrm{d}r = -H^{-1}.$$

Here H is the pressure scale height $= kT / \mu m_H g_{\mathrm{eff}}$.

To solve the equations one has to determine the *outer boundary condition on the pressure*. While this does not offer difficulties in the plane-parallel case, where we find from $\varrho = \varrho_0 \exp(-z/H)$:

$$\tau_{\min} = - \int_R^\infty \varrho\sigma \, \mathrm{d}z = \varrho_0 \sigma H, \qquad (5.4; 23)$$

one encounters a difficulty in the case of a spherical atmosphere, where – analogous to Equation (5.4; 22) – we have $\varrho = \varrho_0 \exp(\xi - \xi_0)$, yielding a finite value at infinity, and a diverging value for the optical depth integral

$$\tau_{\min} = A_\zeta \int_0^{\xi_0} \varrho_0 \exp(\xi - \xi_0) \, \sigma \, \frac{\mathrm{d}\xi}{\xi^2}. \qquad (5.4; 24)$$

Since the density appears to be $\neq 0$ at the star's 'outer boundary' the same applies to the pressure. The cause of this dilemma is related to our assumption that the atmosphere is static: apparently, in that case the total mass of a spherical and iso-thermal gas sphere becomes infinite. By introducing outward motions the density will decrease outward more rapidly than in the case of a purely hydrostatic distri-bution. We thus come to the important *conclusion* that in a static spherical case the mass of the atmosphere should be infinite in order not to have the outward motion. Hence, *consistent* models for spherical *non-expanding* atmospheres *cannot be derived*; outstreaming motions are essential.

For that reason models of static spherical atmospheres must be considered as *approximations* of the real – expanding – case. This approximation is obviously only valid in those parts of the atmosphere where the wind velocities are still small. How should this statement be formulated in mathematical terms?

An often-used approximation is a cut-off procedure introduced by Chamberlain (1963). He showed that for the case $\xi_0 \gg 1$, it is acceptable to take the lower limit in the integral (5.4; 24) equal to $\xi=1$ instead of $\xi=0$.

In that case

$$\tau_{min} = \sigma\varrho_0 H_0 K(\xi_0),$$

with

$$K(\xi_0) = \int_1^{\xi_0} \frac{\xi_0}{\xi^2} \exp[-(\xi_0 - \xi)]\,d\xi \approx 1 + \frac{2}{\xi_0} \quad \text{for} \quad \xi_0 \gg 1.$$

Hence

$$\tau_{min} = \sigma\varrho_0 H_0 \left(1 + \frac{2}{\xi_0}\right) \tag{5.4; 25}$$

an expression that does not differ greatly from the one valid for the plane-parallel case, Equation (5.4; 23).

It is then found that at τ_{min}:

$$(P_g)_{min} = \frac{\tau_{min}}{\sigma_e} \frac{kT_0}{\mu_0 m_H} \frac{1}{H_0\left(1 + \frac{2}{\xi_0}\right)}$$

$$= \frac{\tau_{min}}{\sigma_e} \frac{G\mathfrak{M}}{r_0^2} \frac{1 - \Gamma_F}{1 + \frac{2}{\xi_0}}, \tag{5.4; 26}$$

where suffixes 0 all refer to the τ_{min}-level.

Summarizing, a model of a static spherical stellar atmosphere is obtained by the simultaneous solution of
– the *moment equation of transfer*, Equation (5.4; 9), with boundary values:

$$\text{for } \tau=0: \quad \text{Equation (5.4; 10)},$$
$$\text{for } \tau=\infty: \quad \text{Equation (5.4; 12)};$$

and

– the equation of *hydrostatic equilibrium*, Equation (5.4; 21), with the boundary value for P_{min}: Equation (5.4; 26).

In the case of *radiative equilibrium* these should be completed by the condition

$$\int_0^\infty k_{\nu Q} S_\nu \, d\nu = \int_1^\infty k_{\nu Q} J_\nu \, d\nu, \qquad (5.4; 27)$$

which can be conveniently rewritten as

$$r^2 \int_1^\infty H_\nu \, d\nu = \frac{L}{(4\pi)^2} \equiv H^0 = \text{constant}. \qquad (5.4; 28)$$

A possible computational scheme, developed by Cassinelli, is the following: *assume* values for the stellar mass $\mathfrak{M}$, and for the radius R_{min} at a large τ_R-value, e.g. $\tau_R = 10$. (One has to realize that $R(\tau_\nu)$ is wavelength-dependent.) Assume further first-order approximation functions for $T(r)$, $f(r)$, and solve the equation of hydrostatic equilibrium. With the assumed values of $T(r)$, $P(r)$ and hence $k_\nu(r)$ the optical depths $\tau_\nu(r)$ are computed, and likewise the mean optical depth. Thereafter, by integrating the radiative transfer equations – separately for each frequency – the mean radiation intensity J and the higher moments of I are derived. The equation of radiative equilibrium is tested, and, if not satisfied, the temperature-depth relation is changed. Similarly the boundary conditions are tested.

Hence, with the assumed mass $\mathfrak{M}$, and R we find values for $H_0(r)$, hence L, and $T(\tau_\nu)$, and the stellar spectrum can be computed. We are thus able to verify whether the atmospheric model developed satisfies the parameters of the star one wishes to investigate, like color indexes, $\mathfrak{M}$ and L values. If there is not satisfactory agreement, then the computation should start anew, selecting e.g. another $R_{min}(\tau_R = 10)$-value, for a star with the given mass. Altogether the scheme is rather tedious, and only few stellar atmospheric models have so far been computed.

Detailed iterative methods for performing the computations, and models thus computed were developed by Cassinelli (1971b), Castor (1974a), and Mihalas and Hummer (1974a, b). Cassinelli published six models, for stellar masses of 0.6 and $1.0\mathfrak{M}_\odot$ (intended to represent central stars of planetary nebulae); the luminosity ranges between 2.0 and $3.9 \times 10^4 L_\odot$, and $T(\tau_R = \frac{2}{3})$ between 34 700 K and 92 700 K. The stellar 'radii' are of the order of 1.4 to $6R_\odot$; the photospheric 'thicknesses' are of the same order, so they are undoubtedly to be considered as spherically extended.

Mihalas and Hummer (1974a, b) published seven LTE and seven non-LTE static spherically symmetric models, all corresponding to a star with $\mathfrak{M} = 60\mathfrak{M}_\odot$, $L = 1.25 \times 10^6 L_\odot$, $R = 24R_\odot$, $T_e = 39\,500$ K, and $\log g \approx 3.45$. This would correspond to a spectral type near O6. The models are very close to a balance between g_{grav} and g_{rad}, hence $g_{eff} \approx 0$. The models differ in the assumed values for M (defined in Equation (5.6; 5)) which are tentatively assumed constant (1 ... 1.6) or slightly τ-dependent (decreasing to 0 for $\tau \to \infty$). (As will be shown in Section 5.6 these values may be larger in the outer parts of the atmosphere.) They computed some observable quantities in the theoretical stellar spectra such as the profiles of a few hydrogen lines, the Balmer jump, the continuous spectrum, including the colours. The $L\alpha$ line should occur in emission, the lower Balmer lines and the Balmer jump in absorption.

Castor (1974a) did not publish details of his models but gave some computational results for half a dozen of models for $\mathfrak{M}=30\mathfrak{M}_\odot$. His models have T_e-values ranging from 30 000 K to 50 000 K, Γ-values of 0.95, 0.9, and 0.4.

Along the same line as Mihalas and Hummer (1974b), Kunasz et al. (1975) computed spherical static non-LTE model atmospheres for stars with $\mathfrak{M}/\mathfrak{M}_\odot=30$ and 60, and at various points on the evolutionary tracks. Some improvements were introduced in the computational formalism. At a given τ_λ the temperatures in the non-LTE models are generally lower than in the LTE models, for models with $T_e<45\,000$ K. For models with higher effective temperatures the situation is reverse. The Lyman jump and the Lyman lines are in emission, the Balmer jump and the Balmer lines in absorption. He II $\lambda 4686$ comes in emission in the most extended model atmospheres – see in this connection also our Section 3.2, on Of stars.

Few results have so far been published on model calculations for *cool stars*; the problems encountered in dealing with such objects are formidable indeed. Watanabe and Kodaira (1978) following earlier work by Schmidt–Burgk and Scholz (1977), using the computer code of Hundt et al. (1975) constructed models of static spherical stars in the T_e-range 3200–3800 K, and with $\log g$-values around -0.5. One of the results found from these studies is that the apparent size of the star measured in certain molecular bands may be by 50% larger than the continuum value (Watanabe and Kodaira, 1979), a result that confirms observational data (see Sections 2.5 and 3.11).

One of the purposes of the various model calculations just described, is to examine whether models with extended atmospheres can explain the low colour temperatures (the 'flat' continuous spectra) of Of stars, as already mentioned for the well-studied case of ζ Pup (O4 ef) in Section 3.2 – see in particular Figure 28 there. For ζ Pup: $T_c\approx 30\,000$ K, and $T_e\approx 50\,000$ K. Anticipating the discussion in our next sections it seems, particularly after the work of Castor et al. (1975), that models of *expanding* spherical atmospheres yield a better agreement with the observations than static models, while additional flattening of the spectra is caused by accumulation of fraunhofer lines in the far ultraviolet. Reference is made to Section 5.9, and to a description of the radiation-driven atmospheric model of ζ Pup in Section 3.2.

5.5. Expanding Spherical Stellar Atmospheres; Introduction

Any stellar atmosphere has a certain non-zero rate of mass-loss since at any time there is always a certain fraction of the atmospheric gas that has outward velocity components exceeding the escape velocity $v_{esc}=\sqrt{2GM/R}$. If these particles are at an atmospheric level where the mean free path exceeds the atmospheric scale height ($\approx$ 'effective thickness of the remaining atmosphere') the particles may escape. If the density of the escaping gas is so small that no interaction occurs between the outflowing particles then one is dealing with the case called the *particle-loss model*, or the *exospheric model*. If there is sufficient interaction between the outflowing particles to establish a hydrodynamic pressure, and when the notion 'temperature of the outflowing gas' makes sense, one is dealing with a *stellar wind*; the flow can then be treated from a continuum-hydrodynamics view. In practically all cases of some significance for stellar evolution and structure one is dealing with this latter case.

These stellar winds may lead to a considerable mass-loss of the star. An *upper limit* of the mass flux is obtained on the basis of the assumption of equality of the mass and photon momentum fluxes: $-\mathfrak{M}v_\infty = Lc^{-1}$. Thus, one finds for normal supergiants upper limits of the mass fluxes of a few times 10^{-5} solar masses per year, a value actually reached in Of and WR stars (cf. Section 7.4). The rates of mass-loss derived for extreme objects such as η Car (Section 6.17) and VY CMa (Section 6.16) exceed this limit. They must therefore be considered with caution if they are not due to other, still unknown causes.

Theories on stellar winds have up to now been based on two different kinds of approaches. The first approach follows Parker (1958), who established a detailed theory of the *solar* wind, where the driving force is the gas pressure of the hot solar corona, which gives rise to hydrodynamic flow of gas. Parker's equations, initially developed for the solar corona, are applied to the stellar case (Parker, 1965).

On the other hand, Lucy and Solomon (1970) following ideas initially brought by Milne (1926) have stressed the importance of radiation forces on strong resonance lines in hot stars. Cassinelli and Castor (1973) formulated the problem for an atmosphere under influence of radiation forces. These drive the flow, and thus influence the momentum balance.

In its basic form the theory of stellar winds is described by three equations. The first is that of *conservation of momentum*, which can be derived from the equation of hydrostatic equilibrium, Equation (5.4; 14) or its variant (5.4; 20). This yields

$$v\frac{dv}{dr} + \frac{1}{\varrho}\frac{dP}{dr} + \frac{G\mathfrak{M}}{r^2}\{1 - \Gamma_c(1 + M)\} = 0. \qquad (5.5;\ 1)$$

In the theory of *coronal* winds the term between braces is equal to unity.

The next equations are those of *conservation of mass* (the equation of *continuity*):

$$4\pi\varrho vr^2 = -\mathfrak{M}(=\text{constant}), \qquad (5.5;\ 2a)$$

or

$$\frac{1}{\varrho}\frac{d\varrho}{dv} + \frac{1}{v}\frac{dv}{dr} + \frac{2}{r} = 0, \qquad (5.5;\ 2b)$$

and the equation of *state*:

$$P/\varrho = \mathfrak{R}T/\mu = s^2. \qquad (5.5;\ 3)$$

From these three equations one derives the *velocity equation*:

$$\frac{1}{v}\frac{dv}{dr}(v^2 - s^2) = \frac{2s^2}{r} - \frac{ds^2}{dr} - \frac{G\mathfrak{M}}{r^2}\{1 - \Gamma_c(1 + M)\}. \qquad (5.5;\ 4)$$

This equation has two particular aspects. First, the density does not occur in Equation (5.5; 4) so it can only determine the $v(r)$ law, not the mass-loss! Another aspect is that there may be a point – the *sonic point* (cf. Figure 69) – where $v(r_s)=s$; at that point the left-hand bracket term vanishes. If the righthand term does not vanish too in the sonic point, then dv/dr would approach infinity there, physically an impossible situation. However, a solution in which always $v<s$ would lead to $v\rightarrow0$ at $r\rightarrow\infty$ hence giving $\varrho=\infty$ at infinity. This is also inacceptable. So, for $r\rightarrow\infty$ v must approach a finite value, $\neq0$. Observations of stars, and also of the Sun, show that this terminal value $v_\infty>s$. The only solution that complies with the observation that $v<s$ for small r-values, and with the above considerations is the one going through the sonic point, where v changes from subsonic to supersonic values. In that point the righthand part of Equation (5.5; 4) is zero. That solution of Equation (5.5; 4) gives the velocity law $v(r)$.

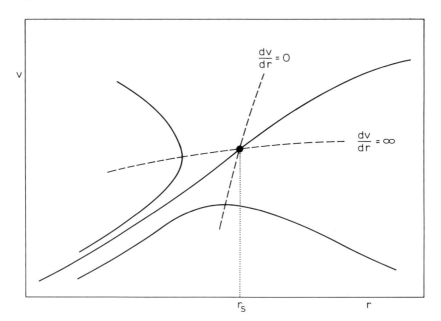

Fig. 69.   A Parker-type (or: X-type) singularity in the solution of the stellar-wind equations in the 'coronal' case.

In order to find the *mass-loss* the equation of conservation of energy has to be added. In the static case the latter equation reduces to the equation of radiative equilibrium but in an expanding atmosphere the gas motion should be included, and the equation may be written as (cf. Cassinelli, 1979b):

$$v\frac{dE}{dr}+Pv\frac{d}{dr}\left(\frac{1}{\varrho}\right)=\frac{1}{\varrho}(Q_A+Q_R-\nabla\cdot\mathbf{q}_c),\qquad(5.5;5)$$

which says that per unit mass the rate of change of the internal energy *plus* the work done by the gas against the pressure *equals* the energy changes due to deposition of mechanical energy (Q_A), by absorption or emission of radiative energy (Q_R) *minus* the loss by conduction (where $\mathbf{q}_c$ is the conductive flux).

The Equation (5.5; 5) applied in conjunction with (5.5; 4) shows the relation between *coronal* or *chromospheric* heating and the *rate of mass-loss*. Hence, in order to be able to predict $\mathfrak{M}$-values the mechanism of outer atmospheric heating and the energy losses should be known; in other words: knowledge of mass-loss demands knowledge of the energy balance of the outer stellar layers.

The physics of coronal heating will be discussed in Chapter 6; cf. Section 6.10; observed rates of mass-loss are summarized in Chapter 7.

In the following sections of this chapter we shall first discuss the force on outer stellar layers by the gradient of radiation-pressure; thereafter we describe models of spherical stellar atmospheres having appreciable, radiatively-driven outward flow of matter. The technique for deriving such atmospheric models was developed by various authors; the first satisfactory models were derived by Castor *et al.* (1975).

5.6. The Radiation Force on Outer Stellar Layers

Large forces can be exerted on stellar atmospheric gas as a result of the absorption and scattering of radiation in the strong ultraviolet resonance lines that are present in the spectra of hot stars (Lucy and Solomon, 1970). A stellar wind may originate as a result of radiation forces, which can thus lead to mass-loss. This idea was taken up by Cassinelli and Castor (1973), Castor (1974b), and was reformulated by Castor *et al.* (1975). In these latter papers and in a paper by Castor *et al.* (1976) the forces on lines other than resonance lines were also considered. It was found that the main contribution is from resonance lines, particularly of abundant ions such as C, N, and O. The spectral range $\lambda < 912$ Å is important for the hottest stars. The total multiplication factor is 10 to 10^2. For ζ Pup it is about 70. Hence, model stellar winds and mass-loss rates calculated including the radiation forces exerted by the many resonance lines have larger velocities and are larger than those derived without. Furthermore, Maciel (1976, 1977) showed that radiation forces on molecular bands of CO, H_2O, and OH can account for stellar winds observed in M-type stars, while Salpeter (1974) stressed the importance of radiation forces on dust particles (cf. Section 7.7).

The radiation force on the gas is related to the solution of the line-transfer problem, and, because the shift of the spectral lines in a frame moving with the gas defines the radiation flux and thus determines the value of the force exerted, the force is a function of the local *v-gradient*. Therefore the stellar wind theory based on this effect differs in this important respect from the theory of stellar winds from a hot corona.

Let us first compute the force produced by a *single line*. Approximate expressions are given by Lucy and Solomon (1970), and by Castor (1974b). The acceleration is

$$g_{\text{rad}, L} = \frac{\kappa_L \pi F_c \, \Delta v_D}{c} \{(1 - e^{-\tau_L})\tau_L^{-1}\}. \tag{5.6; 1}$$

$$= \frac{\pi e^2}{mc^2} \, gf \, \frac{n_i}{\varrho} \, \pi F_c \, \{(1 - c^{-\tau_L})\tau_L^{-1}\},$$

where κ_L is the monochromatic line absorption coefficient per gram, F_C the continuum flux per unit frequency interval at the frequency of the line, $\Delta\nu_D = \nu_0 v_{th}/c$ the Doppler width, and τ_L the optical thickness in the line. In Equation (5.6; 1) the quantity τ_L/κ_L is the column of mass that can scatter the photons, and $(1 - e^{-\tau_L})$ is the probability for scattering to occur. The term between braces can conveniently be approximated by min $(1, 1/L)$, which specifies that the minimum value of the two bracketed quantities is to be taken.

In an expanding atmosphere with a velocity gradient the optical thickness of the shell contributing to the line is approximately

$$\tau_L = \kappa_L \varrho v_{th} \left|\frac{dv}{dr}\right|^{-1} = \frac{\pi e^2}{mc} \lambda_0 \, gfn_i \left|\frac{dv}{dr}\right|^{-1}, \qquad (5.6; 2)$$

since τ_L counts only the number of absorbers in a section of the column along the line of sight towards the observer, across which the velocity changes by v_{th}, the thermal velocity of the atoms. We thus integrate through the line profile as it is seen Doppler-shifted in the observer's frame; microturbulent motions are neglected but could easily be included once that $v_\mu(r)$ would be known sufficiently accurately.

Calling $\beta = k_{FC}/\kappa_L$, where k_{FC} is the flux-mean-opacity due to the continuous spectrum alone (cf. Equation (5.4; 19)), and introducing $t = \beta\tau_L$, we have with Equation (5.6; 2):

$$t = k_{FC}\varrho v_{th} \left|\frac{dv}{dr}\right|^{-1}. \qquad (5.6; 3)$$

In order to obtain the total acceleration produced by *many lines*, one merely sums up the influence of all lines considered. One thus obtains

$$g_{rad} = \Sigma \, g_{rad,L} = \frac{k_{FC}\pi F}{c} M(t), \qquad (5.6; 4)$$

where πF is the total flux, the factor $k_{FC}\pi F/c$ is the acceleration produced by the continuum, and the so-called *radiative force multiplier* $M(t)$ is defined by

$$M(t) = \sum_{\text{lines}} \frac{F_C \Delta\nu_D}{F} \min\left(\frac{1}{\beta}, \frac{1}{t}\right). \qquad (5.6; 5)$$

The factor β^{-1} is the ratio of line opacity to continuum opacity and can be eliminated from the standard expressions for the non-LTE level-populations as:

$$\frac{1}{\beta} = \frac{\pi e^2}{mc} \, gf \, \frac{N_l/g_l - N_u/g_u}{\varrho k_{FC}} \, \frac{1}{\Delta\nu_D}, \qquad (5.6; 6)$$

where N and g are level-populations and statistical weights respectively, and subscripts u and l refer to upper and lower levels. We now consider two extreme cases: If the depth-variable t is as large as unity, which occurs inside the stellar photosphere, then Equation (5.6; 5) shows that the force multiplier is not larger than the fraction of the continuum flux which is blocked by the lines, which is only a small fraction of the total flux. However, $M(t)$ becomes large when $t \to 0$, since then

$$M(t) \to \sum_{\text{lines}} \frac{F_C}{F} \frac{\pi e^2}{mc} gf \frac{N_l/g_l - N_u/g_u}{\varrho \kappa_{FC}}. \qquad (5.6; 7)$$

A maximum value for $M(t)$ can be derived approximately by taking the width of the continuous energy distribution to be about $2.5kT_e/h$ which is $\approx F/F_C$, and by writing X for the fraction of all the atoms in the gas which are able to absorb effectively in the spectral region where most of the flux is emitted. Then

$$M_{\max} \approx 0.7 \times 10^8 \, (10^4/T_e) X. \qquad (5.6; 8)$$

Castor *et al.* (1975) estimate $X \approx 10^{-4}$; hence $M_{\max} \approx 10^3$. In actual practice $1 \leqslant \log M_{\max} \leqslant 2$.

It has been a matter of some research how many and which lines contribute to M, particularly whether subordinate lines contribute too. Since the work by Castor *et al.* (1976) it is clear that *resonance* lines give by far the major contribution. In hot stars the wavelength region below 912 Å is by a factor ~ 5 more important than the region >912 Å.

As was shown in Chapter 1 of this book, the accelerations by radiation forces due to electron scattering is slightly smaller than that of gravity in the *photospheres* of hot supergiants. The above result shows that in the *outer atmosphere* where the strong Fraunhofer lines originate the influence of line-scattering on P_r is one to two orders larger than that of electron scattering, so that mass-flow induced by radiation forces exerted via resonance lines of the abundant ions must determine the instability of the *outer* atmospheres of hot stars.

It is also of importance to note, with Castor *et al.* (1975), that the value of the multiplier M and its depth-dependence, can conveniently be represented by

$$M(t) = kt^{-a}, \qquad (5.9; 6)$$

where k and a are constants to be chosen through fits to detailed calculations. The constant k is called the 'force constant'. If all lines are optically thick $a=1$, while $a=0$ if all lines are thin; so, in general $0 < a < 1$. In any case $a \to 0$ at very large distances from the star, where eventually all lines are optically thin.

5.7. Radiation-Driven Expanding Atmospheres; the Subsonic Region

In the following part of this chapter we have to distinguish between two domains in the stellar surroundings. The first is the *near-stellar expanding atmosphere*, roughly the subsonic region with $r < r_s$, and the second is the stellar wind proper, roughly with $r > r_s$.

The theoretical approach differs for these two regions, and a unified theory has not yet been established. Therefore, we follow that line and discuss in the present section the subsonic region, while the supersonic stellar wind is dealt with in Section 5.8.

As in Section 5.5 our starting equations are those of *conservation of mass* (5.5; 2) and of *conservation of momentum* (5.5; 1). In the latter, as usual, the turbulent pressure term has been omitted. In analogy with Cassinelli and Hartmann (1975) we now define

$$\Gamma^* = \Gamma_C(1+M), \tag{5.7; 1}$$

where Γ_C is defined by Equation (5.4; 19), M by Equation (5.6; 5). (Note that our Γ^* is not equal to the one defined by Cassinelli and Hartmann.) Eliminating the gas pressure and density from Equation (5.5; 1), while using logarithmic derivatives of Equation (5.5; 2), and the equation of state (5.5; 3), one obtains, in analogy with Equation (5.5; 4):

$$\frac{d \ln v}{d \ln r} = \frac{2 - \dfrac{d \ln T}{d \ln r} - \dfrac{G\mathfrak{M}}{r\mathfrak{R}T}(1-\Gamma^*)}{\left(\dfrac{v^2}{RT} - 1\right)}. \tag{5.7; 2}$$

As is the case with Equation (5.5; 4) the solution of this version of the velocity equation gives the velocity law, and not the mass-loss because the gas density does not occur in the equation. The equation contains the temperature gradient, because the flow is driven by radiation forces, which term therefore enters in the momentum balance. Cassinelli and Hartmann (1975) derived the temperature gradient from the equation of radiative equilibrium (5.4; 27), in which the source function S is replaced by the Planckian function B – it was found that this approximation is an acceptable one.

We introduce further the following ratios of opacities

$$\chi_J = k_J/k_P; \quad K_R = k_R/k_{FC}; \quad \chi_F = k_F/k_R, \tag{5.7; 3}$$

where k_J is the opacity averaged over J, and k_P the Planck mean. The Equation (5.4; 27) is then equivalent to

$$\chi_J J = B. \tag{5.7; 4}$$

From Equation (5.4; 7) we derive

$$\frac{1}{q}\frac{d}{dr}(fq\,J) = -k_{F\varrho}H. \tag{5.7; 5}$$

Eliminating J from Equations (5.7; 4) and (5.7; 5) gives (Castor, 1974a)

$$\frac{1}{q}\frac{d}{dr}\left(fq\frac{B}{\chi_J}\right) = -\chi_F K_R k_{FC}\varrho\,\frac{L}{(4\pi r)^2}. \tag{5.7; 6}$$

With this formulation the computation of an expanding stellar model atmosphere goes essentially parallel to the computation of a static model atmosphere where the moment equation of transfer is now Equation (5.7; 6) and the equation of hydrostatic equilibrium is replaced by Equation (5.7; 2). Equation (5.7; 6) is solved iterativerly on the opacity ratios χ_J and χ_F. Both approach unity for $\tau \to \infty$.

In the approach by Cassinelli and Hartmann (1975) the outer boundary of the atmosphere is (arbitrarily) set at the sonic radius r_s, where $v=s$. This is often a reasonable assumption since the more outer layers contribute hardly to the stellar continuous spectrum.

As demonstrated in Section 5.5, Parker's theory of stellar winds shows that a smooth transition from subsonic to supersonic flow over r_s can take place only when the numerator and denominator of Equation (5.7; 2) vanish both at r_s. Hence at that radius, using subscripts s for quantities at the sonic point:

$$s^2 = \Re T_s,$$

and

$$2 - \left.\frac{d \ln T}{d \ln r}\right|_s - \frac{G\mathfrak{M}}{r_s \Re T_s}(1 - \Gamma_s^*) = 0. \qquad (5.7; 7)$$

These equations yield two necessary boundary conditions. The boundary condition of the radiative differential Equation (5.7; 6) is found by the introduction of the 'surface Eddington factor' g:

$$g = H_s/J_s,$$

and becomes, using Equations (5.7; 1) and (5.7; 4)

$$r_s^2 g \left(\frac{\sigma T^4}{\pi \chi_J}\right)_s = \Gamma_s^* \frac{c G\mathfrak{M}}{4\pi k_{FC}}. \qquad (5.7; 8)$$

In a similar way L is eliminated from (5.7; 6), yielding

$$\frac{d \ln \alpha B}{d \ln r} = \frac{-q k_{F\varrho} \left(\frac{r}{r_s}\right)^{-1} \left(\frac{rg B}{\chi_J}\right)_s}{\alpha B(r)},$$

where

$$\alpha(r) = fq/\chi_J.$$

Introducing further $\theta_s = k_{FC}\varrho_s r_s$, we obtain

$$\frac{d \ln T}{d \ln r} = \frac{-q\chi_F K_R \left(\frac{\varrho}{P_s}\right)\left(\frac{r}{r_s}\right)^{-1} \theta(g B/\chi_J)_s}{4\alpha B(r)} - \frac{1}{4}\frac{d \ln \alpha}{d \ln r}. \qquad (5.7; 9)$$

Equation (5.7; 9) together with Equations (5.7; 7) and (5.7; 8) enable one to determine the radius and temperature at the sonic point r_s, with:

$$r_s = \left\{\frac{\Gamma^* cG\mathfrak{M}(\chi_J)_s}{4k_{FC}g\sigma}\right\}^{1/2} \frac{1}{T_s^2} \equiv \frac{C_1 \mathfrak{M}^{1/2}}{T_s^2},$$

and

$$T_s = \left[2 + \frac{1}{4}\frac{\chi_F K_R \theta g}{f} + \frac{1}{4}\frac{d \ln \alpha}{d \ln r}\right]_s \left(\frac{G\mathfrak{M}^{1/2}(1 - \Gamma^*)}{RC_1}\right)^{-1}.$$

Model computations are made by choosing as parameters for the models the stellar mass $\mathfrak{M}$, the sonic temperature T_s and a value for θ_s (which can be regarded as an optical depth for continuous scattering and absorption, if $\varrho(r)$ were proportional to r^{-2}). Initial estimates must further be made of the quantities χ_J, χ_F, f, and g over which iteration has to be taken. Often it is convenient to use data from static models as starting values.

The methods to solve the transfer equation do not differ essentially from the solution in the case of the static problem. For further details on the technique of the computations reference is made to Cassinelli and Hartmann (1975) or to Mihalas et al. (1975b) who dealt with the same problem and gave an elaborate mathematical treatment. Klein and Castor (1978) calculated stellar wind models for $T_e \approx 30000$–50000 K and log $g = 3$–4.5. Lucy (1976a) studied static and expanding models for cool carbon stars, taking $\mathfrak{M} = 1.5 \mathfrak{M}_\odot$, $M_b = -6$, $(C/H) = 1.22 \times 10^{-3}$, $(C/O) = 1.76$, while T_e was varied between 2000 and 3000 K. A weak point in these latter calculations was the assumption of a constant radius, independent of wavelength. The formation of dust-grains was examined and its influence on the emergent energy-wavelength curve calculated.

5.8. Line-Driven Stellar Winds

In one aspect there is no fundamental difference between the stellar wind from a hot corona and that of a radiation-driven wind: in both cases an outward force accelerates the gas. However, the force terms are different in the two cases: it is either the gradient of the gas pressure $P = NkT$ for coronal winds, or else it is the radiation force on the line(s) which is proportional to some power of the velocity gradient for radiation-driven winds.

There is still another difference. As shown in the previous section, the work of Cassinelli and Hartmann (1975) enables one to solve simultaneously for the velocity, temperature and radiation field in the subsonic part of an expanding atmosphere. However, these authors assumed that there is a Parker 'X-type' singularity at the sonic point. Now, the atmospheres of luminous early-type stars are more strongly affected by line-opacity than by continuum opacity. It appears that in such a case the critical point is not an X-type singularity as in Parker's 'coronal' theory; this was shown in the solution given by Castor et al. (1975). Fine and very readable reviews of the various possible solutions of the general stellar-wind equation are given by Holzer (1977), Mihalas (1978, p. 562f), and Cassinelli (1979b).

Our starting equation is again (5.5; 1), that of conservation of momentum. Into this equation we substitute Equation (5.6; 9) for $M(t)$. The velocity equation, corresponding to Equation (5.5; 4), then appears to consist of a linear and a non-linear part as shown below:

$$\left(v - \frac{s^2}{v}\right)\frac{dv}{dr} = -\frac{G\mathfrak{M}(1 - \Gamma)}{r^2} + \frac{2s^2}{r} - \frac{ds^2}{dr} +$$
$$+ \frac{\Gamma_c \mathfrak{M} k}{r^2}\left[\frac{4\pi}{k_{FC} v_{th} |d\mathfrak{M}/dt|}\right]^a \left(r^2 v \frac{dv}{dr}\right)^a. \qquad (5.8; 1)$$

where a is defined in Equation (5.6; 9) and Γ_C by Equation (5.4; 19).

Further analytic simplification is obtained by introducing new variables:

$$w = v^2/s^2,$$

$$u = -2G\mathfrak{M}(1-\Gamma_c)/s^2 r = v_{esc}^2(r)/s^2,$$

and the constant

$$C = k\left(\frac{\Gamma_c}{1-\Gamma_c}\right)^{1-a}\left(\frac{L/c}{|\dot{\mathfrak{M}}|\,v_{th}}\right)^a,$$

while we further write w' for dw/du.

Equation (5.8; 1) then becomes

$$F(u, w, w') \equiv \left(1 - \frac{1}{w}\right)w' + \left(1 + \frac{4}{\mu}\right) - C(w')^a = 0. \tag{5.8; 2}$$

This is the fundamental equation of Castor, Abbott and Klein, describing stellar mass-loss by line-driven winds. Solutions of Equation (5.8; 2) have been investigated by Castor *et al.* (1975).

While it is a property of the coronal stellar-wind equation that there is only one solution, *viz.* the one through the 'sonic' point r_s, where the flow changes from sub-sonic velocities (for $r<r_s$) to supersonic values (see Figure 69, Section 5.5), the present case is different. Castor *et al.* (1975) show that the solution of Equation (5.8; 2) has one *singular point* r_c, which is normally further away from the star than the sonic point. Exact solutions of the differential equation (5.8; 2) at the singular point are given; these can be approximated if one assumes that the temperature follows a power law in r near the singular point: $s^2(:)T(:)r^{-n}$. Then

$$v^2 = \frac{a}{1-a}\frac{2G\mathfrak{M}(1-\Gamma_c)/r_c}{-\frac{1}{2}n + [\frac{1}{4}n^2 + 4 - 2n(n+1)]^{1/2}}, \tag{5.8; 3}$$

$$r^2 v\frac{dv}{dr} = \frac{a}{1-a}G\mathfrak{M}(1-\Gamma_c), \tag{5.8; 4}$$

and

$$\frac{d\mathfrak{M}}{dt} = \frac{4\pi G\mathfrak{M}}{k_{FC}v_{th}}a(1-a)^{(1-a)/a}k^{1/a}\Gamma_c^{1/a}(1-\Gamma_c)^{-(1-a)/a}. \tag{5.8; 5}$$

If $v \gg s$ then, according to Equation (5.8; 1): $r^2 v(dv/dr)$ remains constant until r is as large as $G\mathfrak{M}(1-\Gamma)/s^2$. Then Equation (5.8; 4) is true from the sonic point outward to many stellar radii, and Equations (5.8; 3) and (5.8; 4) can be integrated to fix the integration constant. This yields

$$v^2 = \frac{a}{1-a}2G\mathfrak{M}(1-\Gamma_c)\left(\frac{1}{r_s} - \frac{1}{r}\right). \tag{5.8; 6}$$

and

$$\frac{r_c}{r_s} = 1 + \frac{1}{-\frac{1}{2}n + [\frac{1}{4}n^2 + 4 - 2n(n+1)]^{1/2}}.$$ (5.8; 7)

Since n is approximately between 0 and 0.5, r_s/r_c will be ~1.5 to 1.74. Lucy's (1975) remark that r_s/r_c only appreciably exceeds unity when the thermal motions of the radiation-absorbing ions vanish as compared to the flow-speed remains to be further examined.

In actual practice one starts with a stellar model, characterized by the parameters L, $\mathfrak{M}$, and R. To a first approximation one assumes $r_s = R$ and estimates r_c with Equation (5.8; 7). Then Equations (5.8; 3) to (5.8; 5) or their more precise equivalents published by Castor *et al.* (1975) are used to find the variables at the singular point. Thereupon Equation (5.8; 1) is integrated numerically in each direction from the singular point. The exact relation found between the optical depth and radius can then be used to see if the photospheric radius has the correct value R; if not an improved value for r_c can be found and the integrations repeated.

The condition for obtaining a solution to the equations was formulated in a more simplified case by Marlborough and Roy (1970) as

$$G\mathfrak{M} > \frac{1}{4\pi c} \int_0^\infty k_\nu L_\nu \, d\nu$$

below the sonic point, which is in approximation equivalent to saying that the luminosity should be smaller than the Eddington limit:

$$L < 4\pi c \, G\mathfrak{M}/k,$$

and this is equivalent to the statement [cf. our expressions (1.3; 3) and (1.3; 2)] that no hydrostatic solution is possible of the stellar-wind equation for atmospheric regions that have a net outward acceleration. Another interesting aspect of radiatively-driven stellar winds is that in hot stellar envelopes these winds may not be spherically symmetric but may break up into more complex flows as a result of Rayleigh–Taylor instabilities. This phenomenon is described in some detail in the last part of Section 6.8, from which we summarize that absorption-line-driven amplifications of small sound waves in the supersonic part of winds of early-type stars can led to appreciable disturbances. Observational evidence for such instabilities has indeed been found (see Section 8.5).

5.9. Model Envelopes of Early-Type Stars with Line-Driven Stellar Winds

Cassinelli and Hartmann (1975) made model computations for a central star of a planetary nebula ($\mathfrak{M} = 0.6\mathfrak{M}_\odot$) and for a Wolf–Rayet star with $\mathfrak{M} = 10\mathfrak{M}$. For the latter model $T_s = 30000$ K, and T ($\tau = \frac{1}{3}$) = 43130 K; $\Gamma_F = 0.88$. The continuous spectrum becomes *flatter* than in the case of a plane-parallel atmosphere. However, photospheres of Of stars can be treated as plane-parallel, as follows from the discussion in Section 5.3. But the wind produces an extended halo and we showed in Section 3.2 (cf. Figure 28) that the continuous spectrum of ζ Pup (O4 ef) can be emitted by a plane-parallel atmosphere with a halo of radius $1.33R_*$. However, any discussion of stellar *winds* demands the sphericity approach.

Castor *et al.* (1975) used the theory outlined in Section 5.8 to construct a model for the outer layers of a stellar atmosphere including a stellar wind. The parameters chosen were $\mathfrak{M}=60\mathfrak{M}_\odot$, $L=9.66\times10^5\,L_\odot$, $\Gamma_s=0.4$, $T_e=49\,290$ K, $\log g=3.94$; $R=9.58\times10^{11}$ cm$=13.8R_\odot$, which correspond approximately to an O4f star, like ζ Pup (see Section 3.2). The temperature distribution was based on an assumed LTE model (Castor, 1974a); hydrostatic equilibrium was assumed only in the photosphere, while outside the photosphere a stellar-wind model was developed.

The results found are shown in Figures 70a, b, and c, which give the velocity and density as functions of the radius, and the velocity as a function of the optical depth, respectively. The points P, S, and C denote the photosphere, sonic point, and singular point, respectively. The asymptotic value for the stellar wind velocity is 1515 km s^{-1}. Between the points P and S the temperature decreases from 50 000 to 38 000 K. The rate of mass loss is $-\dot{\mathfrak{M}}=4.2\times10^{20}$ g s$^{-1}=6.6\times10^{-6}\,\mathfrak{M}_\odot$ yr^{-1}. This gives a characteristic time of the star's mass-loss that is comparable with the expected stellar lifetime, the latter being $\sim3\times10^6$ yr. Hence such a star would lose an appreciable fraction of its mass during its lifetime. A comparison of this model of a stellar photosphere including a radiation-driven stellar wind, with a purely hydrostatic model, made by Castor *et al.* (1975) shows that the apparent radius of the flow-model appears to be larger, the difference amounting to about 13% in the visible. However no appreciable influence on the energy versus wavelength distribution-curve results from the inclusion of a stellar flow, because electron scattering is the only significant contributor to the continuum opacity, and it is conservative (i.e. scattered photons are not destroyed).

The calculations described above were later extended – but along the same line – by Klein and Castor (1978) to six spherical radiation-driven stellar wind models of a star of $60\mathfrak{M}_\odot$ with T_e ranging from 30 000–50 000 K. Mass-loss rates were predicted of between 4×10^{-7} and $1.2\times10^{-5}\,\mathfrak{M}_\odot$ yr^{-1}. Line profiles were calculated for several of the fairly strong spectral lines. These results did not yet allow one to make an unambiguous decision as to what model would agree best with a star like ζ Pup, but further progress was thereupon made by Kunasz (1980) who used the models of Klein and Castor to calculate the spectra of H and He II on the basis of statistical equilibrium calculations for the first ten levels of H and He$^+$. The purpose of this investigation was to find under what conditions one would obtain He II $\lambda4686$ in emission and simultaneously He II $\lambda3203$ in absorption, and at the same time Hα in emission and Hβ and higher Balmer lines in absorption – as is seen in early-type Of stars like ζ Pup.

To that end the rates of mass-loss were arbitrarily scaled down with factors ranging between 2 and 4. It is gratifying that the observed spectra could indeed be simulated,

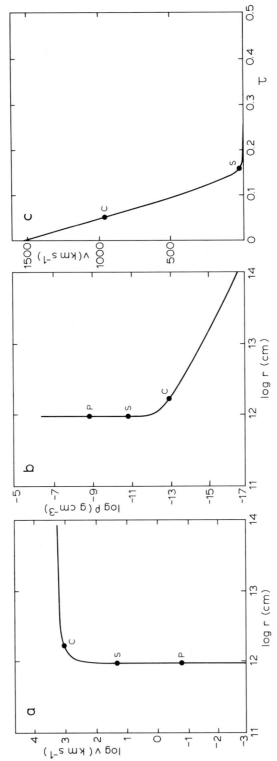

Fig. 70. (a) Velocity distribution in the stellar wind of an O4f star, as a function of the distance r to the stellar center. (b) Same as 70a, for the density distribution. (c) Velocity versus optical depth for electron scattering. (After Castor *et al.*, 1975).

for a model with $T_e = 40\,000$ K, log $g = 3.5$, being one of the models of Klein and Castor, with wind-density scaled down by a factor 2.2. The rate of mass-loss of this model is $5.5 \times 10^{-6}\ \mathfrak{M}_\odot\ \mathrm{yr}^{-1}$ – while the observed value for ζ Pup is $6 \times 10^{-6}\ \mathfrak{M}_\odot\ \mathrm{yr}^{-1}$ (see Section 7.3). It is very remarkable that the observed spectral behaviour appears to occur only in a very *limited* range of atmospheric densities!

5.10. Convection in Supergiant Atmospheres

It is felt almost intuitively that in the bizarre situations that occur in the extremely tenuous and extended atmospheres of supergiants the convective motion field may assume an extraordinary character. Let us therefore examine what theories on atmospheric convection teach us in this respect.

Convective instability occurs, according to the Schwarzschild criterium (4.4; 2), when the adiabatic temperature gradient is in absolute value smaller than the calculated radiative gradient. According to that criterium sub-photospheric convective instability occurs over the *whole* range of stellar effective temperatures, even the highest, where – because of the intense radiation-flow – the adiabatic gradient always takes small values in the ionization regions (cf. Maeder, 1980b). But while instability is there, the convective *motions* do hardly develop or not at all in stars with $T_e \gtrsim 8500$ K, because in such hot stars any volume of matter that would be hotter than its surroundings would radiate its excess energy in a time short as compared to the time required for it to ascend over its own diameter.

Plane-parallel photospheric models including convection were computed by De Loore (1970) for 90 stars with effective temperatures between 2500 and 41 000 K and for five log g-values ranging from 1 to 5. The models were computed on the basis of the mixing-length algorithm.

For the stars cooler than $\sim$8300 K stars De Loore (l.c.) computed quantities such as:
– the optical depth of the top of the convective region;
– the maximum convective velocity v_m;
– the Rayleigh number Ra which is the ratio between the life-time of a convective element and the time for loosing its excess energy; hence Ra is a number defining the efficiency of convection;
– the Reynolds number Re, being the ratio between inertial and viscous forces, and defining the occurrence of turbulence;
– the Prandtl number, Pr, comparing the relative importance of viscosity and thermal diffusivity;
– the Péclet number, Pe = Re . Pr, being the ratio between turbulent and radiative 'conductivities';
– the convective flux emerging from the convective regions.

An example of results of these computations, particularly those relevant to supergiant atmospheres is shown in Figure 71 which gives, for a star with $T_e = 4760$ K and log $g = 1$, values of the various gradients $\nabla = d \log T / d \log P$, where suffixes and indixes denote:

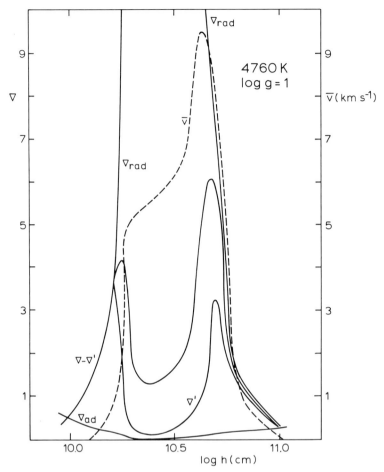

Fig. 71.   The various gradients and convective velocities in a stellar photosphere with $T_e = 4760$ K and $\log g = 1$ (De Loore, 1970).

ad: adiabatic;

rad: radiative;

prime: gradient of a rising convective element.

The physical importance of the function $\nabla - \nabla'$ is that it represents the *driving convective force:* the largest values of $\nabla - \nabla'$ in a stellar atmosphere characterize the region where v_{conv} reaches its largest values and where the main convective elements may be formed. It is interesting to compare these data with those on the *solar* convection, where the function $\nabla - \nabla'$, plotted as a function of depth, is a very peaked one with a width of a few hundred km only. This seems in agreement with the fact that the outer – visible – granulation elements are small. In contrast, the essential aspect of *supergiant* atmospheric convection is that the region where $\nabla - \nabla'$, or $\bar{v}$ (which is more or less equivalent) takes large values, is very extended in depth. In the case of the model $T_e = 4760$, $\log g = 1$ this is approx. 0.5×10^6 km.

The question then arises how to relate this quantity – the thickness of the 'driving region' – to the average size of the convection cells. We now meet two conflicting arguments.

First, Stothers and Leung (1971) advocate that the average size of the convective elements in strongly convective, i.e. late-type, supergiants should be comparable to the whole thickness of the subphotospheric convection zone. The argumentation is based on the *analogy* with the solar case where *giant cells* seem to occur in the photosphere with diameters comparable to the thickness of the convective region, and on the *observation* of long secondary periods (5 to 10 yr) in variable late-type supergiants, which observation was linked to the *assumption* that these periods are the convective overturning times of the giant cells. It was in addition assumed that such giant cells are the main feature in late-type supergiant atmospheres. With this reasoning Stothers and Leung found that in such stars the convective elements should have sizes comparable to the stellar radii.

Schwarzschild (1975), on the other hand, assumed that the sizes of the convective elements should rather be comparable to the thickness of the region where $\bar{v}$ takes large values. To illustrate this assumption he considered an extreme case, a red supergiant with a T_e-value of 3700 K and log $g=-0.15$ (comparable to α Sco of α Ori) and finds for this star a thickness of the region where $\nabla-\nabla'\neq0$ of the order of 60×10^6 km. Since one may infer from *solar* studies of convection that the horizontal diameter of convection elements is approx. three times the depth of the 'superadiabatic region', Schwarzschild's extrapolation of this 'scaling rule' would lead to horizontal scales of the elements of 1.5×10^6 km for a star with $T_e=4760$ K and log $g=1$, and 180×10^6 km for a star with $T_e=3700$ K and log $g=-0.15$. This latter value is nearly comparable to the radius of the star (420×10^6 km). The former case yields elements of ≈0.03 the stellar radius. This leads to the suggestion that any extreme supergiant atmosphere would contain only a small number of convective elements. These elements would have life-times of the order of one day (for the stellar model with $T_e=4760$ K and log $g=1$) to a year (for the stellar models with $T_e=3700$ K and log $g=-0.1$). In view of the small number of these elements, this may result in brightness fluctuations with characteristic times of the order of the lifetimes given above. See for this topic Chapter 8.

Schwarzschild's suggestion is interesting, undoubtedly. But it is based on a rather primitive theory (the mixing length theory of convection) and on a risky extrapolation (from 1000 km elements at the Sun to 200 million km elements in supergiants!). It is therefore important to look for theoretical and/or observational ways to decide between Schwarzschild's and Stothers and Leung's hypotheses. The latter aspect was dealt with by De Jager and Vermue (1979) who compared micro- and macro-turbulent velocities in early type and medium-type supergiants and a hypergiant. The size of the main elements contributing to the velocity field could be derived by means of a diagnostic method based on micro- and macro-turbulent filters (see Section 2.9 and Figure 18), and by introducing the *two* approximations (with $\Phi=f_\mu/f_M$):

$$\frac{\langle v_{\mu}(k)\rangle}{\langle v_M(k)\rangle} \approx \langle \Phi(k)\rangle = \frac{\int\limits_0^\infty \Phi(k)\,F(k)\,dk}{\int\limits_0^\infty F(k)\,dk} \approx \Phi\langle k\rangle,$$

where F is the spectrum of the turbulence. Knowledge of v_{μ} and v_M thus yields the main spatial wavenumber $\langle k\rangle$ of convection and hence the wavelength $l = 2\pi/k$. For OB Ia type supergiants it appears that $l \approx 0.25R$, and for F–G Ib type supergiants $l \approx 0.05R$. The hypergiant HD 217476 (K2–5 Ia$^+$) has main elements of $\sim 0.5R$.

For the intermediate and late-type stars these results agree with the 'predictions'. They seem to show some better agreement with the hypothesis of Stothers and Leung than with that of Schwarzschild, but more research is certainly needed in this field. The occurrence of large moving elements in *early-type* stars cannot be understood on the basis of convection theories because these stars do not have strongly developed convective motions. Tentatively one may think that such elements are related to non-radial pulsations in atmospheres of *early-type* stars (cf. Section 8.5).

CHAPTER 6

CHROMOSPHERES, CORONAE, GAS AND DUST
AROUND LUMINOUS STARS

6.0. Generalities

From a fundamental point of view there is not much reason to distinguish between the various parts of outer stellar envelopes, such as chromospheres, coronae, dust shells, etc. Such a distinction might easily leave the reader with the false impression that these layers are separate physical entities. The contrary is true: the complicated interplay between the mechanical, magnetic, radiative and conductive fluxes, in addition to stellar winds in outer atmospheric regions can lead to the formation of extended envelopes, with a complicated temperature and density distribution. Conventionally, different names are assigned to the various temperature regions – cf. Section 1.8. In the extreme parts of these envelopes where the gas has low temperatures, molecules and even dust particles may form. In other cases youngly formed stars may still partly be embedded in the gas and dust cloud out of which they originated.

The reason why we distinguish in the various sections of this Chapter between the different parts of the outer stellar envelopes that are mentioned in the heading, is certainly not a matter of fundamental physics. The background is merely pragmatic. Different observational techniques are used for observing the various outer layers: strong visual and UV spectral lines for the chromospheres, X-rays for coronae ..etc. But essentially the envelope is a *physical and genetic unity*!

Restriction: As everywhere in this volume we restrict ourselves to envelopes of bright and massive stars. Occasionally we make an excursion to solar observations, but only for heuristic reasons.

References: A recent review on stellar chromospheres is written by Ulmschneider (1979); one on stellar coronae by Mewe (1979). An interesting first attempt to treat all outer stellar layers from one integrated point of view was published by Pecker *et al.* (1973).

6.1. What is a Stellar 'Chromosphere'?

The notion 'stellar chromosphere' is as illogical as, for instance, lunar geology. It would be very hard ever to discover a 'thin sphere around a star characterized by colorful emission' (the solar definition) brilliant enough to deserve the name 'chromosphere'. It is known that the *solar* chromosphere is a region where the temperature increases outwards, due to dissipation of magnetic and mechanical energy. However, it should be realized that this aspect is not *essential* for the solar chromosphere, because even if there was no outward temperature rise, would a chromosphere be visible at *solar* eclipses, albeit with less strong emission lines. It is moreover clear that the solar chromosphere as such would be completely invisible if the Sun would be placed at stellar distances.

For these reasons the notion 'stellar chromosphere' looks not very well chosen. On the other hand, many stars, even dwarfs do show features in their spectra, such as the Hα emission line and the H and K emission cores that are conventionally but not always correctly called 'chromospheric'. Thus, the term has gradually been introduced and settled into the astrophysical literature and clearly needs another definition than the one used for the Sun.

The *solar* chromosphere is that region of the solar atmosphere which, seen tangentially, is transparent for continuous radiation and still opaque for radiation in (strong) spectral lines. A *stellar* chromosphere is defined by Ulmschneider (1979) as 'a stellar envelope (definition: Section 1.8) with a temperature higher than T_e', while Praderie (1973) defines it as "the region of the star giving rise to spectral observations indicating the existence of a mass flux or of non-radiative energy dissipation". It is defined by Linsky (1977) as "that region of a stellar atmosphere where $dT/dr > 0$ and where the energy balance is dominated by radiative and non-radiative (wave-dissipation) heating terms and radiative losses". The three definitions are to some extent equivalent. In proposing a new definition I would refer more closely to the solar one and refrain from including the mechanisms of chromospheric heating in the definition. I thus define a stellar chromosphere as "an envelope with $T > T_e$, which is optically thin in the greater part of the continuous radiation, and optically thick ($\tau \gg 1$) in at least some of the stronger lines". Note that we do not specify *what* temperature is to exceed T_e; normally this should be the kinetic temperature. This definition excludes cases like those described in Section 6.5: the Cayrel effect and the like.

In addition to introducing this definition, two additional remarks, a non-trivial and a trivial one:
– Conventionally a chromosphere is considered to be situated between the chromosphere and the corona, but this need not to be the case with our definition.
– It is clear that the extent of a stellar chromospheres should be considerable to make it detectable, contrary to the solar case.

6.2. Conditions for the Occurrence of Emission Lines; Chromospheric Indicators

In order to detect a stellar chromosphere one should therefore look for observations indicating an envelope with $T > T_e$, while it is also important to search for the origin of the chromospheres by investigating the relation between the T-increase and *mass-flux* or *non-radiative heat dissipation*.

The stellar mass-fluxes will be discussed in Chapter 7. It manifests itself through features such as asymmetric or displaced lines.

Associated with – extended – chromospheric features are also observations of – variable – line splitting (indicating large and changing inhomogeneities), or line asymmetries, and irregular stellar variability (Chapter 8). The most important indication for the occurrence of chromospheres, is the observation of emission lines. Whether a line is observed in emission or absorption depends on two factors: The extent of the envelope and the relative value of the source function as compared to the function representing the photospheric emission. Furthermore, two cases may be considered: a static and an expanding envelope. Additional factors are whether the envelope is optically thick or thin in the wavelength of the line. (We assume in any case, that the envelope is transparent for continuous radiation).

Let us call B_c, B_L, S_L, respectively the source functions of the photospheric continuous and line radiation (core of line) and of the chromospheric line radiation. We assume *no* height-variation of these functions. Let the chromospheric and photospheric (outer) radii be R_2 and R_1 respectively, and the average optical thickness of the chromosphere averaged over the envelope in the line center be τ_L. We assume a homogeneous photosphere (no limb darkening). Then, in this simplified picture (Figure 72a) the photospheric fluxes are $\pi R_1^2 B_c$ in the continuous spectrum and $\pi R_1^2 B_L$ in the line center. The chromospheric contribution in the line center is $\pi R_2^2 S_L \tau_L$ in the optically thin case ($\tau_L \ll 1$) and $\pi R_2^2 S_L$ in the optically thick case ($\tau_L \gg 1$). The relative emission in the line center with regard to the continuous spectrum is then

$$\text{for} \quad \tau_L \ll 1: \quad \frac{B_L}{B_c} + \frac{S_L}{B_c} \left(\frac{R_2}{R_1} \right)^2 \tau_L, \tag{6.2; 1}$$

or

$$\text{for} \quad \tau_L \gg 1: \quad \frac{S_L}{B_c} \left(\frac{R_2}{R_1} \right)^2. \tag{6.2; 2}$$

The question whether the line is observed in emission or absorption, or whether there is just a small emission peak or a (partial) 'filling up' of the line depends in the *static case* (Figure 72a) on the values of S_L/B_c, R_2/R_1 and τ_L. In the case that $\tau_L \gg 1$ the line is in emission when $(S_L/B_c)(R_2/R_1)^2 > 1$. When $\tau_L \ll 1$ the chromospheric emission peak would extend above the continuum if $\pi R_2^2 S_L \tau_L + \pi R_1^2 B_L > \pi R_1^2 B_c$.

In the case of an *expanding chromosphere* (Figure 72b) the situation is different. For simplicity we assume in our schematic picture that the chromosphere is expanding with a constant velocity, large enough for the line radiation emitted by the chromosphere to be well outside the region of the photospheric absorption line. We then obtain two lines in the spectrum, the photospheric absorption line and a chromospheric emission. Restricting ourselves to the emission, it is obvious that the region behind the star (shaded) can not be seen; that the volume around it (unshaded) produces essentially an unshifted line if $\tau_L \ll 1$, and a shortward displaced one if $\tau_L \gg 1$, while the volume in front of the star (dotted) produces a blue-shifted component, in addition to the undisplaced photospheric absorption line.

If $\tau_L \gg 1$ the relative flux in the centre of the shortward displaced emission line is again $S_L/B_c \cdot (R_2/R_1)^2$, and this emission component, at $\Delta\lambda = -v\lambda/c$, has to be *added*

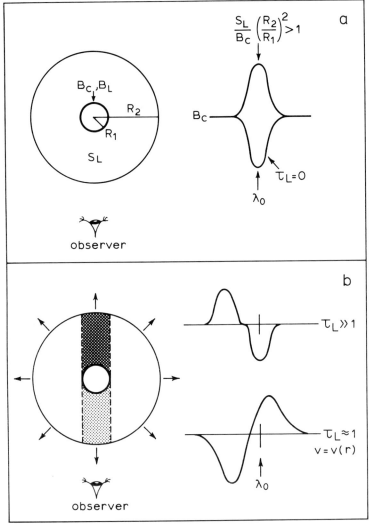

Fig. 72.   To illustrate the origin of emission lines in stars with extended atmospheres; (a) static evelope; (b) expanding envelope.

to the continuous radiation from the photosphere. If $\tau_L \ll 1$ we distinguish again between the (sky-projected) regions in front of the star (area πR_1^2) and around it (area $\pi R_2^2 - \pi R_1^2$). The first area produces an emission component at $\Delta\lambda = -v\lambda/c$ with a flux $\pi R_1^2 S_L \tau_L$. The area around the star has a larger optical thickness: roughly $\sim 2\tau_L$, and produces an emission component with a full width at its basis of $\sim 2\Delta\lambda$, and a total flux of $\sim S_L 2\tau_L \pi (R_2^2 - R_1^2)$.

The case that occurs most frequently is that of $\tau_L \approx 1$ while there is also a velocity-gradient; it is the most complicated case, and it cannot be treated by the simplifying assumptions used here. The resulting line profile is the P Cygni profile which consists of two components:

(1) An emission peak near the wavelength of rest with a relative flux, in the optically thick case approximately equal to

$$\frac{S_L}{B_c}\frac{R_2^2 - R_1^2}{R_1^2};$$

(2) an absorption component, shortward displaced by expanding material in front of the star. With this latter topic we are fully entering the intricate problem of the explanation of P Cygni line profiles, a matter that is further dealt with in Sections 7.3.

The next problem is that of the determination of the source function in the chromosphere. If the chromosphere is hotter than the photosphere, it is so by some form of non-radiative heat dissipation, which shows itself through the occurrence of emission lines, or reversal peaks in absorption lines or by excitation- and/or ionisation anomalies in the spectra.

The strongest lines are most sensitive, because their average level of formation as a rule occurs higher in the stellar atmosphere, where the relative temperature increase may be more than in lower layers. The most important lines are the h and k Mg II lines near 2800 Å, the H and K lines of Ca II near 3900 Å, Hα and the subordinate lines of the Balmer series, and particularly Lα and the other Lyman lines. Other chromospheric indicators are C II 1335 Å, Si II 1808 Å. Some He I lines may also be sensitive indicators of excitation anomalies, while the situation is less clear for He II. In all stars the ultra-violet resonance lines of O VI, N V, and Si IV and other ions *can* be indicators for ionization anomalies, and hence of chromospheres (Linsky and Haisch, 1979). In those hot stars where they are caused by Auger ionization from a corona – or from hot 'coronal' elements – they are obviously not.

Following Ulmschneider (1979) distinction is made here between *type I indicators* which are unambiguous indicators of the existence of chromospheres, and *type II indicators* which are suggestive but not entirely undisputed.

Type I indicators are collision-dominated lines as C II 1335 Å and Si II 1808 Å in medium-T stars, the infrared He II line at 10830 Å in the spectra of a medium-temperature star like Capella or the Sun and late-type stars like α Sco and α Boo, or those of Si IV, N V, O VI (with the reserve expressed above) observable in early-type luminous stars. In late-type (M) stars the UV Fe II lines are type I chromospheric indicators.

Type II indicators are resonance lines with emission cores like Ca II H and K or Mg II h and k. Also the hydrogen Lα line, the Hα line and the infrared triplet of Ca II may be such indicators.

A summary of some well-observed chromospheric indicators, with relevant other data is given in Table XXXVIII.

6.3. Observations in Some Chromospheric Indicators

TYPE I CHROMOSPHERIC EXCITATION INDICATORS

For an enumeration of the main type I indicators we refer to Table XXXVIII and the references given there. A few lines are described here in some detail.

TABLE XXXVIII

Chromospheric indicators for bright stars. The fourth column gives a few remarks on the spectral types in which chromospheric emissions in these lines are mainly observed. General (review) references are Praderie (1976), Linsky (1977), Dupree *et al.* (1979), Ulmschneider (1979), while some *additional references* are given to papers concerned with individual lines or stars.

Indicator type	Ion	λ (Å)	Spectral types	Additional references
I	He I	5876	F and later	
		10830	WR, A and later	O'Brien and Lambert (1979)
	C II	1335	G	Freire (1979)
	C IV	$\sim$ 1550	F, G	Jamar *et al.* (1976b)
	O VI	$\sim$ 1035	(O to A), F to K	Evans *et al.* (1975)
	Si II	1808	G	
	Si III	1206	F to K	
	Si IV	$\sim$ 1400	F to K, (M)	
	Fe II	3150	K, M	
		3300		
II	Ca II	$\sim$ 3950	(A), F and later	
	Ca II	$\sim$ 8500		
	Mg II	$\sim$ 2800	A and later	
	He II	1640	$\sim$F	Jamar *et al.* (1976b)
	Hα	6561	G to K0	
	Lα	1216	(A), F to K	

The *infrared line of He I* at 10830 Å ($^3S-^3P$) is a type I indicator; it occurs as an emission line in P Cyg and in Wolf–Rayet stars (Miller, 1954; Kuhi, 1966). The line, with a lower excitation potential of 20 eV, is undoubtedly an indicator of peculiar excitation conditions and is in most stars of chromospheric origin (Zirin, 1976).

In the far UV-region (1000–2000 Å) the many *strong resonance lines* (often with P Cygnitype profiles) have been described by Morton (1967a, b); see also Morton *et al.* (1968), and were later studied by many others, starting with Smith (1970) and Carruthers (1971), who also noted the large outstreaming velocities, of approximately 1000 to 3000 km s^{-1} in early-type stars.

The *ultraviolet lines of O VI, N V, and Si IV ions* have initially been used by Lamers and Snow (1977) for the study of chromospheres in O- and B-type stars, but it is now clear, since the HEAO–2 (Einstein-Observatory) data have shown coronae to be a common feature of hot stars, that the suggestion (Section 6.4) by Cassinelli *et al.* (1978b) and Olson (1978) is correct; these lines are formed in the stellar winds of hot stars through Auger-ionization by energetic photons from the coronal gas-elements. In intermediate-temperature stars these lines may be considered to be real chromospheric indicators.

In some M-type supergiants the *Fe II ultraviolet emission lines*, detected by Herzberg (1948) are chromospheric indicators. They have excitation energies of up to 5.6 eV, while the ionization energy of Fe I is 7.9 eV. Such energies do not occur in the M-type photospheres. See our description of the spectra of α Sco and α Ori in Section 3.11.

TYPE II CHROMOSPHERIC INDICATORS

The ultraviolet Mg II lines. Nearly all stars, with the exception of the hottest (like ζ Puppis, O4ef; and the Wolf–Rayet stars) show in their spectra the resonance doublet lines (h and k) of Mg II at 2802.7 and 2795.5 Å (3^2S-3^2P). Through a fortunate coincidence this part of the spectrum shows also two lines absorbed from the upper level of that transition (3^2P-3^2D), at 2790.7 and 2798.0 Å. An early comparison of the observed equivalent widths of these lines with predicted values based on non-LTE calculations (Lamers *et al.*, 1973; Snijders and Lamers, 1975; Lamers and Snijders, 1975) was made for spectra of a hundred stars obtained with the Utrecht UV stellar spectrophotometer *S59*, and of 10 stars observed by the *Copernicus* satellite and 7 others observed by Boksenberg *et al.* (1974) and Kondo *et al.* (1972) in balloon experiments. While a comparison of these observations with computed equivalent widths shows fair agreement for main-sequence stars and normal giants, chromospheric indications such as 'filling-up' or large micro-turbulent velocities are observed in spectra of several other groups of stars (Figure 73a). The observed lines are relatively weak in some

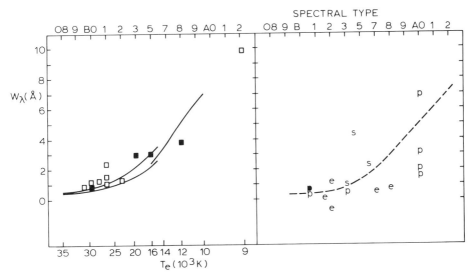

Fig. 73a.   The observed total equivalent widths of the four Mg II UV lines in supergiants (left) and in peculiar stars, shell stars and Be-stars (right) versus spectral type or effective temperature. The filled symbols refer to the high resolution observations and to the most accurate S59 observations. The full lines show the predicted equivalent widths in stars with $\log g = 2.5$ or 3 and microturbulent velocities of 10 and 15 km s^{-1}. In the right-hand figure the peculiar stars are indicated by a symbol '*p*', the shell stars by an '*s*' and the Be-stars by an '*e*'. The broken line shows the average relation for the observed equivalent widths in main sequence stars. (From Snijders and Lamers, 1975).

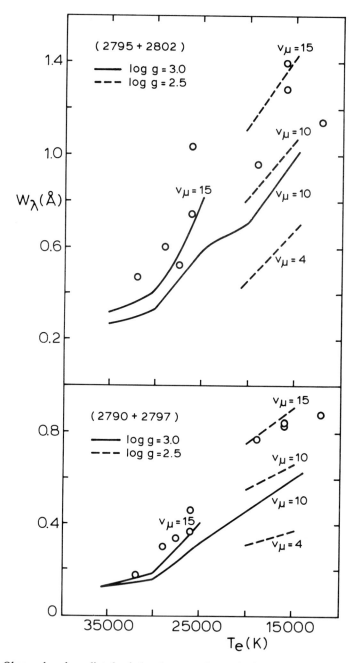

Fig. 73b. Observed and predicted relation between the equivalent widths of the UV Mg II lines (separate for the two pairs of lines in supergiants of luminosity classes Ia and Ib (taken together). The predictions are for two log g-, and various v_μ-values. Best agreement occurs apparently for $v_\mu \approx 10 \dots 15$ km s^{-1}. (From Kondo *et al.*, 1975a).

Be stars (influence of central emission core) and some Ap stars, and they are enhanced in the spectra of some shell stars. A comparison with computations gives for supergiants the best agreement with observations for $v_\mu \approx 10$ to 15 km s^{-1}. In some stars (β Ori, B8 Ia; η CMa B5 Ia) the displaced positions of the emission peaks indicate outstreaming motion.

Later, Kondo et al. (1975a) refined the comparison (Figure 73b) by using observations of the four ultraviolet Mg II lines obtained by the Princeton Telescope Spectrometer on the Copernicus satellite. In general there is good agreement between theory and observation but deviations from the theoretical predictions occur for the earliest spectral types where the rotational effects appear to make the equivalent widths larger. Microturbulent velocities of ~ 10–15 km s^{-1} again give the best agreement with observations for supergiants.

The above discussion refers to the Mg II lines, as they usually appear as absorption lines in medium-dispersion spectra. These lines are useful for giving information on the outermost layers of the stars, simply because of the large absorption coefficients in their centers. However, indicators for an outward increase of the source-function are the central emission peaks in these lines, which are normally – and correctly – identified as 'chromospheric' emissions in the sense of our definition. Such emissions can only be observed in spectra taken with sufficient spectral resolution. They are well observable in giants and supergiants of middle spectral classes, and have been detected with average spectral resolution (0.4 Å) in stars as early as F0 Ib (Evans et al., 1975), and with slightly higher resolution also in early-type stars where they are weakly discernable (Kondo et al., 1976a, 1977). They become more and more conspicious towards later spectral lines, and have been observed in supergiants as late as M2 (α Ori, M2 Iab; Kondo et al., 1972, 1975b; Pagel and Wilkins, 1979); in giants they extend to M5 (α Her, M5 II; Bernat and Lambert, 1978).

Systematic investigations of the Mg II doublet emissions have initially been performed by Kondo and coworkers (Kondo et al., 1975a, b, c, 1976a, b, c). In late-type supergiants the emissions occur against a flat stellar continuum; the 2802 Å emission is symmetric with a central self-reversal; an apparent asymmetry in the 2795 Å emission may be due to Fe I absorption. Detailed measurements for the supergiants α UMi (F8 Ib), α Per (F5 Ib), η Leo (A0 Ib) and ϱ Leo (B1 Ib) were published by Kondo et al. (1976c), and a study of the stars ε Peg (K2 Ib) and α Ori (M2 Iab) was performed by Kondo et al. (1975b). The investigation was extended later by Kondo et al. (1977) to a larger number of stars observed with the Houston–Utrecht instrument BUSS. Weiler and Oegerle (1979) using Copernicus observations, investigated the lines in 49 late-type stars. Pagel and Wilkins (1979) studied them in 12 G–M stars with I.U.E.

The typical behaviour of the K-emission profile of Mg II and Ca II in dependence on log g and on the heating-flux is shown schematically in Figure 74 (after Ayres, 1979).

Several authors, following Wilson and Bappu (1957), derived a relation between the absolute visual magnitude M_V and the emission width (W) (see also Figure 75). It reads: $M_V = -12.45 \log W + 28.78$ (Kondo et al., 1976b, corrected, 1977), or

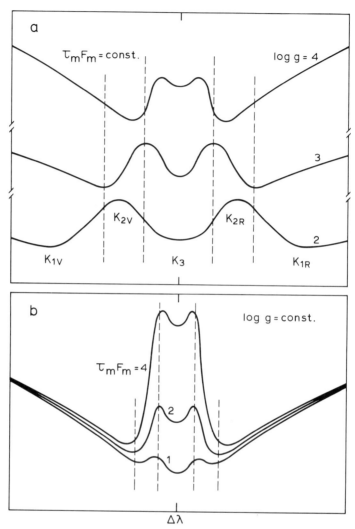

Fig. 74.   Schematic illustration of the dependence of the emission-peak profile in the k lines of Ca II and Mg II on surface gravity and on the chromospheric heating flux, a quantity perhaps to be identified with the mechanical flux F_m times the 'optical depth' for absorption of F_m, hence $\tau_m F_m$. (a) Dependence of $\log g$ for stars with the same T_e and $\tau_m F_m$. The ordinate scales are arbitrary for each of the profiles and the integrated emission between the K_1-features should be equal in all cases. The broadening of the emission for decreasing $\log g$ (increasing luminosity) is the Wilson–Bappu-relation. (b) In this case g and T_e are taken constant but $\tau_m F_m$ varies. The integrated intensity of the peak changes markedly; the K_1-distance increases; the K_2-distance reduces. (Partly after Ayres, 1979).

$= -15.15 \log W + 34.93$ (Weiler and Oegerle, 1979) from a larger material. W is the full width at half maximum of the emission peak, measured in km s^{-1}.

   The H and K lines of Ca II. A great deal of research has been put into describing and explaining the central emission cores of the H and K lines in stellar spectra.

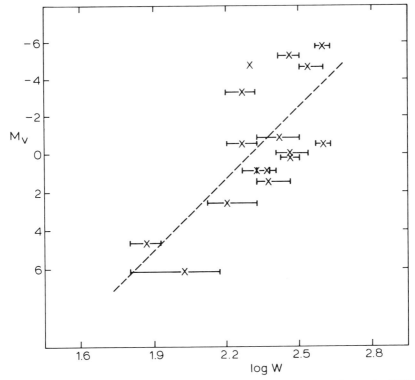

Fig. 75.   The Wilson–Bappu relation for the Mg II lines, (after Kondo *et al.*, 1975b, 1976b, 1977); *W* is the width of the emission in Ångström units times c/λ_0.

Wilson and Bappu (1957) and Wilson (1959), established their well-known relation

$$M_V = -14.94 \log W_0 + 27.59, \qquad\qquad (6.3;\ 1)$$

where $W_0 = W - 18$ km s^{-1} (corrected for slit-width). The relation is valid for G, K, and M stars over a range of 15 mag and has been extended later by Warner (1969) to southern stars. In further investigations Wilson (1970a, b, 1976), presented more detailed considerations on the absolute calibration of the law, based on a larger material (about 700 medium- and late-type subgiants, giants and supergiants with well-known parallaxes). No clear reason was found to change expression (6.3; 1).

High-resolution spectra of the cores and wings of the H and K lines in 43 stars of various luminosities, and calibrated in absolute flux units were described by Linsky *et al.* (1979a). They tabulated radiation temperatures in the K_1 emission core and information on the line widths and line asymmetries. A comparison between the measured line-widths, those of Wilson and Bappu, and those of Stencel (1977) showed agreement but only roughly so for supergiants. This latter aspect was tentatively ascribed to supergiant variability.

There are other possible ways to express the fact that the central emission features in the Ca II lines increase in width with increasing stellar absolute luminosity. Thus, Engvold and Rygh (1978) found relations between W_0 and the distance ΔK_1 (km s^{-1}) between the center of the K-line and the K_1 intensity minima, and with ΔK_2, being the distance (km s^{-1}) between the K-line center and the K_2-peak intensity:

$$\log \Delta K_1 = 0.78 \log W_0 + 0.34,$$

$$\log \Delta K_2 = 1.06 \log W_0 - 0.56.$$

It is one step further to relate W_0 to the physical properties of the atmosphere through a discussion of such properties for the stellar atmospheres involved; this is often done *via* model studies. Thus Reimers (1973) reformulated Equation (6.3; 1) as

$$W (:) g^{-0.2} T_e^{1.1}, \tag{6.3; 2}$$

and Lutz and Pagel (1979) found from data on 55 stars

$$\log W_0 = -0.27 \log g + 1.65 \log T_e + 0.10\, [F_e/H] - 3.69 \tag{6.3; 3}$$

a relation that is valid if $-2.7 \leq [F_e/H] \leq 0.44$, with a standard deviation of ± 0.06.

Ayres (1979) found for $\Delta \lambda_{K_1}$ (the wavelength distance between K1$_V$ and K1$_R$) and for $\Delta \lambda_{K_2}$ (similarly for the K_2-features) the following relations:

$$\log(\Delta \lambda_{K_1}) = \tfrac{1}{4}[F_e/H] + \tfrac{1}{4} \log F_{ch} - \tfrac{1}{4} \log(g/g_\odot) +$$
$$+ \tfrac{7}{4} \log[T_e/(T_e)_\odot] + \text{const.} \tag{6.3; 4a}$$

and

$$\log(\Delta \lambda_{K_2}) = - \tfrac{1}{4}[F_e/H] - \tfrac{1}{4} \log F_{ch} - \tfrac{1}{4} \log(g/g_\odot) -$$
$$- \tfrac{5}{4} \log[T_e/(T_e)_\odot] + \tfrac{1}{2} \log v_t + \text{const.} \tag{6.3; 4b}$$

Here F_{ch} is the rate of chromospheric heating, a quantity not further specified by Ayres. We may interprete it as the product $\tau_m F_m$ (cf. Figure 74) where F_m is the mechanical flux incident on the chromosphere, and τ_m the chromospheric 'optical thickness' for F_m.

The near-infrared Ca II triplet. Related in behaviour to the Ca II H and K lines are those of the near-infrared Ca II triplet near 8500 Å. The lines are strongest in medium-*T* spectral types, and although they do not show emission features comparable to the Mg II h and k, or the H and K lines, they are filled-in in spectra of stars with active chromospheres. The chromospheric radiative loss-ratios calculated for the λ8542 line correlate well with those derived for the Mg II and Ca II H and K lines, and with the Wilson–Bappu relation (Linsky *et al.*, 1979b).

For the *interpretation* of these relations several possible effects may be considered. The emission-peak widths and their intensities can be related to the thickness of the column of enhanced line-source-function S_L. and to the value of S_L. These are connected to the rate of dissipation of mechanical energy and perhaps also to the mass-flux. The widths of the peaks may be caused by stochastic motions, micro- or macro-scopic. It is true that the v_t-values do not occur in a marked way in the relations (6.3;2) through (6.3; 4) – it appears only in Equation (6.3; 4b) – but this may be due to the imperfect knowledge of v_t in stars rather than to a physical lack of relation.

A first attempt to explain the law in terms of the existence of a stellar chromosphere was made by De Jager (1958). A mechanical flux F_m, originating from the stellar convection zone dissipates energy in the higher layers and heats the chromosphere. The dependence of chromospheric heating on stellar magnitude can be estimated. This defines the intensity of the emission peak. The width of the emission peak is thereupon defined as the wavelength distance between the two points in the profile where the optical depth is unity, assuming Doppler broadening. The relation $(W(:) g^{-1/2})$ thus obtained between the emission peak width and the absolute stellar magnitude was at that time not in disagreement with the observed relation. Later, Thomas (1973) on the basis of similar arguments with an improved quantitative reasoning found a similar agreement between emission width and mass-flux. Lutz and Pagel (1979) interpreted their empirical relation [our Equation (6.3; 3)] in terms of dissipation of an acoustic energy flux and Doppler broadening. Ayres (1979), similarly, related the Wilson–Bappu law to a chromospheric optical thickness increasing with decreasing g.

The above authors tried to explain the effect as dependent on the heating flux or mechanical flux F_m and assuming Doppler broadening. Such explanations should then also yield a dependence of the total emission flux *in* the emission peaks on W_0, but these quantities appear to be *uncorrelated*! Another group of explanations assumes therefore that W_0 is related to the velocity field in the chromosphere. This v-field naturally is related to F_m, but in such a case there need not be correlation between W_0 and the emission flux in the peaks. Thus, Scharmer (1976) assumed that the emission peak width is due to supersonic turbulence, which in turn would be caused by the star's mechanical flux, as being dissipated in its chromosphere. Although a functional relation could be derived that was in reasonable agreement with Reimers's, the weak point in this treatment is that very large supersonic microturbulent velocities v_μ would be needed to explain the observed width, while it seems doubtful if such large microturbulent velocities will originate: in supersonic turbulent velocity fields shock wave dissipation would become very important, and would thus prevent v_μ to exceed the sound velocity s by an appreciable amount. Nevertheless it seems allowed to *conclude* that the widths of the emission peaks are due to stochastic motions related to the dissipation of a mechanical energy flux in the chromospheric layers.

The Wilson–Bappu relation has been established for medium- and late-type stars. Earlier-type stars do not seem to obey that relation (e.g. the study of α Cyg, A2 Ia by McClintock and Henry 1977) which suggests that the chromosphere of an early type star is different from that of a late type one, for example by not showing signs of stellar activity, perhaps common to stars with convection zones.

It is important that the emission peaks in Ca II and Mg II are split by a central absorption. The two peaks do not have the same intensity. This is shown in Figure 76 for the supergiants ε Gem, G8 Ib and ε Peg, K2 Ib. The profiles show an asymmetrically placed central inversion. By mirroring the righthand-side of the profile the

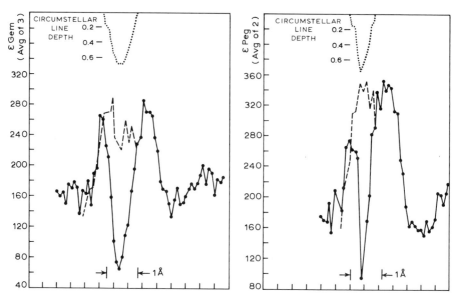

Fig. 76.   The central K emission cores in the supergiants ε Gem, G8 Ib (left), and ε Peg, K2 Ib; after Vaughan and Skumanich (1970); attempt to explain these profiles by the superposition of a displaced circumstellar absorption line on a (less/un-) displaced emission core.

dashed profiles are found. It is then assumed that the absorption profile is caused by circumstellar gas. That profile is obtained if the observed profile is subtracted from the dashed line. The measured wavelength shifts appear to indicate outstreaming velocities in these supergiants; they are of the order of 20 km s^{-1} (Vaughan and Skumanich, 1970). Generally, for Ca II and Mg II the ratio V/R is defined (Wilson, 1976) as the intensity ratio between the $K2_V$ and $K2_R$ features. Apart from short-time variability, which perhaps is to be attributed to stellar activity – similar to solar activity – the V/R-ratio varies smoothly from <1 in late K and M giants, to >1 in G giants (Stencel, 1978, 1979). It is perhaps significant that the transition from >1 to <1 coincides with that between stars with negligible to significant mass-loss.

The Lα line. After the launch in 1972 of the *Copernicus* satellite with the Princeton spectrometer, data became gradually available on the intensity and profile of the Lα line in stars. The importance of these observations is that the Lα line is formed at greater height than the Mg II lines. McClintock *et al.* (1975) made a compilation of their own and other observations of the line-width of the Lα and Mg II emission peaks in K-type stars of different luminosities, and found a clear dependence of the line width on the stellar luminosity, quite similar to the Wilson–Bappu (1957) relation found for the Ca II emission peaks (see above). Their relation for Lα and Ca II is shown in Figure 77. It is clear from the foregoing that for each of the Ca K, Mg II and Lα lines a 'Wilson–Bappu' relation exists. The various widths reflect the variation of the broadening mechanism with height in the stellar chromospheres. Lα emission has not yet been detected in supergiants because of the interstellar blocking for these – mostly – distant stars.

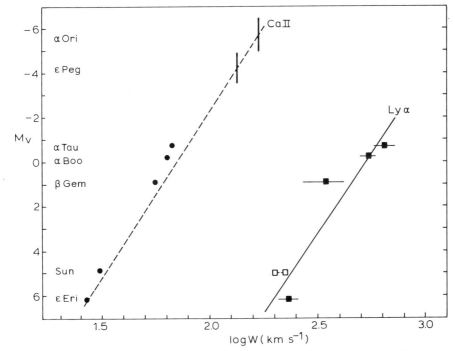

Fig. 77.   The relation between the width of chromospheric emission lines and stellar luminosity. The ordinate is the absolute visual magnitude. The abscissa is $\log W$ measured in km s^{-1}. Squares represent Lα (full half-widths). Filled dots are data for Ca II widths (Wilson and Bappu, 1957). The dashed line is the relation reported by Wilson (1959) for the Ca II K line. For two stars (vertical lines) the absolute magnitudes were computed from the Ca II widths. Open squares are solar values. (After McClintock *et al.*, 1975).

6.4. Properties of Chromospheres and 'Warm Envelopes' in Various Types of Stars as Derived from Spectral Investigations

We review in this section how the investigation of stars through the various chromospheric indicators has given information on stellar chromospheres around different groups of stars.

Bright O- and B-type stars. In the spectra of the brightest of these stars Lamers and Snow (1978) found lines of O VI, N V, and Si IV ions, which originate in expanding envelopes, as can be judged from their velocities.

Figure 78 gives the location of the stars with such lines in the HR diagram. The filled symbols refer to those stars where these lines are present or probably present. The hatched area shows the regions in the HR diagram where such lines could occur if they were due to radiative ionization in a gas with a kinetic temperature of $0.8T_e$ (i.e. the approximate 'boundary temperature' of a grey stellar model) and a mean intensity of radiation

$$J_\nu = W(r)\, F_\nu,$$

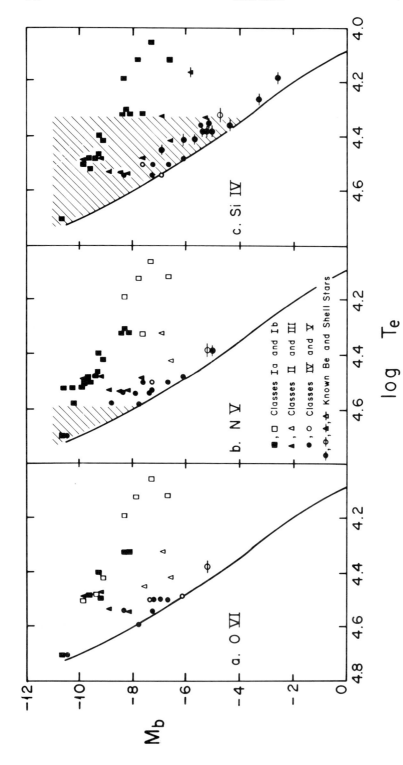

Fig. 78. The location of the stars which have O VI, N V or Si IV ions in their envelopes in the temperature-luminosity diagram. Filled symbols refer to those stars where the lines of these ions are present or probably present. Open symbols refer to those stars where the lines are absent or probably absent. The hatched area shows the location of the stars for which one would expect to observe the lines if there was no other ionization source than photo-ionization from the photosphere. (After Lamers and Snow, 1978).

where F_ν is the monochromatic flux, derived from stellar models, and $W(r)$ the geometric dilution factor.

It is obvious from Figure 78 that a relatively high degree of ionization does not only occur for stars in those regions of the Hertzsprung–Russell diagram where such is possible by radiative processes but also in those parts where the star's temperatures are too low for that. This was initially considered as an important indication that the *kinetic temperature* of the star's envelopes must be sufficiently high to produce such ionization, but this hypothesis was rendered obsolete for the following reasons: First: X-ray observations made aboard the Einstein Observatory (HEAO-B) showed that many luminous early-type stars are accompanied by X-ray-emitting plasmas that are of high T_k, conventionally assumed to be a corona. Secondly, prior to this discovery, Hearn (1975a) had already advocated that such stars could have a corona, and thereupon Cassinelli (1979a) suggested that the regions where the high-ionization lines are formed are lying *above* the corona, and would be due to Auger ionization by X-rays emitted from the (supposedly: thin) corona just around these stars. See also Cassinelli and Olson (1978) where this suggestion is further elaborated: the consequence of this idea would be that the existence of a thin corona would make a fairly hot chromosphere with temperatures of $\sim 4 \times 10^5$ K unnecessary; but the excitation conditions would be anomalous. Hence, Figure 78 demonstrates that in the non-hatched areas coronae occur; if they are not there, the lines must be emitted by chromospheres with $T_{kin} \approx 10^5$ K.

Be and shell stars. We mentioned already that the 'chromospheres' of these stars have probably the shape of a flattened disc. Whether in these cases it is still appropriate to speak of chromospheres is mainly a matter of definition.

Figure 79 gives a number of Hα profiles in different Be-stars and a model of the star's envelope, after Boyarchuk (1973). For an observer at O_1, who sees the star pole-on, the profile would show a faint stellar absorption line over which is superimposed a fairly strong emission, due to the extended parts of the envelope. For an observer at O_3 the strong absorption by the extended envelope would mark a central absorption line with slight emission peaks, due to the remainder of the envelope. In O_2 an intermediate case is observed.

The computation of the spectral Balmer decrement (variation in intensity over the lines of the Balmer series) has to take into account that the shell is optically thick in most of the Balmer lines. Boyarchuk assumed a velocity of expansion for the stars, varying with distance to the surface and considered the probability for photons, emitted in lower layers, to escape outwards. The ratio between the intensities Hδ/Hβ and Hγ/Hβ, computed for moving envelopes as well as for a static atmosphere is given in Figure 80. Other observations (Section 3.6) suggest a small rate of mass-loss from Be stars. The present observations do not allow a decision on the question whether one is dealing with moving envelopes or whether the Be-disks can be treated as a static atmosphere.

The hypergiants of medium spectral types have some kind of chromospheres, as was already shown by the early study of the chromospheric parameters of the hyper-

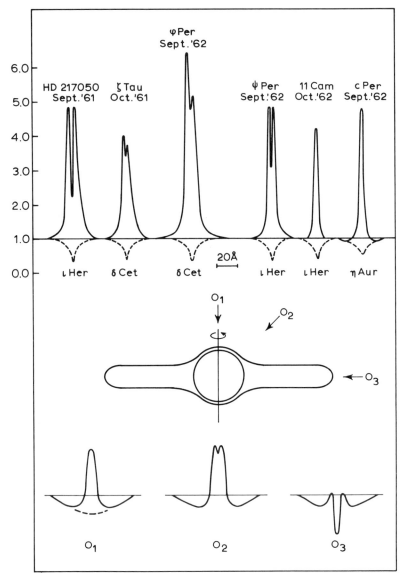

Fig. 79.   Observed Hα profiles in Be stars (above); model of Be star envelopes (middle); expected
         line profiles seen from different directions (below). (After Boyarchuk, 1973).

giant ϱ Cas (F8 Ia$^+$) by Sargent (1961). The observed weakness of the Balmer lines
of this star is explained by their being filled in by emission. The explanation of this
emission seems to offer some difficulties since the star is surrounded by a circumstellar
shell (see Section 3.8) with excitation and ionization temperatures as low as approx.
4000 K. Hence the Balmer line emission must be due to recombination in a hotter
chromosphere, presumably situated between the photosphere and the expanding shell.
Sargent showed that the observed filling-in of the Hα line cannot be due to absorp-

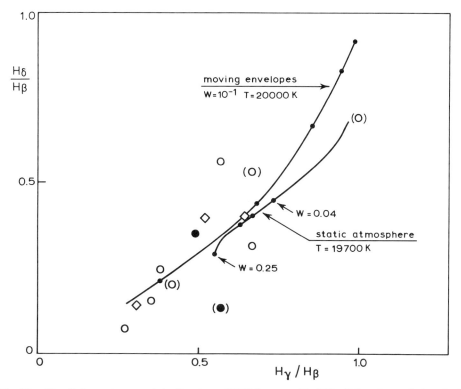

Fig. 80. The Balmer decrement in Be stars: $H\delta/H\beta$ versus $H\gamma/H\beta$. (After Boyarchuk, 1973).

tion and downward cascade of Lyman-continuum photons (the Zanstra-mechanism) originating from the photosphere, and therefore suggests heating by a mechanical energy flux, originating in the photosphere for which a value $F_m = 10^7$ erg cm^{-2} s^{-1} seems needed, equal to the energy output observed in $H\alpha$.

Chromospheres of other hypergiants have been studied by various authors – see Section 3.8 – for example by Wolf (1972, 1973) and by Wolf *et al.* (1974). The latter study, dealing with HD 160529 is the most detailed one, yet is still largely qualitative. Apart from emission features such as the Balmer lines, particularly $H\alpha$ and $H\beta$, which show P Cygni profiles, variable in character, there is, for instance, the variable absorption line He I $\lambda5876$ line which is much stronger than could be expected in this star, for which $T_e = 8800$ K and log $g = 1.1$. A similar feature occurs in HD 7583 in the Small Magellanic Cloud (Wolf, 1973). The variability of the lines observed in HD 160529 could be due to shock waves travelling through the atmosphere. The characteristic time (a few days) is compatible with a shock velocity of approx. 50 km s^{-1} and the expected extension of the chromosphere (some 20 million km).

Zeta Aurigae stars. These stars have fascinating atmospheres indeed! The occultations of the main sequence star behind the extended atmosphere of the supergiant offer a magnificent possibility to probe the atmosphere of the primary with nearly a point-source (Figure 39) (A review was given by Wright, 1970).

It was actually in these stars that stellar chromospheres have been detected. The chromospheres show evidence for high excitation; for instance 31 Cyg has a chromosphere with a temperature of 15000 K; $n_e \approx 10^{10}$ cm^{-3}. Micro- and macroturbulent velocities have been measured of the order of 10 to 20 km s^{-1}.

Late-type supergiants. Here the pioneering work was done by Deutsch; see his review (1970) and the one by Vardya (1972). The observers stress the instability of the atmosphere, the occasional occurrence of high-exitation lines such as these of He I. First indications of an excited envelope were given by Herzberg's (1948) observation of Fe II lines in the spectrum of α Sco.

The availability of high-resolution ultraviolet observations opened the way to a quantitative treatment. The Mg II h and k lines around 2800 Å, which appear as strong emission lines in the supergiants α Ori (Betelgeuse; M1.5 Iab) and α Sco (Antares; M1.5 Iab) were investigated in Copernicus spectra by Bernat and Lambert (1976a). They remarked the asymmetry of the k-line against the virtual symmetry of the h-line – see Figure 81.

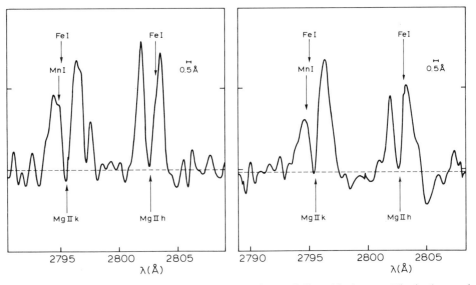

Fig. 81.   *Copernicus* scan of the Mg II doublet in Betelgeuse (left) and in Antares. The background (dashed) represents the noise level and not the stellar continuous spectrum. The ordinate scale is in arbitrary units. Positions of the Mn I and Fe I resonance transitions are shown above the spectrum. (From Bernat and Lambert, 1976a).

This asymmetry can be explained by overlying lines of Mn I and Fe I which is remarkable at first sight since these lines should be too weak to produce an appreciable asymmetry. However, selected Fe I lines in the visible appear to be weakened by fluorescent emmission from the Mg II lines which supports the original assumption. The absorption by Mn I and Fe I cannot be due to matter in the assumed circumstellar shell – the total density of matter in this slightly expanding shell is too small, but it could occur in a cool, stationary, but turbulent region ($v_\mu \approx 10$ km s^{-1}) between the top of the chromosphere and the base of the circumstellar shell. This picture is supported by Kondo *et al.* (1977) from their *BUSS* observations. They found that in α Ori the h and k emission peaks have absorption cores which are asymmetrically placed, but the *envelope* of the emission peaks is close to a (symmetric) Doppler profile. Yet, the variability of the envelope of α Ori (cf. Section 3.11) may cause variable asymmetries of the emission peaks and their central absorption.

6.5. Outward Increase of the Source Function or the Temperature in Near-Photospheric Layers

In this and the following sections, we wish to show that:

(a) There are cases in which, even without dissipation of a mechanical flux, an outward rise of the temperature or the source function can occur;

(b) In stars with effective temperature below approx. 10 000 K dissipation of mechanical energy from (sub-)photospheric convection regions can produce sufficient heating to yield a chromosphere or even a stellar corona;

(c) Amplification of waves by radiation pressure may be a mechanism to produce heating in hotter supergiant atmospheres.

(d) Dissipation of magnetic energy can contribute to the formation of a hot envelope.

First we discuss case (a) (cases (b), (c), and (d) are described in subsequent sections). Cayrel (1963) has considered the following problem: with increasing height, hence with decreasing density in a stellar atmosphere the number of ionizations by collisions decreases and photo-ionizations take over, in a way depending on the major opacity source (cf. also Feautrier, 1968). The relevant radiation field is defined by the temperature variation in the deeper layers of the stellar photosphere, in a way depending on the transparency of the atmosphere for the radiation involved. The temperature, thus assumed by the outermost layers is defined by the part of the radiation field that is responsible for the photo-ionizations, and may therefore increase outward in dependence on the ratio of photo- over collisonal-ionizations to reach a value, about equal to the brightness temperature at that wavelength at which the medium is most transparent. Feautrier (1968) and Mihalas (1970) have published models of stellar atmospheres in radiative equilibrium showing an outward temperature increase at very small optical depths.

The situation, of course, gets more complicated when Fraunhofer lines are also taken into consideration (cf. Frisch, 1966; Athay, 1970).

A similar case occurs when in certain stellar atmospheres the ratio σ/κ increases strongly outward, a case that occurs primarily in hot supergiants (De Jager and Neven, 1975). Since the source function is partly defined by the scattering term and partly by the absorption term (the latter acting only on the local radiation field; the former on the radiation field over a certain depth range) the influence of the surroundings and particularly of the deeper layers on the value of the local source function becomes stronger and stronger in more and more outward regions. In certain supergiants atmospheres this can give rise to a notable outward increase of the continuum source function.

6.6. Chromospheric and Coronal Heating – the Solar Case

The identification of the solar coronal lines, as due to forbidden transitions in highly ionized atoms, was initiated by the discovery of Adams and Joy (1933) of five coronal lines in the spectrum of the recurrent nova RS Oph (see Sections 9.2 and 9.5). This led Grotrian (1939) to suggest that the coronal lines are forbidden transitions in highly ionized atoms; a suggestion that was followed up by Edlén (1941, 1942). He made it clear that the solar corona is a hot plasma ($T \approx 10^6$ K) surrounding a cooler star with a photospheric temperature of $\sim$6000 K. Thus the problem arose of how such a plasma can derive its high temperature from the cooler Sun, a process that seemingly conflicts with the second law of thermodynamics. Apparently *thermodynamic upgrading of energy* is needed to explain the corona. The breakthrough came when Biermann (1946) found that solar coronal heating is due to *loss of mechanical energy* by shock dissipation of wave energy. It is only in the *shock fronts*, which are extremely thin, that the velocity gradient is large enough for viscous and conductive heating. Elsewhere in stellar chromospheres viscosity is unimportant, as is thermal conduction. Semi-quantitative considerations in the line of Biermann were given by Schwarzschild (1948) and were put in more quantitative shape by Schatzman (1949). A few years later another step forward was made when Lighthill (1952) and Proudman (1952) were able to derive quantitative expressions for the generation of a mechanical energy flux by thermal convection. Since these authors still assumed an incompressible non-stratified medium the application of their results to stellar cases is not without risks.

On the basis of these data the first succesful attempts to compute the physical parameters of the solar corona and of the transition region to the chromosphere were made for the non-magnetic case by De Jager and Kuperus (1961), and for a magnetized solar plasma by Osterbrock (1961). They were followed later by more refined investigations (see Sections 6.10 and 6.11).

Semi-quantitatively, the theory of solar-type coronae may be described as follows: In the convection region, extending in the Sun up to the level with an optical depth $\tau_{0.5} \approx 0.8$, with several elements overshooting up to $\tau_{0.5} \approx 0.1$, the large-scale convective motions generate turbulent motions on smaller scales; large turbulent elements in turn generate smaller elements, and so on. Thus, a spectrum of turbulent motions is generated, whereby the *large-scale elements*, or better: the *motions with large wavelengths in the Fourier transform of the velocity field*, transmit their energy of motion into the motions with smaller wavelengths through the non-linear terms in the equations of motion. There is a limiting wavelength l_1 where the motion energy is finally dissipated in shocks. The turbulent waves may have the character of sound waves, while gravity waves are generated by the convective motions in the radiative region just above the convective region. In magnetic regions they are Alfvén-waves. The mechanical or magnetic energy thus created, propagates partly or wholly outward, yielding a flux of mechanical or magnetic energy

$$F_m = \theta \varrho v_t^2 s. \tag{6.6; 1}$$

Here $\theta \approx 1$, v_t is the average velocity amplitude of the waves, still to be specified; in non-magnetic regions s is the velocity of sound. Since $s(:)T^{1/2}$, and since T does not increase greatly near the base of the stellar photospheres, v_t should increase proportionally to $\varrho^{-1/2}$, if there is no dissipation. Hence, the velocity amplitude of the sound waves increases upward, until the non-linear terms in the hydrodynamic equations become so important that they lead to the formation of shocks. Thus, if the amplitude of the wave is large enough, that is if v_t is comparable to s (it need not be strictly equal and may well be smaller than s), shock dissipation of the wave energy takes place, resulting in much more efficient energy dissipation. Solar chromospheric observations show that $v_t \approx s$ over the greater part of the upper chromosphere. Hence the dissipation of mechanical energy is there of the order of $\varrho s^3/L$ erg cm^{-3} s^{-1} where L is a characteristic dissipation-length. On the other hand the radiation of the chromospheric plasma is approximately $2 \times 10^{-21} T^{-1/2} n^2$ erg cm^{-3} s^{-1}. With $s=15$ km s^{-1}, $L=10^3$ km, $\varrho=1.6 m_{||} n_e$ and $T=6000$ K dissipation equals radiation for $n_e=4\times 10^9$ cm^{-3}. Higher up there is more energy dissipated than can be radiated so that necessarily the plasma temperature takes such high values that part of the dissipated energy can eventually get lost conductively over the steep temperature gradient which is formed between the chromosphere and the corona, and a stable situation is reached again. So, in this simplified picture the level with $n_e=4\times 10^9$ cm^{-3} would correspond with the 'top' of the chromosphere.

A very rough estimate of the coronal temperature could be obtained, in principle, when the mechanical flux at the basis of the corona is known, and assuming further that the greater part of that flux is dissipated in the corona, and conducted downward (hence neglecting radiative losses and solar-wind losses), by integrating the conduction equation [cf. Equations (6.10; 3) and (6.10; 4)]

$$F_{\text{cond}} = 5 \times 10^{-7} T^{5/2} \, dT/dz.$$

This semi-qualitative review on solar-type chromospheres and corona allows us to *summarize* a few obvious *problems and conclusions*.

It is apparently essential to know the various photospheric wave-modes and their upward propagation, in brief: similarly as the structure of stellar *photospheres* is defined by the equations of *radiative* and/or *convective energy* transport, the structure of the *outer layers* is defined by the equations describing the *transport of mechanical and/or magnetic energy*.

The obvious difficulty in determining coronal structure is that knowledge of the mechanical flux at the base of the corona presupposes knowledge of the losses of mechanical energy in the chromosphere, which involves detailed knowledge of the dissipation and radiation mechanisms. These are very complicated in chromospheric conditions, and this is the b a s i c difficulty in the computation of stellar c o r o n a e. Another difficulty is that by far the largest part of F_m is dissipated in a stellar chromosphere so that it is difficult to determine with sufficient accuracy the small fraction of F_m that is transmitted to coronal regions.

An aspect, that will be amplified in the next sections is that it seems sure that the solar-type *chromosphere* is heated by acoustic and gravity waves, but it is doubtful whether the mechanical energy-flux remaining at the top of the chromosphere is sufficient for heating the *corona*. Skylab-investigations of the corona have shown the importance of *magnetic structures* for heating at least large parts of the corona. A comparative evaluation of the mechanisms of solar-type coronal heating by magnetic or non-magnetic wave-modes is highly desirable (cf. Section 6.10.3).

6.7. Wave Modes in a Stellar Photosphere

The study of possible wave modes in a free atmosphere has made considerable progress in the years after 1960, particularly by rocket experiments in the E a r t h's atmosphere producing along the rocket's trajectory a luminescent or illuminated gas trail, which thereafter changes in shape by atmospheric winds and waves. The knowledge thus acquired for the terrestrial atmosphere could be transformed to stellar atmospheres. Possible wave modes in an atmosphere are defined by the forces restoring a deviation of an element of matter from the equilibrium configuration. If this force is the gas pressure one is dealing with c o m p r e s s i o n (or: a c o u s t i c) w a v e s. If it is gravitation, with buoyancy as the related effect, we have i n t e r n a l - g r a v i t y w a v e s (Figure 82). The waves are A l f v é n w a v e s, or m a g n e t o - a c o u s t i c w a v e s if the restoring forces are magnetic tension or pressure, respectively. R o s s b y w a v e s are due to the Coriolis force.

We next briefly describe a few properties of these various types of waves, and their possible role in heating stellar chromospheres and coronae.

A parcel of fluid which is displaced from equilibrium and oscillates in pressure equilibrium with its surroundings represents a wave mode, called an *internal-gravity wave*; it oscillates at the Brunt–Väisälä frequency

$$\omega_{BV} = (\gamma - 1)^{1/2} g/s,$$

where γ is the ratio of specific heats.

Compression waves propagate by the adiabatic increase or decrease of the gas-pressure during contraction or expansion of the gas. Since the phase-velocity (λ/P) depends on the period, the waves show dispersion. Hence, vertical compression is not possible for waves with periods exceeding the time necessary for a sound wave of infinitely high frequency to propagate over the scale-height of an atmosphere. To that period corresponds the acoustic cut-off frequency

$$\omega_A = \gamma g/2s,$$

which is nothing else than the vibration frequency of the entire atmosphere.

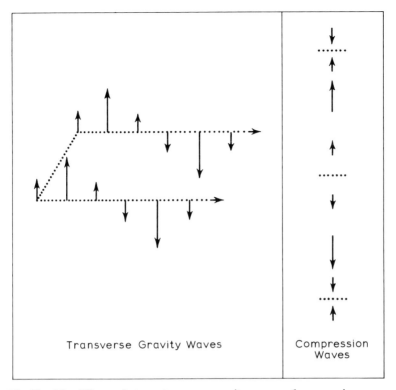

Fig. 82.   The difference between transverse gravity waves and compression waves.

Gravity waves can only exist in the radiative region of the atmosphere; in the convection region they would be unstable. Actually, internal-gravity waves and convection may be considered as stable and unstable manifestations of the same wave-mode (Stein and Leibacher, 1974).

If gas pressure and gravitation are both acting as restoring forces the modes of the possible *acoustic-gravity waves* can be described by a diagnostic dispersion diagram, which gives the boundaries of the regions where acoustic or gravity waves can propagate – see Figure 83 and for an extended discussion Stein and Leibacher (1974). The hatched areas show regions where vertical propagation is possible. It is important to realize that with changing T and P the boundaries change also, so that waves that propagate upward in one part of a stellar atmosphere may arrive higher up in a region where propagation is no longer possible.

Gravity waves as a means for heating the solar corona were initially proposed by Whitaker (1963) because sound waves with wavelengths of the same size as the solar granulation elements would not propagate upward. Doubt arose when Souffrin (1966) showed that gravity waves of these wavelengths would damp out in a time scale of seconds by radiative relaxation.

But B. Mihalas (1979), who made detailed numerical calculations of the propagation of internal-gravity waves in a realistic model of the solar atmosphere, found that a

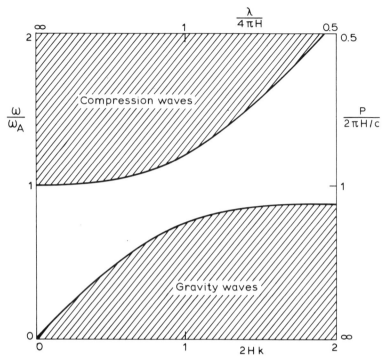

Fig. 83. Diagnostic diagram showing the 'transmission' of an atmosphere for compression waves and gravitation waves. H is the scale-height of the atmosphere; λ and k are wavelength and wave-number, P the period.

significant flux of internal-gravity wave energy can survive radiation damping and can propagate into the solar atmosphere. There, the propagation is mainly limited by effects of non-linear wave breaking as the wave amplitude increases with height. The fluxes that can penetrate highest are those with the smallest group-velocities, closest to the limiting boundary of the gravity-wave domain in Figure 83. The calculations show that waves carrying a flux of $\sim 10^8$ erg cm^{-2} s^{-1} in photospheric levels, can carry a flux of $\sim 10^5$ erg cm^{-2} s^{-1} to heights of ~ 1400 to 1500 km above the Sun's limb. Hence, gravity waves can be efficient in heating the solar *chromosphere*, not the corona.

In addition, the chromosphere is heated by acoustic waves of higher frequency, above the photospheric cut-off frequency for sound waves (0.02 s^{-1}). These appear able to produce a sufficient mechanical flux, according to computations by Ulmschneider (1974). A fair approximation of the properties of the field of acoustic waves (Ulmschneider *et al.*, 1977) is by assuming monochromatic waves with the acoustic period P, and a velocity amplitude $u_0 = (2\pi F_m/\varrho s)^{1/2}$. Such waves correspond approximately to the maximum in Stein's (1968) computed acoustic spectrum (Section 6.8). These waves are also able to penetrate to, and heat the chromosphere. Very approximately one finds that the base of stellar chromospheres, assumed identical to the level of temperature minimum, lies where the velocity amplitude is roughly equal to the sound velocity. As a rule this approximation is valid for those stars where the shocks do not form in the radiative damping zone. In stars where shock formation occurs in that zone (Schmitz and Ulmschneider, 1979a, b) the temperature minimum occurs at the top of the radiative damping zone irrespective of where shock formation occurs.

6.8. The Generation of Mechanical Fluxes in Stellar Photospheres

The theory of the generation of a flux of acoustic waves in a homogeneous turbulent medium has been developed by Lighthill (1952, 1960), Proudman (1952), and others. An early summary is given by Kuperus (1965, Section II.2), a later one by Stein and Leibacher (1974).

In this approach the equation of motion is written as

$$\varrho \frac{d\mathbf{v}}{dt} + (\mathbf{v} \cdot \mathbf{grad}) \, \mathbf{v} = -\mathbf{grad} \, p + \varrho g + \frac{\mu}{4\pi} \, \text{curl } \mathbf{B} \times \mathbf{B},$$

and that of continuity:

$$\frac{d\varrho}{dt} + \text{div}(\varrho \mathbf{v}) = 0.$$

Here g is the effective acceleration of gravity, B the magnetic field strength, and the other symbols have their usual meaning. The equations are linearized by introducing perturbations p_1, ϱ_1, and B^1 to the undisturbed values p_0, ϱ_0, and B^0. This yields an inhomogeneous wave equation for one of the variables:

$$\frac{\partial^2 \varrho_1}{\partial t^2} - s^2 \nabla^2 \varrho_1 = -\frac{\partial F_i}{\partial x_i} + \frac{\partial^2 S_{ij}}{\partial x_i \, \partial y_j},$$

where F_i is the force term

$$F_i = \varrho_1 g \delta_{i3} - \frac{\mu}{4\pi} \, B_j^0 \, \frac{\partial B_i^1}{\partial x_i},$$

and δ_{ij} is the Kronecker function. Further, S_{ij} is the stress tensor due to velocity shear and magnetic shear:

$$S_{ij} = \varrho v_i v_j + \frac{\mu}{8\pi} \, (B^1)^2 \, \delta_{ij} - \frac{\mu}{4\pi} \, (B_i^1 B_j^1).$$

In most cases the magnetic field is assumed negligible ($B=0$). The linear terms yield the wave propagation operator, and the non-linear terms are a quadrupole-type source function. As in the theory of electromagnetic radiation retarded potential solutions are then sought of the inhomogeneous wave equations

$$\varrho(\mathbf{x}, t) - \varrho_0 = -\frac{1}{4\pi s^2} \frac{\partial}{\partial x_i} \int F_i\left(\mathbf{y}, t - \frac{|\mathbf{x} - \mathbf{y}|}{s}\right) \frac{d\mathbf{y}}{|\mathbf{x} - \mathbf{y}|} +$$

$$+ \frac{1}{4\pi s^2} \frac{\partial^2}{\partial x_i \, \partial y_j} \int S_{ij}\left(\mathbf{y}, t - \frac{|\mathbf{x} - \mathbf{y}|}{s}\right) \frac{d\mathbf{y}}{|\mathbf{x} - \mathbf{y}|},$$

where x and y are respectively points inside and outside the turbulent region. The integration is performed over the turbulent region taking the retardation into account.

In Lighthill's treatment the atmosphere was considered homogeneous. The next step should be the extension of the discussion to a gravitationally stratified atmosphere, where under the influence of gravity and pressure as restoring forces gravity waves and pressure (acoustic) waves originate. A successful discussion of this case has not yet been published; an early attempt by Stein (1967) appeared to be erroneous (Stein and Leibacher, 1974) since it was based on the assumption that internal-gravity waves exist in the convection region.

The frequency dependence of the acoustic flux was calculated by Stein (1968) and Milkey (1970), who found that the maximum flux would occur at a frequency$=s/H$, where H is the pressure scale height at the point of maximum sound generation. To this spectrum of acoustic waves a low frequency 'tail' may be added through the mechanism of convective overshooting: the penetration of convective elements from the convection zone into the overlying radiative-equilibrium part of the atmosphere. These penetrating convective elements, when decelerating, will be able to generate internal-gravity waves in the nearly adiabatic regions near the temperature minimum (Stein and Leibacher, 1974). The theory, like the convection theory, is very phenomenological and largely qualitative.

The case of the generation of a mechanical flux in a m a g n e t i c r e g i o n was treated by Kulsrud (1955) along the lines described above. He found for the generated acoustic power per unit of volume and time:

$$P_{\mathrm{mag}} = a_1 \varepsilon M^5,$$

where $\varepsilon = \varrho \langle v^2 \rangle^{1/2} M$ is the turbulent energy dissipation per unit volume and time, and

$$a_1 = 13.5 + A(\beta)\, \gamma^2 + 12.3\beta\gamma^4.$$

Here

$$\gamma = \frac{\langle (\mathbf{B}^1)^2 \rangle}{4\pi\varrho} \Big/ \langle \mathbf{v}^2 \rangle$$

is the ratio between the energies of the turbulent magnetic field and of the turbulent motions, while $A(\beta)$ is a very slowly varying function of the scale factor β. Figure 84, taken from Kuperus (1965) gives the ratio between P_{mag} and P_0, the power output in the absence of magnetic fields, as a function of γ, for $\beta=1$, 2, and 4.

According to the conventional mixing-length formalism, thermal convection would not develop in stars earlier than approximately A5. Convectively instable subphotospheric regions do exist however in such stars but convective *motions* will hardly

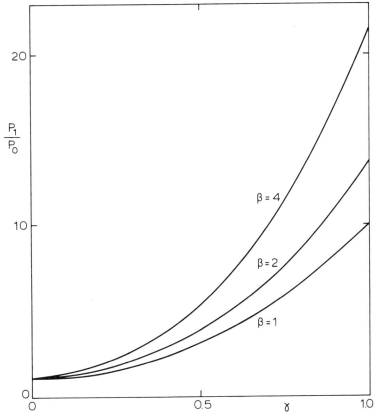

Fig. 84.   Amplification of the total acoustic power output in the presence of a turbulent magnetic field. (After Kulsrud, 1955).

develop there because radiative equalization of temperature fluctuations will occur in a short time-scale. A more refined theory of stellar convection based on a discussion of atmospheric instability, using the equations of compressible fluid dynamics (La-tour *et al.*, 1976; Toomre *et al.*, 1976) seems to indicate that the mechanical fluxes calculated for A-type stars with the mixing-length approach are too small, but work is still in progress and final results have to be awaited. In any case this theory too does not seem to work for O- and B-type stars. How then to explain that turbulent motions exist in all early-type stars? How are these motions generated if not by convection?

Several possible solutions have been proposed, starting with a suggestion by Hearn (1972, 1973) that in hot stars chromospheric heating could occur by a mechanical flux originating from sound waves amplified by continuous radiation forces. Minor velocity perturbations in a hot stellar atmosphere may be amplified by the interaction of the perturbations of the absorption coefficient with the radiative forces, i.e. the gradient of the radiation pressure. The difference in the absorption coefficients in the 'top' and 'bottom' of a sound wave interact with the gradient of the radiation pressure, which is strong in hot stars. The wave is amplified till eventually stability is reached through the action of second order terms, causing viscous dissipation. (It is interesting to note in passing that Mestel *et al.* (1976) derived the same radiative-driven instabilities for expanding envelopes of quasars.) Although this mechanism seems promising it does not work in its initially proposed form: Berthomien *et al.* (1976) found that the *e*-folding distance for the amplification of radiation-driven sound waves is much greater than the pressure-scale-height of the atmosphere. Therefore the total amplification of a perturbation in a hot stellar atmosphere cannot be very great, typically by a factor of 2 only.

Therefore, further elaborations of these considerations by Nelson and Hearn (1978), Martens (1979), and MacGregor *et al.* (1979) are of interest. The first authors showed that a Rayleigh–Taylor instability with a short *e*-folding time may arise in hot stars, being driven by the radiative forces related to the (UV) resonance lines of ions in an expanding atmosphere at the basis of the stellar wind. An element of gas that would receive a small perturbation to the expansion velocity of the atmosphere may thus be further accelerated. In addition, Martens (1979) calculated that in the super-sonic part of an expanding atmosphere sound waves are strongly amplified by res-onance-line opacity forces. He calculates for these perturbations periods of 15 min to several hours. These waves would produce an acoustic flux of 8×10^9 erg cm^{-2} s^{-1} for ζ Pup (O4 ef), and 7×10^8 erg cm^{-2} s^{-1} for ε Ori (B0, Ia). Using a linearized theory, applicable to optically thin waves, parallel research was done by MacGregor *et al.* (1979). They found that sound waves in the supersonic wind regions tend to grow, with *e*-folding distances for wave-growth that are smaller than the distances over which the flow properties vary. It is thought (but not proved) that such disturbances may be a source for coronal heating in luminous stars with well-developed stellar winds.

For atmospheres of cool stars where hydrogen is mainly neutral, a mechanism for the amplification of sound waves exists, based on the 'kappa-mechanism' of hy-drogen ionisation: in a sound wave originating in the hydrogen convection zone and propagating outward, the degree of ionisation will be different in the regions of com-pression and of dilution, and so will be the absorption coefficient for radiation. Thus, amplification of sound waves may result – a mechanism that has analogies to Hearn's mechanism of radiation-driven sound waves. Ando (1976) has investigated the effect numerically and found that κ-amplification is important for certain modes in a region at the low-temperature side of the Cepheid instability strip in the HR-diagram. From the numerical results an interpolation formula can be derived for the maximum growth rate η of the excited modes:

$$\eta \ (:) \ g^{-0.5} \ T_e^6. \tag{6.8; 2}$$

This relation can be understood in a semi-quantitative way as follows: the growth rate should be of the same order of magnitude as the ratio of the pulsational time-scale to the thermal time-scale of the driving zone:

$$\eta(:)\tau_{\text{puls}}/ \int c_p T \, dm/L.$$

Now, $\tau_{\text{puls}} \approx s/g$ and

$$\int c_p T \, dm/L \approx \langle c_p T \rangle P/Fg,$$

where P and F are the pressure and the heat flux at the hydrogen ionisation zone. Since the average value $\langle c_p T \rangle$ in the ionisation zone can be considered almost constant for all late-type stars, we have $\eta(:)F/P$, and assuming that $\kappa(:)P^{0.7}T^{10}$ we obtain

$$\eta(:)g^{-0.6}T^{10},$$

a relation that agrees roughly with Equation (6.8; 2).

6.9. Comparison of Predicted and 'Observed' Stellar Mechanical Fluxes

Modern computations of stellar mechanical fluxes have been made by Ulmschneider (1967). His Figure 9 gives lines of equal mechanical fluxes in the Hertzsprung–Russell diagram. This was followed by work by De Loore (1970) who computed F_{mech}-values for 90 model photospheres. This work was repeated by Renzini et al. (1977) but still based on the mixing-length convection theory. They find overall acoustic fluxes that are approximately one order of magnitude smaller than those of De Loore, which shows the enormous influence of slight changes in the convection calculations on the results. Figure 85 gives lines of equal acoustic fluxes in a $(\log T_e; \log g)$-diagram. The period P_{max} of the acoustic waves for which the monochromatic flux has its maximum value is about $0.1P_a$, where P_a is the acoustic cut-off period at the point of maximum flux generation: $P_a = 4\pi s/\gamma g$. Renzini et al. (1977) find as a good fit to their numerical results:

$$\log P_{\text{max}} = 5.8 - \log g.$$

We next summarize the scanty observational indications for the mechanical energy flux; we try to estimate its value along the spectral sequence, and compare these values with theoretical predictions.

If the mechanical flux originates in (sub-)photospheric layers of the star, and progresses outward, an observable manifestion for the value of the mechanical flux in photospheric layers is yielded by observations of the photospheric random velocity field.

Obviously, we then immediately meet a fundamental problem: that of determining what part of the 'turbulent' motion field of a stellar photosphere contributes to the mechanical energy flux F_{mech} of the star. And since the star's photospheric motion field is mostly only known by its two asymptotic values, the micro- and macro-turbulent velocity components, the problem is that of the relation between the micro-

256 CHAPTER 6

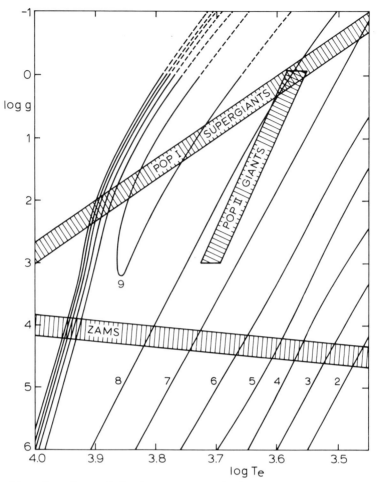

Fig. 85. Lines of equal acoustic flux $\log \pi F_m$ (erg cm^{-2} s^{-1}) as a function of $\log g$ and $\log T_e$, for a ratio a (=mixing length over scale height) equal to unity. The location of the Zero Age Main Sequence (ZAMS), of the population II giant branch, and of the population I supergiants is indicated. (After Renzini *et al.*, 1977).

and macro-turbulent velocity components and the velocity parameter defining the mechanical energy flux. An associated problem is that of the velocity of propagation of the mechanical energy flux.

Attempts to arrive at an answer to this complex of questions were made by Lamers and De Loore (1976) and Ulmschneider (1979). These authors determined the 'observational' mechanical energy flux for a number of supergiants with well-determined microturbulent velocity components (see Table XI). They assumed for the outward component of the mechanical energy flux of the star (cf. Equation (6.6; 1)):

$$F_m = \tfrac{1}{2}\varrho a^2 s,$$

where s, the velocity of outward propagation of the mechanical flux is assumed equal to the local velocity of sound $s=(\gamma\Re T/\mu)^{1/2}$, with $\gamma=\frac{5}{3}$. The velocity amplitude a of the waves that define the mechanical flux was assumed to be equal to the micro-turbulent velocity component ζ_μ as defined from the line widths, in the way described in our Section 2.9, and given in Table XII. The values of ϱ and T were taken from model atmospheres at $\tau_R=0.1$; for the F- and G-type atmospheres the model values at $\tau_R=0.2$ (Parsons, 1967) were taken. The resulting values are given in Figure 86, where these 'observational' data are also compared with theoretically predicted values of F_m. These latter values are derived from convective atmospheric models (Renzini *et al.*, 1977) for stars with $\log T_e<3.9$, and for stars with higher T_e-values F_m is calculated with Hearn's value based on the concept of radiation-driven sound waves. The agreement appears to be satisfactory, which is curious since we know that Hearn's initial expressions for F_m are overestimated!

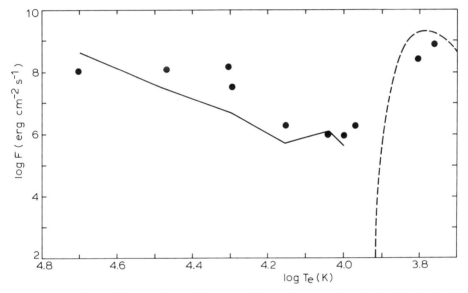

Fig. 86.   Comparison of the mechanical flux derived from the observed microturbulent velocities (*dots*) with predicted mechanical fluxes, for supergiants. *Solid line*: flux predicted from radiation-driven sound waves (Hearn); *dashed line*: flux calculated for convection (Renzini *et al.*, 1979). There is a fair agreement! (After Lamers and De Loore (1976), as modified by Ulmschneider (1979)).

Another way to observationally determine the value of mechanical fluxes is by determining the radiative output of a star in chromospheric lines. This value may in reasonable approximation be identified with the mechanical flux at the basis of the chromosphere, because – as will be shown in Section 6.10 – nearly the whole mechanical flux as it emerges from the top of the photosphere is dissipated in the chromosphere; only a very small fraction of it remains for the heating of the – very rarified – outer coronal regions.

Thus, De Jager (1959) and later Linsky and Ayres (1978) attempted to set a quantitative basis to the theory of late-type stellar chromospheres. They assumed that the net chromospheric emission in a line is the difference of the observed flux of (the emission core of) the line with the flux computed for a photospheric model without a chromosphere. By adding up contributions from all lines for which observations are available (mainly Mg II h and k; Ca II H and K; Lα) in 32 stars including the Sun, chromospheric fluxes are found as are shown in our Figure 87a, where these results are compared with theoretical predictions of F_m by Renzini *et al.* (1977) and Ulmschneider *et al.* (1977). Apparently there is fair agreement for main sequence stars; for supergiants the fluxes determined from the observations are approximately an order of magnitude too small. A solution to this problem was suggested by Schmitz and Ulmschneider (1979a), who found that the contribution of H⁻ emission (neglected by Linsky and Ayres!) is for *giants* about ten times the summed up contribution of the lines. This is *corrected* in our Figure 87b where the H⁻ contribution is taken into account in a comparison of empirical (E) and theoretical (T) chromospheric fluxes. As is clear, this comparison has so far been made for main-sequence stars and giants, but not yet for supergiants. Little information on chromospheric fluxes of the most luminous stars is yet available! An additional contribution to solve the problem of the apparent F_m-deficiency in supergiants is from Cram (1978) who found that in a non-LTE atmosphere a given amount of mechanical heating will produce a stronger temperature-rise than in a LTE-atmosphere. The two solutions given here have yet to be combined and critically compared.

In later investigations (Musielac and Sikorski, 1979; Basri and Linsky, 1980; Linsky *et al.*, 1979b) part of the radiative losses were determined for a larger material. In several cases considerable differences remain between the radiative fluxes and calculated mechanical fluxes.

The pronounced variability of the chromospheric indicators in long-period variables, as well as the feeling that the existence of chromospheres in such stars can not be explained by a dissipation of a mechanical flux – any flux originating from subphotospheric layers seems to be damped in the photosphere – has led Bychkov and Panchuk (1977) to the tentative hypothesis that such transient are due to magneto-hydrodynamic effects such as the reconnection of field lines above the photosphere.

6.10. The Computation of Stellar Coronal Parameters

6.10.1. HISTORICAL AND GENERAL REMARKS

After the first more or less successful attempts to understand the structure of the solar corona the first predictions of stellar mechanical fluxes and of the physical parameters at the bases of stellar coronae were made by De Jager (1960). The computations were based on Böhm–Vitense's (1958) theory of stellar convection, Lighthill's (1952) theory of the generation of acoustic noise and on some first order considerations on the transfer and dissipation of a mechanical energy flux through the solar chromosphere, for the non-magnetic as well as the magnetic cases. Values for the electron density and temperature at the base of the corona were given. These results were used

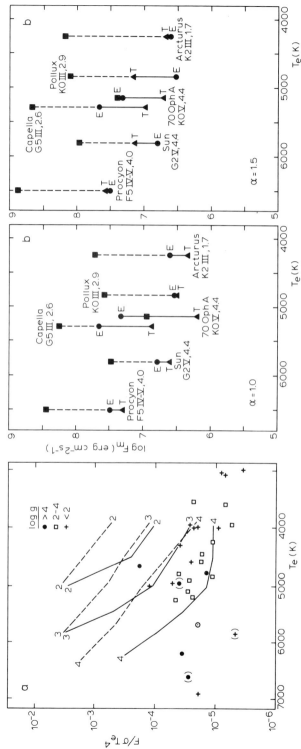

Fig. 87. (a) Measured radiative losses in Mg II h and k normalized to the surface luminosity σT_e and plotted versus effective temperature. Stellar gravities were crudely estimated according to luminosity class and spectral type. Solar Mg II fluxes are plotted for the average Sun (⊙); the two stars well above the mean are λ And (●) and β Dra (+). Data which are underestimates are in parenthesis. Also shown are the total acoustic flux generation curves of Renzini et al. (1977; dashed) designated by log g, and the acoustic flux available at the temperature minimum after photospheric radiative damping losses as computed by Ulmschneider et al. (1977; solid). (From Linky and Ayres, 1978). (b) A comparison of empirical (E) and theoretical chromospheric fluxes for six stars. Filled squares are acoustic fluxes by Renzini et al. (1977) revised upward by a factor of 2. Triangles are Ulmschneider et al.'s (1977) calculated fluxes. The quantity α is the ratio between mixing length and pressure scale height. (From Ulmschneider, 1979).

later, in somewhat modified form, to publish the first estimates of the expected ultra-violet and X-ray emission from stellar chromospheres and coronas (De Jager and Neven, 1961). Later, predictions of stellar coronal temperatures were published by Kuperus (1965), who computed lines of constant coronal temperatures in the Hertz-sprung–Russell diagram. For main sequence stars of spectral types later than early A (only for such stars convection developes and were calculations made) the highest coronal temperatures were predicted for stars with $T_e \approx 7000$ K. Solar type super-giants should have $T_{cor} \approx 1$ to 1.5×10^6 K. A more detailed treatment of the problem, in which models of the transition regions to the stellar coronae were derived for four stellar models (the Sun, G2 V, and K1 III, G3 III, and G7 V stars) was thereafter given by Ulmschneider (1967, 1971a, b) who used improved radiative-loss functions and also included gas-flow (the 'stellar winds') in his equations. For the four types of stars mentioned above, maximum coronal temperatures are found of respectively 3.2, 2.8, 3.5, and 1.7 MK.

Stellar coronal parameters were computed over a large part of the Hertzsprung–Russell diagram by De Loore (1970), for 90 model photospheres, and while com-puting the consequent coronal electron densities and temperatures, he also included in these computations the stellar winds in the subsonic regions.

For stars with spectral types later than A5 model coronae have been computed by Landini and Fossi (1973). They predicted an increase of the coronal temperature and of the radiation flux when the star, during its evolution, moves off the main se-quence. When, however, the giant and supergiant stages are reached the acoustic energy flux will be dissipated mainly in the stellar chromospheres, and radiated there. Hence, in line with earlier results they find that supergiant stars would not have hot coronae but at most a warm envelope that could better be called a 'chromosphere' (although the *name* of the envelope is a matter of definition rather than of physics!).

This latter conclusion can easily be proved. Extreme supergiants often have large radii. Hence the velocity of escape, $v_{esc} = (2G\mathfrak{M}/R)^{1/2}$ is small. When this velocity would be smaller than v_m (the average coronal particle velocity in assumed Max-wellian distribution) the corona would escape almost immediately and would therefore not exist. Even for $v_m \approx \frac{1}{2} v_{esc}$ a corona would disappear in a time span short as com-pared with the life time of most stars. We write

$$v_m = (8\mathfrak{R}T/\pi\mu)^{1/2} \approx 1.5 \times 10^4 \, (T/\mu)^{1/2} \text{ cm s}^{-1},$$

and require for a corona to exist that

$$v_m < \tfrac{1}{2} v_{esc}. \tag{6.10; 1}$$

With $g = G\mathfrak{M}/R^2$ the condition (6.10; 1) reduces to

$$T < 160g \, (R/R_\odot). \tag{6.10; 2}$$

For the Sun, with $g = 2.7 \times 10^4$ cm s^{-2} the maximum possible coronal temperature would therefore be 4×10^6 K. For a red supergiant such as α Sco, with $g \approx 1$ and

$R/R_\odot \approx 600$, $T_{max} \approx 10^5$ K. The same data apply to the F8 super-supergiant ϱ Cas (see Section 3.8 for pertinent data). But the O4ef star ζ Pup ($g=6\times10^3$; $R/R_\odot=20$) (Section 3.2) has $T_{max}=20\times10^6$ K.

Hence, hot coronas will not exist in the upper right part of the Herztsprung–Russell diagram.

After this introduction we describe the numerical methods for the computation of stellar coronae, and the equations for estimating the approximate values of several of the coronal parameters.

We have now to distinguish between two different approaches; they refer to the magnetic and the non-magnetic cases respectively. Both are inspired by solar studies.

The *non-magnetic approach* is the conventional one. The star is assumed spherically symmetric and the equations for the transfer of mechanical energy are solved. This case will be treated first, below.

In the *magnetic approach* the star could again be treated as spherically symmetric and an amplification factor of noise-generation like the one suggested by Kulsrud (cf. Figure 84) could be introduced in the equations. But, again inspired by the solar observations, one might as well assume that coronal heating occurs only in localized regions of the stellar atmosphere, by some process of transformation of magnetic into kinetic energy. This case, far less developed than the first, will be treated secondly.

6.10.2. THE NON-MAGNETIC APPROACH

The basic equations for the *non-magnetic approach* are again the equations of:
– continuity (conservation of mass): Equation (5.5; 2);
– conservation of momentum: Equation (5.5; 1);
– state: Equation (5.5; 3); and
– conservation of energy: Equation (5.5; 5).

For the study of coronal parameters the treatment is sometimes simplified by neglecting the stellar wind. In any case the wind-velocity does not seem to be large in coronal regions. For the present problem the most important equation is (5.5; 5), that of conservation of energy. In this equation Q_A=heat input due to dissipation of mechanical energy [erg cm^{-3} s^{-1}];

$$\nabla \cdot q_c = \text{heat input by conduction} = \frac{\mathrm{d}}{\mathrm{d}z}\left(K\frac{\mathrm{d}T}{\mathrm{d}z}\right) \qquad (6.10; 3)$$

where K is the coefficient of thermal conduction:

$$K = \kappa T^{5/2}, \quad \text{with} \quad \kappa = 5 \times 10^{-7} \text{ erg cm}^{-1}\text{s}^{-1}\text{deg}^{-7/2} \qquad (6.10; 4)$$

(Braginskii, 1965); hence

$$\nabla \cdot q_c = \frac{\mathrm{d}^2\eta}{\mathrm{d}z^2} \quad \text{with} \quad \eta = \frac{2\kappa}{7} T^{7/2}, \qquad (6.10; 5)$$

$$Q_R = \text{radiative energy loss} = \varrho^2\varphi(T) \text{ erg cm}^{-3}\text{s}^{-1}, \qquad (6.10; 6)$$

where $\varphi(T)$ is a function initially computed by Cox and Tucker (1969) for an optically thin plasma. For coronal conditions this function was recomputed by Mewe (1972). Later calculations are by McWhirter *et al.* (1975) and Raymond *et al.* (1976).

The left-hand side of Equation (5.5; 5) can be written as

$$Q_{\text{wind}} = \frac{-\dot{\mathfrak{M}}}{4\pi R^2} \left[\frac{\mathrm{d}}{\mathrm{d}z} \left(\tfrac{1}{2}v^2 + \tfrac{5}{2} \frac{RT}{\mu} \right) + g_{\text{eff}} \right], \tag{6.10; 7}$$

where $5RT/2\mu$ is the enthalpy of a mono-atomic gas. The most difficult and uncertain term of Equation (5.5; 5) is the mechanical energy-flow term, Q_A, for which one may write, just formally

$$Q_A = -\frac{\mathrm{d}F_m}{\mathrm{d}z} \equiv \frac{F_m}{L} = \frac{F_0}{L} e^{-z/L}, \tag{6.10; 8}$$

where F_0 is the mechanical energy-flux at the basis of the transition layer, and L the e-folding distance of the mechanical flux. Loss of mechanical energy-flux by reflection against the steep temperature gradient of the transition region is formally included in this treatment, but it makes L strongly height-dependent.

With the above equations those of conservation of momentum and of energy can be rewritten as follows:

$$\frac{1}{v} \frac{\mathrm{d}v}{\mathrm{d}z} = \frac{g_{\text{eff}} + \frac{R}{\mu} \frac{\mathrm{d}T}{\mathrm{d}z}}{\mathfrak{R}\frac{T}{\mu} - v^2}, \tag{6.10; 9}$$

$$\frac{\mathrm{d}^2\eta}{\mathrm{d}z^2} = -\frac{F_m}{L} + M\frac{\mathfrak{R}T}{\mu} \left(\frac{3}{2T} \frac{\mathrm{d}T}{\mathrm{d}z} + \frac{1}{v} \frac{\mathrm{d}v}{\mathrm{d}z} \right) + Q_R \tag{6.10; 10}$$

and

$$-\frac{\mathrm{d}F_m}{\mathrm{d}z} = \frac{F_0}{L} \left(\frac{T_0}{T} \right)^{1/2} \exp\left\{ -\int_{x_s}^{x} \frac{T_0}{T} \mathrm{d}x/L \right\}, \tag{6.10; 11}$$

with $L = 1.0 \times 10^4 \, PT_0^{1/2}$ (Lamers and de Loore, 1975). These expressions define the distribution of temperature, velocity and of F_m with height z. They show the well-known (cf. Section 5.5) singularity in the sonic point, typical for solar-type coronae and winds. In that point

$$g_{\text{eff}} + \frac{\mathfrak{R}}{\mu} \frac{\mathrm{d}T}{\mathrm{d}z} = 0. \tag{6.10; 12}$$

In an atmosphere with $g_{\text{eff}} > 0$ the sonic point occurs at a level where $\mathrm{d}T/\mathrm{d}z < 0$, but in atmosphere with $g_{\text{eff}} \approx 0$, as in many supergiants or other stars with extended atmospheres, the sonic point occurs at the level of maximum temperature.

One of the crucial aspects of the problem area discussed here is that of the value of Q_A, the local heat input by dissipation of mechanical energy. So far this quantity has been *formally* described [in Equation (6.10; 8)] by introducing a 'dissipation length' L. Knowledge of this quantity is important because the actual value of L determines both the magnitude and the position of the coronal temperature maximum, as shown by computations of Endler *et al.* (1979). The actual value of L and its variation with height depends on the mechanism(s) responsible for coronal heating. So far this problem has only been studied for the solar corona, where shock-wave heating was conventionally assumed. However Athay (1976) suggested that thermal conduction from the hot compressed parts of the wave towards the cooler diluted parts may be important. D'Angelo (1969) proposed Landau damping, being absorption of wave energy by ions moving in the wave's electric field – it will occur mainly when the mean free path of the ions becomes comparable to the wavelength of the oscillations.

At this time no general recipe can be given for the calculation of heating of a specific corona: for each case the possible heating mechanism should be compared separately.

An interesting attempt to calculate the parameters of stellar coronae is due to Hearn (1975b; see also Hearn, 1979) on the basis of the assumption that the coronal parameters are *defined* by the condition that the total energy flux, necessary to maintain the corona against *heat losses*, due to *thermal conduction* towards the photosphere, to *radiative* and to *mass-loss* should be minimized.

The *minimum-flux corona* thus obtained, appears to have temperatures and base-densities which are single-valued, monotonic functions of the total energy flux; this corona also appears to have the largest pressure of possible coronal models; it is a *maximum-pressure* model.

This approach has the mathematical advantage that the theory can be set up in a nearly closed form; physically, however it is not certain that the minimum-flux assumption is correct, because it is the specific heating law which defines the structure of the corona. It has not been proved that any heating law would correspond to the minimum-flux corona (Endler *et al.*, 1979).

6.10.3. THE ROLE OF MAGNETIC FIELDS IN CORONAL HEATING

This role has been underestimated by most authors. Yet it seems that in the solar case the main source of coronal heating is to be found in the magnetic regions. An estimate of the relative efficiency of the magnetic and non-magnetic solar areas in generating a mechanical flux can be found as follows (De Jager and De Loore, 1970): We know the relative fractions of the solar active and 'quiet' regions that are covered with magnetic fields; we also know the dependence of coronal electron density on the F_m-value, and the average coronal electron densities above active and quiet coronal regions. A comparison of these three sets of data leads to the result that per unit area the solar magnetic regions generate a mechanical flux, about seven times as large as the flux from non-magnetic regions.

In addition, the discussion in Section 6.8 and Section 6.10.2 has shown that in the non-magnetic areas the gravity-wave-flux will heat the chromosphere but will bring a flux of $10^5 \, \mathrm{erg \, cm^{-2} \, s^{-1}}$ only to about 1500 km above the solar photosphere.

Higher-up they are still stronger damped. A flux of acoustic-wave-energy may penetrate further but the flux, available at the top of the chromosphere is small. Observations of solar areal oscillations in the middle-chromosphere give an estimated acoustic-wave-flux $\sim 1 \times 10^4$ erg cm^{-2} s^{-1} (Athay and White, 1978), which is about a factor ten less than is thought necessary for heating the corona. This result was amplified by Bruner (1978) who measured the energy contained in oscillations in the solar transition zone between chromosphere and corona. The flux he found is 9×10^3 erg cm^{-2} s^{-1}, in good agreement with Athay's result. As to order of magnitude these discrepancies agree with the above-described earlier result of De Jager and De Loore (1970).

The question is then how and what mechanical flux is generated in magnetic regions. One line of approach is to keep the concept of wave-heating but to apply the amplification factor as calculated by Kulsrud (1955), as shown in Figure 84. This figure shows that the required amplification of $\sim 10 \times$ may occur in regions where $\gamma (=E_{mag}/E_{turb}) \approx 1$. In a typical photospheric region at the top of the convection zone where $v_{conv} \approx 2$ km s^{-1} this would correspond to field-strength fluctuations $\Delta B \approx 300$ G requiring fields of at least that order of magnitude. This is large – such fields occur only in small and rather localized areas. In addition, the dissipation-rate of magneto-hydrodynamic waves seems too small to sufficiently heat the corona. Several investigations to find wave-modes that dissipate better are still inconclusive. Therefore one has sought for other mechanisms.

Rosner *et al.* (1978a, b; see also Vaiana and Rosner, 1978) suggested that *coronal* loops may be a source of coronal heating, perhaps by anomalous current dissipation. Van Tend (1980) proposed that random motions in prominences would generate magneto-hydronamic waves in the corona; heating would therefore occur most efficiently in the coronal loops overlying prominences. Although semi-quantitative estimates of the amount of flux generated seem to indicate a fair agreement with the flux required, this line of research is only at its very beginning.

The obvious difficulty in appying these considerations to the stellar case is that very little information is available on stellar magnetic fields, and no information at all on localization and shape of magnetized stellar plasma. Nevertheless, it is my feeling that this avenue has to be followed in order to improve understanding of stellar coronal heating. This statement would not only apply to what are conventionally called 'solar-type stars' but also to hot stars in the winds of which important disturbances can easily develop. There is no reason why such stars should not have sufficiently strong ($\sim 10^2$ G) but still unmeasurable magnetic fields, perhaps of fossil origin. Without doubt magnetic fields can also be important in cool giants with their large convective regions.

6.11. The Observation of Coronae of Luminous Stars

Spatially resolved observation of a stellar corona is so far only possible for the Sun. For sta r s one has to resort to indirect methods, *viz.* the observation of:

– stellar X-radiation;

– stellar spectral lines of high degrees of ionization; these occur mainly in the UV spectral region;

– a radiation excess in the infrared/sub-mm/microwave spectral regions.

After the discovery of *X-radiation* from α Aur (Capella) (Catura *et al.*, 1975) and of Sirius (Mewe *et al.*, 1975a, b, 1976) more stars emitting X-rays were detected by means of the satellites SAS-3 and HEAO-1 and particularly the Einstein Observatory (HEAO-2). A summary of the pre-Einstein data is given in Figure 88a, taken from a review by Mewe (1979). In this diagram, which gives the situation as it was in the summer of 1979, X-ray fluxes or upper limits are given for the energy range 0.16–0.28 keV. Significant fluxes were measured only for a ten stars, among which, remarkably enough, the late-type supergiant 12 Peg (K0 Ib) and the giants ζ Cyg (G8 II) and α Aur (G5 III). The latter, however, is a binary, and it may be that the X-radiation is related to the binary nature of the star.

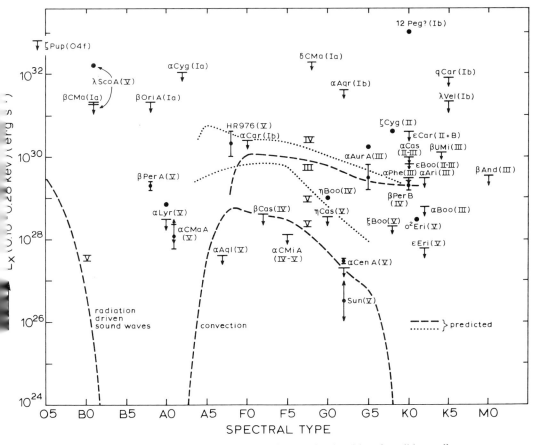

Fig. 88a. Comparison of observed and predicted soft X-ray luminosities of candidate stellar coronae; *dashed lines*: calculations on the basis of the minimum-flux coronal formalism; *dotted lines*: predictions of Landini and Fossi (1973). (From Mewe, 1979).

In order to compare such data with results of theoretical predictions one has to know, besides the predicted coronal parameters, the emission of a coronal gas of given composition and temperature. The *emissivity data* are summarized for the temperature range 10^5–10^8 K in Figure 88b (Mewe, 1979, review; with reference to previous work, mainly Mewe, 1972, 1975a). The most important characteristic of this graph is the large contribution from lines, mainly Si and Fe, near 10^6 K. The predicted coronal data were taken from De Loore (1970), Lamers and De Loore (1975), and from Landini and Monsignori Fossi (1973); the resulting predicted X-ray fluxes for stars of various luminosity classes are given in Figure 88a.

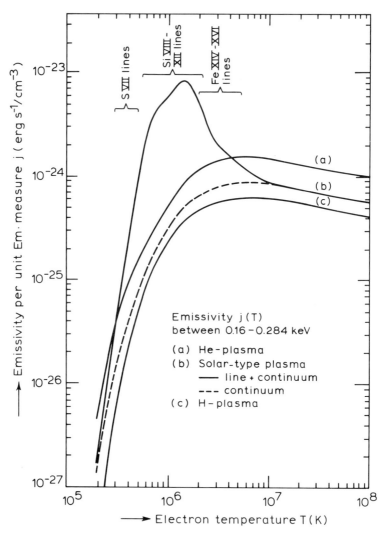

Fig. 88b.   Emissivity $j(T)$ between 0.16 and 0.284 keV *versus* coronal temperatures, calculated for (a) a He plasma; (b) a solar type plasma (————: line and continuum; – – –: continuum); (c) a H plasma. (From Mewe, 1979).

After these initial discoveries the break-through came when the Einstein Observ-atory satellite was pointed at various parts of the sky and discovered X-radiation from a number of bright stars. It now appear that X-ray fluxes are associated with most main-sequence stars, and particularly with M-type dwarfs and giants (not super-giants), and with luminous O- and B-type stars. Harnden *et al.* (1979) who investigated the association VI Cygni, discovered X-radiation sources associated with early-type stars. Figures 89b and 89a show the optical and X-ray images of the association. Of the five X-ray sources detected four can be identified with early-type stars, of spectral types O6 Ib, O5 If, B5 Ia, and O7 Ia. The last of these stars is the contact binary BD+40.4220 (cf. Tables VIII and XI in Chapter 2), one of the most massive couples known. In the energy-range 0.2–4.0 keV these sources have typical X-ray luminosities of $\sim 5 \times 10^{33}$ erg s^{-1}, and temperatures of $\sim 10^{6\cdots}$ K.

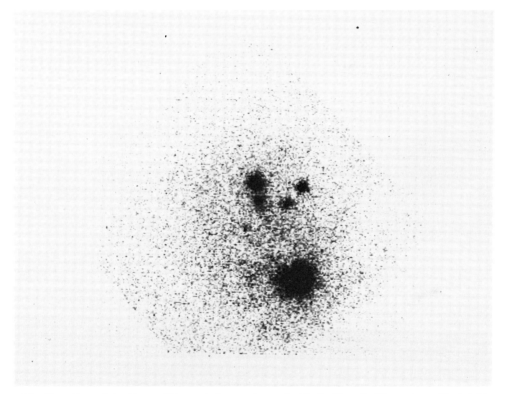

Fig. 89a.   Part of the Cygnus VI association as observed with the Einstein Observatory. The bright source in the lower right is Cygnus X-3.

Another area rich in stellar X-ray sources is the region around η Car (Seward *et al.*, 1979) where a number of individual sources could be detected, with luminosities $\sim 10^{33}$ erg s^{-1}. Two X-ray pictures of this region are shown in Figure 99 (Section 6.17). Apart from η Car itself, these sources could be identified with a cluster of

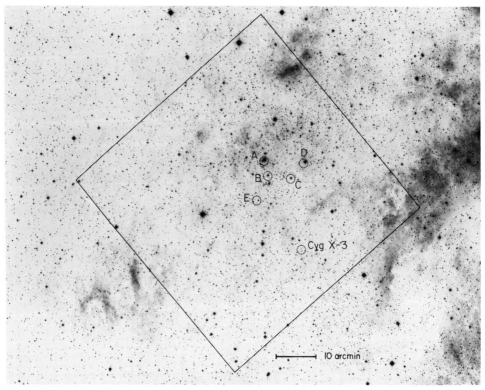

Fig. 89b. The same region reproduced from the Palomar Sky Survey (red point) with the X-ray locations indicated. North is up and east is to the left. (From Harnden *et al.* (1979). Original print: *Astrophys. J.* Photo courtesy of Harvard-Smithsonian Center for Astrophysics).

O-type stars, some individual O-type stars, and a Wolf–Rayet star. Among the identified objects are the group of O3 stars HD 93128, 93129A, 93129B and HD 93250, O3 V((f)), described in Section 3.3; they belong to the most luminous stars of the Galaxy.

The detection of coronal features of stars, on the basis of *spectral lines of high degree of ionization* started when Morton *et al.* (1968) found fairly highly excited resonance lines in the spectra of several supergiant stars: C III 1175 Å, Si III 1206 Å, N v 1240 Å, Si IV 1394, 1403 Å, and C IV 1550 Å. The presence of these lines indicates ionization temperatures T_{io} of up to above 10^5 K. It was initially assumed that $T_i \approx T_{el}$, so that the first conclusion was that these stars would be surrounded by a warm envelope. It now seems more probable that the ionization of at least some of these ions is due to Auger-ionization by photons emitted by a hotter plasma, usually called a corona (Cassinelli, 1979a). In Section 7.3 we describe how a study of the Hα-profile of ζ Pup and ζ Ori A enabled Hearn and Cassinelli *et al.* to establish the existence of coronae around these stars.

Another indirect way to detect stellar coronae is from *infrared flux measurements*. Cassinelli and Hartman (1977) and Cassinelli *et al.* (1978b) have shown that a slight excess in brightness of ζ Pup in the 1–10 μm range, as well as the Hα line profile (see Section 7.3) can be explained by the assumption of a coronal model in which the corona is fairly thin (between $r/R_* = 1.1$ and 1.5) and has a temperature of 2×10^6 K, and a mass-loss rate of $(8–11) \times 10^{-6} \, \mathfrak{M}_\odot \, \text{yr}^{-1}$. The fairly small vertical extent of the corona should be due to the fact that the density in the stellar wind is high enough to cool the flow by recombination radiation where mechanical energy deposit ceases. This coronal model predicted an X-ray flux in accordance with the ANS-upper limit.

6.12. Circumstellar Gas – (a) Circumstellar Spectral Lines

There are cases in which the surrounding gas is so far away from the star or relatively speaking so cool that one should speak of circumstellar (CS) gas or of envelopes, rather than of a chromosphere or corona. In all such cases the gas has either been emitted by the star and is related to the stellar mass-loss (a topic to be dealt with in Chapter 7), or are star and envelope genetically related: the case of contracting proto-stars.

The existence of CS gas is found in various ways: either by circumstellar absorption lines (see our discussion of α Sco, Section 3.11), or by emission lines apparently emitted by an extended tenuous envelope – often emitting forbidden lines, or else by an infrared emission excess, or by microwave emission. Important are also the maser effects observed for many red giants and supergiants, or in contracting protostars. CS absorption lines are primarily observed around the coolest stars, the same applies to the maser effects, while thermal microwave emission can originate from extended envelopes of mass-losing stars of all spectral types.

In this section we describe observations of *circumstellar absorption or emission lines*. Pioneering research was done by Deutsch (1960) who found that circumstellar lines are common in M0-type and later giants and supergiants; they are stronger for later types and increased luminosity. Figure 90 after Reimers (1975b, 1977a) summarizes the locations of the late-type stars with circumstellar spectral lines in the Hertzsprung–Russell diagram. In virtually all cases are the Ca II, H and K lines the strongest of the circumstellar lines in the visible spectral region. The K_3 absorptions are as a rule superimposed on K_2 circumstellar emissions. Both are violet-shifted and there is a correlation between the shifts (Δ) of K_2 and K_3 (Reimers, 1975a, Figure 2). Roughly $\Delta(K_2) = 0.26 \, \Delta(K_3)$. In the G and K supergiants the K_3 absorption cores are often variable, in strength and in velocity. In a star like γ Aql but also in some others, a transient absorption component may occur at the violet side of the K_2 emission. It is denoted by K_4, and was studied by Wilson and Bappu (1957), Deutsch (1960), and Reimers (1975a). Other circumstellar lines, such as Na D, Ca I 4227, Sr II 4077, Hα, ... etc. occur in the spectra of *M* giants and in all late-type supergiants.

Before discussing the interpretation of circumstellar lines, we describe a few typical objects.

The classical prototype is α^1 Her (M5 II) of which Deutsch (1956) disocvered that

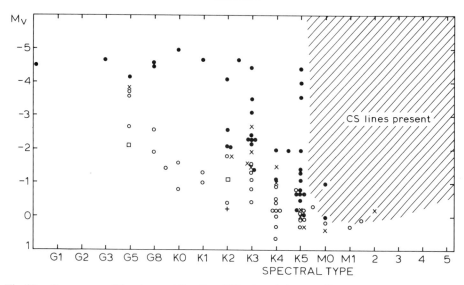

Fig. 90.   Occurrence of the circumstellar Ca II K line in red giants earlier than M0 in an HR-diagram.
Spectral types and absolute magnitudes are usually from Wilson (1976), some are from Reimers
(1977c). Symbols: ● stars with always visible CS lines; × stars which show sometimes CS lines;
○ no CS lines have been found; □ HD 138481 and HD 13480; + α Boo (CS Mg II lines). In the
hatched area all stars show CS lines. (After Reimers, 1977c).

the strongest violet-shifted lines were also observable in the companion a^2 Her (G0
II–III) at 4″.7 distance. This suggests a very extended envelope. A list of circumstellar
lines was published by Reimers (1977a). In the spectrum of α Her B no lines of excitation $>$zero eV could be detected (Weymann, 1962b), thus yielding $T_{ex}=0$. From
the radiation properties of the main star it can be argued that the inner $10R_{star}$
should contain mostly Ca^{++} ions, so that the circumstellar lines, in any case those
of Ca II, should be formed outside $10R_{star}$. The rate of mass-loss for this star is given
in Table XLII; it is approx. $0.1 \times 10^{-6} \, \mathfrak{M}_\odot \, yr^{-1}$. Other interesting cases are the M
supergiant binary α Sco (Deutsch, 1960; Kudritzki and Reimers, 1978; Van der Hucht
et al., 1979c) which has been described in Section 3.11; and the long period variable
Mira (o) Ceti (Deutsch, 1960), see Section 3.11.

In the spectrum of α Ori, Weymann (1960, 1962a) observed a hundred circumstellar lines, most of them of low excitation potential, between 0 and 1 eV. The
(micro-)turbulent velocity component was $\approx 4 \, km \, s^{-1}$ as determined from curve-of-
growth studies. Further $T_{kin} \approx 1000$ K. The electron density was 300 cm^{-3}, and the
stellar mass-flux $\approx 10^{-8}$ particles cm^{-2}.

Systematic investigations of the physical parameters of the circumstellar envelopes
in late type giants and supergiants were made by Reimers (1975a, b), by Sanner
(1976b), the latter on the basis of high resolution scans of strong resonance lines
in the visible spectral region obtained at the 2.7 m telescope at McDonald Observatory, and by Hagen (1978). In the interpretation of the observations it is assumed
that the shell is detached from the photosphere. Hence a clear distinction can be made

between the radiation from the stellar photosphere and the observed line spectrum. The photospheric radiation acts as some kind of background 'continuum' which is incident on the inner part of the shell, where it is scattered, so that the circumstellar lines are formed by redistribution of the photospheric line cores, in some cases into P Cygni type profiles. Hence, for a distant observer, the number of photons in net emission relative to the stellar 'continuum' should be equal to the number in net absorption. Furthermore, it is assumed that the photospheric lines are symmetric. These two assumptions relate the profiles of the circumstellar lines to the photospheric data.

It is found that the shells lie within a few stellar radii from the star, with increasing distance and size for increasing luminosity. For the supergiants the separation is between 5 and 15 stellar radii. The values of n_H range between approx. 10^6 and 2×10^8 cm^{-3}, while $n_e/n_H \approx 10^{-6}$, independent of spectral type and luminosity. The total column densities range from 2 to 4×10^{21} cm^{-2} for the late giants to 5×10^{21}–2×10^{22} cm^2 for late supergiants. For giants later than M5 metals are deficient in the circumstellar envelope, presumably because they are condensed onto grains.

From a review of Ca II observations, partly in binaries, Reimers (1975a, b), confirming Weymann's result for α Her, concluded that early M giants have an envelope which contains Ca^{++} particles up to $10R_{star}$. For later M types this sphere reduces in diameter. The Ca II lines are formed outside this sphere, where most Ca is Ca$^+$. There are clear indications that the velocity of expansion increases in this sphere with distance to the star; in the earliest M-stars up to $\sim$25 km s^{-1} and in the later types up to $\sim$15 km s^{-1}.

6.13. Circumstellar Gas – (b) Thermal Microwave and Infrared Emission from Bright Stars

Apart from the continuous microwave emission, due to *free-free radiation* from a – generally also: expanding – envelope, various circumstellar gas clouds emit *microwave* (mainly cm) *line radiation*, in all cases of molecular origin. These emissions occur mainly around cool stars. One distinguishes between the weak *thermal emission* (CO, SiO, H$_2$O, and other molecules) and the non-thermal *maser lines* (OH, H$_2$O, SiO). In this section we mainly discuss the continuous *thermal microwave and infrared emissions*, in their relation to stellar winds, and we discuss briefly the few observations of thermal line emissions.

First, *microwave emission of the hottest stars*. The, already 'classical', example is P Cyg, the first star for which radio emission was observed (Wendker *et al.*, 1973) at 5 and 10.68 GHz, later confirmed and extended by others; see Section 3.5 and notably Figure 33. The spectral index a (defined from $S(v) = v^a$ const.) was defined at 0.7 from these early observations.

From a comparison of the radio emission at these two wavelengths with the spectra expected for an optically thin or thick gas respectively, one infers that the gas must be optically thick at 5 GHz, and have an optical thickness of approx. unity at 10 GHz. With these data and the known fluxes at the two frequencies one may derive upper and lower limits for the radius of the radio emitting region, which turns out to be 2400 and 260 AU respectively.

Assuming a temperature of the emitting gas of 10^4 K this gas should have $n_e \geq 2 \times 10^5$ cm^{-3} and $\mathfrak{M} \geq 2 \times 10^{-2}$ $\mathfrak{M}_\odot$. Since P Cygni is losing mass at a rate of 2×10^{-5} $\mathfrak{M}_\odot$ yr^{-1} (see Section 7.4) this amount of mass would have been lost by the star in 10^3 yr. This time span is much shorter than the expected life time of the star's present configuration, and so it seems reasonable to assume that the microwave emission of P Cyg is indeed emitted by a large cloud of circumstellar gas lost by the star in fairly recent times.

Since the fundamental discovery of the continuous microwave radiation from P Cyg such emission was later discovered for a larger number of stars including O, B, Oe, Be, Of, WR, T Tau stars, (proto-)planetary nebulae, and H supergiants. Summaries are given by Barlow (1979) and Schwarz (1979); we refer also to Morton and Wright (1979). In all these cases the microwave radiation is emitted by a fairly dense *stellar wind*, and a critical discussion of the flux-frequency function $S(\nu)$ – also extended to the infrared spectral region – should enable one, in principle, to derive the $n(r)$ and hence the $v(r)$ relation, and hence $\dot{\mathfrak{M}}$. To that end the flux should be known at a fairly large number of microwave (and infrared) frequencies, which is the case only for a small, but happily rapidly increasing number of stars. The ways to analyse such data is explained later in this Section; the $\dot{\mathfrak{M}}$-values are given in Section 7.4.

We first describe a few interesting objects for which the microwave radiation has been measured.

The well-known *Wolf–Rayet binary γ^2 Vel* also emits microwave radiation (measured at 5.00, 6.27, and 8.87 GHz: $\alpha = 0.6 \pm 0.3$), ascribed to free-free emission from a circumstellar gas cloud extending to $r \gtrsim 10^{10}$ km (10^2 AU); farther out than the orbital radius of the binary (Seaquist, 1976). Other Wolf–Rayet stars emitting microwave radiation are HD 192163 (Wendker *et al.*, 1975), and HD 193793 (WC7+05) (Florkowski and Gottesman, 1977).

The first Of star for which radio-emission was detected is ζ Pup (O4ef), with a flux of 7.2 ± 1.1 mJy at 14.7 GHz. Morton and Wright (1979) deduced from this observation $-\dot{\mathfrak{M}} = (3.9 \pm 0.7) \times 10^{-5}$ $\mathfrak{M}_\odot$ yr^{-1}. Thereafter Abbott *et al.* (1980), using the highly sensitive Very Large Array of the National Radio Astronomy Observatory (U.S.A.), detected radio emission at 6 cm for a number of the brightest O- and B-type stars: 9 Sgr, ζ Pup, λ Cep, ζ Ori A, ε Ori, and P Cyg – all stars with a high rate of mass-loss.

Altenhoff *et al.* (1976) made a search for radio-emission of stellar objects with the 100 m radio-telescope at Effelsberg (F.R.G.) and with the Westerbork Synthesis Radio Telescope (Netherlands), at frequencies between 1.4 GHz and 10.69 GHz. A detectable flux around *single* stars was recorded for a few *emission-line stars*, or *stars embedded in nebulae*. An interesting object is Lk Hα 101, a faint and highly reddened Hα emitting star, illuminating the reflection nebula NGC 1579. The star is at a distance of 800 pc and has an effective temperature of approx. 10^4 K. The luminosity is 5.5×10^3 $L_\odot$ (Altenhoff *et al.*, 1976) or $> 2.5 \times 10^4$ $L_\odot$ (Cohen and Woolf, 1971). The infrared spectrum fits a black-body temperature of 800 K, which indicates that *this* constituent of the radiation is from circumstellar *dust*. The circumstellar *gas* cloud, responsible for the *radio*-emission, has a diameter of 400 AU, with $n_e \approx 3 \times 10^5$ cm^{-3}. The spectrum shows many bright emission lines of H, O I, and Fe II, and is sometimes called F or Fe (Schwarz, 1979), but most likely the central star is an O- or B-type supergiant. Infrared spectra of the Br β line show a P Cygni-type profile with a blue-shifted absorption indicating mass-loss with $v_\infty \approx 400$ km s^{-1} (Thompson *et al.*, 1976).

A comparable object, but being a Be star, is MWC349 of which radiodata were compiled by Olnon (1975). Optical observations by Kuhi (1973) show the central star to have $M_v = -7.9$, $T_{\text{Zanstra}} \gtrsim 45000$ K, while Greenstein (1973) gives possible T_e-values of 16000 and 28000 K. Combining optical and radio-data Olnon finds (with $D = 2.1$ kpc): $R = 11$–170 AU and $n_e = n_0 (r/R)^{-2.1}$, with $n_0 = 9 \times 10^8$–3×10^6 cm^{-3} (from Kuhi's or Greenstein's data respectively). The electron temperature of the shell, found from a combination of radio-interferometric data with the observed radio-spectrum is estimated at 10^4 K. The mass-loss is $-\dot{\mathfrak{M}} = 3 \times 10^{-7} v$ km s^{-1} $\mathfrak{M}_\odot$ yr^{-1}, with $v =$ velocity of expansion in km s^{-1}. From the Hα line profile one infers, rather crudely, $v \approx 1500$ km s^{-1}, yielding the high value $\dot{\mathfrak{M}} \approx -4 \times 10^{-4}$ $\mathfrak{M}_\odot$ yr^{-1}. However Schwarz and Spencer (1977), from a rediscussion of the radio-data, find the more plausible value 7×10^{-5} $\mathfrak{M}_\odot$ yr^{-1}.

Another case of a hot star surrounded by an extended circumstellar cloud may be the object HBV475, of which Altenhoff and Wendker (1973) detected a flux at 10.68 GHz of 0.008 ± 0.002 f.u. The broad strong emission features, attributed to H, He and several times ionized C, N, and O were for Crampton *et al.* (1970) a reason to assume a Wolf–Rayet central star, surrounded by clouds of ejected gas. It may be that this object is in a stage developing into a planetary nebula (see Section 3.12). In this connection we should also mention the *proto-planetary nebulae* V1016 Cyg and HM Sge (Section 3.12), and also the *probable* proto-planetaries Vy2–2 and M1–92, described in Section 6.14. All these objects are radio-emitters, which may be a significant observation, perhaps indicating dynamic phenomena in the (newly formed?) expanding envelopes of these objects.

About 40% of the *planetary nebulae* are known to be weak radio emitters; a case is NGC2438, being optically thin at 21 cm. Goss *et al.* (1973) assumed $T_{e1} = 12000$ K and derived $n_e = 2.1 \times 10^2$ cm^{-3}. The mass of the ionized hydrogen would be $0.1\mathfrak{M}_\odot$. Other well studied cases are the planetary nebulae M1–11 (Feldman *et al.* (1973) 10.6 GHz; Terzian *et al.* (1974), 2.7 and 8.1 GHz; Altenhoff *et al.* (1976) 1.4 GHz), and M1–70 (Altenhoff *et al.*, 1976). Khromov (1976) finds approximately

$$\log S_{6630} = -25.8 \ (\pm 0.8) + 1.67 \log n_e,$$

where n_e is the average of the electron densities in the nebula derived from intensities in some forbidden lines, and S_{6630} is the radio-flux at 6630 MHz. Different ions yield different relations, however, which indicates a stratification effect in the nebula, with n_e decreasing outward.

Turning to the *coolest stars*, continuous microwave emission has so far been detected for a bright supergiant α Ori (M2 Iab), and for the Mira stars R Aql (M6.5e)
and R Aqr (M7e+?). A summary of the observations at various frequencies is given
by Bowers and Kundu (1979). The observations of microwave radiation of α Ori are
also collected in our Section 3.11.

Summarizing the available data on stellar *continuous microwave emission* we find
that such emission occurs chiefly around stars which have a large rate of mass-loss,
and hence a dense expanding envelope which is moreover of fairly high temperature ($\gtrsim 10^4$ K). The hydrogen is this envelope must be mainly ionized, so that the
source has the properties of a H^+-region. The radio emission is caused by free-free
emissions of electrons in the fields of protons. In that case the emission coefficient
$\varepsilon(:)\nu^{-2}$. Panagia and Felli (1975), and Wright and Darlow (1975) assumed a density
decreasing as $n = n_0 r^{-\beta}$, and found that then the radio spectral index would be

$$\alpha = (4\beta - 6.2)/(2\beta - 1).$$

The radiation transfer equation was thereupon solved for the case $\beta = 2$ (which means:
constant expansion velocity, with v generally assumed equal to the velocity at 'infinity' v_∞), and assuming an isothermal wind. Then the flux is

$$S(\nu) = 23.1 \frac{(\gamma g \gamma)^{2/3}}{d^2} \left(\frac{Z}{\mu v_\infty} \frac{-\mathfrak{M}}{\mathfrak{M}_\odot}\right)^{4/3} \tag{6.13; 1}$$

hence

$$-\dot{\mathfrak{M}} = \frac{0.095 \mu v_\infty S_\nu^{3/4} d^{3/2}}{Z \gamma^{1/2} g^{1/2} \nu^{1/2}} \quad \mathfrak{M}_\odot \text{ yr}^{-1}, \tag{6.13; 2}$$

where most symbols have their usual meaning, and further:
γ = number of electrons per ion;
g = the gaunt factor = $(3^{1/2}/\pi) \{\ln[(2kT)^{3/2}/(\pi Z e^2 \nu m_e^{1/2})] - 1.44\}$;
Z = the average charge of the ions.

The (weak) dependence of g on ν tends to flatten the spectral index α from $\frac{2}{3}$ to
0.6, as is indeed observed in the few available cases. This shows that the basic assumptions in deriving Equation (6.13; 1) are apparently correct. It is useful to repeat
these assumptions: *constant velocity*, and *isothermal wind*. Can it be shown *a priori*
that these assumptions are correct? Taking $v = v_\infty$ is based on the fact that the coefficients of absorption and emission vary as λ^2, so that for long wavelengths optical
depth unity is reached in a fairly distant envelope, of which one may think that it
is farther away from the star than the region of emission of the 'violet edge' of the
P Cygni profiles of the strong UV resonance lines, normally used for deriving v_∞.
The belief in this assumption would be strengthened when a sufficient number of
VLBI measurements would be available of stellar microwave emission, because this
would give unambiguous information on the extent of the microwave emitting region.

The assumption that the distant stellar winds are *isothermal* was investigated by Barlow (1979): the expanding gas will start to cool when the adiabatic cooling rate is comparable to the radiative cooling rate. From this criterium Barlow derived the radius r_c beyond which adiabatic cooling becomes important:

$$r_c = 4.3 \times 10^{10} \left(\frac{-\dot{\mathfrak{M}}}{10^{-6}\,\mathfrak{M}_\odot}\right) \left(\frac{v}{10^3\,\text{km}}\right)^{-2} \text{km}.$$

Inserting numbers one finds for ζ Pup: $r_c = 4 \times 10^{10}$ km (300 AU), which is undoubtedly beyond the region from where microwave emission is expected.

Infrared excess of radiation of stars. If stars have significant emission in the microwave range one may also expect some contribution in the (far) infrared: free-free emission goes according to λ^2. The detection of this emission often leaves some ambiguity, for two reasons. First, in most cases the stellar photosphere also emits some radiation in the infrared (contrary to the radio-case). Next, strong stellar winds are often associated with the formation of dust particles at larger distances to the star. The emission of these dust particles (typical temperature $\sim$600 K – see Section 6.15) adds to those of the other components. So, in order to know whether there exists an *infrared excess* emitted by a stellar wind, the *photospheric infrared spectrum* must be known (from model computations) and the contribution of a possible *dust component* must also be subtracted. This can only be done from detailed studies of the infrared spectrum: the dust component is assumed to have a black body spectrum, the temperature of which can be determined, while the emission of the gas is described by Equation (6.13; 1). For the case that the gas temperature is *not* constant Hartmann and Cassinelli (1977) found that with $n(:)r^{-\beta}$, $T(:)r^{-m}$, the spectral index is

$$\alpha = (4\beta - m - 6)/(2\beta - 1.5m - 1), \tag{6.13; 3}$$

which would again reduce to (6.13; 1) for $m=0$.

So far, clear and unambiguous 'infrared excesses' emitted by a stellar wind have been detected for Be stars (Woolf *et al.*, 1970; Gehrz *et al.*, 1974); for P Cyg (see our Section 3.5); for Wolf–Rayet stars, and for a few early-type supergiants (see the various contributions in Conti and De Loore (1979), and particularly Barlow's review there). It is most difficult to identify an infrared excess with sufficient degree of certainty in late-type stars, because of the complicated shape of their infrared spectra with the many molecular bands. Swings and Klutz (1976) report an infrared excess of radiation in the spectrum of the red variable RX Pup (spectral type Ne). The star's 1972 and 1975 spectra have H emission lines showing P Cyg profiles, furthermore forbidden lines of neutral and singly ionized atoms, and around 1940 also high ionization lines like [Fe x] (Swings and Struve, 1941a).

Generally speaking *thermal microwave emission from molecules* is observed in Mira stars, late-type supergiants and C-stars. It is mostly observed in the infrared spectral range. Thermal emission from SiO molecules was discovered by Buhl *et al.* (1975) in several oxygen-rich long-period variables, such as R Cas, W Hya, R Leo,

VY CMa. The thermal CO emission was found by Zuckerman *et al.* (1977) in some carbon- and oxygen-rich objects. The SiO lines are ground state lines ($v=0$) and have a width of $\gtrsim 20$ km s^{-1}. The center of the line has the same v_R-value as the average of the maser lines (see Section 6.14) so they must be emitted by the same regions of the clouds, and the widths must give information on the expansion velocity and on the mass-flux (Morris and Alcock, 1977; Morris *et al.*, 1979).

Infrared excesses of 0.5 and 1.0 mag between 5 and 8 μm in μ Cep (M2 Ia) and R Cas (M7e) is ascribed to thermal emission of circumstellar H_2O gas (v_2-fundamental) by Tsuji (1978b). The molecules would occur at $6R_*$ and $3R_*$ resp. Of the HC_3N molecule two lines were discovered by McGee *et al.* (1977) in the spectrum of the Mira star W Hya.

In contrast to these emissions are the broad molecular absorption features at 3.05 μm and 3.9 μm, later found in the spectra of all C-stars, – with (C/O)>1 (Merrill and Stein, 1976; Noguchi *et al.*, 1977). They do not occur in spectra of oxygen-rich and normal H-type giants. The 3.05 μm feature, which has a FWHM of 0.3 μm, has fine structure, and is due to HCN and C_2H_2 (Ridgway *et al.*, 1976), while that at 3.9 μm is attributed to a combination of CS and C_2H_2 (Bregnam *et al.*, 1978). Another absorption observed near 10.5 μm in C-star spectra may be attributed to silicon-carbide SiC (Jones *et al.*, 1978).

6.14. Circumstellar Gas – (c) Maser Effects

Certain late-M-type supergiants, late-type semi-regular, and Mira type variables as well as a few peculiar emission line stars, surrounded by gas and dust are characterized by intense line emissions in the microwave region. The non-thermal character of these emissions is apparent from a comparison of their high brightness temperatures (typically $\gtrsim 10^8$ K) with the line widths, the latter indicating kinetic temperatures of the order of 100 K. A pumping mechanism must be responsible for the excitation of the – assumed – masering effect. This effect is certainly related to the gaseous envelopes around all such stars, but also the dust, which occurs around many of them may play a role in the maser mechanism.

The first maser lines have been detected of the OH molecule, at 1612 MHz (Wilson and Barrett, 1968; 1972; see also Wilson *et al.*, 1972). Later, Schwarz and Barrett (1970) discovered H_2O emission at 1.35 cm for four M stars, and Buhl *et al.* (1974) found SiO maser emission at a few frequencies in the mm-region. A catalogue of the about 350 objects, mostly evolved stars, showing maser line radio emission (Engels, 1979) contains 320 OH/IR objects, of which 171 are identified with stars, and 56 SiO objects identified with late-type infrared stars.

In this section we discuss observations relevant to the emission in these three molecules; we describe a few characteristic objects, and deal briefly with some proposed explanations.

Maser lines of OH are observed at 1612 MHz (satellite line), 1665 and 1667 MHz (main lines). The OH-maser line sources have been listed in some detail by Reid *et al.* (1977). Systematic searches were undertaken by Baud *et al.* (1979a, b), Rieu *et al.* (1979), Olnon *et al.* (1979). The sources are mainly restricted to the reddest giants and supergiants; among these are about fifty Mira variables and very long-period infrared variables; a relatively small fraction of the M-supergiants is associated with substantial OH emission at 1612 MHz (Bowers and Kerr, 1977, 1978; Lépine and Paes de Barros, 1977; Le Squeren *et al.*, 1979; Silvergate *et al.*, 1979). No carbon-rich S stars are known to be OH maser sources. Two peculiar early-type emission line stars, perhaps proto-planetary nebulae have OH maser lines (Davis *et al.*, 1979).

A characteristic of the emission is that the line is mostly double: it consists of two emission peaks separated by velocities ranging from a few to some 40 km s^{-1}, – see as an example Figure 91. Detailed spectra of the 18 cm OH lines in long-period vari-ables show in some of them emission between the lines, and relatively small values of the maximum-to-minimum flux ratio, which suggests that in these cases the lines are produced in saturated or nearly-saturated masers (Fix, 1978). Jewell *et al.* (1979) who monitored eight IR/OH sources found that the satellite line is mostly saturated. Positions of 11 OH masers determined to an accuracy of $\pm 5''$ show that eight of them correspond in position with H$_2$O masers (Evans *et al.*, 1979); about half of them cor-respond to protostars and compact H$^+$-regions.

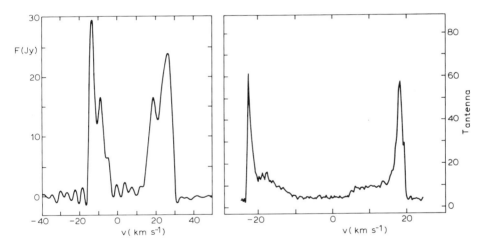

Fig. 91.   Profiles of the OH-maser line at 1612 MHz for VX Sgr – a type III source (left; from Reid *et al.*, 1977), and from the type II source OH 30.1–0.7. (From Winnberg, 1977, 1978).

According to a classification of Wilson and Barrett (1972) the OH maser sources fall into three groups (see also reviews by Winnberg, 1977a, 1978):

Type I: One of the OH lines (usually that at 1667 MHz) is the strongest in the ground-state hyperfine quartet. The line profiles are split in two, separated by 7 to 20 km s^{-1}. The association is with late M-type Mira stars, or semi-regular variables, often also with 1.35 cm H$_2$O maser emission. Type I OH-masers are also often related to compact H^+ regions (Matthews *et al.*, 1977). For type I masers the intrinsic luminosity ranges between $\sim 5 \times 10^{15}$ and 5×10^{17} W (Rieu *et al.*, 1979).

Type II: The 1612 MHz satellite line is the strongest. It is always split in two doppler components separated by 20–40 km s^{-1}. There may be weaker emission in the main lines at 1667 and 1665 MHz. This group is less common than the type I objects. Association: late M-type Miras which show an excess emission beyond 3 μm. Johansson *et al.* (1977) studied the kinematic properties and the galactic distribution of a number of unidentified objects, and confirmed identification with Mira variables. For these objects the intrinsic luminosity is between $\sim 10^{17}$ and 5×10^{18} W (Rieu *et al.*, 1979).

Type III: The sources are a rather inhomogeneous group with the common characteristic that they are associated to supergiants. The members are: PZ Cas (M3), VY CMa (M3–5e I) (Hyland *et al.*, 1969; see Section 6.16), VX Sgr (M4e), S Per (M4e Ia), AH Sco (M5 Ia-Iab), NML Cyg (M6), MY Cep and IRC 30308. All stars have excess radiation beyond 3 μm; they are also associated with H_2O and SiO emission; the main line radiation can be polarized.

The objects S Per and AH Sco were discussed by Baudry *et al.* (1977), and were found in a systematic search for OH sources in M supergiants in three clusters. Another list of OH Miras was published by Bowers and Sinha (1978). It appears from these investigations that, statistically, less than approx. 10% of the infrared sources show OH maser emission. Apparently, at least 10^{17} to 10^{18} OH molecules cm^{-2} seem needed to make OH emission observable at 1612 MHz (Litvak, 1972). Such molecules may be formed after photo-dissociation of H_2O by interstellar UV photons (Elitzur *et al.*, 1976) or via sticking of O atoms to circumstellar grains (Watson, 1974). There seems no correlation with the *photospheric* OH abundance, neither with the (circumstellar) H_2O radio emission flux. Some OH maser sources (10%) are also H_2O masers (Marques dos Santos *et al.*, 1979).

So far a few *systematic properties* have been found that help to understand the origin of the OH emission in Miras and M supergiants:

• The emissions are displaced symmetrically about the stellar radial velocity (Reid, 1976; Reid and Dickinson, 1976).

• The OH intensity in masers associated with long-period variables shows a time-lag of 10–50 days with respect to the optical variation (Jewell *et al.*, 1979). This suggests that:

– the maser lines are emitted by an expanding shell around the star;

– the lines are pumped by infrared radiation of the stellar source;

– the maser shell region has a size of a few times 10^{11} km ($\sim$1000 AU).

At supergiants the separation between the OH peaks is generally large ($\Delta v \approx 30$ to 60 km s^{-1}). The separation is $\sim$4 times smaller for OH-Mira's than for OH–IR sources. The OH luminosity increases with the infrared colour index; this quantity is smallest for type I Mira's and highest for unidentified OH–IR sources. In the cases of NML Cyg and VY CMa the OH feature as a whole was 'red'-shifted before 1973; between 1973 and 1975 it moved to the central OH velocity (Wallerstein, 1977). Interferometric observations indicate that the sources are rotating (Masheder *et al.*, 1974; Davies *et al.*, 1972). For other sources the stellar velocity seems to be very close to that of the centre of the OH 1612 MHz emission features (Wallerstein, 1977).

Maser lines of SiO at 86.243 GHz are attributed to the ($v=1$; $J=2-1$) transition (Snyder and Buhl, 1974); other transitions occur at 43.122 GHz ($v=1$; $J=1-0$; Snyder and Buhl, 1975), and at 42.820 GHz ($v=2$; $J=1-0$; Buhl *et al.*, 1974). The ($v=0$; $J=2-1$) transition is between thermally populated levels.

The lines have been observed in various red or infrared giants, mostly long-period Mira variables; the most pronounced case is R Cas (Schwartz, 1977). Other objects include stars such as o Cet, NML Cyg, R Leo, W Hya, VY CMa, VX Sgr, NML Tau, and the (symbiotic?) Mira star R Aqr (Zuckerman, 1979). Cahn and Elitzur (1979) investigated seventeen SiO-Mira-masers with known distances and (hence) known luminosities. At the time of writing 56 SiO maser sources are known and listed by Kaifu *et al.* (1975), Snyder and Buhl (1975), Blair and Dickinson (1977), Ballister *et al.* (1977), Lépine *et al.* (1978), Morris *et al.* (1979), and Engels (1979). Practically all of these correspond to late type stars, and most of them are associated with H_2O sources, a number also with CO sources.

The $v=1$ and $v=2$ lines of R Cas have the main velocity components symmetric with respect to the same radial velocity, but the $v=2$ lines are shifted from that central velocity by about 1 km s^{-1}. This pattern suggests a shell-like, at least a more or less symmetric, structure in the SiO-emitting region. The various observations can be combined to one picture of the $v_{\text{rad}}(R/R_*)$ function for R Cas, as shown in Figure 92, after Schwartz (1977). In all stars which show the SiO maser lines, with the exception of NML Cyg and R Cas, the lines have a weak underlying pedestal emission, and it seems that the central velocity of the pedestal may be identified with the stellar radial velocity (Snyder *et al.*, 1978). There is a clear, virtually linear relation between the envelope ('terminal') expansion velocity as measured from the $v=0$ lines, and the stellar period of pulsation, according to $v=-7.7+0.04P(d)$, for $200d<P<800d$.

For the *interpretation* of the SiO maser mechanism a few observations may be significant.

First, there is a good correlation between the SiO luminosity and that of the associated star at the wavelength of the $v=0$ to $v=2$ vibrational transition. This correlation suggests very strongly that *radiative pumping* must be at the base of the maser phenomenon. This relation also suggests that the masering region occurs at all stars at about the same distance to the star's centre.

Secondly, the SiO maser emission shows a fairly strong linear polarisation. For a sample of ten maser sources associated with variable stars linear polarisation has been measured in the $v=1$, $J=2\rightarrow1$ transition, with no indication of circular polarisation (Troland *et al.*, 1979). There seems no reason why this finding would not apply to most SiO sources, at least to those associated with variable stars. For the well-studied object R Cas the Stokes parameters showed significant changes over characteristic periods of the order of a day.

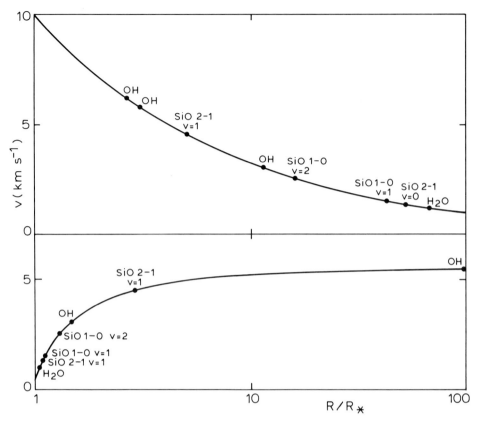

Fig. 92.   Radical velocities in R Cas as measured in several maser lines (labeled to the curve) for an assumed symmetric model with (*upper figure*) assumed pulsation driven flow, and (*lower figure*) assumed stellar wind flow driven by radiation pressure, with $v_\infty = v_{circ}$. (After Schwartz, 1977).

The third molecule for which maser emission is known, is H_2O (Schwarz and Barrett, 1970; Wilson and Barrett, 1972), which emits at 22 GHz. Lépine and Paes de Barros (1977) list 40 H_2O masers, including 34 regular Mira type variables with $264d \leq P \leq 561$, and $\langle P \rangle = 372$; and six semiregular variables. The absolute radio fluxes of the H_2O line at 1.35 cm at maximum are 4×10^{43} photons s^{-1} with a very small dispersion. Notably, the flux is independent of the period P. Kleinmann *et al.* (1978) list 82 known H_2O masers known to be associated with stars. The emission shows in most cases only one line component or several closely spaced components with some degree of symmetry. The radial velocities are always within the velocity range of the OH lines. For six long-period stars for which lines of both molecules were observed, Dickinson and Kleinmann (1977) found that the component separation (in km s^{-1}) is the same for OH and H_2O. On the other hand the *intensity* of the maser components does show irregular variations on time scales of days to years (White and Macdonald, 1979). The M-type supergiant S Per has an H_2O line complex at 22 GHz which shows at least five well-discernable peaks over a wavelength range of 20 km s^{-1}. They vary during the period of light variation, but with phase differences up to a few hundred days between the various components and the optical radiation (Cox and Parker, 1978). No radial-velocity *changes* larger than 0.5 km s^{-1} have been observed for the H_2O maser line observed in *long period variables*, which suggests that in such stars the circumstellar gas cloud is in constant expansion, associated with mass-loss (see Section 7.4).

It makes sense to *divide the masing envelopes in two kinds*. In a gas, in which the abundance ratio (C/O) > 1, the CO molecule will be abundant, but if (C/O) < 1 molecules like OH, H_2O, SiO will be more abundant. The former envelopes are called graphite-like, the others silicate-like. This division, naturally, is also valid for the thermal emissions.

The large intensity of the maser lines makes it possible to investigate the *detailed structure of the sources* by the technique of Very Long Baseline Interferometry. Observations made in the OH-line at 1612 MHz (Reid *et al.*, 1977) have shown that the structure of the sources can be interpreted in a 'core-halo' model. A small core ($\lesssim 0\overset{''}{.}03$) of radio brightness temperature 10^9 K, is surrounded by a large ($\gtrsim 0\overset{''}{.}5$) halo with a radio brightness temperature $\lesssim 10^8$ K. In some cases there is more than one core distributed over a small area ($\lesssim 0\overset{''}{.}2$). Assuming a 'typical' distance of 200 pc one thus obtains the following rough numbers: core area $\lesssim 7 \times 10^8$ km; maser sources area $\lesssim 6 \times 10^9$ km; halo $\gtrsim 1.5 \times 10^{10}$ km. We compare this with other data: For a number of Mira stars with known or reliably estimated distances the size of the H_2O maser sources emission region was determined at $\lesssim 10^9$ km (Spencer *et al.*, 1979). This value should be compared with the radius of a typical Mira star (Section 3.11), being $500R_\odot \approx 3.5 \times 10^8$ km. So most maser features occur 'close' to the star ($\lesssim 20R_*$).

In addition to the gaseous component the shells also contain dust particles, detectable by their infrared emission. The size of these *dust halos* correspond to linear dimensions of $\gtrsim 3 \times 10^{10}$ km, or 20 AU (typically $100R_\odot$). The dust halo in IRC+10011 is densest at a radial distance of 5×10^9 km and extends outward till beyond 5×10^{10} km. Between these distances the dust temperature decreases from 600 K to 200 K. The halo must probably be interpreted as an optically thick expanding circumstellar shell, or a disk (see e.g. Elmegreen and Morris, 1979) most probably driven out by radiation pressure of the luminous stars. Like in all such cases the expansion velocity should increase with distance, till an asymptotic velocity is reached. The final gas velocities appear to be correlated with the mean M_b-value. They may be put equal to half the velocity distance between the two maser components, and range between 3 and 22 km s^{-1} for Mira stars with $\log(L/L_\odot)$-values between 3.7 and 4.3 (Cahn and Wyatt, 1978b). The OH emission should be produced by those parts of the envelope where the Doppler velocity to the observer is nearly constant, i.e. from two small caps in the front- and back-parts of the expanding shell. This is so because the paths of maximum maser gain must be those in which the radial velocity gradient has a minimum, because in a gas with a v-gradient masering would not start. This explanation also rules out the shock front hypothesis for the maser model. The primary input to the maser radiation (at 18 cm) can hardly be the star itself, but should rather be located near the shell boundaries, where it should be due to 'spontaneous' emission, perhaps in high temperature regions near shocks.

Thus one may arrive at a *global picture*, in which a star ($R \approx 500 R_\odot$) is surrounded by a gaseous expanding envelope; most sources occur at distances $\lesssim 20 R_*$. At about 10 to 10^2 R_* from the star's center dust condensation can occur. In subsequent shells at increasing distance to the star, maser lines of SiO, H_2O, and OH are generated, where mostly the OH line is the strongest and would be emitted from regions at distances of 10–100 stellar radii (Olnon, 1977). The line is assumed to be pumped by 3 μm photons emitted by the dust particles. One of the strongest sources, that can be explained by this model, is OH 26.5+06, with $-\mathfrak{M} \approx 3 \times 10^{-5} \ldots 10^{-4} \mathfrak{M}_\odot \ \mathrm{yr}^{-1}$.

While this model accounts for the overall picture of the observations related to average stars, *detailed measurements* show again that there are cases in which nature is more complex than our models! Spatially, detailed structure is sometimes observed in the emission complexes with maser velocity components spread over a range of 5 to about 10 km s^{-1}. This observation may show that the circumstellar shells are not homogeneous, or/and the shell may have been ejected in various 'puffs' at different times. With Mira stars it may also be that the formation of dust particles is linked in some way to the cycle of variability of the star.

The spatial structure of H_2O-masers, as determined from VLBI in which five stations participated in one case, has shown the complexity of such sources in which a hierarchy of scales is apparent (Walker *et al.*, 1978). There are tight groups of low velocity sources with a probable size of 5×10^9 km for individual masers; the strongest of these have H_2-densities of 10^9–10^{10} cm^{-3}. These groups of masers, which have typical sizes of 4×10^{10} km, are often surrounded by high-velocity sources occupying a larger area (Walker *et al.*, 1977; Genzel and Downes, 1977; Genzel *et al.*, 1978). The total extent of a H_2O maser emission region is of the order of 10^{12} km. We compare this with measurements of type II OH maser regions, for which an average diameter of $(4 \pm 2) \times 10^{11}$ km was found (Reid *et al.*, 1977; Schultz *et al.*, 1978). It may further be significant that in one well-studied case (the H_2O maser source W49) several spectral components show correlated variability with characteristic periods of a week to several years. This would suggest that the input of the maser line energy should be the same source for all components. This source may be identified with a variable star cocooned in a dust shell $\gtrsim 4 \times 10^{11}$ km (2700 AU) (White, 1979). Such correlated variations, over shorter periods, were also observed for VY CMa (Cox and Parker, 1979) – see Section 6.16. Interesting are also the high-velocity components. These can not be gravitationally bound, and it was suggested (Genzel and Downes, 1977a) that these sources are remnants from the star formation process. To be further investigated!

A rather different case is that of U Ori, an oxygen-rich ((O/C) > 1) Mira-type variable which has been known to emit maser radiation of H_2O, SiO, and OH. However, in the first months of 1974 the main OH emission (1667 MHz) disappeared, while the 1612 MHz emission increased strongly in intensity. Later the 1612 MHz emission intensity decreased strongly, and 1667 MHz reappeared. Can this be explained by a short-lived disturbance in the stellar photosphere? (Cimerman, 1979).

We describe a *few of the best studied objects*. A classical object among the non-Mira M-type supergiants is the infrared source NML Cygni (Neugebauer *et al.*, 1965). At 10 μm it is one of the apparently brightest objects, and it is the strongest 18 cm-OH emitter known. It is commonly believed to be a heavily obscured late M-type supergiant (Hyland *et al.*, 1972), at an estimated distance of 500 pc. Interferometric observations (Davies *et al.*, 1972; Benson and Mutel, 1979) show that the OH emission comes from the outer regions of a cloud with an angular size of $3\rlap{.}''3 \times 2\rlap{.}''3$ (corresponding, at 500 pc, with a size of 1650×1150 AU) which is expanding and slowly rotating. The strongest OH radiation is emitted by the outer regions. To explain that there is no OH emission from the far side of the cloud it is postulated that the cloud must be optically thick at 1612 MHz, perhaps because of free-free absorption by a partly ionized gas. If this is true, free-free radio emission should be observable. The source *seems* to emit a weak continuous microfrequency radiation at 2.8 cm (Goss *et al.*, 1974b), with a flux of 12 ± 2 mJy, but this has been questioned by Gregory and Seaquist (1976). At 86 GHz Wilson (1971) found an upper limit of 270 mJy. Gregory and Seaquist detected radiation at 10.5 GHz, the flux density is 27 ± 4 mJy. But it is uncertain whether this radiation is emitted by NML Cyg or by a faint nebulosity, adjacent to the object, at an estimated distance of ~ 2 kpc. Very high resolution radio observations with high positional precision are desirable; VLBI observations have high resolution but relatively bad positional accuracy.

The best known example of a star showing at the same time a *gas component* with many *thermal lines* as well as *maser lines*, and a *dust component* with low-temperature infrared continuous emission is the late type Carbon star CW Leo (IRC+10216) (Herbig and Zappala, 1970; see review by Winnberg, 1978). At 5 μm it is the brightest object at the sky.

The spectrum of the dust component peaks near 5 μm and has $T_c \approx 650$ K (Becklin *et al.*, 1969). This component is apparently optically thick in visual wavelengths and re-radiates the stellar radiation in the infrared. An emission peak at ~ 11.5 μm suggests that SiC is an important constituent of the dust (Treffers and Cohen, 1974). At wavelengths of 2.2, 3.5, and 4.8 μm the brightness distribution is an uniform circular disk of $\sim 0\rlap{.}''4$ diameter (Tooms *et al.*, 1972; Foy *et al.*, 1979). The emission peak at ~ 11 μm has a more complicated structure: First, there is a compact, unresolved core of diameter $< 0\rlap{.}''2$. In addition there is an extended component which contains 91% of the total flux and has an e^{-1}-diameter of $0\rlap{.}''90$. This extended component has also circular symmetry (Sutton *et al.*, 1979), in contrast to the brightness distribution and polarisation measured at shorter wavelengths (1 to 5 μm) where the ellipticity is ~ 2 (references in Sutton *et al.*, 1979).

The molecular lines from the envelope include those of CO, CN, CS, HCN, C_2H, SiS, NH_3, SiO, and HCN_3. The volumes emitting the molecular lines are much larger than the 'stellar' source, and have different sizes for the various lines; it is $\lesssim 40''$ for HCN, and $\sim 2\rlap{.}'1$ for CO (Wannier *et al.*, 1979). The NH_3 observations demand a certain H_2-density (on the basis of dissociation equilibrium calculations) and this, in turn, yields an estimated value for the mass of the *gaseous envelope*: $\mathfrak{M} \approx 0.1 \mathfrak{M}_\odot$

(Betz *et al.*, 1979). This is qualitatively consistent with the strong observed mass-loss (Morris, 1975; Kwan and Hill, 1977).

Another well-known object is the supergiant IRC+10420, of spectral type F8 Ia. It is of much earlier type than all other known stellar masers, which are of spectral type M3 or later. It is a strong OH maser source and has similar maser properties as OH/IR supergiants such as NML Cyg and VY CMa (see Section 6.16), in that the main emission is at 1612 MHz and doubly peaked. The infrared spectrum is very similar to that of η Car (see Section 6.17). Detailed information on the masering gaseous shell is obtained from maser observations of OH at 1612 MHz (Mutel *et al.*, 1979; Benson *et al.*, 1979). These show some eight individual sources with a *core* of $0''015(\sim 50\ \mathrm{AU})$ and a *halo* of $0''15\ (500\ \mathrm{AU})$, extending over an area of $\sim 5000\ \mathrm{AU}$. Since 1975 a new blue-shifted emission was observed at $\Delta v \approx 9\ \mathrm{km\ s^{-1}}$. This emission was spatially separated from the other sources by $0''75$. This observation and the high circular polarisation of the OH maser lines may be interpreted as an expanding shell with $v_{\mathrm{exp}} \approx 30\ \mathrm{km\ s^{-1}}$ and a radius of about 2500 AU to 4000 AU. The star does not show appreciable SiO or H_2O maser emission. One may wonder whether this is a star, moving leftward in the HR-diagram, either in a late evolutionary phase or else as a pre-main-sequence object.

Two other intriguing objects are the early-type emission line objects Vy 2–2 (M1–70) and M1–92, described by Davis *et al.* (1979, with references to earlier work). The first is a strong infrared source; $T_c \sim 1800\ \mathrm{K}$ between 1.6 and 2.2 μm, and 190 K between 10 and 18 μm which suggests a dust component. It emits a single asymmetric peak in the satellite OH line at 1612 MHz; this line may be pumped by the intense far infrared stellar radiation. It seems likely that the observed emission comes from the approaching part of a nebula containing dust and neutral gas, while the receding part, which might produce the other maser component is supposed to be occulted by an optically thick ionized nebula closer to the star. It may be that Vy 2–2 is in the intermediary phase between a Mira object and a planetary nebula. The other object, M1–92, also called 'Minkowski's Footprint' for its peculiar shape, consists of two emission lobes with on the 'bridge' in-between, a faint red star, perhaps responsible for the nebular emission. Maser lines are observed in 1667 and 1612 MHz; the first shows four sharp components with velocities between -15 and $+20\ \mathrm{km\ s^{-1}}$ with respect to the local standard of rest, and a broader component at zero velocity. The visible spectra of the two lobes contain sharp-lined low-excitation emission lines, as well as a few broader P Cygni-type lines. The model suggested for this object is that of a central star surrounded by a dense flattened dust disk, and a much larger tenuous gas cloud. The two lobes should be the polar regions of the cloud, illuminated through the more transparent poles of the flattened disk. The broad OH-line at $v=0\ \mathrm{km\ s^{-1}}$ should originate in the low-density outer dust halo; its width, suggesting $v_{\mathrm{exp}}=20\ \mathrm{km\ s^{-1}}$ fits well with the velocities derived from the visual spectra of the lobes. Also this object is suggested to be a proto-planetary nebula.

For the *interpretation of the emitted maser radiation* in terms of the physical parameters of the masing region and of the star, the most simple case to be considered is that of a saturated maser. If then V is the volume of the emitting region, Ω the solid angle into which the maser radiation is beamed, and ΔP the (positive) difference in pump rate between the upper and the lower of the maser levels, then the photon emission rate Φ is (Kwan and Scoville, 1974)

$$\Phi = \tfrac{1}{2}\Delta P V \frac{\Omega}{4\pi},$$

while for pumping of an optically thick line at the frequency ν the pump rate P is:

$$P(\nu) = \frac{I(\nu)}{hc} \frac{\pi R^2}{r^2} \frac{d\nu}{dr}.$$

Here, R and r are the radii of the star and of the masing region respectively; $I(\nu)$ is the stellar radiation intensity and $d\nu/dr$ the gradient of the radial velocity in the masing region.

It is convenient to write $\Delta P = \eta P$, where η is defined as the pumping inversion efficiency. An estimate of η for a number of well-studied Mira stars by Cahn and Elitzur (1979) yielded a typical value of 10%.

Next, some *physical aspects of the population inversion*. It is obvious that for maser action to be efficient, the upper level of the transition involved must be overpopulated, so that stimulated emissions can play an important role. The conditions under which such a 'population inversion' occur can be calculated for given molecular species. Physically, population inversions can occur when during collisions the system is not only excited to the adjacent level ($\Delta J = 1$) but also to the next ($\Delta J = 2$). Thus, the OH maser lines at 18 cm may be explained by collisions of H and H_2 with OH; this results in excitation of the upper ($\varepsilon = +1$) Λ levels of $j = \tfrac{5}{2}$ manifold, followed by radiative decay to the $j = \tfrac{3}{2}$ levels. Furthermore collisional transitions in $j = \tfrac{3}{2}$ should not result in restoring thermal equilibrium in the non-thermal distribution of level populations induced by radiative decay from the non-equilibrium populated $j = \tfrac{5}{2}$ level. This process occurs for $n(\text{H})$-values from 10^3 to 10^{11} cm^{-3} (Kaplan and Shapiro, 1979).

Varshalovich and Khersonskij (1978) considered the rotational levels of CO, SiO, and CS molecules for interstellar and circumstellar conditions ($n = 1 \dots 10^7$ cm^{-3}; $T_k = 10 \dots 100$ K; $T_r = 2.7$ K, characterizing the radiation field, defining the population of the rotational levels) and found that population inversion occurs for SiO in a narrow range of densities, near $n = 10^5$, in the temperature range considered. For the other molecules the results are similar. For establishing the inversion the occurrence of dust particles – apparently an essential element in most masing objects – may also prove to be important. Goldreich and Kwan (1974) showed that the cold dust particles, with a temperature T_d, low in comparison with the kinetic gas temperature T_k, can remove resonance photons from the gas while hot dust particles, with $T_d > T_k$ can contribute to the radiative excitation of the molecules. Strel'nitskii (1977) and Bolgova *et al.* (1977) discussed the level population distribution and radiative transfer in a gas-dust mixture, on the basis of the homogeneous, isothermal, plane-parallel approximation. They found that absorption and emission of the continuous radiation can contribute considerably to the level population. Further, if $T_d \ll T_k$ the population inversion can indeed strongly increase for the relevant transition in the molecule. When $T_d \approx T_k$ the inversion would be annihilated.

Such consideration do not help to solve the curious problem of the different phase lags between the 22 GHz H_2O components and the optical radiation in S Per as observed by Cox and Parker (1978). The phase difference may suggest that the H_2O masers in this star are pumped by a collisional process.

Reference is further made to the monograph and review paper by Cook (1977, 1978).

6.15. Circumstellar Dust

Circumstellar dust has initially been detected by *stellar reddening*, or – on larger scale – by the observation of *reflection nebulae*. Later, the opening of the infrared spectral region to observational astrophysics has made it possible to detect a dust cloud near a star from its *continuous emission spectrum*, which shows itself as an excess radiation, added to that of the star, and characterized by the colour temperature T_c of the dust cloud's radiation. However, still earlier there was the discovery (Woolf and Ney, 1969) of an *emission feature* near 10–11 μm in the spectra of cool supergiants and Mira variables.

The 10–11 μm feature is fairly broad, extending over a few μm. High-resolution spectra do not show any spectral fine structure so that the identifications with dust particles of various sizes, in a cloud or shell near the star seems appropriate.

A weaker excess, announced in the range 0.7–8 μm by Humphreys (1974) particularly in variables such as S Per, VX Sgr and VY CMa, and attributed by her to a chromospheric H^+ emission, was not found by Fawley (1977) – the discrepancy may be due to overcorrection for interstellar extinction in the first case.

Another way to detect dust around stars is by the *polarization of starlight* scattered by the dust particles. While there are cases in which the dust is discovered from the observation of the direct polarisation of the starlight, there are also cases in which the star appears to be surrounded by a halo of polarized light (McMillan and Tapia, 1978).

Some late-type supergiants show an intrinsic linear polarisation, which is moreover wavelength-dependent. Such measurements were presented for μ Cep (M2 Ia) by McLean (1979) and for α Ori (M1.5 Iab) and α Sco (M1.5 Iab) by Tinbergen *et al.* (1980).

The observations show that the polarisation changes in angle and value with characteristic times of the order of one year (α Ori), as if it were due to an asymmetric cloud or asymmetric illumination with that characteristic time. The wavelength-dependence is partly Rayleigh-like (increasing towards shorter wavelengths, roughly $(:)\lambda^{-4}$), and may be ascribed to scattering at bound molecules. However, a deviation from the λ^{-4}-law at wavelengths $\lesssim 0.48$ μm may indicate the presence of grains of sizes ≈ 0.1 μm.

The *occurrence of dust shells* around stars is not restricted to cool stars, neither is it directly related to the stellar luminosity: dust shells appear around stars of all spectral types that have a sufficiently strong rate of mass-loss. So it appears that all O–B3 stars with $M_v < -4$ are redder than stars of the same spectral types but intrinsically less bright (Pecker, 1962; Isserstedt, 1978). This is ascribed to dust envelopes around such stars. Such envelopes occur particularly in the vicinity of young pre-main sequence stars such as the Herbig Ae–Be objects. A number of these, studied by Harvey *et al.* (1979) have an infrared excess that can be attributed to thermally radiating dust (presumably left over from the star formation?) with colour temperatures of 30–70 K.

In Section 3.4 we mentioned that dust shells, sometimes variable, occur around the late members of the WC sequence.

Dust clouds occur also around Be stars: Allen and Swings (1976) found that about 10% of all Be stars have infrared excesses near 2–3 µm. Such an excess must be due to a fairly hot dust cloud ($\sim 10^3$ K). The presence of the dust cloud radiating at 2–3 µm is more common for the relatively lower-temperature emission-line stars. In these clouds the simultaneous presence of a tenuous gas is shown by forbidden lines such as [O II] and [S II]; the n_e-value should be $> 10^5$ cm^{-3}.

Apart from the Be-stars many other emission-line objects of greatly varying spectral types have dust clouds. Thus Cohen and Barlow (1973) measured the infrared excess near 20 µm of the emission nebulae NGC 7635 (central star: BD+60°2522; O6.5 IIIf), and IC 1470 (spectral type of central star: O7f). The excess indicates colour temperatures of 180 K and 150 K respectively, and should be attributed to a dust nebula. Harvey *et al.* (1979) detected dust clouds around six out of eight investigated, mainly cool emission line objects, such as LK Hα 101, R Mon, MWC 1080. The T_c-values are low, ~ 20 to 100 K. Some clouds are spatially extended.

Lebofsky and Rieke (1977) drew attention to a number of dust-surrounded carbon stars from the AFCRL Survey at 4, 11, and 20 µm of which CRL 3068 seems to be the most extreme case. The dust shell must be of large optical depth. The object was discovered as an infrared emission object in an 11 µm survey. Narrow-band infrared continuum spectroscopy yields a colour temperature of 350 K. The southern objects of the Survey were investigated in other wavelength ranges by Allen *et al.* (1977): most of the stars surrounded by dust are C or late M stars, but in addition there are a few emission nebulae, WC and Be stars, and sources lying in or near reflection nebulae.

The size of the dust cloud is either measured by direct optical scanning – if the cloud has a large angular extent, such as Lk Hα 101, R Mon and MWC 1080, where the cloud is $> 40''$ (Harvey *et al.*, 1979). For other cases spatial heterodyne interferometric measurements such as those made at 11 µm by Sutton *et al.* (1977), applied to four late-type stars, give the following results:

– α Sco (M1.5 Iab), optically thin dust cloud, mainly for $r > 12R_*$;

– α Ori (M1.5 Iab), optically thin, dust mainly for $r > 12R_*$, for this star McCarthy *et al.* (1977) found that half of the 11 µm radiation is emitted in a shell with radius of $30R_*$;

– VY CMa (M3–5I), optically thick shell, gaussian intensity distribution with e^{-1}-intensity at $0\rlap{.}''58$ distance;

– R Leo (M6.5e–9e), uniform spherical shell, optically thin, diameter $0\rlap{.}''28$ ($\approx 5R_*$).

However, without doubt, dust may extend to far greater distances, as has been proved for two of the stars mentioned above: In Section 6.16 we describe VY CMa; particularly Figure 97 gives details on the extended dust cloud, detected from the polarisation halo around the star. Also, polarisation measurements of the sky around α Ori show the existence of a shell where the plane of the electric vector of the polarized component is at right angles to the direction to the star. Barely significant (1.7σ) polarisation is still detectable at 90″ distance (17000 AU $\approx 6000R^*$), while a distinct shell exists in any case up to 30″.

It therefore seems that further, refined observations of skylight polarization around cool supergiants may reveal the existence of more extended dust clouds around such stars.

After this brief review of the observations we deal with the *astrophysical measurements* aiming at identification of the properties of the dust particles, and of the shells.

The region in the HR diagram occupied by stars showing the infrared silicate excess near 11 μm appear to contain all stars with luminosities above a line defined by the spectral types M6 III, M1 Iab, G8 Ia, G0 Ia$^+$ (references in Reimers, 1975a).

Some insight in the properties of the dust shells and the particles therein can be obtained by using results of four-colour photometry, in a way developed by Gilman (1974). If the dust emission comes from a fairly spherical shell away from the star, the flux at wavelength λ is

$$F_\lambda = F_\lambda^* \left\{ \left(\frac{R_d^2 T_d}{R_*^2 T_*} - 1 \right) \left(1 - e^{-\tau_\lambda} \right) + 1 \right\}, \qquad (6.15; 1)$$

where R_d and R_* are shell and stellar radii, F_λ and F_λ^* are the respective fluxes, T_d and T_* the temperatures, and $\tau_\lambda = \kappa_\lambda d$ the optical depth of the shell, where d is the shell's effective thickness. Assume the opacities $\kappa_1 \ldots \kappa_4$ known for the four colours, then, a two-colour diagram may be constructed, involving three wavelengths, – such as $(F_1 - F_2):(F_2 - F_3)$ – see Figure 93: Any given value of $(R_d/R_*)^2 (T_d/T_*)$ defines a curve in the diagram for increasing optical depths. Stars with negligible or small infrared emission are found close to the origin of axes. The dust-temperature T_d is found from a comparison of two colours measurements, while the position of the star in the diagram yields T_λ and, after some computations: R_d. So far, using observations at wavelengths 3.5, 8.4, 11, and 18 μm average shell temperatures were found of 400–1000 K (Gehrz and Woolf, 1971; Gilman, 1974), the majority of determinations being fairly close to 1000 K. For 60 infrared sources Dorschner *et al.* (1978) and Gürtler *et al.* (1979) determined the values of the optical depths of the center of the 10 μm absorption (0.4 … 6) the equivalent widths (1.4 … 8 μm) and the consequent dust shell temperature: 200 … 700 K, median value $\sim$300 K. This temperature is, however, the result of a selection effect: objects with $T \approx 300$ K have maximum radiation at 10 μm.

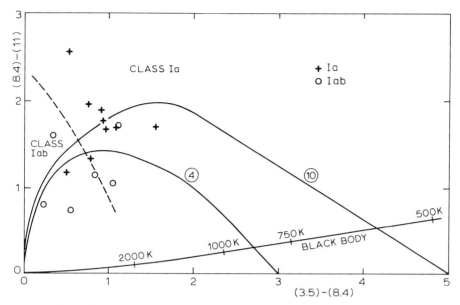

Fig. 93.   Colour-colour diagram for late-type supergiants. The absicca and ordinate are marked with the wavelengths (in μm) of the relevant flux measurements. The two upper *solid curves* are two-colour lines for fixed values of $(R/R_*) (T/T_*)^{1/2}$; these values are labeled to the curves. The *dashed line* separates the luminosity classes Ia and Iab. (After Gehrz and Woolf, 1971).

Assuming for the time being a temperature of 700 K, optical depths and shell radii R_s can be derived from the results of the two-colour diagrams of Gehrz and Woolf (1971). For a wavelength $\lambda = 11$ μm one thus finds for various groups of stars:

	τ (11 μm)	R_s/R_*
Mira variables	1	5
type M Iab supergiants	0.2	10
type M Ia supergiants	0.1	20 (Humphreys *et al.*, 1972)
type G Ia supergiants	0.8	20
type G Ia⁺ supergiants	0.4	40 (Humphreys *et al.*, 1971)

Along the same lines Bergeat *et al.* (1976a, b, c) analysed 22 cool C-stars (essentially C4–C9), of which photometric data was available in many wavelengths up to 3.4 μm. The infrared emission component longward of 1.2 μm comes from an optically thin dust shell for which the inner and outer radii, expressed in that of the photosphere are on the average 1.61 and 1.86. The dust shells may predominantly consist of large graphite grains of radii $a \approx (2.0 \pm 0.3)$ μm. The grains should have temperatures of approx. 1350 K. These temperatures are considerably higher than those measured by others, who used the spectral distribution of the dust-emission: in the present case where the observations do not extend far enough into the infrared to allow for such a treatment, the grain temperature was simply computed with $T_{gr} = T_e (R_{ph}/2R_{sh})^{1/2}$, where R_{sh} is the average shell radius. This high temperature may also explain the fairly small optical depths found.

Temperatures of the dust particles were further estimated by Thomas *et al.* (1976), who measured the IR excess near 10 μm of 16 stars, in 5 wavelengths between 8.4 and 12.5 μm ($\lambda/\Delta\lambda \approx 50$). Since the excess flux may be assumed to be proportional to the blackbody emission function and the optical depth τ_λ, the temperature of the shell can be estimated by comparing the excess function with theoretical calculations. The best agreement would be obtained for $T \approx 250$ K.

We conclude that the various determinations of dust temperatures, mainly based on infrared multi-colour observations yield dust temperatures over a wide range, between about 200 and 1000 K. If an average value should be adopted, one should perhaps take 600 ± 300 K.

The question may readily arise whether the large scatter in the dust temperatures reported reflects a real scatter in the temperatures or errors in measurements and/or interpretational technique. It is useful to realize that most dust shells are opaque ($\tau \gg 1$); the precise discussion of the transfer of radiation through an opaque dust shell is of importance for enabling a reliable interpretation of the observations. Unno and Kondo (1977) and Haisch (1979) have developed the relevant theory: the latter assumed multiple particle sizes – yielding multiple temperature distributions!

The *relation* between the existence of a (circum)stellar *dust cloud*, its mass-loss, and the *fundamental stellar parameters* was investigated by Reimers (1975). We already mentioned that for hot stars it is not so much the temperature or L, but $\mathfrak{M}$ that defines whether circumstellar dust exists around a star. In Section 7.4 (last part) we refer to the linear relation existing between $\log(-\mathfrak{M})$ and $\log(L/gR)$, the quantity between brackets being the ratio between the stellar luminosity and its negative potential energy; so also the existence and extent of dust clouds should be related to that quantity. Gehrz and Woolf (1971) have found that

$$\Delta F = F(3.5 \; \mu m) - F(11.4 \; \mu m) = 2.5 \log\left(-c_1 \frac{R_d \, T_d}{R_* \, T_*} \dot{\mathfrak{M}} + 1\right), \quad (6.15; 2)$$

assuming $-\dot{\mathfrak{M}} = c_2 L/gR$ the second part of expression (6.15; 2) becomes

$$2.5 \log\left(c_3 \frac{R_d \, T_d}{R \, T} \frac{L}{gR} + 1\right).$$

Plots of $\log(L/gR)$ against ΔF enable one to identify various values of the constant c_3 – see Figure 94, where there is a ratio of five between the c_3-values for the uppermost and lower of the two curves. This figure shows too that a star must be cooler and more luminous than a certain limit in order to produce an observable infrared excess. Reimers found that this limit corresponds with $L/gR \geq 1.5 \times 10^5$ (where all three parameters are in solar units).

Another aspect of the genetic relation between the occurrence of a stellar wind and the existence of a circumstellar cloud of gas and/or dust particles was brought forward by Kuan and Kuhi (1975). They found that a *gas* cloud occurs only for stars with

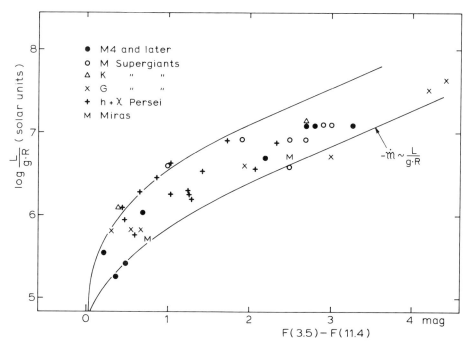

Fig. 94. Values of L/gR (in solar units) versus the infrared silicate excess (from various sources). The abscissa gives the flux differences between 3.5 and 11.4 μm in magnitudes. The solid lines represent infrared excesses calculated from the assumption $-\mathfrak{M}(:)L/gR$. (After Reimers, 1975a).

a large momentum flux ϱv^2; for smaller values of ϱv^2 there would be a *dust* cloud. Figure 95 gives the main result; in this figure the ratio $F(10\ \mu\mathrm{m})/F(2.2\ \mu\mathrm{m})$ characterizes the occurrence of gas or dust. In this vein Cahn and Wyatt (1978a) assumed that in the stellar wind around cool giants the temperature decreases with distance; beyond approx. $2R_*$ the gas becomes cool enough for grains to form.

The *composition of the dust grains* is hard to determine since the available data such as the intensity profile of the 10 μm excess, the polarisation, ... allow for explanations with various models. Comparison of the spectral data near 10 μm with laboratory spectra, and theoretical considerations mainly based on the cosmic atomic abundances and on discussions about particle formation make probable that the particles are essentially silicates, perhaps of sub-μm size (Gilman, 1974).

Proposed grain models include *hydrated silicates* (Larimer and Anders, 1967): below 500 K silicate grains become hydrated; *dirty silicates*: pure silicate grains have a much smaller absorption coefficient and for that reason can not fully reproduce the 10 μm excess profile; and more recently, particles consisting of olivine, $(\mathrm{Mg, Fe})_2\mathrm{SiO}$ or enstatite $(\mathrm{Mg, Fe})\mathrm{SiO}_3$ (Stephens and Russell, 1979), or phyllosilicates such as occurring in carbonaceous chondrites (Zaikowski and Knacke, 1975; Dorschner *et al.*, 1978b; Friedemann *et al.*, 1979) are suggested as possible candidates. The latter do show absorption at the wavelength of the observations, but the laboratory bandwidth is narrower than the observed value.

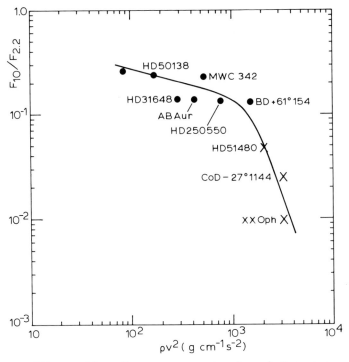

Fig. 95.   Ratio of 10 μm to 2.2 μm flux versus momentum flux ϱv^2. *Dots*: dust; *crosses*: free-free emission. (From Kuan and Kuhi, 1975).

Day (1976a, b) found from laboratory measurements, that magnesium silicates pro-duce the observed absorption at 10 μm and the observed optical extinction, but that they should also be responsible for an excess at 20 μm, upon cooling to 80 K. Others agree that the laboratory absorption increases upon cooling (Robinson and Hyland, 1977). Runciman *et al.* (1973) found no such effect for olivine in the 10 μm band, but Dorschner *et al.* (1977), from a comparison with laboratory measurements, found that the 10 μm band could indeed be due to olivine, while Duley *et al.* (1978) found laboratory arguments for SiO: the shape of the interstellar extinction band at 9.6 μm is very similar to the laboratory extinction of SiO particles of μm diameter.

Theoretical computations of the expected (continuous) spectrum of spherical dust clouds around stars were made by Jones and Merrill (1976). Radiative equilibrium was assumed in the cloud; parameters that were varied were shell optical depths, shell radii and mass distribution functions forh both silicate and graphite grains. The out-come was that it appears hard to produce spectra similar to those observed unless the shells are assumed to contain, 'dirty' silicate grains (which absorb stronger for $\lambda < 8$ μm than 'clean' grains).

The dust particles can also produce a *polarisation* at optical wavelengths by scat-tering, if the particles are not spherical and aligned, or if the shell is asymmetric. A review was published by Zellner and Serkowski (1972). Some 18 very cool stars have been investigated for polarisation in seven wavelength bands between 0.3 and 1 μm by Kruszewski and Coyne (1976) who noted the strong polarisation (up to 15 and 20% near 0.3 μm) of a great part of these objects. The polarisation may vary with wavelength, $(dp/d(1/\lambda) \approx 5\%$ per μm^{-1} for the object CIT 6), and with the phase of variability, the latter with time scales of the order of days. There is a good correlation between the mean degree of polarisation and the intensity of the 10 μm silicate excess.

Very interesting is also the relation between the average degree of polarisation p (in the visual spectral region) and the separation W of the components of the 18 cm OH emission feature, the polarisation increasing with increasing W-value. The approximate relation is $p \approx -2 + 0.1W$ ($W > 10$ km s^{-1}), with p in % and W in km s^{-1}, where $p \approx 0$ for $W \lesssim 10$ km s^{-1}.

The average particle radius, derived from a comparison of scattering models with polarisation observations is 0.08 μm for silicates and slightly smaller for iron or graphite grains. For the dust around η Car Warren–Smith *et al.* (1979) find $d \approx 0.1$ μm.

Summarizing the present, partly still conflicting results, it is allowed to *conclude* that the circumstellar dust particles are *composed* of silicon-oxides and dirty silicates; they have *diameters* of the order of 0.1 μm, and average *temperatures* of ~600 K.

6.16. Circumstellar Gas and Dust: the Red Supergiant VY CMa

The star VY CMa (M3–5e I), $T_e = 2740$ K, is an irregular variable, surrounded by extended *dust clouds*, as appears from the fact that it is embedded in a small reflection nebula ($8'' \times 12''$), and by the observed infrared excess (Hyland *et al.*, 1969; Gillett *et al.*, 1970). That it is also surrounded by a gaseous envelop follows from its emission of *microwave radiation*, as observed in the 3.5 mm region by Wilson (1971) with a flux level of 0.2 ± 0.07 f.u. Furthermore, the star is a well-known maser-source, and emits lines from the molecules OH, H_2O, and SiO. It is one of the brightest known sources of infrared radiation in our Galaxy. Many data on its radiation properties are found in a paper by Rosen *et al.* (1978).

The star is situated at the interface between a large (~15 pc) molecular cloud described by Lada and Reid (1978) and the large (~5°) H^+ region Sharpless 310. The latter region is excited by the bright O-star τ CMa and perhaps also by the massive binary UW CMa (O7f+O8.5 If – cf. Table VIII); it contains a number of very young objects, and the cluster NGC 2362. There is a bright rim at the edge of S 310 near VY CMa (Herbig, 1972), perhaps responsible for the microwave emission. This rim coincides with a rim in the CO emission of the molecular cloud. Since VY CMa has a velocity similar to those of the molecular cloud complex and the stars of the cluster NGC 2362, these various coincidences strongly suggest a physical relation between VY CMa, the molecular cloud, and the cluster. Hence the distance of the star is known (1.5 kpc; Herbig, 1969), and its luminosity ($5 \times 10^5 L_\odot$). With a view to its location in the HR-diagram we therefore estimate a mass of ~$20\mathfrak{M}_\odot$. The uncertainty in this value (factor 2?) is related to the uncertainty about the evolutionary phase of the star: is it a contracting protostar or an evolved star? The association with so many young objects suggests the former.

There must be a gaseous envelope close to the star, detectable by the thermal SiO emission lines at 8.3 μm (Geballe *et al.*, 1979). At that wavelength McCarthy *et al.* (1977) measured at ~10 μm a continuum radius of the star of $0\overset{''}{.}21\pm0\overset{''}{.}06$, but Geballe *et al.* think that the emission must originate from closer to the star (estimated: ~$0\overset{''}{.}15$), in view of the fairly high molecular rotation and vibration temperatures (~600 K), while at $0\overset{''}{.}2$ one would expect $T_{rot}\approx T_{vib}\approx400$ K. Since the abundance ratio $(SiO/H)\approx10^{-6}$, hence $[SiO/H]\approx-1.5$ most of the silicon must be in the form of dust grains, already at that distance to the star.

At longer infrared wavelengths that inner dust shell is directly observable. In Section 6.15 we referred to measurements of the size of the inner dust shell by Sutton *et al.* (1977) at 11 μm, who found for an assumed gaussian distribution an e^{-1}-intensity at $0\overset{''}{.}58$ distance from the center. Upper limits at 1.25 μm and 3.7 μm of $1\overset{''}{.}05$ and $0\overset{''}{.}52$ were communicated by Bensammar *et al.* (1978).

It seems that the inner dust shell around this star has also been observed in the visual spectral range. Ever since 1927, when Van den Bos announced the detection of a starlike 'companion' to VY CMa at $0\overset{''}{.}56$ distance, the existence of a companion has been claimed, apparently moving at a regular angular speed (~0.8 deg yr^{-1}) with a separation that increased to more than $0\overset{''}{.}7$ in the seventies. The fact that the motion apparently does not obey to Kepler's law of areas, while masses computed for a circular orbit (assuming a stellar distance of 1500 pc) would be impossibly high ($4600\mathfrak{M}_\odot$), brought Wallerstein (1978) to assume that the star is closely surrounded by a rotating dust cloud with holes, through which a bundle of light shines on a more distant part of the cloud. This is supported by the observation that the brightness variations of star and 'companion' tend to go parallel. Also, the distance of the 'companion' ($0\overset{''}{.}6$) agrees well with Sutton's (1977) determination of the radius of the dust cloud mentioned earlier in this Section. For an assumed mass of $20\mathfrak{M}_\odot$ for VY CMa the distance of the holey dust cloud to the star should be $0\overset{''}{.}15$ or 230 AU ($\approx16R_*$) in order that it rotates with the observed angular speed.

Very Long Baseline Interferometry by various authors (see e.g. Rosen *et al.*, 1978; and Rosen and Mutel, 1979), has shown that the masing region has an extent of about $25\times1''$, which corresponds with 5×10^{10} km to 2×10^{10} km or $7\times10^5\,R_\odot$ to $1.5\times10^5\,R_\odot$. There are at any time about ten individual sources with diameters ~5×10^8 km ($700R_\odot$). There are indications that the masing region rotates around the star with a period of up to 10^4 yr (Masheder *et al.*, 1974), which would explain the long-term variability of the sources if these may be assumed to be highly directional. The individual sources, and the resulting maser line profile vary with time (Robinson, 1970) while also the positions and the number of the sources vary continuously. This is shown very clearly in Figures 96 (which show the maser sources in January 1974, and September 1976) and Figure 97 showing among other things the sources in the first days of April 1973. Substantial changes seem to occur in periods of a few months, and are correlated, consistent with an assumed central pump source (Cox and Parker,

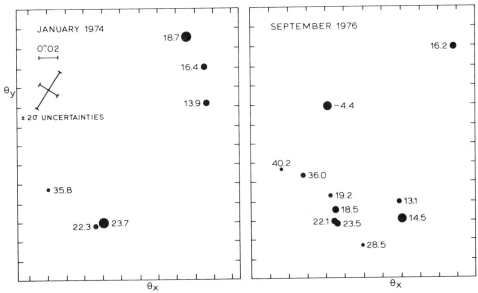

Fig. 96.   Maps of the spectral components in VY CMa constructed from fringe-frequency data. The spot sizes in each map are proportional to the flux densities (not angular sizes) of the features. Each spot is labeled by the velocity of peak of the corresponding spectral component in the cross-power spectrum. The greater number of features mapped in 1976 is due primarily to the increased velocity coverage in that experiment. The continuously changing pattern is apparent, and becomes still more obvious, when compared to the April 1973 observations of Moran *et al.*, shown in Figure 97. (After Rosen *et al.*, 1978).

1979). There seems to be a tendency for the low-velocity maser components to occur close to the centre of the region. The apparent symmetry of the $v=1$, $v=2$, $J=1-0$ line complex of SiO (Buhl *et al.*, 1974) can be explained if these molecules are found in a circumstellar ring system (Van Blerkom, 1978b). This is a very probable suggestion: evidently, circumstellar gas finds its most stable position in an equatorial ring system.

The total amount of gas in the maser region is 10^{-1} $\mathfrak{M}_\odot$, the total amount of the H_2O gas is 10^{-4} $\mathfrak{M}_\odot$; the gas density in the masing regions is $n_{\rm gas} \approx 10^9$ cm^{-3}. From the P Cygni profiles of the SiO lines at 8.3 µm Geballe *et al.* (1979) derive an out-streaming velocity at $0\rlap{.}''15$ ($\approx 3 \times 10^9$ km $\approx 10^5$ R) of 15 km s^{-1}. Assuming for the masing region an expansion velocity of ~ 20 km s^{-1}, a radius of 10^{10} km, one obtains a stellar rate of mass loss $-\mathfrak{M}=2\times 10^{-3}$ $\mathfrak{M}_\odot$ yr^{-1}. This is a very high value, which may be too high, if one would take into account that the masers occur in the relatively dense parts of the region. It should be noted that this mass-loss-rate is by a factor of about 10^3 higher than in red giants or supergiants of comparable spectral types – see Table XLII in Section 7.4. A reason to doubt this large rate of mass-loss is that it needs only 10^4 yr to make the star lose $20\mathfrak{M}_\odot$, an amount equal to its present mass. However, in order to influence the bright rim (at ~ 3 pc distance) by gas out-streaming with 20 km s^{-1} mass-loss must have occurred already 1.5×10^5 yr ago (Lada

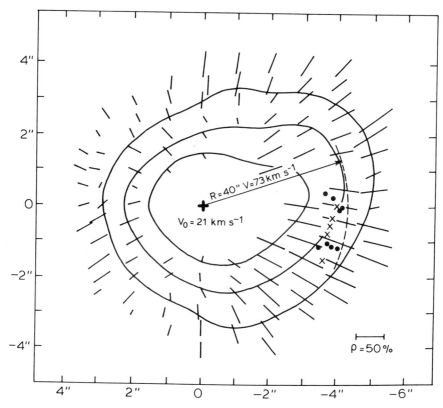

Fig. 97.   Map of the positions of the 13 principal 1612 MHz OH features in VY CMa as observed early April 1973 (Moran *et al.*, 1977a). The contours connect points of equal unpolarized flux at 17, 50, and 150 arbitrary units above the background. The length of the straight lines is proportional to the degree of linear polarization according to the scale in the lower right corner and the direction of the lines is perpendicular to the E-vector. The centre of expansion in the model been brought to coincidence. (After Winnberg, 1978).

and Reid, 1978). Hence a rate of mass-loss of a few times $10^{-4} \, \mathfrak{M}_\odot \, \mathrm{yr}^{-1}$ seems a more reasonable estimate.

Finally, we compare the VLBI-pictures of VY CMa with polarisation measurements of the source. This indicates a distinct relation between the degree and direction of polarisation – presumably due to dust particles – in dependence on the position around the star (Figure 97).

6.17. Circumstellar Gas and Dust: the Object η Carinae

This southern object, located in the Carina nebula (NGC 3372) at $\alpha = 10^h 41^m$; $\delta = -59°$ is intriguing indeed, and has a remarkable history. It is one of the brightest objects known – see Figure 2. Prior to 1837 it was of the 2nd to 4th apparent magnitude, but in that year it brightened to first and zeroth magnitude, and remained so for some time (Figure 98a). With a repetition time of about 20 years four light-maxima occurred, but gradually the object faded down till seventh and eight magnitudes. (Innes, 1903; Walborn and Liller, 1977). At present it is of the sixth magnitude, and consists of a *central condensation* with a stellar appearance, and with at a wavelength of 4 μm a diameter of $\leq 1\overset{\prime\prime}{.}4$ (Aitken *et al.*, 1977). It is surrounded by a somewhat elongated *outer envelope* often called (after Gaviola, 1950) the *homunculus*, after the visual appearance of the isophotes. The homunculus has an axial ratio of approx. 2 : 3, with average half-flux diameters ranging between $2\overset{\prime\prime}{.}7$ at 3.7 μm to $6\overset{\prime\prime}{.}5$ at 12.2 μm (Koornneef, 1978). In the infrared there appear to be two intensity maxima with separation increasing with wavelength from $\sim 1\overset{\prime\prime}{.}1$ at 3.6 μm to $\sim 2\overset{\prime\prime}{.}2$ at 11.2 μm (Hyland *et al.*, 1979). There is a steep density decrease beyond that distance (Harvey *et al.*, 1978). The complex morphology of the homunculus can be best studied from the polarisation of its dust component, and was described by Walborn (1976b) and Warren–Smith *et al.* (1979). In addition there is an extended *nebula*, the Carina Nebula, of $\sim 0\overset{\circ}{.}3$ diameter (see the frontispiece of this volume). This nebula has been the subject of investigation of many astronemers; a fairly extensive description was given by Walborn (1976b).

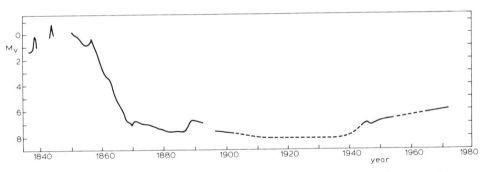

Fig. 98a.   Visual light curve of η Carinae. (After the compilations by Innes (1903), O'Connell (1956), and Feinstein and Marraco (1974)).

Its brightest part, at about $1\overset{\prime}{.}5$ distance from the star is certainly a reflection nebula, since its spectrum is similar to that of η Car. Hence at least that part of the Nebula must consist of dust particles, and it seems that most of the Nebula is reflecting the star's light. In its central part, close to the homunculus features have been discovered with tangential velocities $\gtrsim 1000$ km s^{-1} (Walborn *et al.*, 1978).

The visual and near infrared spectra of the central condensation show at present many emission lines – the Balmer series, [Fe II]-lines (see: Johansson, 1977), [Mn II]-lines (Thackeray and Velasco, 1976), ionic emission lines of higher ionization such as [Ne III], N III, [S III] and Fe III, and the Bracket α line at 4.05 μm (Aitken *et al.*, 1977). The latter authors also found that at 4 μm the $1\overset{\prime\prime}{.}4$ central condensation shows a marked infrared excess, much larger than that of any 'normal' star on the HR diagram.

In the course of time the spectrum – like the star itself – has shown large variations. An 1893 spectrogram shows a spectrum rich in absorption lines (Whitney, 1952; Walborn and Liller, 1977), and with some resemblance to that of novae. It is – as with novae – similar to the spectra of F supergiants with a few emission lines, and may therefore be emitted by an expanding shell ejected shortly before. This hypothesis is supported by the measured expansion velocities, and by the P Cygni profiles of the ultraviolet resonance lines of C II, C IV, Si II, Si III, Al II, Al III, Mn II, Mg II, Fe II (Giangrande et al., 1979). Such spectra show two systems of ultraviolet absorptions, a shell absorption at -90 km s^{-1} and a large absorption probably due to an expanding envelope with v ranging from about -200 to -700 km s^{-1}. The terminal velocity is about 715 km s^{-1} (Cassatella and Viotti, 1979).

Since the distance d of η Car is estimated to be between 1.5 and 3 kpc, with 2 kpc as the most probable value (Thé and Vleeming, 1971), the total luminosity should be approx. 10^{40} erg s^{-1}, yielding $5 \times 10^6 L_\odot$ (Aitken et al., 1977), thus making η Car one of the most luminous objects of the Galaxy. This is confirmed by Viotti (1969) who found $M_v = -10.5 \pm 1$ mag and with a bolometric correction of -1.1 mag (with $T_e = 29\,000$ K, see below):

$$M_b = -11.6 \pm 1, \quad \text{or} \quad \frac{L}{L_\odot} = 1.4 \times 10^6 \quad \text{to} \quad 9 \times 10^6.$$

Davidson (1971) determined the main parameters of the central condensation. The determination is difficult, because only a small part of the continuous spectrum can be attributed to the star, and this is strongly attenuated by the surrounding gas and dust from the envelope and nebula. At 3300 Å the extinction may reduce the continuum intensity by about a factor 200. With a Zanstra method the effective temperature of the star is estimated; if one assumes $T_c = T_e$ one finds $T_e \approx 29\,000$ K. With the known luminosity ($L \approx 10^{40} (d/2 \text{ kpc})^2$ erg s^{-1}) one then finds a radius

$$R = 4.4 \times 10^7 \left(\frac{d}{2 \text{ kpc}}\right) \text{ km} \approx 60 \, (-20; +40) \times \left(\frac{d}{2 \text{ kpc}}\right) R_\odot.$$

From the mass-luminosity relation of Stothers (see Section 2.4) the mass would then be

$$\mathfrak{M}/\mathfrak{M}_\odot \approx 115 \left(\frac{d}{2 \text{ kpc}}\right)^2,$$

a fairly large value, at the limit of possibilities. Further, $g \approx 800 \, (d/2 \text{ kpc})^{-0.5}$ cm s^{-2}.

The *circumstellar gas* occupies a region of a few arc sec diameter in the inner part of the envelope. Its properties were studied by Rodgers and Searle (1967) and Davidson (1971) who, from the relative intensities of several forbidden transitions, found respectively: $n_e \approx 3 \times 10^6$ cm^{-3} and $T \approx 2 \times 10^4$ K (R and S), and $n_e \approx 4 \times 10^6$ cm^{-3}, $T \approx 7500$ K (D). The mass of the H$^+$ region is $\approx 0.025 \mathfrak{M}_\odot$.

The *dust envelope* and *nebula* can be best studied in the infrared. It must be composed of metallic silicates, observable through the emission hump at approx. 10 μm (Robinson et al., 1973). Angular size measurements by Gehrz et al. (1973) and Koornneef (1978) show that the diameter of the 10 μm dust shell is about three times that at 2 and 3 μm.

Figure 98b gives the infrared spectrum of the envelope, between 2 and 200 μm, mainly from the observations of Joyce (1975) and Harvey *et al.* (1978), who also included earlier observations (Gehrz *et al.*, 1973; Sutton *et al.*, 1974) in their compilations. The main part of the spectrum has a colour temperature of 250 K, which may be identified with the temperature of the emitting dust particles. Thomas *et al.* (1976) find 225 K from observations over a shorter wavelength range. For the envelope Robinson *et al.* (1973) find dust temperatures of 375 and 875 K. In the spectrum given in Figure 98b the silicate emission hump is clearly visible near 10 μm. There is apparently no emission or broad absorption at shorter wavelengths.

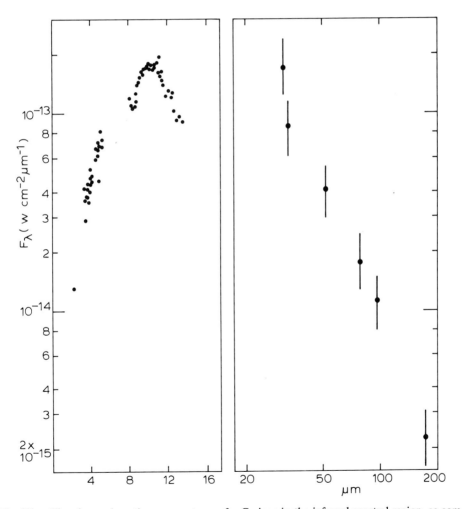

Fig. 98b. The observed continuous spectrum of η Carinae in the infrared spectral region, as compiled by Joyce (1975) and Harvey *et al.* (1978). The wavelength scale of the left-hand diagram is linear, the other is logarithmic!

The linear polarisation measurements of the homunculus around the central (stellar) condensation and the intensity-distribution show some asymmetry which can be interpreted as due to a toroidal circumstellar dust ring with radius $\sim 10^4$ AU, with an axis inclined $\sim 70°$ to the solar direction (Warren–Smith *et al.*, 1979; Hyland *et al.*, 1979). Assuming a certain geometry for the nebula, Carty *et al.* (1979) find a distribution function for the grain sizes such that grains of 0.15 μm contribute most to scattering in optical wavelengths while 0.1 μm grains produce maximum emission at 10 μm wavelength.

The star and the surrounding nebula are X-ray sources, as was found by Einstein-Observatory observations (Seward *et al.*, 1979). The high-resolution observations of Figure 99a show that η Car itself is an extended source and shows moreover that also hot O-type stars in the Carina complex are X-ray sources. This is again demonstrated in the low-resolution image (Figure 99b), which shows a few more sources and in addition a large X-ray emitting area. Finally, X-ray isophotes are shown by Figure 99c.

The *evolutionary scenario of the last 150 years* has been composed by Davidson (1971), who stressed that the fact that the star was around 1840 some six magnitudes

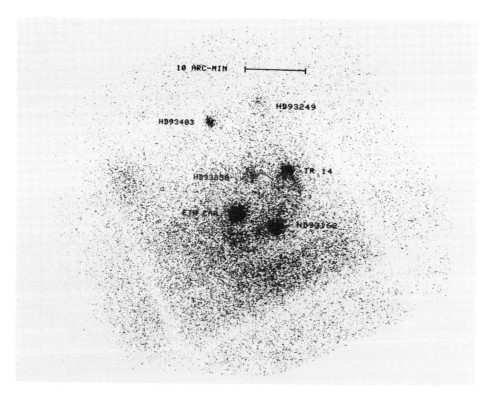

Fig. 99a.   The η Carinae region as observed by the Einstein Observatory. The instruments window support structure, 38′ wide, shadows part of the field of view.

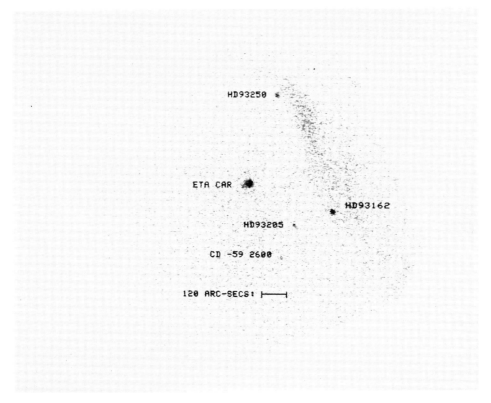

Fig. 99b. The inner part of the η Carinae region as observed by the Einstein Observatory.

brighter than now, need not mean that the object was intrinsically brighter at that time: extinction by dust may since have greatly increased, and the bolometric correction may have changed. If the star's effective temperature would have been 10^4 K at that time, the bolometric correction was zero, while it is -1 to -2 mag now. The extinction is some 2 mag. Assume that before 1840 the star was at magnitude 4. Then, by a sudden instability, gas ejection started, which may also have led to the toroidal dust ring, now surrounding the central star. The apparent large brightness after the ejection could have been caused by the increased size of the gaseous nebula around the star (see for similar considerations Section 9.2 – on nova spectra). Later (>1856) when the gas arrived at larger distances from the star, dust particles started to form; hence the brightness of the star decreased by extinction. As a consequence of the increasing size and thickness of the dust cloud the star continued to fade till approximately 1890. Thereafter a more stationary situation set in.

The *mass $\mathfrak{M}_d$ of the dust cloud* and its rate of increase, $-\dot{\mathfrak{M}}_d$ were estimated by Andriesse *et al.* (1978). They assumed that dust can only exist for distances to the star's center $> R_i$ (closer to the star dust would evaporate), and $< R_0$, the outer boundary of the dust shell. It is assumed that R_0 increased linearly with time since 1856,

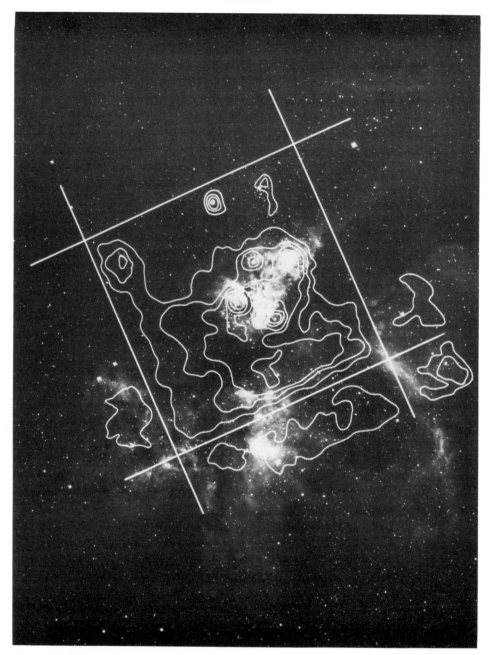

Fig. 99c. Contours of constant X-ray counting rate in the η Carinae region have been overlaid on a near-UV photograph. The square indicates the position of the instrument support structure and is 38′ wide. The strengths of contours in arbitrary units are ∼3, 5, 7, 12, 18, and 30. Cf. also the frontispiece of this book. (Figures 99a, b, and c are from Seward *et al.* (1979). Original print: *Astrophys. J.* Photo courtesy of Harvard-Smithsonian Center for Astrophysics.)

apparently the time when dust formation started, as can be inferred from the decreasing stellar brightness. Assume

$$R_0 = R_i + vt, \qquad (6.17; 1)$$

where v=the velocity of expansion of the envelope, ≈ 900 km s^{-1} (deduced from the present diameter of the dust cloud assuming that expansion started in 1840), $R_i = 3.7 \times 10^{11}$ km (corresponding to the radius – $\sim 1''$ – of the inner circumstellar gas cloud around the star). Let $m(R)/m(R_i) = (R/R_i)^a$, with $a = 1.5$ (Andriesse, 1976), where $m(R)$ is the mass in a shell with radius R and thickness unity. Write for the visual optical depth of the nebula:

$$\tau = \pi a^2 Q \int_{R_i}^{R_0} \frac{m(R)}{4\pi a^3 \varrho / 3} \, dR$$

where $Q = a\omega/c$ is the visual extinction coefficient, ω the visual frequency, a the average radius of a dust particle, and ϱ its average mass density. Writing for the total mass of the nebula:

$$\mathfrak{M}_d = 4\pi \int_{R_i}^{R_0} R^2 \, m(R) \, dR, \qquad (6.17; 2)$$

one obtains

$$\mathfrak{M}_d / \tau = \frac{16\pi}{3} \frac{a+1}{a+3} \frac{\varrho c}{\omega} \frac{R_0^{a+3} - R_i^{a+3}}{R_0^{a+1} - R_0^{a+1}}, \qquad (6.17; 3)$$

and

$$-\dot{\mathfrak{M}}_d = \dot{\tau} \frac{16\pi}{3} \frac{\varrho c}{\omega} R_i^2.$$

Inserting for $\dot{\tau} = 1.75 \times 10^{-8}$ s^{-1} (referring to the initial fading of 0.57 mag yr^{-1}) one finds $-\dot{\mathfrak{M}}_d = 7 \times 10^{21}$ g s^{-1}. With (6.17; 1) and (6.17; 3) one may find $\mathfrak{M}_d$ by integration. This yields the present value $\mathfrak{M}_d(1980) \approx 5 \times 10^{31}$ g. Andriesse et al. (1978) deduce from this value a total mass of the expelled gas of 2×10^{34} g ($\approx 10 \mathfrak{M}_\odot$), a value for which they find support by the *interstellar* dust (:) gas ratio (a weak argument in *circum*-stellar environments), and from the fact that the particles that are able to form chemical bonds have an abundance by mass less than 0.01 of the H-mass, yielding $\mathfrak{M}_g \geq 5 \times 10^{33}$ g. We *conclude* that there are (rather indirect) indication for a total shell-mass of a few to ~ 10 solar masses. Assuming that all mass has been expelled continuously from 1856 onward one finds an average mass-loss of $(2-8) \times 10^{-2} \mathfrak{M}_\odot$ yr^{-1}. This is the largest rate of mass-loss found for any (semi-)stable star – so large that it certainly needs independent confirmation. One consequence of this large mass-loss is that the often suggested identification of η Car with a nova or nova-like object loses ground: novae do not lose more than approx. $10^{-4} \, \mathfrak{M}_\odot$ and this happens in one event, quite contrary to the fairly continuous, much larger mass-loss rate of this, apparently very unstable object.

CHAPTER 7

MASS-LOSS FROM BRIGHT STARS

7.1. Outline

In Sections 5.5 through 5.8 the theory of expanding spherical stellar atmospheres, and resultant model atmospheres were outlined. This Chapter deals with *observations* relevant to stellar mass-loss; we describe the way one derives *mass-loss values* from such observations, summarize and compare 𝔐-values, and consider the proposed *mechanisms* that produce the observed phenomena.

Historically, two possible mechanisms were initially considered. Following the fundamental research on the solar wind by Parker (1958) stellar mass-loss by *evaporation from a hot corona* was first discussed by de Jager (1960) and Rubbra and Cowling (1960) and has later found many advocates among which Hearn (1975a, b). *Radiation forces* were introduced in astrophysics by Milne (1926) and their effects on stellar mass-loss shown by Lucy and Solomon (1970); it is important only in hot stars. Later developments are from Cassinelli and Castor (1973), Castor (1974a, b), Castor *et al.* (1975). A star does not need to be surrounded by a corona in order to have appreciable mass-loss, but most hot stars do have one!

The above considerations refer to individual stars. In binaries mass-transfer from one of the components may exceed the rate of loss in individual stars by orders of magnitude, when in the process of evolution of the binary one of the components overflows its Roche lobe and transmits matter to its companion. This case was discussed in Sections 4.10, 4.11, and 4.12.

7.2. Observational Indicators and General Properties of Mass-Flow

In most cases of stellar mass-loss the wind velocity is small close to and in the photosphere, but continuous acceleration generally makes the velocity increase with distance from the star. At large distances the escaping gas typically appears to have line source functions with *excitation temperatures* decreasing with increasing distance from the photosphere, because with increasing distance the rates of photo-ionisation, photo-excitation, and collisional excitation and ionisation decrease. In seemingly contrast is that certain *strong UV resonance lines* of relatively high stages of ionisation (like Si IV, C IV, N V, O VI) originate at fairly *large* distances from the star. This is because the *ionisation temperatures* do *not* decrease with increasing distance: The photo-ionisation rates depend essentially on the ion density and the stellar colour temperature, while the recombination rates – dependent on the electron temperature and on $n_e n_i$ – decrease faster. Hence the ionisation temperature increases with increasing distance. Combined with the decreasing T_{ex}-values, this explains why the strong UV resonance lines are formed at large distances from the star. (To this effect should be added that the thermodynamic state of the gas can be 'frozen-in' by the flow and thus carried to larger distances from the star). In this picture one would expect that essentially all subordinate lines, of any appreciable excitation potential, are formed fairly close to the stellar photosphere.

Hence, appreciable mass-loss will show itself by:
– *asymmetric* or slightly-*displaced* lined of photospheric excitation;
– larger *wavelength-shifts* for lines of lower excitation; a velocity-excitation relation derived by Hutchings (1978) is shown in Figure 100;

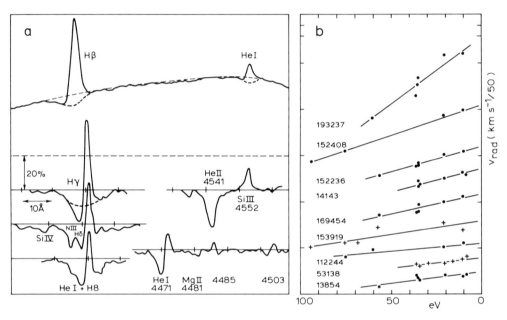

Fig. 100.   (a) Line profiles in the spectrum of the Of star HD 108. The lines with highest excitation show less emission. The lines $\lambda\lambda 4485$, 4503 are unidentified emission-lines found in many Of stars. (b) Velocity-excitation relation for various O-type stars, identified by their HD number. (From Hutchings, 1978).

– the '*Balmer-line velocity progression*' which shows a decrease of expansion velocity for higher members in the Balmer series, indicates the same phenomenon; – see Figure 101;
– strongly-*displaced UV resonance lines*, often showing P Cyg profiles, are usually the best indicators for the largest (up to terminal) velocities in the stellar winds – Figure 102;
– some *emission lines* indicate the presence of extended envelopes with a source function $S > B_{phot}$; often, such envelopes are outstreaming; particularly the strength of

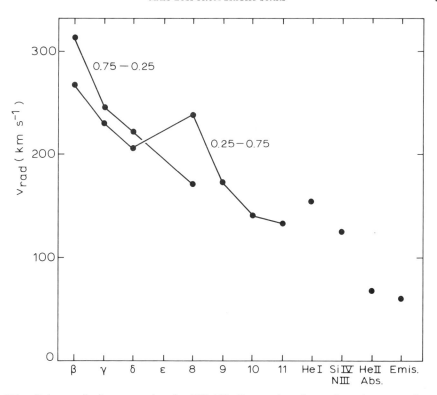

Fig. 101.   Balmer velocity-progression for HD 108. Progression shows dependence on phase in binary orbit with phases as indicated. Mean values for ions are also shown. Anomalous values from H8 indicate significant contribution from He I λ3888. (From Hutchings, 1978).

the Hα emission is a sensitive indicator of the mass-loss rate (Conti and Frost, 1977; Klein and Castor, 1978);

– *radio emission and infrared radiation excess* is ascribed to free-free transitions in a dense stellar wind, at fairly large distances from the star. In Section 6.13 we described how mass-loss values can be derived from this kind of data. The essential expression is Equation (6.13; 2).

The *history of the study of stellar mass loss* is short and interesting. Adams and MacCormack (1935) were the first to observe that the cores of strong resonance lines in some M-type stars were systematically blue-displaced by 5 km s^{-1}, suggesting out-streaming motions. Deutsch (1956) observed sharp, displaced absorption lines in the spectrum of the G0 III companion of the visual binary α Her; these lines showed velocities equal to the heliocentric velocities of the strongest circumstellar lines of the primary (see also Section 6.12). The fundamental significance of this discovery is that even at a distance of several hundred AU from the primary the outstreaming velocity is still appreciable and in fact larger than v_{esc}. This was the first uncontro-versial proof of stellar mass-loss. Similarly, Sargent (1961) found blue-displaced lines in the spectrum of ϱ Cas (F8 Ia^{+}), indicating $v \approx 25$ km s^{-1}. An important further

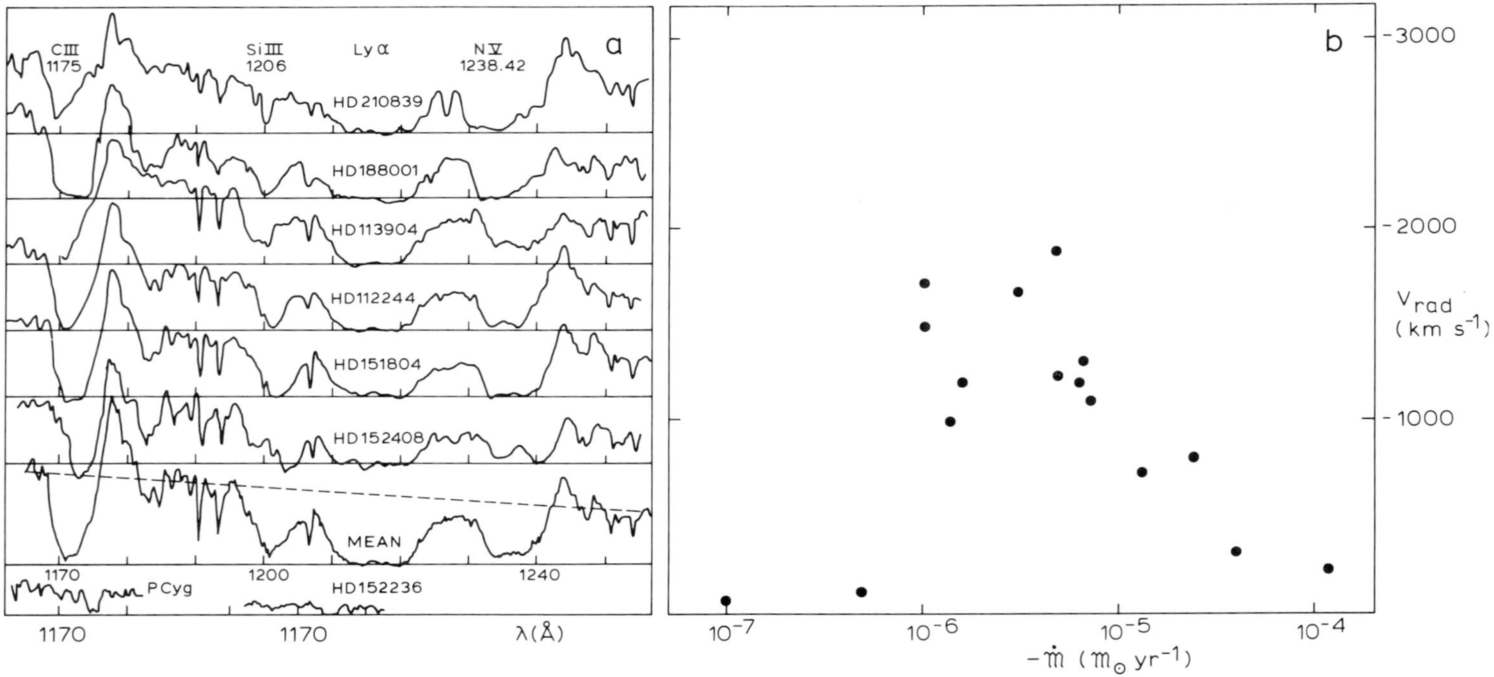

Fig. 102. (a) *Copernicus* scans of the UV spectra of stars with extreme rates of mass-loss. Attention is drawn to the P Cygni profiles of C III and N V. It is also noteworthy that the stars with strong mass-loss (P Cyg and HD 152236 – bottom) do not show P Cyg-type profiles, but just photospheric line profiles at λ1175 – see also Section 3.5. (b) The (absolutely) largest velocities as derived from UV resonance lines, in their relation to the stellar mass-loss rate —𝔐. (From Hutchings, 1978).

step was the discovery by Morton (1967a) of large wind velocities ($\sim$2000 km s^{-1}) in ultraviolet rocket spectra of δ Ori (O9.5 III), ε Ori (B0 Ia) and ζ Ori (O9.5 Ib); see also Morton *et al.* (1972). Again, this was a *direct* observation of wind-velocities greatly exceeding v_{esc}. Hutchings (1968a) analysed the spectra of two Of and one B1 Ia star and found displaced lines of P Cygni-type profiles in all three stars. The observations allowed him to deduce the velocity field $v(r)$ in the outer parts of the stellar atmospheres – the inner regions of the stellar wind. In all three cases $v(r)$ rises from fairly small values at $r=R_*$ (R_* being the star's radius, r being measured from the stellar centre) to much larger nearly-asymptotic values, at larger distances, e.g.:

$$\text{HD 152408, O8f, } M_v = -7; v \approx 30 \text{ km s}^{-1} \quad \text{at} \quad r = R_* \text{ to}$$
$$\sim 600 \text{ km s}^{-1} \quad \text{at} \quad r = 6R_*,$$
$$\text{HD 151804, O8f, } M_v = -7; v \approx 30 \text{ km s}^{-1} \quad \text{at} \quad r = R_* \text{ to}$$
$$\sim 200 \text{ km s}^{-1} \quad \text{at} \quad r = 3R_*,$$
$$\text{HD 152236, B1 Ia}^+, M_v = -8; v \approx 10 \text{ km s}^{-1} \quad \text{at} \quad r = R_* \text{ to}$$
$$\sim 180 \text{ km s}^{-1} \quad \text{at} \quad r = 1.5R_*.$$

In the far-ultraviolet observations the large mass-loss was evident from the conspicuous P Cygni profiles of lines from twice, three and four times ionised atoms of C, N, O, Si, and other ions. A list of P Cygni profiles in the UV spectra of eight other stars was published by Hutchings (1976c), and since they have been observed with *I.U.E.* in spectra of many other stars. The short-wavelength edge of the P Cygni profiles is often identified with the 'terminal velocity' v_∞ (Snow and Morton, 1976; cf. Table XIII, Chapter 3), and assumption that is correct when v_∞ is reached in still fairly dense expanding layers: for the Sun, for example, this velocity (-300 km s^{-1}) is not observable in the spectrum because the solar wind is optically transparent. Lamers *et al.* (1976) found $v_\infty \approx 2.8 v_{esc}$ from a study of 8 bright stars; this relation, which can be predicted theoretically, was confirmed by Abbott (1978) from a larger material, 40 stars, who found $v_\infty \approx 3 v_{esc}$.

Another property of mass-loss in fairly early-type stars is that *appreciable mass-loss* is restricted to the *stars with the largest luminosities*. Thus, in O- and B-type stars, clear P Cygni profiles are shown by the ultraviolet resonance-lines of C IV and Si IV of *all* stars with absolute bolometric magnitudes brighter than -8, which indicates that significant outflow of matter occurs in all supergiants, or at least in those sufficiently hot to have intense far-ultraviolet spectra (Henize *et al.*, 1976). A similar result was obtained earlier by Rosendhal (1973b) who investigated the visual spectral region of 62 supergiants of spectral types between O7 and A8. Particular attention was paid to the Hα line profile, which showed P Cygni characteristics in many spectra, but radial velocities were also measured for a few other lines (C II, Si II, Fe II) in order to see whether any velocity gradient exists. One of the main results was that conspicuous emission in Hα occurs only for the brightest stars, e.g. for $M_v \lesssim -6$ ($M_b \lesssim -8$) for B0–B1 stars, and for $M_v \lesssim -7$ ($M_b \lesssim -7.5$) for B8–A3 stars; see Figure 103. The transition magnitude cannot be determined very sharply and has naturally no real physical significance; it rather reflects the (low) dispersion of the instruments used

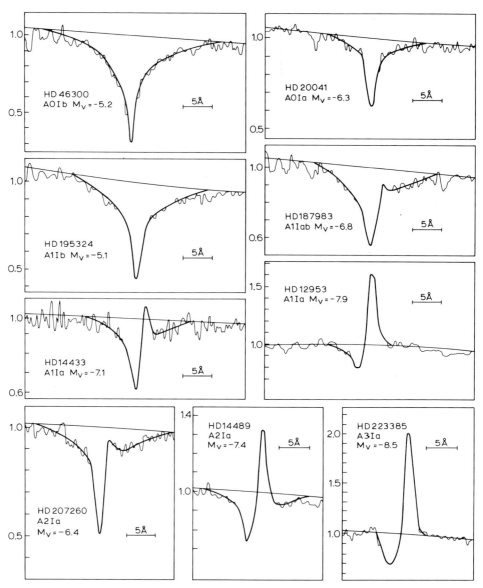

Fig. 103.   Representative Hα profiles in A-type supergiants. The profiles show that conspicious displaced components occur only in stars brighter than about $M_v \approx -7$. (After Rosendhal, 1973b).

in these surveys. At $M_b \approx -6$ (and -7 for medium-T stars) another division line can be drawn: below that luminosity for most stars no appreciable mass-flow has been detected. This limit is confirmed from the large survey of mass-loss data obtained with the *Copernicus* satellite by Snow and Morton (1976).

   A next general result is that the *terminal velocities decrease* with *increasingly later spectral type*. This does not mean that the $-\dot{\mathfrak{M}}$-values depend in the same way on the spectral type: there is no clear dependence, and for some late-type stars very large rates of mass-loss have been determined. Comprehensive summaries of terminal expansion-velocities in early-type stars, as measured from the ultraviolet resonance lines of C III, IV; N III, IV, V; O VI; Mg II, Si III, IV; P V; S IV, VI; Fe IV are given by Hack (1976) and by Snow and Morton (1976). For Wolf–Rayet stars, Of stars, and O-type supergiants the average expansion-velocity is -1400 km s^{-1}, for B-type supergiants it is -250 km s^{-1}, for the A-type supergiant a Cyg (A2 Ia) it is -80 km s^{-1}. A sharp decrease occurs near spectral type B0: for ε Ori (B0 Ia) it is still -800 km s^{-1}. This apparent dependence on spectral type may reflect the M_b-dependence: the later-type stars have smaller luminosities. A transition in the expansion velocities apparently occurs near $T_e \approx 13000$ K.

This luminosity- and T_e-dependence is reflected in the atmospheric velocity gradient as shown by a comparison of relative velocities of lines (Rosendhal, 1973b), Figure 104.

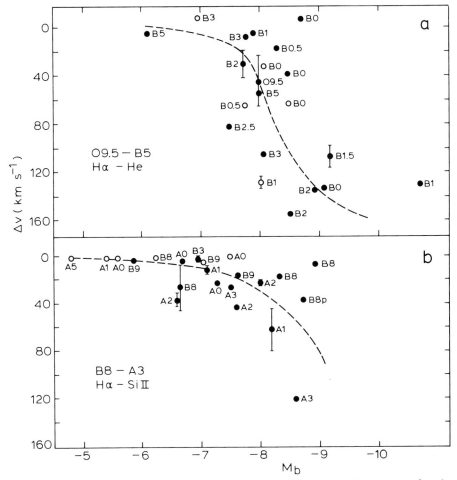

Fig. 104. (a) The velocity difference (Hα absorption – He I) in O9.5–B5 stars as a function of absolute bolometric magnitude. (b) The velocity difference (Hα absorption – Si II) in B8–A3 stars as a function of absolute bolometric magnitude. (After Rosendhal, 1973b).

In F and G supergiants Reimers (1979) finds velocities of $\sim$20 to 50 km s^{-1} only. This trend continues toward cooler stars:

In late-type supergiants 1.5(e.g. α Ori, M 1.5 Iab) and Mira stars (χ Ori, G7e, and R Leo, M7e) the Na I and K I resonance lines show P Cygni profiles at maximum and post-maximum phases (Sanner, 1976a, b). This shows that the stars are surrounded by an envelope, which must be quite extended – because the emission component is well-developed. The expansion velocity is rather small, about 3 to 7 km s^{-1}. Such velocities are usually observed for late-type supergiants. We mentioned already that Adams and McCormac (1935) found velocities of $\sim$5 km s^{-1} in some M-type stars. For the maser sources, as a rule late Mira variables and red giants (Section 6.14), the value of v_∞ may be put equal to half the velocity-difference between the two maser components; one thus finds values of a few to a 20 km s^{-1}.

An interesting and rapidly developing field of research deals with the *variability of stellar winds*; this aspect is dealt with in Section 8.5.

7.3. The Expected Shape of Spectral Lines in an Expanding Atmosphere; Some Well-Studied Cases

The first contributions to this field were made by Chandrasekhar (1934), Kosirev (1934), and Sobolev (1947a, b) when it was realized that outstreaming motions, often with a considerable velocity gradient, are characteristic for certain types of bright stars such as P Cygni, Wolf–Rayet stars, novae and supernovae. A review of the state of the art around 1969 was given by Rybicki (1970), who added a list of the most important references published until then; Kalkofen (1970) also gave a useful summary. From among the investigations now published we mention those of Castor (1970), Lucy (1971), Noerdlinger and Rybicki (1974), Kuan and Kuhi (1975), Mihalas *et al.* (1975b, 1976). Most of this is summarized in the important review paper by Hummer (1976), and in Mihalas's (1978) book.

While the basic aspects of the problem were already formulated by Chandrasekhar (1934) and Kosirev (1934), the analytical solution, even for an assumed fairly simple $v(\tau)$-function, poses a considerable problem and rather unrealistic approximations have often to be introduced if one is to cope with the simultaneous solution of the equation of radiative transfer and that of statistical equilibrium between the atomic levels involved. To handle a more appropriate velocity field (e.g. a solution of the stellar wind equations) requires the use of numerical techniques. When, in addition, reasonably sound assumptions are introduced for the radiative transfer and statistical equilibrium of the various atomic levels, the problem becomes of such a high level of complication that it can be handled only by the largest existing computers. Yet additional complications arise when realistic assumptions about the nature of re-distribution in the line profile are to be treated.

In this section we first give the basic formalism, and thereafter we describe some aspects of the solution for the two extreme cases that are normally considered, those of *slow* and *fast* flows (i.e. with respect to the mean thermal speed).

The velocity gradient changes the frequency of the radiation as seen from a local frame of reference moving with the matter. Assume complete redistribution, hence a frequency-independent source function; then in the non-relativistic case the profile of the absorption coefficient, which would be $\varphi(\mathbf{r}, \nu)$ at rest, transforms into $\varphi(\mathbf{r}, \mathbf{l}, \nu)$ $=\varphi(\mathbf{r}, \nu - \nu_0 \mathbf{l} \cdot \mathbf{v}(\mathbf{r})/c)$, in which ν_0 is the frequency at the line centre, and $\mathbf{l}$ the vector specifying the direction of photon propagation. We normalize φ such that $\int \varphi(\mathbf{r}, \nu) \, d\nu = 1$.

The level populations of the specific atom considered, n_u and n_l (where subscripts u and l stand for 'upper' and 'lower') are defined by the equations of statistical equilibrium which state that for any level the number of ingoing transitions equals the number of outgoing ones. The number of radiative transitions between two such levels are then

$$R_{ul} = A_{ul} + B_{ul}\bar{J}_{ul},$$
$$R_{lu} = B_{lu}\bar{J}_{ul},$$

where A and B are the Einstein probability coefficients, and

$$J_{ul} = \frac{1}{4\pi} \int_0^\infty d\nu \, d\mathbf{l} \, \varphi[\mathbf{r}, \nu - \nu_{ul} \mathbf{l} \cdot \mathbf{v}(\mathbf{r})/c] \, I_\nu(\mathbf{r}, \mathbf{l}), \tag{7.3; 1}$$

where $I_\nu(\mathbf{r}, \mathbf{l})$ is the specific intensity of radiation with frequency ν at the point $\mathbf{r}$ moving along the direction $\mathbf{l}$.

Then, the equation of transfer for radiation in the transition $u \rightarrow l$ is:

$$\mathbf{l} \cdot \frac{\partial}{\partial \mathbf{r}} I_\nu(\mathbf{r}, \mathbf{l}) = k(\mathbf{r}) \, \varphi[\mathbf{r}, \nu - \nu_{ul} \mathbf{l} \cdot \mathbf{v}(\mathbf{r})/c] \, [S(\mathbf{r}) - I_\nu(\mathbf{r}, \mathbf{l})] +$$
$$+ k_c(\mathbf{r}) \, [B_c(\mathbf{r}) - I_\nu(\mathbf{r}, \mathbf{l})], \tag{7.3; 2}$$

where the *line source function* is

$$S(\mathbf{r}) = \frac{2h\nu_{ul}^3}{c^2} \left(\frac{g_u N_l(\mathbf{r})}{g_l N_u(\mathbf{r})} - 1 \right)^{-1}. \tag{7.3; 3}$$

Here ν_{ul} is the frequency of the line at rest, g_i the statistical weight of the level i, k the line opacity

$$k = h\nu_{ul} (N_l B_{lu} - N_u B_{ul})/4\pi,$$

k_c the continuous opacity and B_c the continuum black body-radiation.

The above expressions can be written somewhat more simply by introducing the following quantities

$$x = (\nu - \nu_{ul})/\Delta_{ul}, \tag{7.3; 4}$$

with

$$\Delta_{ul} = \nu_{ul} v_{th}/c, \quad \text{and} \quad u \equiv v/v_{th},$$

where v_{th} is the atmospheric thermal velocity.

Then, Equations (7.3; 1) and (7.3; 2) become

$$\bar{J}_{ul} = \frac{1}{4\pi} \int_0^\infty dx \int dl \; \varphi[\mathbf{r}, x - \mathbf{l}\cdot\mathbf{u}(\mathbf{r})] \, I_x(\mathbf{r}, \mathbf{l}) \qquad (7.3; 5)$$

and

$$\mathbf{l}\cdot\frac{\partial}{\partial r} I_x(\mathbf{r}, \mathbf{l}) = k(\mathbf{r}) \, \varphi[\mathbf{r}, x - \mathbf{l}\cdot\mathbf{u}(\mathbf{r})] \, (S - I_x) + k_c(\mathbf{r}) \, (B_c - I_x). \quad (7.3; 6)$$

In Chandrasekhar's approach one now writes Equation (7.3; 6) as:

$$\frac{\partial I_x(\xi)}{\partial \xi} = \kappa_x(\xi) \, [S_x - I_x], \qquad (7.3; 7)$$

where ξ measures the path-length along a trajectory from a given point.
Further

$$\kappa_x = k(\xi) \, \varphi(\xi, x) + k_c(\xi) \qquad (7.3; 8)$$

is the total opacity, and

$$S_x(\xi) = \frac{k(\xi) \, \varphi(\xi, x) \, S(\xi) + k_c(\xi) \, B_c(\xi)}{k(\xi) \, \varphi(\xi, x) + k_c(\xi)} \qquad (7.3; 9)$$

is the total source function.
Equation (7.3; 7) has the *formal solution*:

$$I_x(\xi) = I_x(0) \exp\left[- \int_0^\xi \kappa_x(\xi') \, d\xi' \right] +$$

$$+ \int_0^\xi \kappa_x(\xi') \, \varphi_x(\xi') \, S_x(\xi') \exp\left[- \int_{\xi'}^\xi \kappa_x(\xi'') \, d\xi'' \right] d\xi'. \qquad (7.3; 10)$$

For any given set of function values for $\kappa(x)$ and $S(x)$ the emergent intensity along a ray can be computed. Approximations of various kinds are often introduced to deal with the problem in certain specific cases. Numerical methods are described by Hummer (1976).

For *slow flows* Kalkofen (1970) made a set of computations for a semi-infinite atmosphere with the velocity law $v(\tau)=10(1+\tau/T)^{-1}$, where the parameter T has been set at values 10, 10^2, 10^3, 10^4 and where τ is the line optical depth at the specific wavelength. A constant temperature was assumed for $\tau > 10^3$; above that an increase was assumed as shown in Figure 105. The resulting source functions appear to depend only very slightly on the flow, which apparently is a result of the fact that the velocity-gradient is taken to be small so that the shortward shift of line opacity does not reduce the core opacity sufficiently to allow it to carry appreciable flux. Mostly, the reduction of radiation in one wing does not prevent the radiation from escaping in the other wing.

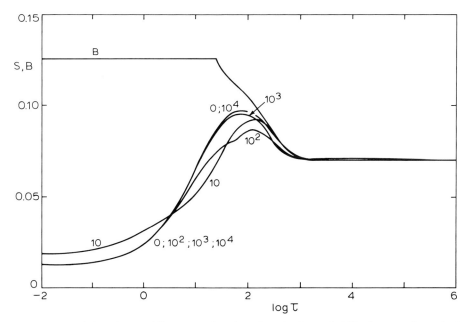

Fig. 105. Planck functions and line source functions for ε (=the probability for true absorption)= 10^{-2}, and $v(\tau)=10/(1+\tau/T)$ with $T=10$, 10^2, 10^3, and 10^4. The static case is labeled by 0. (From Kalkofen, 1970).

However, the effect on the emergent flux is considerable, as shown by Figure 106. Although in the present case we are dealing with planar atmospheres, imposing a temperature rise outward produces P Cygni-like profiles when the maximum of the source function occurs in a region of the atmosphere for which the flow velocity is sufficiently small.

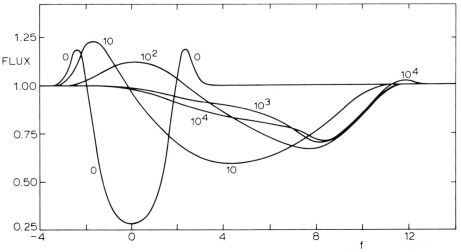

Fig. 106. The emergent monochromatic flux corresponding to the source functions of Figure 105. The frequency f is measured from the line center in thermal doppler units. (From Kalkofen, 1970).

Fast flows ($v \gg v_{\text{th}}$), including large velocity gradients, facilitate the solution of the problem. Assume for a moment that all photons are actually emitted in the line centre; then all photons seen at a specific frequency x_1 in the stationary frame are emitted from the surfaces for which

$$v(\mathbf{r})\,\mu = x_1,$$

where $\mu = \cos\theta$. These surfaces are called the 'constant-velocity surfaces', and the use of this concept – initially due to Sobolev (1947a) – has proven to be very useful. An example of the course of such lines in a given v-field is given in Figure 107. See further, for reviews: Hummer (1976) and Mihalas *et al.* (1975b, 1976). Olson (1978), using the Sobolev approximation, produced a grid of resonance-line profiles for two velocity laws, several assumed radial dependences of the ionization, and a range of ionic abundances. These were used by him for the analysis of the UV spectrum of ζ Pup. An atlas of theoretical P Cygni profiles was computed by Castor and Lamers (1979) for different velocity laws, again using the Sobolev approximation. Similar but less

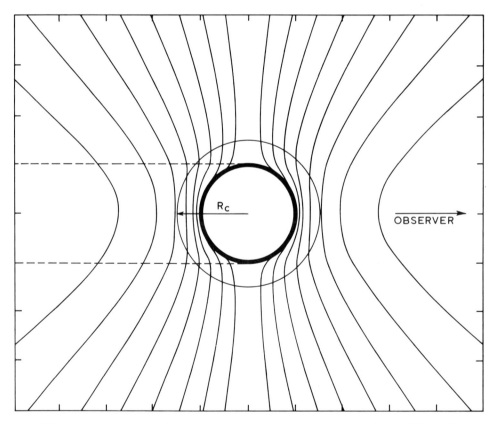

Fig. 107. Contours of equal line-of-sigh-velocity component for a model of ζ Ori, with the velocity law of Equation (7.3; 12), with $v_1 = 10v_0$. The successive contours are for $v_z/v_\infty = 0.1\,(0.1)\,0.8$ where $v_z = v(r)\cos\theta$. R_c is the radius of the outer boundary of the coronal zone. The thick line gives the position of the stellar photosphere. (From Cassinelli *et al.*, 1978).

extended calculations were published by Surdej (1979). Such atlases are obviously important instruments for the diagnosis of P Cygni-type line profiles, and even when their comparison with observations does not give a definite answer, the diagnosis is helpful in restricting the possible range of values for the most important parameters whose values are sought.

Empirical determination of the velocity law v(r). It would be helpful when an empirical method could be found to establish the velocity-distance law $v(r)$ in a stellar wind. Unambiguously this has not yet been done, but first steps have been made. The radial optical thickness τ_r of a point in a spectral line, in an expanding stellar atmosphere, that can be approximated in Sobolev's way, is:

$$\tau_r(r) = \frac{\pi e^2}{mc} \lambda_0 f n_i (\mathrm{d}v/\mathrm{d}r)^{-1}, \tag{7.3; 11}$$

where λ_0 is the rest-wavelength of the transition, n_i (cm^{-3}) the number density of the absorbing ions at distance r, and $\mathrm{d}v/\mathrm{d}r$ (s^{-1}) the gradient of the velocity law. Study of profiles of a number of resonance lines of various ions, guided by an assumed model, should allow one to separate the effects of $\tau(v)$ and $v(r)$ on the line profiles, because the ratio of the amount of absorption to the amount of emission depends mainly on $v(r)$, while the shape of the absorption part of the profile, corrected for the contribution of emission, depends mainly on $\tau(\mathbf{r})$. By comparing the derived values of $\tau(\mathbf{r})$ for different ions at the same value of the velocity v, the ratios between the values of n_i can be derived for the ions observed. Thus the two unknown relations are found. Lamers and Castor (1979) described how $v(r)$ could be derived, in principle, from accurately observed P Cygni profiles.

As examples of some investigations we describe below the analyses of P Cygni profiles and resulting rates of mass-loss for two early-type stars. These stars are among the best investigated and can serve as *standard stars* for mass-loss determinations.

The star ζ Ori A (O9.5 Ib) has an interesting Hα profile with P Cygni characteristics. Figure 108 gives the accurate photo-electric line profile as observed by Ebbets (1980) and for comparison earlier photographic observations by Conti and Leep (1974). The new observations apparently constitute an important observational improvement. The figure gives also a theoretical profile, calculated by Auer and Mihalas (1972) for a photosphere with $T_e = 31\,000$ K and $\log g = 3.2$. Hearn (1975a) considered the algebraic sum between the areas of the observed (Conti and Leep) and the predicted profile, and assumed that this resulting area can be identified with that of the P Cygni profile caused by the stellar wind. He thus derived, after an elaborate analyses including a discussion of the H-line populations, a rate of mass-loss of $1.8 \times 10^{-6}\,\mathfrak{M}_\odot\,\mathrm{yr}^{-1}$. Use of Ebbet's new data would obviously have changed the result, but – we may expect – not to a large amount. This value compares well with earlier results which range between 1 and 2×10^{-6} (Morton, 1967b; Hutchings, 1970a).

Cassinelli *et al.* (1978b) while using the same observational material, made one further step, by not restricting attention to the equivalent widths of the line but by calculating the detailed profile, assuming a hot and fairly thin 'coronal' zone close

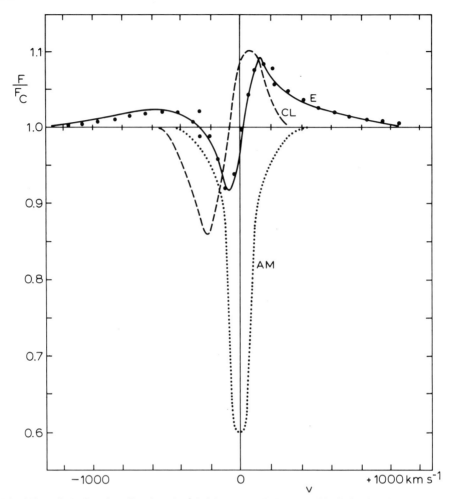

Fig. 108.   The P Cygni profile of Hα for ζ Ori A measured photographically by Conti and Leep (1974; broken line) and photo-electrically by Ebbets (1980; filled circles and full-drawn line). In addition a theoretical photospheric profile is given (dotted line) calculated by Auer and Mihalas (1972). The conspicuous P Cygni-type profile indicates mass-loss.

to the stellar surface. The rationale behind this latter assumption is the following: the absence of a very intense emission component at low velocities demands that low-velocity material in the stellar wind does not contribute to the profile, which is satisfied by postulating a low ($\sim 1.1 R_*$) and hot ($\sim 3 \times 10^6$ K) corona, related to a plateau in the velocity distribution, followed at about $2.5 R_*$ by a (second) rapid acceleration of the stellar wind to the terminal speed. The computation of line profiles in this rapidly expanding atmosphere is simplified because Sobolev's approximation can be used. Figure 107 shows the iso-velocity surfaces for the initially assumed $v(r)$-law:

$$v(r) = v_0^2 + v_1^2 \left(1 - R_*/r\right). \tag{7.3; 12}$$

The remainder of the discussion concerned the computation of the level populations, the choice of a temperature and velocity model, and of its influence on the Hα profile. An important other aspect is the discussion of the ionization in the stellar wind, based on the observed equivalent widths of the strongest UV resonance lines. It is shown that these ions are caused by Auger ionization by X-ray photons from the corona (Olson, 1978, 1979). This yields for the main region of the wind (where the UV resonance lines are formed, up to the region where v_∞ is reached): $T_e \approx 35\,000$ K.

From a further analysis of the (Conti-Leep) Hα profile, a rate of mass-loss was found of $-\mathfrak{M} = 2.9 \times 10^{-6}\,\mathfrak{M}_\odot$ yr^{-1}. The rate of mass-loss has also been determined by Barlow and Cohen, using infrared photometric data and by Klein and Castor (1978) who used again the Hα emission-line profile. These data were summarized and adjusted (cf. Section 7.4) by Conti and Garmany (1980) who concluded to a rate of mass-loss of $1.8 \times 10^{-6}\,\mathfrak{M}_\odot$ yr^{-1}. This value does not disagree with the one found by Abbott *et al.* (1980) from measurements of the 5 GHz radioflux ($2.3 \times 10^{-6}\,\mathfrak{M}_\odot$ yr^{-1}).

Summarizing, we find a rate of mass-loss for ζ Ori of $2 \times 10^{-6}\,\mathfrak{M}_\odot$ yr^{-1}, with a 20% uncertainty, apparently an accurate result.

The expanding atmosphere of ζ Pup (O4ef) is one of the best studied stellar winds, and offers a fine example of a stellar rate of mass-loss determined by different means. The best way to determine the mass-loss rate is by using the *microwave-radiation* since this involves the smallest number of hypotheses (cf. Section 6.13). Thus Morton and Wright (1978) using observations at 14.7 MHz determined a mass-loss of $(6.3 \pm 1.5) \times 10^{-6}\,\mathfrak{M}_\odot$ yr^{-1}. This value was corrected later (1979) to $(7.2 \pm 1.1) \times 10^{-6}\,\mathfrak{M}_\odot$ yr^{-1}. Abbott *et al.* (1980) from observations at 5 GHz found $(3.5 \pm 1.8) \times 10^{-6}\,\mathfrak{M}_\odot$ yr^{-1}.

The ultraviolet resonance lines were discussed by Lamers and Morton (1976), Lamers (1976), and Olson (1978) using ultraviolet spectra from the *Copernicus* satellite. The study of the shapes of the profiles gives information on the outward streaming velocity, and its variation with distance to the star. Figure 109 shows, as typical cases, the profiles of the N v lines at 1238.82 and 1242.80 Å, and those of the O vi lines at 1031.94 and 1037.62 Å, in ζ Pup and – for comparison – in the B0 V star τ Sco. Lamers has compared these profiles with theoretical line profiles, using Lucy's (1971) theory for the formation of absorption lines in an expanding atmosphere. The profiles depend on the velocity law $v(r)$ with $v_\infty = 2660$ km s^{-1} (see Table XIII) and on the optical depth parameter $\tau(v)$ according to Equation (7.3; 11). Values of $v(r)$ as well as the various n_i-values are derived in the way as described in the 2nd paragraph of page 317. These n_i-values can then be compared with the ionisation degree as calculated for different assumed values of the temperature, density and radiation fields, and assuming a collisional ionization balance. For ζ Pup a mass-loss rate of $(7.2 \pm 3) \times 10^{-6}\,\mathfrak{M}_\odot$ yr^{-1} was thus found. The amount of momentum required for accelerating the atmosphere cannot, so it seems, be provided by the radiation pressure on the resonance lines in the near and middle UV; rather it appears to come from the resonance lines in the extreme UV (228–912 Å) (Castor *et al.*, 1976).

A subsequent analysis of *the Hα profile* of ζ Pup by Cassinelli *et al.* (1978b), along the same lines as described above for ζ Ori, but assuming a photospheric model with

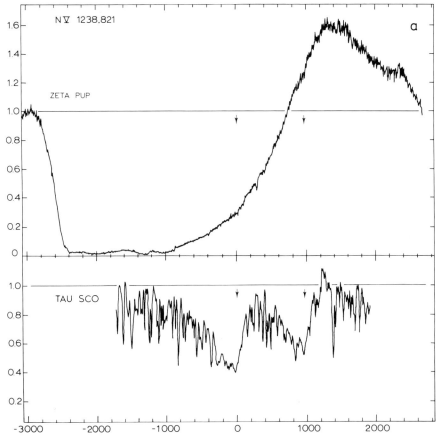

Fig. 109a.   The profiles of the N v $\lambda\lambda$1238.821, 1242.796 lines in the spectra of ζ Pup and τ Sco. The horizontal axis gives the velocity (km s^{-1}) in the frame of the star. The arrows indicate the laboratory wavelength of the lines.

T_e=50 000 K, log g=4 (Auer and Mihalas, 1972), T_{e1}=30 000 K, and the velocity distribution of Lamers and Morton (1976) yielded a mass-loss of $(8–11)\times10^{-6}\,\mathfrak{M}_{\odot}$ yr^{-1}.

Olson (1978), like Lamers, used *Copernicus* observations of *UV resonance lines*, which were analyzed with his grid of theoretical resonance-line profiles. It appears that the observed profiles are not incompatible with a model consisting of a cool wind and a thin but hot ($\approx5\times10^6$ K) corona at the base of the flow. This corona produces enough hard X-ray photons to ionize the cooler outer layers to produce O^{5+} ions, thus explaining the O vi lines observed in the UV spectrum. The predicted X-ray emission is also not incompatible with the upper limit observed by *ANS* (Mewe *et al.*, 1976), and agrees also with *Einstein* observations. The model predicts a mass-loss of $8\times10^{-6}\,\mathfrak{M}_{\odot}$ yr^{-1}.

Summarizing, it seems that ζ Pup is surrounded by a thin, fairly hot corona, at the basis of a dense stellar wind. The results of the various determinations of the rate

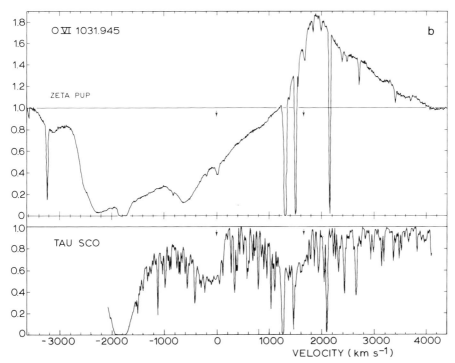

Fig. 109b.   The profiles of the O VI $\lambda\lambda$1031.945, 1037.619 lines in the spectra of ζ Pup and τ Sco. The arrows indicate the laboratory wavelengths of the lines. The strong line at -1900 km s^{-1} is Lβ. The sharp lines in the spectrum of ζ Pup are interstellar lines. *Copernicus* observations. (After Lamers, 1976).

of mass-loss cluster around the average value $-\dot{\mathfrak{M}}=(6\pm1.5)\times10^{-6}\,\mathfrak{M}_\odot$ yr^{-1}, a fairly well determined average. Part of the scatter in the data may be physically real: Conti and Niemelä (1976) and Snow (1977) found long-term variations in the rate of mass-loss of ζ Pup; the former estimate that $|\dot{\mathfrak{M}}|$ has decreased by 25% in a three-years interval. Therefore the above given value for $-\dot{\mathfrak{M}}$ should never be considered as a canonical one!

7.4. Summary of Observationally-Determined Values for Mass-Loss

For individual stars or groups of stars we summarize the $\dot{\mathfrak{M}}$-values, often semi-quantitative, that are found in literature:

η *Carinae*: We first mention η Car, a truly outstanding case, being one of the most massive stellar objects known in our Galaxy, and described in some detail in Section 6.17. There it was found that η Car loses mass at a rate of $(2–8)\times10^{-2}\,\mathfrak{M}_\odot$ yr^{-1}, a remarkably *high* value which needs further inspection.

P Cygni: For P Cygni De Groot (1969, 1973) reports a mass-loss rate of 10^{-5} to $10^{-4}\,\mathfrak{M}_\odot$ yr^{-1}, while Hutchings (1969, 1970a) found $-\dot{\mathfrak{M}}\approx5\times10^{-4}\,\mathfrak{M}_\odot$ yr^{-1}. The smaller result was confirmed by a discussion of the radio *and* infrared spectra of P Cyg

(Wright and Barlow, 1975): $-\mathfrak{M} \gtrsim 1.2 \times 10^{-5} \, \mathfrak{M}_\odot \, yr^{-1}$. Later, Barlow and Cohen (1977) and Schwarz and Spencer (1977) on the basis of improved but similar data found 1.5 and $2 \times 10^{-5} \, \mathfrak{M}_\odot \, yr^{-1}$, while, based on the study of the whole continuous spectrum, Nugis et al. (1979) found $8 \times 10^{-5} \, \mathfrak{M}_\odot \, yr^{-1}$. Abbott et al. (1980), from 5 GHz microwave data, found $-\mathfrak{M} = 2 \times 10^{-5} \, \mathfrak{M}_\odot \, yr^{-1}$. Kunasz and Van Blerkom (1978), and Van Blerkom (1978a), who studied Balmer lines profiles, found (1.5 to 3)$\times 10^{-5} \, \mathfrak{M}_\odot \, yr^{-1}$. We conclude that for P Cygni: $-\mathfrak{M} \approx 2 \times 10^{-5} \, \mathfrak{M}_\odot \, yr^{-1}$ seems the best value.

Wolf–Rayet stars: For the WC8 component of γ^2 Vel Seaquist (1976) derived $-\mathfrak{M} = 1.2 \times 10^{-5}$ and $3.1 \times 10^{-5} \, \mathfrak{M}_\odot \, yr^{-1}$ from the radio and infrared free-free emission respectively. Morton and Wright (1978) found on the basis of their own microwave measurements: $-\mathfrak{M} = (3.9 \pm 0.7) \times 10^{-5} \, \mathfrak{M}_\odot \, yr^{-1}$. Noerdlinger (1979) constructed a number of models of the stellar wind, the results of which were compared with the observed C III] λ1909 P Cygni profile and with radio data. This yielded $-\mathfrak{M} = (1.1$ to $1.4) \times 10^{-4} \, \mathfrak{M}_\odot \, yr^{-1}$. Willis et al. (1979a), also using the C III] line, observed with IUE found 1×10^{-4}. So there is a substantial difference between the micro-wave- and the intercombination-line-data. The 'best' value apparently must be somewhere between 3 and $10 \times 10^{-5} \, \mathfrak{M}_\odot \, yr^{-1}$.

Some fifteen WR stars with spectral types between WN4 and WN8, WC5 and WC8 were investigated by Willis et al. (1980) in a semi-statistical way, using 10 μm infrared fluxes (Barlow, 1979) and measured or estimated v_∞-values. The Equation (6.13; 2) was then used to estimate the rate of mass-loss. Values found range between $\sim 6 \times 10^{-5}$ (for the earliest type stars) to $\sim 4 \times 10^{-4} \, \mathfrak{M}_\odot \, yr^{-1}$ for the later types. The Wolf–Rayet star HD 192163 (WN6; $M_v = -4.8$) is associated with a filamentary nebula (NGC 6888) visible optically (notably in Hα) and in radio-waves (Wendker et al., 1975). The nebula is situated asymmetrically with respect to the star and has dimensions $18' \times 12'$ (7.6$\times$5.0 pc). The mass-loss was determined by Hackwell et al. (1974) at $-\mathfrak{M} = 2 \times 10^{-5} \, \mathfrak{M}_\odot \, yr^{-1}$.

We conclude that Wolf–Rayet stars lose mass by stellar winds at a rate of approx. a few times 10^{-5} to a few times $10^{-4} \, \mathfrak{M}_\odot \, yr^{-1}$. Late spectral types loose more than earlier ones. Since the duration of the Wolf–Rayet phase is about 10^4 to 10^5 yr (Kippenhahn, 1969) the total loss in that period may have some – but not outstanding – evolutionary significance.

An important aspect of the mass-loss of Wolf–Rayet stars is that at least one of them shows strong variability, which sometimes takes dramatic values. Wood (1977) summarized radio-observations of HD 193793 (WC7+O5): these decreased slowly in intensity since 1970 and by a factor 5 between 1975 and March 1977, but almost returned to the 1970 value only two months later.

O- and bright B-type stars; comparison with B-type main-sequence stars, Be-stars and binaries: Historically, Morton (1967b) found for early B-type supergiants a rate of mass-loss of about $10^{-6} \, \mathfrak{M}_\odot \, yr^{-1}$, from the profiles of ultraviolet resonance lines. A first extensive study of the O- and B-type stars was made by Hutchings (1970a) who found that mass-loss is a common feature for the brightest hot stars; indications were found that the rate of mass-loss increases with L and T_e. Since that time more determinations of the mass-loss rate of early-type stars were made. A symposium devoted to mass-loss of O-stars describes the state-of-the-art in 1978 (Conti and De Loore, 1979).

The results of the various investigations differ, according to the observational material used. For the O- and B-type stars this material consists of visual emission-lines particularly Hα, the strong ultraviolet resonance-lines, and of infrared and microwave emission data. From the specific investigations we mention that of Conti (1976) who compared line profiles in the spectra of *O3f and O4f type stars* with emission-line profiles calculated by Klein and Castor (1978) and concluded there is a mass-loss of $\sim 10^{-5} \, \mathfrak{M}_\odot \, \mathrm{yr}^{-1}$. This was later refined and confirmed in a more detailed study by Conti and Frost (1977) and Conti *et al.* (1977b). Abbott *et al.* (1980) used microwave-emission data at 5 GHz to determine $-\dot{\mathfrak{M}}$-values for six stars.

Barlow and Cohen (1977) derived the mass-loss from broadband infrared photometry of 34 OBA and 10 Of and Oe stars at wavelengths of 2.2 and 3.6 μm. The infrared excess normally observed for these stars is assumed to arise in an expanding wind emitted by the star, as in the case of P Cyg (see Section 3.5 and Figure 33). By assuming for all stars investigated the same shape for the velocity-distance relation in the stellar wind, namely that of P Cyg, rates of mass-loss of the stars (apart from the Oe stars for which a different situation may apply) can be derived. This approach suffers from the apparent weakness that – as we now know – the $v(r)$-relation for P Cyg seems to *differ* from that of all other early-type stars! Lamers *et al.* (1980) rectified this situation by a first-order correction: namely, by adjusting all determinations of $\dot{\mathfrak{M}}$ of Barlow and Cohen such that their value for ζ Pup agrees with the value $(6.3 \times 10^{-6} \, \mathfrak{M}_\odot \, \mathrm{yr}^{-1})$ found by Morton and Wright (1978) from their radio data. This value practically equals our assumed one – Section 7.3. They applied the same procedure to the Hα-determinations of Conti *et al.*, described above, and of Klein and Castor (1978). The adjustment factors are in no case larger than a factor 2. The same corrections were applied by Conti and Garmany (1980) in reviewing $\dot{\mathfrak{M}}$-values for O-type stars. Their values together with those of others are given in Table XXXIX.

Table XL gives the rates of mass-loss for bright B-type stars.

In these tables we do not include the less accurate results of investigations of a more statistical character, such as the investigation made by Hutchings (1976a, b) who studied 65 luminous galactic O- and B-type stars. The flow-rate was estimated on an arbitrary scale from an examination of the shapes of spectral lines, and the results were subsequently calibrated by using a subset of the same stars with well-determined mass-loss rates. The bolometric absolute magnitudes of the stars were found by estimating M_v from a $W(\mathrm{H}\gamma)$-M_v relation (Hutchings, 1976b), and by adding the bolometric correction. A relation between spectral type and effective temperature was assumed and thus a theoretical HR-diagram could be obtained, which shows for each individual star the mass-loss; from these results curves of equal mass-loss can be drawn in the HR diagram. Similar summarizing diagrams were later made by De Loore (1977), Cassinelli (1979), and Hutchings (1980). An important aspect of mass-

TABLE XXXIX

Rate of mass-loss for single bright O-type stars

HD(E)	Other designation	Spectral type	$-\mathfrak{M}$ (in $10^{-6}\,\mathfrak{M}_{\odot}\,yr^{-1}$)	References
93129A		O3f	21; 15	5, 8
93250		O3 V((f))	0.2	8
190429		O4f	11; 8	5, 8
66811		O4f	5	6
16691		O4f, O5 I	8; 5	5, 8
15570		O4f	18; 13	5; 8
	ζ Pup	O4ef	6	Section 7.3
168076		O4((f))	9.7	5
	9 Sgr	O4 V((f))	0.3; 25	8; 9
14947		O5f	8; 5	3, 5, 7, 8
15558		O5(f)	3	7, 8
228766		O5.5f	10	4
	λ Cep	O6ef	6	9
210839		O6f	3	6
153919		O6f	15	6
152233		O6(f)	2	7, 8
47129		O7	5	2
148937		O7f	8	6
135591		O7 III(f)	0.06	8
152408		O8 If	12	7, 8
167771		O8 I((f))	1.5	7, 8
151804		O8–9f	8; 10; 9	6, 1, 7, 8
	ζ Ori A	O9.5–9.7 Ib	2	Section 7.3
	δ Ori	O9.5 I	1; 1.7	3, 7, 8
16429		O9.5 I	1.6	7, 8
152424		O9.5 II	3	7, 8

References:

1. Hutchings (1969, 1970a).
2. Hutchings and Cowley (1976).
3. Barlow and Cohen (1977).
4. Massey and Conti (1977).
5. Conti and Frost (1977).
6. Conti *et al.* (1977b).
7. Klein and Castor (1978).
8. Conti and Garmany (1980).
9. Abbott *et al.* (1980).

loss in O-type stars was noted by Conti and Garmany (1980): rates of mass-loss can be very different for stars of the *same* bolometric magnitude: Of stars have larger mass-loss rates than O and O((f)) stars with the same M_b. At $M_b \approx -10$ this factor is ~ 60.

To sum up the above results, one finds for *Of-stars, O-, and B-supergiants and giants* average rates of mass-loss of the order of 10 to $2 \times 10^{-6}\,\mathfrak{M}_{\odot}\,yr^{-1}$, with a few exceptions. O9.5 I, II stars have $\sim 2 \times 10^{-6}\,\mathfrak{M}_{\odot}\,yr^{-1}$. It is interesting to compare these values with the much lower ones found for *main-sequence stars* such as τ Sco (BoV): $4 \times 10^{-8}\,\mathfrak{M}_{\odot}\,yr^{-1}$ (Rogerson and Lamers, 1975) and the O6.5 V star HD 54662: $8 \times 10^{-9}\,\mathfrak{M}_{\odot}\,yr^{-1}$ (Conti and Garmany, 1980). Also, *Be-type stars*, although surrounded

TABLE XL

Rate of mass-loss for bright B-type stars

HD(E)	Other designation	Spectral type	$-\mathfrak{\dot{M}}$ (in $10^{-6}\,\mathfrak{M}_{\odot}\,\mathrm{yr}^{-1}$)	References
37128	ε Ori	B0 Ia	2; 3	1, 2
38771	κ Ori	B0.5 Ia	1	1
169454		B1 Ia$^+$	3.2; 1.8	1, 4
152236	ζ^1 Sco	B1 Ia$^+$	2.1; 10	3, 4
2905	κ Cas	B1 Ia	1.4	1
269700		B1 Iae	2.3	4
91316	ϱ Leo	B1 Iab	0.9	1
–	BD $-14°\,5037$	B1.5 Ia(+?)	0.4	4
190603		B1.5 Ia$^+$	3	1
194279		B1.5 Ia	1.3	1
14143		B2 Ia	1.2	1
41117	χ^2 Ori	B2 Ia	2.2	1
14134		B3 Ia	0.7	1
53138	o^2 CMa	B3 Ia	1.7	1
198478	55 Cyg	B3 Ia	0.6	1
58350	η CMa	B5 Ia	0.4	1
36371	χ Aur	B5 Iab	1.4	1
164353	67 Oph	B5 Ib	0.1	1
15497		B6 Ia	0.8	1
34085	β Ori	B8 Ia	1.1	1
166397	μ Sgr	B8 Ia	1.1	1
21291		B9 Ia	0.4	1

References:

1. Barlow and Cohen (1977).
2. Abbott *et al.* (1980).
3. Hutchings (1969, 1976a).
4. Sterken and Wolf (1978a).

by extended 'chromospheric' discs, presumably consisting of gas that streamed out of the equatorial regions, have a surprisingly small rate of mass-loss. Values $\sim 10^{-8}$ $\mathfrak{M}_{\odot}\,\mathrm{yr}^{-1}$ are the largest reported (Hutchings, 1970a; Poeckert and Marlborough, 1978a, b and earlier), and most communicated results are smaller.

Snow and Marlborough (1976) list a dozen *Be stars* which all show evidence of outstreaming motions: asymmetries in ultraviolet resonance lines, observable especially in the Si IV doublet at 1400 Å indicate maximum outstreaming velocities of up to 1000 km s^{-1}. Yet, for one of the most 'active' Be stars, 59 Cyg (B1.5 Ve) the rate of mass-loss is estimated to be only of the order of 10^{-10} to $10^{-9}\,\mathfrak{M}_{\odot}\,\mathrm{yr}^{-1}$. For the Be star φ Per (B1 III–Vpe) Bruhweiler *et al.* (1978) found from line asymmetries in the ultraviolet spectrum a rate of mass-loss of $5 \times 10^{-11}\,\mathfrak{M}_{\odot}\,\mathrm{yr}^{-1}$. For the well-studied object Pleione (HD 23863; B8 IV–Ve – see also Section 3.6 and Figure 34) the mass-loss rate was determined from the increase of the mass of the shell. This must yield a lower limit because the shell, in turn, may lose mass: $-\mathfrak{\dot{M}} \geq (4\ \mathrm{to}\ 5) \times 10^{-11}\,\mathfrak{M}_{\odot}\,\mathrm{yr}^{-1}$ (Hirata and Kogure, 1978; Higurashi and Hirata, 1978). For γ Cas (B0.5 IVe) and X Per (O9.5 III–IVep) Hammerschlag-Hensberge *et al.* (1979b) find

$-\dot{\mathfrak{M}}=7\times10^{-9}$ and $1\times10^{-8}\,\mathfrak{M}_\odot$ yr^{-1}. Hence, the *early*-type object X Per has the strongest mass-flux of the Oe–Be–Ae group of stars. It is noteworthy that all these values are even *smaller* than the one derived for the main-sequence B-type stars, a matter that deserves further investigation.

In *binaries*, especially in close systems, mass-loss and mass-exchange may interact in a complicated way. For the *binary* UW CMa (HD 57060; O8 If+O7) McCluskey *et al.* (1975), McCluskey and Kondo (1976) determined the rate of mass-loss from ultraviolet resonance lines of doubly, triply and fourfold ionized atoms of various abundant elements. The lines show P Cygni profiles; the absorption lines yield radial velocities of -200 to -800 km s^{-1}, while the peaks of the emission components indicate velocities of $+400$ to $+800$ km s^{-1}. The emission is produced by a circumbinary shell, which does not participate in the orbital motions. The average rate of mass-loss is estimated at $3\times10^{-6}\,\mathfrak{M}_\odot$ yr^{-1}, but for this system, which has an orbital eccentricity of 0.09, the value of $-\dot{\mathfrak{M}}$ of the primary varies during the cycle between 10^{-7} and $1.5\times10^{-5}\,\mathfrak{M}_\odot$, hence over a range of factor 150 (Hutchings, 1977). For the highly eccentric systems HD 187399 and AZ Cas ($e\geq0.4$) mass-loss occurs only at periastron (Hutchings and Redman, 1973; Cowley *et al.*, 1977). Hutchings (1977) shows that the ratio $\dot{\mathfrak{M}}_{max}/\dot{\mathfrak{M}}_{min}$ decreases with decreasing e:

System	$-10^{-6}\,\dot{\mathfrak{M}}(\mathfrak{M}_\odot$ yr$^{-1})$		max/min	e
	max	min		
29 CMa (HD 57060)	15	0.1	150	0.09
HD 163189	15	1	15	0.08
HD 47129	15	2	7	0.04
HD 108	40	10	4	?

A-type supergiants: A 'classical' object is α Cyg (A2 Ia; $M_{bol}=-8.6$; $T_e=9170$ K; $\mathfrak{M}=17\mathfrak{M}_\odot$). It was first studied by Lamers (1975) who used spectra obtained with the Utrecht ultraviolet spectrometer *S59* aboard the ESRO TD1A satellite. Later studies are by Kondo *et al.* (1975b), Snow and Morton (1976), Lamers *et al.* (1978), and Praderie *et al.* (1980); these authors used higher-resolution *BUSS* and *IUE*-spectra. The resonance lines of Mg II and Fe II show violet-shifted components. This indicates that the material is accelerated to a terminal velocity of 250 km s^{-1}. From the profile of the Fe II resonance lines a mass-loss rate of $1.1\times10^{-8}\,\mathfrak{M}_\odot$ yr^{-1} is derived. From the UV resonance lines Praderie *et al.* (1980) find $-\dot{\mathfrak{M}}=(1.1$ to $7)\times10^{-8}\,\mathfrak{M}_\odot$ yr^{-1}. The 'best' value seems therefore a few times $10^{-8}\,\mathfrak{M}_\odot$ yr^{-1} with a considerable uncertainty. This uncertainty may have a physical base: Some Fe II lines in α Cyg show components at -125 and -195 km s^{-1}, which indicates that variable shells of higher density exist: mass-loss is apparently not a stationary phenomenon, but takes place in 'puffs' spaced irregularly in time, and not necessarily emitted in spherical shells.

TABLE XLI

Mass-loss of A-type supergiants

Star	HD	Spectral type	M_b	$-\dot{\mathfrak{M}}$ (in $10^{-6}\,\mathfrak{M}_\odot$ yr^{-1})	References
	21389	A0 Ia	-7.1	0.42	Barlow and Cohen (1977)
HR 1040		A0 Ia		0.01	Praderie *et al.* (1980)
η Leo	87737	A0 Ib	-5.5	0.047	Barlow and Cohen (1977)
				$<10^{-3}$	Praderie *et al.* (1980)
				0.0003	Kondo *et al.* (1976c)
	12953	A1 Ia	-8.0	0.80	Barlow and Cohen (1977)
9 Per	14489	A2 Ia	-7.5	0.52	Barlow and Cohen (1977)
α Cyg	197345	A2 Ia	-7.9	0.69	Barlow and Cohen (1977)
				0.011	Lamers *et al.* (1978)
				0.01 to 0.07	Praderie *et al.* (1980)
	17378	A5 Ia	-7.8	0.32	Lamers *et al.* (1978)
HR 2874		A5 Ib		0.003	Praderie *et al.* (1980)

In conflict to the above value for $\dot{\mathfrak{M}}$ are determinations by Barlow and Cohen (1977), based on the interpretation of the excess-radiation near 10 μm, assuming this to be due to outstreaming matter, and taking a velocity-law scaled from that of P Cyg (but it is known that this $v(r)$-law deviates strongly from that of most other stars!). For α Cyg they find a value of $69\times10^{-8}\,\mathfrak{M}_\odot$ yr^{-1}. This large difference is partly due to the erroneous $v(r)$-law but may lead further to the consideration that the envelope may be much hotter than assumed by Lamers *et al.*, so that less Fe is in the first stage of ionisation than calculated. This problem is still open. A similar problem applies to η Leo (A0 Ib) for which Kondo *et al.* (1976c) found $-\dot{\mathfrak{M}}=3\times10^{-10}\,\mathfrak{M}_\odot$ yr^{-1}; while Lamers *et al.* (1978) found no evidence for mass-loss. Barlow and Cohen found $4.7\times10^{-8}\,\mathfrak{M}_\odot$ yr^{-1}. Obviously more work is needed!

Results for A-type supergiants are collected in Table XLI. The enormous differences between values derived from spectral studies and those from infrared radiation excess are striking and demand further investigation! These differences make it impossible at present to conclude what is the most probable rate of mass-loss of A-type supergiants.

F–G–K-type supergiants: Few investigations have been made for these kinds of stars. For the hypergiant ϱ Cas (F8 Ia$^+$) Sargent (1961) found a rate of mass-loss of 2.5×10^{-6} to $2.5\times10^{-5}\,\mathfrak{M}_\odot$ yr^{-1}, and Sanner (1976a, b) found $-\dot{\mathfrak{M}}>1.2\times10^{-5}\,\mathfrak{M}_\odot$ yr^{-1}. These values are large, and certainly indicative – together with the observed variability of luminosity and spectrum, – of the extreme instability of the class of hypergiants. The difference between the two published values for $\dot{\mathfrak{M}}$ does not so much represent the uncertainty in the determination but rather real physical fluctuations, which occured over a period of less than a year.

Another hypergiant, HR 8752 (G0 Ia$^+$) has a rate of mass-loss of $10^{-5}\,\mathfrak{M}_\odot$ yr^{-1} (Lambert and Luck, 1978) on the basis of radio-data. Reimers (1979) finds $7\times10^{-6}\,\mathfrak{M}_\odot$ yr^{-1} from Hα emission.

A systematic investigation of mass-loss in F and G-type supergiants is started by

Reimers (1979) on the basis of Hα emission profiles. The G0 Ia stars R Pup and BS 4441 have $-\dot{\mathfrak{M}}(\mathrm{H}^+) \approx 4 \times 10^{-7} \, \mathfrak{M}_\odot \, \mathrm{yr}^{-1}$. The true rates may be higher by one or two orders of magnitude when H is mainly neutral. For 89 Her (F2 Ia) Sargent and Osmer (1969) found $-\dot{\mathfrak{M}} = 10^{-8} \, \mathfrak{M}_\odot \, \mathrm{yr}^{-1}$. The supergiant ζ Aur (K4 Ib) was investigated by Saito (1970, 1973) who found in two investigations mass-loss rates of 10^{-8} and 10^{-7} $\mathfrak{M}_\odot \, \mathrm{yr}^{-1}$. The circumstellar cloud should have a number-density of $2 \times 10^{11} \, \mathrm{cm}^{-3}$ at $10R_*$ and $2 \times 10^{10} \, \mathrm{cm}^{-3}$ at $100R_*$. Turbulent velocities in the cloud range between 20 and 50 km s^{-1}.

For the supergiants HD 216946 (K5 Ib), HD 17506 (K3 Ib), HD 48329 (G8 Ib) and HD 208606 (G8 Ib) Reimers (1973) derived mass-loss rates of 20, 20, 4 and $4 \times 10^{-8} \, \mathfrak{M}_\odot \, \mathrm{yr}^{-1}$, respectively.

So, while keeping in mind the lack of data one may summarize for F–G–K-type supergiants:

luminosity class: Ia–Ib: $\dot{\mathfrak{M}} = 10^{-7}$ to $10^{-8} \, \mathfrak{M}_\odot \, \mathrm{yr}^{-1}$

$$\mathrm{Ia}^+ \qquad (10 \text{ to } 2.5) \times 10^{-6}.$$

While dealing with stars of these types we should bear in mind that they occupy a region of the HR-diagram within which (for stars with $M_v \lesssim 0$) a transition line can be drawn separating stars *with* circumstellar lines from those *without*. The transition between the two can be drawn fairly sharply, as is clear from Figure 90 (Section 6.12). It is obviously interesting to pose the question whether this transition line is related to the stellar mass-loss process. Mullan (1978) has advanced arguments that it divides also the regions where the stellar wind at the coronal base changes from sub- to super-sonic (above the division line). Verification of this interesting suggestion has to await additional observational data on stellar coronas for intermediate-type stars.

Late-type giants and supergiants: This field was pioneered by Deutsch (1956, 1966) who found for the giant α^1 Her (M5 I-II) a rate of mass-loss of $3 \times 10^{-8} \, \mathfrak{M}_\odot \, \mathrm{yr}^{-1}$, and who gave as an average for M5–M8 supergiants the – too low – value $4 \times 10^{-9} \, \mathfrak{M}_\odot \, \mathrm{yr}^{-1}$. The mass-loss rate for α^1 Her was redetermined by Reimers (1977a) on the basis of observations of circumstellar absorption lines in the spectrum of the visual companion α^2 Her. He found

$$-\dot{\mathfrak{M}} = 1.1 \times 10^{-7} \, \mathfrak{M}_\odot \, \mathrm{yr}^{-1} \text{ (error } \approx \text{ factor 2)},$$

where

$$\mathfrak{M} = 1.7 \mathfrak{M}_\odot, \, M_b = -3.5 \, (L/L_\odot = 3.3 \pm 0.2).$$

Weymann (1962b) derived for the supergiant α Ori (M1.5 Iab) a much higher value, *viz.* of $4.5 \times 10^{-6} \, \mathfrak{M}_\odot \, \mathrm{yr}^{-1}$. The order of magnitude of this latter value was confirmed by Gehrz and Woolf (1971) who found for α Ori the somewhat smaller value of $0.7 \times 10^{-6} \, \mathfrak{M}_\odot \, \mathrm{yr}^{-1}$. A redetermination by Reimers (1975b) yielded $10^{-6} \, \mathfrak{M}_\odot \, \mathrm{yr}^{-1}$.

A similar object is α Sco (M1.5 Iab) for which Van der Hucht *et al.* (1980c) determined the mass-loss rate from a discussion of the UV emission-line spectrum, yielding $7 \times 10^{-6} \, \mathfrak{M}_{\odot} \, \mathrm{yr}^{-1}$. This value, significantly higher than a value given by Kudritzky and Reimers (1978; 7×10^{-7}) seems more reliable because its determination included a discussion of the ionization equilibrium in the shell. However, it may be too large when the spectral line used (Zn II) would have a strong interstellar component.

Gehrz and Woolf (1971) listed $\mathfrak{M}$-values determined from IR radiation excesses for a number of M-type stars that were observed by them and by others. For the supergiants in their list mass-loss rates of 0.7 to $27 \times 10^{-6} \, \mathfrak{M}_{\odot} \, \mathrm{yr}^{-1}$ are found. Hagen (1978) determined mass-loss rates for 9M-type giants and supergiants from gas column-densities and expansion velocities (from Hα P Cygni line profiles). Sanner (1976b) investigated twelve M-type giants; their spectral types range from K5 – M0 Iab – Ib (ψ^1 Aur) to M5 Ib – III (α^1 Her). The physical parameters of the circumstellar shells were determined from spectral scans of strong resonance-lines in the visible spectral range. Resulting mass-loss rates range from $-\dot{\mathfrak{M}} = 10^{-9}$ to $10^{-8} \, \mathfrak{M}_{\odot} \, \mathrm{yr}^{-1}$ for stars with $M_v \approx -1$, to values of 10^{-6} for $M_v \approx -7$. The data fitted to the empirical relation (α Sco and VV Cep were excluded):

$$\log(-\dot{\mathfrak{M}}) \, [\mathfrak{M}_{\odot} \, \mathrm{yr}^{-1}] \approx -10 - \tfrac{1}{2} M_b.$$

A re-analysis of the circumstellar gas shells and mass-loss rates of four M-type supergiants by Bernat (1977) yielded much higher values than the ones given above. The difference was ascribed to the conclusion that the extents the circumstellar shells exceed the values assumed in earlier work. The observed line profiles were fitted to a model consisting of a shell, detached from the photosphere, haivng a constant expansion-velocity, and hence a density decreasing as r^{-2}. Bernat's values differ very substantially from all others, and these differences demand further investigation, and prudence in dealing with the latter data.

For Mira-type variables with OH-emission (for which typically $M_b = -4$ to -5) the mass-loss was estimated by Lépine and Paes de Barros (1977). The terminal expansion-velocity (3.5 km s^{-1}) is equalized to half the separation, in velocity-units, between the OH-emission peaks, and with a gas density $N(OH) \approx 10^6 \, \mathrm{cm}^{-3}$ (Litvak, 1969) and a shell-radius of 10^{10} km, a mass-loss rate of $-\dot{\mathfrak{M}} = 2 \times 10^{-7} \, \mathfrak{M}_{\odot} \, \mathrm{yr}^{-1}$ is derived. For these stars the $-\dot{\mathfrak{M}}$-value appears to increase with increasing pulsational period (Katafos *et al.*, 1977).

A *summary* of determinations of mass-loss rates for M-type supergiants – with various degrees of accuracy! – is given in Table XLII. The large differences between Bernat's determinations and the others make one hesitant to accept the former's values. In Section 6.16 we already have called attention to the remarkable case of the very red supergiant VY CMa, and we suggested that the deviating $\mathfrak{M}$-value may indicate that this star is a contracting protostar: Remember that the rate of mass-loss for this star was estimated assuming the shell around the star to have been *expelled* by VY CMa.

Summary of mass-flux data: comprehensive representations. The foregoing sections, and especially Tables XXXIX through XLII and Figure 110 review presently

TABLE XLII

Mass-loss rates of M-type supergiants and a few other M-stars. After Deutsch (1956), Weymann (1962b), Gehrz and Woolf (1972), Reimers (1975b, 1977a, b, 1978) Sanner (1976a, b), Bernat (1977), Schwarz and Spencer (1977), and Hagen (1978). Determinations from these authors are usually designed by the initial, apart from those of Gehrtz and Woolf. A few other (individual) determinations are referenced by footnotes (a, b, c, ...).

Star	Spectral type (at maximum)	$-\dot{\mathfrak{M}}$ (unit: $10^{-6}\ \mathfrak{M}_\odot\ yr^{-1}$)
ψ Aur	K5–M0 Iab	0.5
6(BU)Gem	M1 Ia	2.1
TV Gem	M1 Iab	1.2
α Ori	M1.5 Iab	5.0 (W)
		0.7
		0.17 (S)
		34 (B)
		1 (R)
		0.03 (SS)
		0.15 (H)
α Sco	M1.5 Iab	0.1 (S)
		2.2 (B)
		0.7[a]
		7[b]
		0.13 (H)
μ Cep	M2 Ia	10
		1.0 (S)
		420 (B)
		0.49 (H)
VV Cep	M2 Iab	0.13
RW Cyg	M2 Ia–ab	5
119(CE)Tau	M2 Ib	0.24
VY CMa	M3e I	200[d]
BC Cyg	M3 Ia	7
BI Cyg	M3 Iab	7
S Per	M4 Ia	27
UY Sct	M4 Ia–ab	6
α^1 Her	M5 Ib–II	0.03 (D)
		0.9
		0.67 (B)
		0.11 (R)
		0.028 (H)
Y Lyn	M5 Ib–II	0.23
o Cet (Mira)	M5	0.2[c], 0.6[e]
RS CnC	M6 Ib–II	0.4

Additional references:

[a] Kudritzki and Reimers (1978).
[b] Van der Hucht *et al.* (1980).
[c] Yamashita and Maehara (1978).
[d] Section 6.16.
[e] Cahn and Wyatt (1978b); see also Section 3.11.

available information on mass-loss rates. The material shows a dependence on luminosity class and on spectral type. Several authors have tried to establish empirical relations between $\dot{\mathfrak{M}}$ and stellar parameters. These will be reviewed here, first

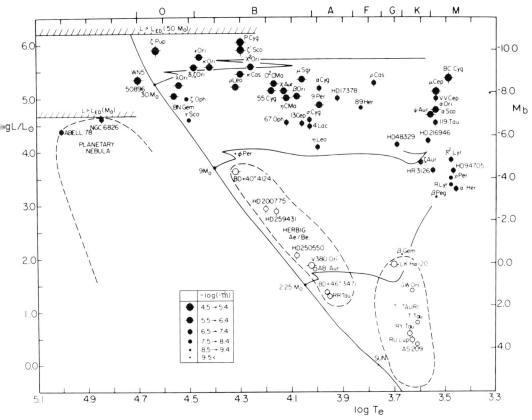

Fig. 110. Mass-loss rates and their distribution over the Hertzsprung-Russell diagram after a synthesis by Cassinelli (1979b). Locations of stars and rates of mass-loss do not always correspond with data given in this volume (see e.g. α Cyg, P Cyg), but the diagram is illustrative in showing the general trend of mass-loss rates.

for early-type stars: Empirically, Barlow and Cohen (1977) derived from their discussions of mass-loss from early-type stars:

For the O-stars (excluding HD 108):

$$- \dot{\mathfrak{M}} = 6.8 \times 10^{-13} L^{1.10 \pm 0.06} \mathfrak{M}_\odot \text{ yr}^{-1};$$

for the B- and A-supergiants:

$$- \dot{\mathfrak{M}} = 5 \times 10^{-13} L^{1.20 \pm 0.08} \mathfrak{M}_\odot \text{ yr}^{-1};$$

where L is in units of $L_\odot$.

It is understandable that the numerical values in these early determinations may change greatly when more information is available. The expressions are consistent with the assumption that the mass-loss is caused by radiation forces.

Klein and Castor (1978) have established for early-type stars an empirical relation between the stellar mass-loss rate, the star's mass $\mathfrak{M}$ (expressed in solar masses), and the amount of energy emitted in the Hα line: $L(H\alpha)$, expressed in solar units $L_\odot$, namely:

$$\log[L(H\alpha)/L_\odot] = 2\log(-\dot{\mathfrak{M}}) - \log(\mathfrak{M}/60) + C,$$

where the parameter C depends on the star's effective temperature:

$$
\begin{aligned}
C &= 11.691 \quad \text{for} \quad T_e = 30\,000 \text{ K},\\
&\quad\ 11.448 \qquad\qquad\ \ 40\,000,\\
&\quad\ 11.300 \qquad\qquad\ \ 50\,000.
\end{aligned}
$$

As an aid in computing $L(H\alpha)$ one may use, as did Conti and Frost (1977), Allen's relation (1973) viz.:

$$\log L_\lambda(V) = -0.4M_v + 31.63,$$

where $L_\lambda(V)$ is the stellar luminosity in erg s^{-1} Å^{-1} in the V-band. The total Hα emission equivalent-width is determined (in a not fully satisfactorily) way by adding the photospheric absorption equivalent-width taken from model computations to the observed emission equivalent-width (i.e. from the part of the profile above the continuum level).

The importance of mass-loss for the earliest O-type stars is obvious: with mass-loss rates of $\sim 10^{-5}\,\mathfrak{M}_\odot$ yr^{-1}, and with, for a $60\mathfrak{M}_\odot$ star a life-time of the Of phase of between 10^6 and 3×10^6 yr, such a star will lose $\frac{1}{6}$ to $\frac{1}{2}$ of its mass during that period.

For the *cool stars* Reimers (1975, 1977a, 1978) has tried to set up a unifying representation of mass-loss as a function of fundamental stellar parameters. The simplest possible quantity with dimension [mass/time] may be L/gR. Writing all three quantities in solar units, the empirical fitting relation:

$$
\left.
\begin{aligned}
-\dot{\mathfrak{M}} &= (5.5 \pm 1) \times 10^{-13}\ L/gR\ [\mathfrak{M}_\odot \text{ yr}^{-1}]\\
&= (5.5 \pm 1) \times 10^{-13}\ LR/\mathfrak{M}\ [\mathfrak{M}_\odot \text{ yr}^{-1}]
\end{aligned}
\right\}
\qquad (7.4;\,1)
$$

satisfies the observations reasonably well, although in one case (89 Her) a deviation by a factor 10^2 occurs. Since gR is the negative potential energy of the star, the assumption that the ratio between $\dot{\mathfrak{M}}$ and L/gR is constant is equivalent to assuming that the same fraction of the luminosity is used to provide the necessary potential energy for the escaping material.

For *stars of intermediate temperature* Reimer's formula (7.4; 1) was modified by Fusi-Pecci and Renzini (1975) and by De Loore (1977), such that the radiation flux L is replaced by the mechanical flux F_m, yielding

$$\dot{\mathfrak{M}} = \eta F_m/gR, \qquad (7.4;\,2)$$

where η is an efficiency factor, equal to -8×10^{-4} in the case of the Sun. A reasonable agreement with observational values is apparently obtained in De Loore's investigation.

7.5. Stellar Mass-Flux Theory Based on Fluctuation Theory

We should keep in mind, following Andriesse (1979, 1980a, b) that stellar mass-loss is not a constant and steady phenomenon but – as shown clearly for such stars for which sufficient observations are available, like the Sun and several others – that the mass-flow fluctuates regularly over large ranges. The source of the fluctuations must be in the photosphere, which is not in complete thermal equilibrium. To gain insight into the nature of the fluctuations one can compare two characteristic relaxation times, *viz.* the *dynamical time-scale*, being the time that determines the dynamics at the stellar surface, or the time needed to reach stability after a surface perturbation:

$$\tau_d = (\tfrac{4}{3}\pi R^3/G\mathfrak{M})^{1/2} = (\varrho G)^{-1/2}, \qquad (7.5; 1)$$

and the *thermal (Kelvin–Helmholtz) time-scale* which is the time needed to radiate the stellar potential energy, hence the time for reaching thermal equilibrium [cf. Equations (4.1; 9) and (4.1; 10) and Equation (7.4; 1)]:

$$\tau_t = G\mathfrak{M}^2/RL = \mathfrak{M}gR/L. \qquad (7.5; 2)$$

In all stars $\tau_d \ll \tau_t$ as it should be: otherwise the star would not be stable. For example for the Sun $\tau_d \approx 3 \times 10^3$ s and $\tau_t \approx 10^{15}$ s; and for a star with $L = 10^6$ $L_\odot$, $\mathfrak{M} = 40\mathfrak{M}_\odot$, $R = 30R_\odot$, $\varrho = 2 \times 10^{-3}$ g cm^{-3}, one would have $\tau_d = 10^6$ and $\tau_t \approx 10^9$ s. A supergiant of the latter type seems closer to instability than the Sun: this too is a wellknown fact!

In order to study mass-loss with the thermodynamic-fluctuation theory Andriesse (1979) selected the dimensionless parameter

$$\lambda = -\frac{G\mathfrak{M}}{LR} \dot{\mathfrak{M}}. \qquad (7.5; 3)$$

The significance of this quantity is that, if P is the total power generated in the stellar interior, energy balance requires that $\lambda = (P-L)/L$. Complete thermalisation would mean $\lambda \to 0$, while in a real star, in which surface perturbations occur continuously, λ would approach an average value $\bar{\lambda}$. Since we assume $\mathfrak{M}$ to be a fluctuating quantity, λ is so too, and one may write:

$$\dot{\lambda} = -\lambda/\tau_t + A(t).$$

This is the Langevin equation, where the change of λ is expressed in a systematic term and in an independent non-systematic or stochastic term. The quantity $A(t)$ is elaborated and specified in Andriesse (1980b). It follows from this equation that the r.m.s. average $\bar{A}$ of $A(t)$ is the normalized strength of the gravitational perturbations, which then represents an average power of $\bar{A}$ units of the gravitational energy, hence $\bar{A}(G\mathfrak{M}^2/R)$. We now *adopt* a linear relation between these perturbations – assumed to be small – and the thermal luminosity L of the star:

$$\bar{A}(G\mathfrak{M}^2/R) = \chi L, \quad \text{or} \quad \bar{A} = \chi/\tau_t, \qquad (7.5; 4a)$$

while it can be shown that for the upward fluctuations

$$\langle A_+ \rangle = \bar{A}/v\pi. \tag{7.5; 4b}$$

Next the important *assumption* is made that χ is the same for all stars. A further discussion then leads to the result that

$$\langle \lambda^2 \rangle = \tau_t \langle B^2 \rangle / 2\tau,$$

where

$$\langle B^2 \rangle = \int_0^\tau dt \int_0^\tau dt' \langle A(t) \, A(t') \rangle$$

is the integral of the cross-correlation of $A(t)$ over a fixed time τ chosen in between τ_d and τ_t. The ratio $\langle B^2 \rangle / 2\tau$ is independent of this choice, however. With the reasonable assumption that the cross-correlation of $A(t)$ decays in the dynamical time-scale τ_d it follows that

$$\langle B^2 \rangle = \tau \langle A_+ \rangle (2\pi\tau_t/\tau_d)^{-1/2},$$

and hence

$$\lambda = \chi(\tau_d/\tau_t)^{1/2}, \qquad \langle \lambda_+ \rangle = (\tau_d/2\pi\tau_t)^{1/2}.$$

From Equations (7.5; 1), (7.5; 2), and (7.5; 3) one then finds

$$-\overline{\mathfrak{M}} = 0.40 \, L^{3/2} \, (R/\mathfrak{M})^{9/4}/G^{7/4}, \tag{7.5; 5}$$

where the constant (0.40) is equal to $[2\pi \, (3\gamma - 4)^{1/2}]^{-1/2}$.

Figure 111, constructed by Andriesse (1979) shows a comparison between the observed mass-fluxes (see also our tables XXXIX through XLII), where the straight solid line refers to main-sequence stars for which $\mathfrak{M}(:)L^{0.29}$ and $R(:)L^{0.208}$, yielding

$$\overline{\mathfrak{M}}/\overline{\mathfrak{M}}_\odot = (L/L_\odot)^{1.32}, \tag{7.5; 6}$$

a relation that seems well obeyed for main-sequence stars. The other lines in the diagram refer to luminosity classes III and I, and are found from Equation (7.5; 5) by introducing the relevant relations between $\mathfrak{M}$, L, and R for such stars. It is with these groups, particularly the supergiants, that considerable discrepancies occur between observations and expectations, in certain cases even amounting to a factor 10^3, larger than the uncertainty in the data. Andriesse (1980b) gave arguments that actually the *core* radius of supergiants should be used in the formula; this would remove the discrepancy.

We *conclude* that the approach to the problem of stellar mass-loss using the concept of fluctuation theory may constitute an interesting basis for more research. There are a few ad-hoc assumptions in the theory, and these are a weak aspect. Fundamental in the theory is, that the fluctuations at the basis of the proposed mechanism are of a sub-photospheric nature, but there is little observational evidence, so far. The

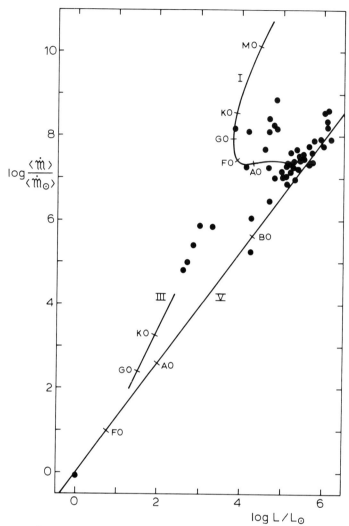

Fig. 111. Mass-flux as a function of the luminosity as predicted by Equation (7.5; 5) (full lines). The labels I, III, and V refer to supergiants, giants and main-sequence stars, respectively; in the latter case the straight line follows equations (7.5; 6). Markings are given for the different spectral types. (After Andriesse, 1979).

approach described in this section shows how fairly general considerations, which do not deal with the specific mechanism of mass-loss, enable one to establish a relation between stellar parameters and $\mathfrak{M}$. The next obvious question to discuss is what *physical mechanisms* produce mass-loss.

7.6. The Mechanisms of Supergiant Mass-Loss

First of all a word of *caution*: the only star for which the physics of the mass-loss process can be studied in sufficient detail is the Sun, an object which, if placed at the distance of an average star, would have an *unobservable* magnetic field. Yet it appears certain that the solar magnetic field is all-dominant in determining the rate of solar mass-loss, which, surprisingly enough at first sight, seems *not* to originate in the dense regions of the solar corona, where the gas appears to be constrained by the fields of the solar active regions. On the contrary, the wind seems to originate from the enigmatic *coronal holes*, regions where the particle density is much less than in the average corona, and with an unclear, but perhaps 'open', magnetic-field pattern (Figure 112). What then is one to think of the various refined and detailed theories set up to explain stellar mass-loss?...

Apparently one has to be modest in the appraisal of what has been achieved so

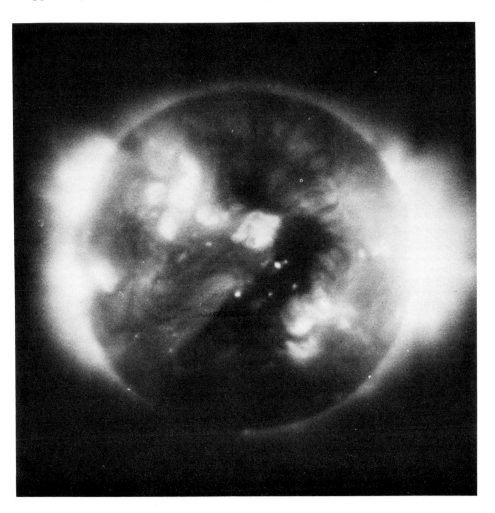

Fig. 112. The Sun seen in soft X-rays by Skylab. The corona is densest above the active regions where coronal gas is apparently confined by closed magnetic fields. No observable X-ray flux is emitted by the coronal hole, which occupies a fair fraction of the disc. The coronal holes seem to be the main source of the solar wind. (American Sci. & Engineering photograph).

far; in all respects we are still at the beginning, and surprising new developments are likely.

The next point is that for many of the objects discussed in this book, particularly early- and late-type giants, the existence of a *corona and/or a chromosphere* seems to be a rule – see Section 6.11. It is likely that the A- and the late-type supergiants have chromospheres rather than coronas: the low g_{eff}-values in the outer layers of the latter would prevent the formation of a corona, while only little mechanical flux seems to develop in A-type supergiants.

The *origin* of chromospheric or coronal heating in the brightest stars is unclear. Since the turbulent elements in supergiants appear to be extremely large, with sizes many times the local density scale-height (Section 5.10), the heating is presumably not related to dissipation of sound-wave energy, but rather to dissipation in features that have analogy to gravity waves, probably related to Rayleigh-Taylor instabilities in the atmospheres of such stars.

Through these 'chromospheres' or 'coronae' matter is *streaming outward*. In this connection it is perhaps crucial that the point where the velocity changes from subsonic to supersonic values is situated close to (and sometimes in) the atmosphere of many giants and supergiants: this result emerges from computation, but empirically it is also apparent from the supersonic values found for the macroscopic velocities observed in many early-type supergiants (Section 2.9). This remark applies also to certain late-type supergiants (cf. Mullan, 1978).

The question we have to answer after these introductory remarks is whether the stellar winds find their origin in the coronae or chromospheres, or whether reversely coronae are caused by the winds. Or could both features – winds and coronae – have the same origin, for example a mechanical flux, and be unrelated otherwise? A critical discussion of these problems was written by Cassinelli *et al.* (1978a).

The answer differs for the type of star considered. *Solar-type* coronae are caused by a mechanical or magnetic flux. The solar wind originates in the corona and is accelerated by the pressure of the hot coronal plasma in the solar regions with an 'open' magnetic field pattern. It seems likely that the same comments apply to most of the cooler giants and supergiants *viz.* also there the stellar winds may originate in the coronae or chromospheres. In addition we will show in the next section that the coolest supergiants owe their winds to radiation forces on dust molecules since it may be that these drag the gas along, but that latter problem needs further research.

Coronae of *early-type stars* (O and B) cannot account for the strong winds observed in these objects: first because application of the coronal-wind theory to such stars – for which the terminal velocities are a few times 10^3 km s^{-1} – would demand coronal temperatures of several times 10^7 K (Lucy, 1975). Therefore the corona plays a negligible role in accelerating the stellar wind, and the strong radiation flux will be the dominant factor. There is another way to show this: the energy-flux F_w carried per second by early-type stellar winds is comparable to the total thermal energy content of the stellar coronae, roughly and by order of magnitude. Hence if the winds would obtain their driving energy from the early-type stellar coronae, the corona should be 'refilled' with energy about once every few seconds or minutes, a highly unlikely situation. In the contrary, disturbances in the stellar winds of early-type luminous stars, amplified by radiative forces associated with the strong UV resonance lines (Martens, 1979; MacGregor *et al.*, 1979) can yield mechanical energy to heat the supersonic parts of the stellar wind. This may give rise to hot elements in this part of the winds, certainly not to a neat spherically symmetric coronal layer (but there is no reason why the coronae of early-type stars should be spherically symmetric!). Hence: early-type stellar winds make coronae.

A-type stars particularly the supergiants form an interesting case in this connection: it seems that their wind-fluxes as well as their coronae and chromospheres are only weakly developed (cf. α Cyg). This is related to the small mechanical fluxes expected in these stars, and this finding supports again the view that at least for certain stars the origin of coronae and winds is linked – probably by a mechanical or magnetic flux.

A last aspect, that has not been discussed yet – not because it would not be important but rather because it appears very difficult for a satisfactory treatment – is the influence of *stellar rotation* on the winds. A few brief and semi-qualitative remarks on this subject were made by Castor (1979) to whom we refer.

7.7. Expanding Dust Shells

Many stars, particularly those with fairly large $-\dot{\mathfrak{M}}$-values are surrounded by shells of dust particles that have condensed from the gas molecules. These dust shells must also expand because of radiation forces on the grains, or they move outward because of momentum-coupling between the stellar wind and the dust grains (Gilman, 1972; Lucy, 1976c). Let us examine this question for Mira-type stars! The theory, essentially similar to the considerations given in Section 1.3 which led to the Eddington-limit (Equation (1.3; 3)), was given by Salpeter (1974), who found, upon integration of the time-independent hydrodynamic equation, that

$$v_\infty = v_{\text{esc}} \, (L/L_{\text{lim}} - 1)^{1/2}, \qquad (7.11; 1)$$

where L_{lim} is given by Equation (1.3; 3) but in which κ is the opacity by grains and gas combined. For the dust particles κ is difficult to estimate but $\kappa \approx 1$ may be a good value. Equation (7.11; 1) may be combined with the expressions for: the velocity of escape $v_{\text{esc}} = (2G\mathfrak{M}/R)^{1/2}$; the luminosity equation $L = 4\pi R^2 \sigma T_e^4$, and the pulsation equation (4.7; 1). Thus, following Michalitsianos and Kafatos (1978) one obtains the quadratic equation

$$R^2 - \left(\frac{\pi F P^2 \kappa}{cG\beta}\right) R + \frac{v_\infty P^2}{2G\beta} = 0. \qquad (7.11; 2)$$

Here, P is the pulsational period of the star, $\beta = \frac{4}{3}\pi Q^2 \varrho_\odot$ and the other quantities have their usual meaning. If we take $Q = 0.04d$, hence $\beta = 0.0094$, assume for v_∞ half the largest splitting of the maser line components (km s^{-1}), and take $\kappa = 1$, then for any given Mira star with known total flux πF (which is one of the difficulties!) and period P the radius can be derived. If πF is not known, Equation (7.11; 2) yields for any period P a relation between πF and R. Michalitsianos and Kafatos (1978), taking $Q = 0.03d$, found $\langle \mathfrak{M} \rangle$-values of 1 to $2\mathfrak{M}_\odot$, in fair agreement with other determinations or estimates. A change of Q to $0.04d$ would not sensibly change the resulting masses.

Essentially this result means that the basic assumptions of the present discussion are correct: hence the region of the envelope from where the maser lines are emitted must be expanding because of *radiation forces on circumstellar dust* and simultaneous momentum-coupling with the gas.

CHAPTER 8

VARIABILITY OF SUPERGIANT ATMOSPHERES

8.1. Introduction; History

Many if not all supergiants show signs of variability, which is another indication of the instable character of these very extended atmospheres. Most of the variations are of a small amplitude and have a more or less erratic character, but a 'characteristic period' is often detectable. Apart from the supergiants, there are in the upper part of the Hertzsprung–Russell diagram several other groups of variable stars. These are, in order of decreasing temperature, the β-Canis Majoris- (or: β-Cephei-) stars, the (Delta) Cepheids, the semi-irregular (UV Tauri) variables and the long-period red variables of the Mira-type. The β CMa stars have luminosity classes II, III, IV; Cepheids reach III and II and the same is true for Mira stars. These stars are not supergiants. In this chapter we restrict ourselves to the variability in supergiant atmospheres.

Before giving in the next sections a review of the observed variations of brightness, colour and spectrum we briefly review the history: Among the supergiants α Cyg (A2 Ia) is perhaps the best studied spectrum variable; it was known as such since 1896. For the radial velocity Paddock (1935) found some indication of periodicity, with a 'characteristic time' of about $12d$. His observations were studied anew by Lucy (1976a) who found no less than sixteen significant periods between 6.9 and $100.8d$, – see Section 8.4.

Variability of a supergiant in an extragalactic system was detected almost simultaneously with that of α Cyg: the star S Dor (HD 35 343, A2–5 eq, – see Section 3.9) in the LMC was found to be variable in 1897 (Pickering, 1897). Since, it has shown several deep minima, the last one around 1965 (Van Genderen, 1979a).

Variability of other supergiants of intermediate spectral type ($\sim$A–F) in extragalactic systems was discovered in the spiral system M33, independently by Duncan (1922) who found three variables, and by Wolf (1922) for one of these three stars. This latter star, named Var 2 by Duncan, was later studied by Humphreys (1975); it shows He I emission lines on a type A absorption system. More than thirty and fifty years after these initial discoveries the variability of five other A–F supergiants in extragalactic systems was found by Hubble and Sandage (1953), and confirmed by Rosino and Bianchini (1973), in M31 and M33, from observations covering a period of more than 50 yr. Variables of the kinds described here are sometimes called Hubble–Sandage variables, or S Dor-variables, but the types are not clearly defined, and the names not generally used.

Tammann and Sandage (1968) found five blue irregularly variable supergiants in NGC 2403.

In the LMC, Rosendhal and Snowden (1971) found light variation in five supergiants. The variations from night to night ranged between 0.02 and 0.07 mag. In another study of the LMC, Appenzeller (1972) found that nearly all out of 18 bright O and B stars varied on a time scale of weeks; later (1974) he even found photometric variations on time scales of *hours* in S Dor and HD 37836.

Underhill (1966) remarked already that 'it is very probable that most, if not all of the 'bright' supergiants are variable in light and/or colour and/or radial velocity'. This is now generally accepted: all Ia supergiants are somewhat variable. Further questions are how this variability depends on the star's properties such as brightness and mass, and what physical laws regulate the variability. The first question can now – roughly – be answered: information is available on the dependence of variability on the stellar parameters – Sections 8.2 through 8.4 –, but the physics of the variability is not yet very well understood (Sections 8.6 and 8.7).

8.2. Brightness Variations

All supergiants of class Ia show irregular brightness variations. A very well-known case is the hypergiant ϱ Cas (F8 Ia$^+$) – see Section 3.8 – which normally shows irregular fluctuations in brightness, with a typical amplitude of 0.2 mag, and a period of about 400 days. Sudden brightness jumps can also occur, e.g. in 1946–47 the brightness of this star decreased by 1.5 mag, to recover years later.

All other well-studied hypergiants show brightness variability, see e.g. HD 160529 (Wolf *et al.*, 1974; Sterken, 1976a), HD 33579 in the LMC (Wolf, 1972), and HD 7583 in the SMC (Wolf, 1973). Particularly the fairly large brightness variations in the first star (HD 160529), amounting to 0.3 mag are interesting, since they are apparently related in various ways with the star's velocity field: there is a relation between the brightness variations and the velocity variations, as if progressive waves through the atmosphere influence the star's continuous radiation, *and* there is a correlation between the brightness variations and the line splitting, in the sense that the star becomes fainter with increasing line splitting.

Systematic analyses of the brightness variations of supergiants have been made by Maeder and Rufener (1972; cf. Maeder, 1980a), Sterken (1976a, b), Rufener *et al.* (1978), Burki *et al.* (1978), Burki (1978), Van Genderen (1979, 1980a, b). Sterken analysed light variation in a dozen A- and B-type supergiants with $-9.0 \le M_v \le -5.7$, while Rufener *et al.* (1978) dealt with seven Ia-type supergiants near the clusters h and χ Persei. Maeder (1980a) described the large Genève photometric program. Van Genderen (1979) made photometric observations, extending over a period of five years, of three supergiants and one hypergiant (HD 33579, – see Section 3.8) in the Large Magellanic Cloud. He also investigated (1980b) three supergiants in the prolongation of the Cepheid strip.

These observations show different kinds of variability:

Short-term fluctuations, on time scales of a few hours, and with amplitudes of a few hundredths of a magnitude were only detected for two of the three (or four) hypergiants, hence for the *most extreme objects*, viz.: HD 160529 (A2 Ia$^+$; $M_v = -8.9$), and HD 168607 (B9 Ia$^+$; $M_v = -9.0$). For the objects HD 152236 (B1 Ia$^+$; $M_v = -8.8$) and BD$-14°5037$ (B1.5 Ia$^+$ (?); $M_v = -7.2$) no such short-term variations were detected. The two objects showing variability have Hα lines with marked P Cyg profiles; the widths of the emission peaks being $\sim$20 to 30 Å (Sterken, 1976a).

Similarly, Mendoza (1970) and Appenzeller (1974) found short-term variability in S Dor, and Wolf (1975) in HDE 269006.

Intermediate-term variations (from one night to the other – on a time scale of the order of days to months occur in all objects investigated by Sterken, and by Rufener *et al.* (1978), but for the least bright one: HD 167264 (B0 Ia; $M_v = -5.7$). Figure 113 gives some examples. There is no strict periodicity but a 'semi-period' P_s can be established by an autocorrelation procedure (Burki *et al.*, 1978). The value of P_s

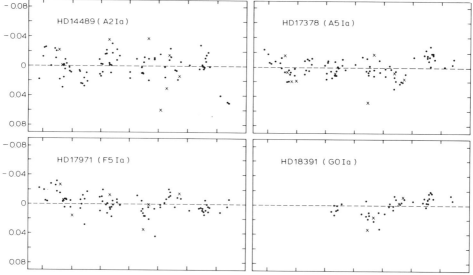

Fig. 113.   Medium-term variability of the visual brightness m_v in four Ia-type supergiants. Abscissa: time; each of the graphs covers a period of 150d; ordinate: Δm_v. Crosses denote uncertain values. (After Rufener *et al.*, 1978).

appears to increase from 15 to 20 days for B-supergiants to more than 80 days for the G0 Ia star. If we define 'spectral type' by the symbol S, and assume one 'unit of spectral type' equal to the difference A0–B0, ... etc., then roughly: $\log P_s = 1.2 + 0.25$ $(S-B0)$. The average amplitude of the light variation is 0.05 mag (Sterken, 1976a; Burki *et al.*, 1978). Larger y-amplitudes were found for the stars HD 91619 (B5 Ia) – $\sim$0.09 mag – and HD 96 919 (B9 Ia) – $\sim$0.11 mag (Wolf and Sterken, 1976). The characteristic times of the variability of the two latter stars was estimated at $\sim$15 to 20 days, but additional completely irregular fluctuations of $\Delta y \approx 0.05$ may occur.

A systematic search for the brightness fluctuations of stars in the whole upper part of the HR diagram by the Genève group resulted in Figure 114a (Maeder, 1980a, b) which gives for all stars that have been observed more than 3 times, the square root of the fluctuations multiplied by 3.3. This value simulates the 'peak-to-peak amplitude' of the brightness fluctuations.

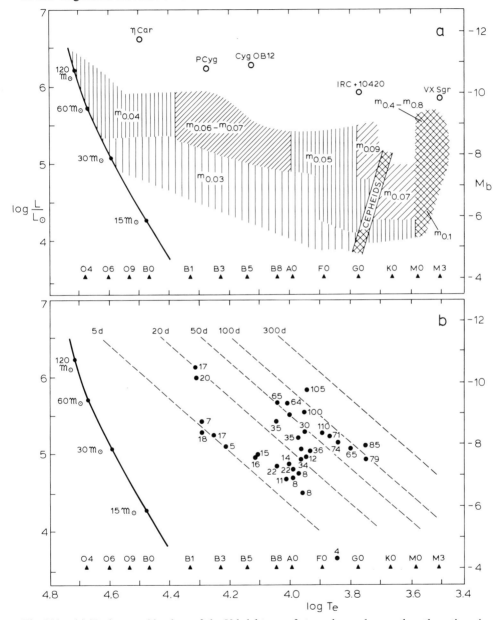

Fig. 114. (a) 'Peak-to-peak' values of the V brightness of stars observed more than three times in the Genève photometric program, and their distribution over the HR diagram (Maeder, 1980a). (b) Approximate lines of constant P_s-values in the Hertzsprung–Russell-diagram. After Burki (1978), as modified by Maeder (1980a). P_s-values are in days.

Important is in this diagram: the increase of amplitude with luminosity; the interesting variation of amplitude with T_e for the brightest supergiants: there are maxima for O, B-type and very late-type stars. Maeder (1980a, b) found a relation between the rate of mass-loss and the 'amplitude'. He also remarked that the variation of the 'amplitude' with T_e is to a large extent similar to the dependence on T_e of the limit of deep convection in the outer layers of massive stars.

Important is the systematic relation between period and luminosity. Abt (1957a) remarked already – from v_R observations – that there is a fair correlation between P_s and the absolute visual magnitude M_v; – this leads to a *statistical Period-Luminosity relation*. This was confirmed for the brightness variation by Sterken (1976a, b), Burki (1978) and Maeder (1980a) – see the latter's compilation in Figure 114b. We refer further to Figure 115, where values for two LMC supergiants HD 33 579 (Van Genderen, 1974) and HDE 268 907 (Maeder and Rufener, 1972), as well as values from colour observations are included. Apparently, the P_s–M_v relation depends on the spectral type.

If *long-term variations* are defined as differences between the average magnitudes taken over successive months to years, then all investigated objects but the absolutely faintest ones show such variations in brightness or colour. Sterken (1976a, b) found that such variations are characterized by Δm- or Δc-values of the order of one or two hundredths of a magnitude but this is not true for all objects. For instance, the four LMC super- or hypergiants investigated by Van Genderen (1979) have the following largest Δv-values and 'semi-periods' (eye-estimates from lightcurves):

HD 33 579	(A3 Ia$^+$ (e))	$\Delta v \approx 0.04$ mag	$P_s \approx 100\ d$
S Dor – HD 35 343	(A2–5 eq)	$\begin{cases} 0.5 \\ 2 \end{cases}$	$\begin{matrix} 2\ \text{yr} \\ 10\ \text{yr} \end{matrix}$
HDE 269 006	(B2.5 Iep)	$\begin{cases} 0.5 \\ 1 \end{cases}$	$\begin{matrix} 0.5\ \text{yr} \\ 5\text{–}10\ \text{yr} \end{matrix}$
HDE 268 757	(G5–8 Ia)	$\begin{cases} 0.08 \\ 0.3 \end{cases}$	$\begin{matrix} 1\ \text{yr} \\ 3\ \text{yr} \end{matrix}$

This brief summary shows already that there is a fourth kind of variability:

Very long-term variations are only discovered by observations extending over several years or several decades. They consist of irregular fluctuations of fairly large amplitudes, particularly for the absolutely brightest objects. Thus, similarly to Van Genderen (1979), Gottlieb and Liller (1978) found brightness variations of the order of tenths of magnitudes and more, over periods of tens of years for several luminous stars, among which a few with η Carinae characteristics.

8.3. Colour Variability

A large investigation of colour variability, including 660 stars, and based on a long series of photometric measurements was published by Maeder and Rufener (1972).

Also Van Genderen (1979) measured colour variations for the four objects studied by him. While giant stars of class II do not statistically show more colour variability than normal dwarfs of luminosity classes V to III the variability is more pronounced for the class I type stars: Intrinsically bright supergiants show clear variations in their colours: about $\frac{1}{3}$ of the supergiants of luminosity classes Iab/Ib and all Ia-type stars seem to be variable in colour. These conclusions are not fully verified by Sterken's (1976a) analysis of brightness and colour index changes of A- en B-type supergiants, the same applies to Campusano's (1977) investigation of intermediate time-scale (days) variations of the continuous spectra of six A- and B-type supergiants of luminosity classes Ia$^+$, Ia and Ib: the object HD 167838 (B5 Ia) did not show variations in Campusano's investigation – which covered only seven nights, however. While the colour variation too is not strictly periodic there is in the colour variation some regularity, similar to the one found for radial velocities by Abt (1957b) and for the brightness variation by Sterken (1976b) and Burki (1978). As with the brightness variations there is not so much question of a real period but rather of a characteristic semi-period P_s of the light variations. A summary of data from the literature till 1972 was collected by Maeder and Rufener (1972) and is shown by the squares in Figure 115. The agreement with values derived from the brightness variations is good. We *conclude* that for a broad range of spectral types there exists for the colour variations a Period-Luminosity relation that is very similar to the one found for the brightness variations, and that is parallel to the relation for Cepheid variables.

As is the case with the brightness variations, colour variability is strongest in *hypergiants*. Sterken (1976a) found (*u-v*) amplitudes up to 0.19 mag (HD 160529). The well studied star HR 5171 (=HD 119796; G8 Ia$^+$) is interesting because it appears to show *two* modes of variability (Harvey, 1972), one with a long characteristic time, of ~10 yr, and with a slope of 1.6 in the (*B*, *B-V* plane), and another with a time scale of about one year and a slope of 4.2.

While all hypergiants and most supergiants show colour-variability, this is not the case for less bright stars: there, variability was only found for bright O- and early B-type stars. (As is stated earlier – Chapter 2 – the distinction between the various luminosity classes nearly loses meaning for very early-type stars.) In Figure 116 crosses denote stars for which the dispersion in the colours exceeds the value of 0.0075; dots are stars with smaller dispersions. The accumulation of crosses in the left upper part of the diagram (spectral type earlier than B4) is obvious. This is supported by radial velocity observations of stars of these types, for which Underhill (1966) has remarked that they show irregular variations over a range of about 30 km s^{-1}.

8.4. Spectrum Variations

The brightest supergiants, of luminosity classes Ia and Ia$^+$ all show spectral variations to a certain extent. The *long-term variations of the spectral type* are in some cases conspicuous. Morgan and Keenan (1973) noted that the luminosity class Ia$^+$ stars

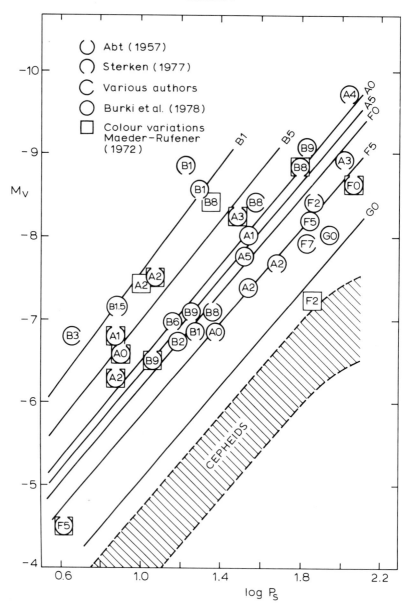

Fig. 115.   Relation between semi-period P_s (for the brightness and colour variations) and luminosity for supergiants. Data from brightness variations (circles) are from Burki's (1978) compilation, while those from colour variations (squares) are from Maeder and Rufener (1972).

HDE 269953 and HDE 269723, classified by them as G0 Ia$^+$ and G4 Ia$^+$ were given the type G0 Ia$^+$ by Feast and Thackeray (1956), apparently indicating long-term spectral changes of the second star. An extreme case is the hypergiant ϱ Cas, actually of type F8, but which changed from a K-type in 1930 to G+MI in 1945–46 (Beardsley,

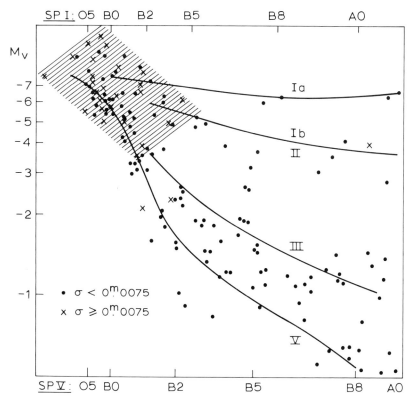

Fig. 116. Early-type part of the Hertzsprung–Russell diagram with positions of stars observed by the Genève photometric program. Dots denote stars with an intrinsic colour scatter less than 0.0075 mag; crosses refer to stars with an intrinsic scatter exceeding this value. The variability is pronounced in the hatched area, where the brightest O and B-type stars occur. (After Maeder and Rufener, 1972).

1961) and returned to F8 Ia$^+$ around 1950 (Sargent, 1961). Another is BS 8752, which slowly changed from G0 Ia$^+$ to G5 Ia$^+$ between 1950 and 1970 (Keenan, 1971). The spectral type of the star FG Sge changed from A3 Ib in 1968, via F7 Ib in 1973 (Chalonge *et al.*, 1977) and B5 Ia in 1975 to G2 I in 1976 (Smolinski *et al.*, 1976). Parallel to the spectral changes the colour index *B-V* varied, to reach a maximum value in 1977 (Kupo, 1977). Section 3.12 gives a discussion of the possible relation between this star and nuclei of planetary nebulae. Rosendhal (1973a) found that all early-type supergiants with emission lines are probably intrinsic spectrum variables. See for more examples on hypergiant spectrum variability Section 3.8.

That *radial-velocity changes* are quite common in Ia-type supergiants of early and intermediate type was already stated by Abt (1975b) in an extensive study of supergiant spectra. A list of references to investigations of this kind was published by Aydin (1979) who, himself, studied 6 Cas (A3 Ia) and discovered radial-velocity variations of

Apart from the variations in spectral type described above, and colour variations described in the preceding Section there are also variations in the *radial-velocities*, the *line strengths* or *line profiles*.

a semi-regular nature with amplitudes in the range of 5 to 10 km s^{-1} and on time scales as small as one hour. Generally, lines of different elements may have different radial-velocities which vary with time. The amplitudes of the radial-velocities are of the order of 4 to 8 km s^{-1}. No simple periodicities exist but in some cases characteristic periods are present. In line with the results presented in previous sections these periods seem to be longer for stars with higher luminosity and of later spectral types. Radial-velocity changes are much more common than light variability.

A classical case is α Cygni (A2 Ia), already mentioned in the introductory Section 8.1. On the basis of 447 accurate measurements of the radial velocities of the star, made during the years 1927–1935 at the Lick Observatory by Paddock (1935), Lucy (1976a) made a search for periodicities in v_R. Fluctuations in time spans of one night appear to have a random character, caused by observational errors. There appears to be a long-period variation with an amplitude of approx. 6 km s^{-1} and a period of about 800 days, but it is not clear whether this periodicity may be due to a possible binary character of the star. After subtraction of this long-period variability the velocity variations in 1931 (the year during which most observations were made), being subjected to a periodicity analysis, show sixteen significant periods with values between 6.9 and 100.8 days and amplitudes ranging between 0.29 ± 0.08 km s^{-1} (for $P=6.9d$) and 1.02 ± 0.09 km s^{-1} (for $P=11.4d$). The six terms of largest amplitude (>0.78 km s^{-1}) have P-values ranging between 10.0 and 18.9 days.

It is interesting to compare these periods with those expected to occur for radial pulsations. Assuming, with Lucy (1976a) for α Cyg $M_b=-8$; $\mathfrak{M}=12\mathfrak{M}_\odot$, $R=140R_\odot$ and hence: $\langle \varrho \rangle = 4.4 \times 10^{-6} \varrho_\odot$, one finds with the pulsation parameter $Q_0=0.04$ days a fundamental period of $P_0=19d$, just compatible with the six terms of largest amplitude. The longer periods, however, can apparently not be ascribed to radial pulsations and Lucy forwarded the suggestion that these pulsations are *non-radial*. This suggestion can be supported theoretically and confirmed empirically.

Theoretical support was given by the fact that the non-radial so-called g-mode instabilities have a very broad spectrum of eigenvalues towards pulsations with large wave numbers i.e. small vertical wavelengths. We refer to a discussion of this aspect by Maeder (1980b).

In addition there is some empirical confirmation. Line profiles were calculated for α Cyg, assuming the star to pulsate non-radially. Taking 12 km s^{-1} for the Doppler broadening by thermal and microturbulent motions, it appears that the observed line widths – usually assumed to be caused by 'macroturbulence', and typically $\sim$28 km s^{-1} (Groth, 1961) – can be reproduced for oscillations with surface harmonics of degree 3 to 4 (Lucy, 1976b). There is some analogy between this piece of research and the one by De Jager and Vermue (1979) who found that the observed ratio between the microturbulent and macroturbulent velocity components in supergiants indicates the presence of very large photospheric 'elements' moving coherently.

In addition, stratification effects have been observed in the motion field of the same star. The metallic lines, assumed to be of photospheric origin, have a typical amplitude of velocity variation $\Delta v_R \approx 6$ km s^{-1}. The higher layers, from where the lower Balmer lines and Mg II $\lambda 4481$ are emitted, have $\Delta v_R \approx 15$ km s^{-1}, and the highest layers (central parts of Hα and Mg II h and k) have Δv_R ranging between 50 and 150 km s^{-1} (Inoue, 1979).

These observations demand a thorough discussion of the motion fields of supergiants, including non-radial pulsations and stratification effects. It may very well be that, once that a complete theory of stellar non-radial pulsations has been set up, detailed investigations of radial velocities of giants and supergiants may prove to become a new and promising tool for the investigation of the structure of the vibrating envelopes of such stars. These observations also ask for a reconsideration of Fath's (1935) finding that the light variability would be 270° out of phase with the v_R-variability: the extremes in the light-curve would precede those in v_R by approx. 2.5d. Assuming pulsation to be the cause, Groth (1972) then finds $\Delta R \approx (3 \text{ to } 4) \times 10^6$ km $\approx$ 0.04R_*.

In the most extreme stars depth-dependent variations like those described above for α Cyg are conspicious. Wolf *et al.* (1974) have measured radial velocities in the hypergiant HD 160529 (A2 Ia$^+$) and found that the hydrogen lines, as well as the metal-lines (see Figure 117) show simultaneous changes of v_R with amplitudes of approx. 20 km. There are clear indications for depth-dependence of v_R (e.g. v_R(H9) − v_R(Hγ) ≈ 25 km, with the other Balmer lines in between), as well as multiplicity in the other stellar layers (e.g. the Na D lines show sometimes three components).

Next, the other aspect of spectral variability: the number of supergiants that show *variability of line strength* or *line profile* is about half of those having a variable radial velocity. From Sterken's (1976a) investigation of brightness variations of early-type supergiants it appears that all B and A supergiants with appreciable Hα emission are photometric variables, and vice versa.

Erratic or quasi-periodic changes in spectral line profiles occur quite often in supergiants. In the visual spectral range the *Hα line* is the best to study such variations, because of its strength: like in the Sun the outer parts of the atmosphere seem to respond stronger to atmospheric instabilities than the deeper, photospheric parts. Since also radial velocity measurements often suggest that in supergiants and in hypergiants Hα is formed in an outer envelope, Hα variations like those listed by Smolinski (1971) indicate changes in these envelopes. Rosendhal (1973a, b) who investigated some 20 early-type stars of high luminosity found that about $\frac{2}{3}$ of the investigated stars show large variations in the profile and strength of Hα, as well as of other strong lines in the red part of the spectra of these stars. In the supergiants HD 91619 (B5 Ia) and HD 96919 (B9 Ia) strong variations were found in the Hα profiles including various phenomena such as only emission, normal and inverse P Cygni type profiles, asymmetric absorption, multiple splitting, and broad emission with central absorption (Wolf and Sterken, 1976). The radial velocities measured in Hα indicate variable large-scale velocity fields with relative velocities occasionally up to 100 km s^{-1}.

Another interesting line is He II $\lambda 4686$ of which Brucato (1971) investigated the variability in five Of stars, and found it to vary on time scales as short as 10 min, particularly in the earliest spectral types. Similarly, variations on a day-to-day scale were observed for the He II $\lambda4686$ line in λCep (Of) by Hutchings (1979b).

Variations in the Ca II, H and K *emission peaks* have so far only been detected for the G2 Ib supergiant α Aquarii (Hollars and Beebe, 1976). The variations refer to the relative strengths of the K_2 emissions and occur on time scales as short as a day (the shortest interval for which observations were available).

A fine example of spectral line changes in supergiants is shown in Figure 117a, which gives the profile of the *Hβ-line* in the A-type hypergiant HD 160529. While not all the variations shown in the three different tracings may be real: – part of it is certainly due to plate noise – the interesting variation of the profile of the Hβ-line is noteworthy. At one time the profile shows a distinct P-Cygni type and at another time (29 April 1970) the line has a more symmetric character. The same star also shows irregular variations in its integrated brightness.

In stars of luminosity classes Ia and Ib the *Hγ-line* and its variations were measured by Hutchings (1967b) who found small ($<10\%$) erratic changes with time scales of

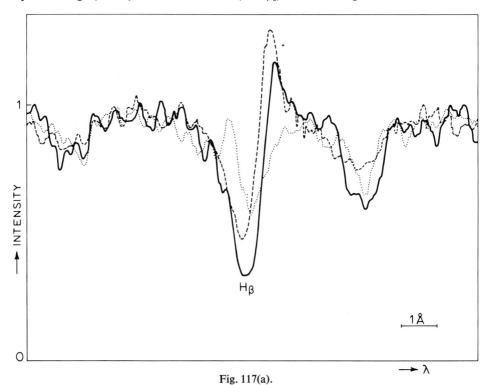

Fig. 117(a).

Fig. 117.   Variations of line profiles in the hypergiant HD 160529 (A2 Ia$^+$). (a) The diagram shows three Hβ-spectra: one taken on 29–IV–197 0(dotted), another on 27–III–1972 (dashed) and on one 21–IV–1972 (drawn). The two latter spectra, less than a month apart, show the large variations that took place in the atmosphere of the star over that short interval of time. A comparison with the earlier observations shows a still more dramatic change of the atmospheric parameters (Wolf, 1973). (b) Two spectra taken on 21–III–1972 and 20–III–1973 show variable line splitting with $\Delta v=15$ km s^{-1}. (Wolf *et al.*, 1974).

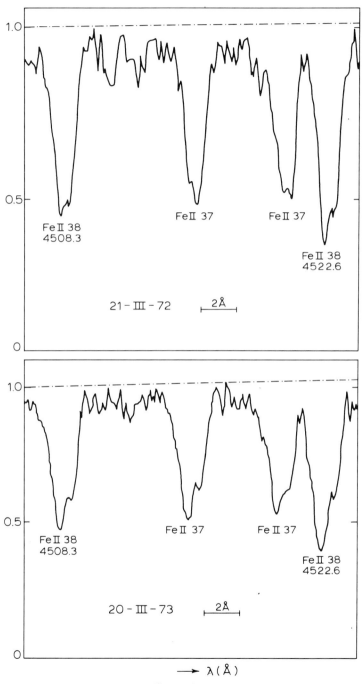

Fig. 117(b).

days. A systematic investigation of the spectrum variations in A-type supergiants was made by Rosendhal and Wegner (1970), who investigated the variations in radial velocity and equivalent width in three stars: ϱ Cas (F8 Ia$^+$), 9 Per (A2 Ia) and ν Cep (A2 Ia). Both the radial velocities and the equivalent widths appear to change rather quickly. The characteristic periods are not always easy to determine, since the time difference between successive observations is typically of the order of one to a few days. A clear correlation between the radial velocity variations and those of the equivalent widths was apparent for two stars, but for ν Cep rather an anti-correlation was observed.

The question then arises how to interpret such and similar equivalent width observations. Rosendhal and Wegner (1970) ascribe the line broadenings to microturbulence, but Wolf *et al.* (1974) could study the same phenomenon for the more extreme case of the hypergiant HD 160 529 (A2 Ia$^+$), and in *this* case it became clear that the broadening of the lines is apparently due to variable line splitting, hence – as in the case of the radial velocity changes of α Cyg, described before – to *variable mass motions on very large scales* in the atmospheres of these stars. Two spectra, about one year apart, show velocity differences of 15 km s^{-1} – see Figure 117b.

The rapid line profile variations on time scales of minutes to hours, which occur in many intermediate-temperature-supergiants have not (yet?) been detected in the visual spectral region of stars that are closer to the main sequence, or that just have higher T_e-values: Be, Of, Wolf–Rayet stars (Lacy, 1977).

Conclusions. In this section we investigated variations related to the stellar photospheres and the lower parts of the envelopes. It appears that variations of the radial velocities of lines are common for most supergiants, with stronger variations for more extreme supergiants. Semi-period-luminosity relations describe the character of the variations. These are ascribed to what is conventionally called 'non-radial pulsations' but what should rather be seen as irregular and incoherent slow pulsational motions of large parts of the atmospheres of these big stars. A smaller number of supergiants, particularly those which emission lines, show variations in the line strengths and profiles. These variations are important for strong lines such as Hα, and He II λ4686. These are formed in the *outer layers*, in the basis of the stellar winds. They indicate rather irregular variations in the envelopes of the stars. The relation with the pulsations and with the overall brightness variations are not yet sufficiently investigated; neither has the periodicity of the spectral-line changes been investigated well enough to draw general conclusions.

8.5. Variability of Stellar Winds

It was only after the opening of the far ultraviolet spectral region to repeated regular observations with observatory-type spacecraft (*Copernicus* and *I.U.E.*) that it became possible to observe variations in stellar winds, because these can be studied best through the strong far-ultraviolet resonance lines. Before dealing with the *stellar* case we should have a look at the best studied stellar wind – that of the Sun. The *solar wind* shows very important spatial and temporal fluctuations over a broad range of scales, down to very small ones. For *stars* one measures the integrated effect only, so that a great deal of the fluctuations cancel out, and only the gross effects can be observed. But even with that restriction important variations have been noted:

Long time-scale variations were found by Conti and Niemelä (1976) who studied the Hα emission-line profile in ζ Pup between 1974 and 1976 and found a gradual change of the profile that can be interpreted as a decrease of density in the envelope, and of the rate of mass-loss: about 25% in three years. Snow (1977) observed the ultraviolet spectra of 15 Of, O, B0 and B1-type stars that had also been observed some 2 to 4 years before. In most cases the P Cygni-profiles in the spectra showed small but distinct variations, and in two cases, ζ Pup and δ Ori A, these changes confirm that significant variation of the rate of mass-loss can occur over a period of several years.

The observations described here show that considerable changes can occur in stellar winds over time intervals of years. Another kind of variations was recorded by Taylor and Münch (1978) who observed small cloudlets with enhanced Hα and [N II] emission in some areas of the H⁺ emission nebula M42 near θ^2 Ori A; some of these move with velocities of 100 to 200 km s⁻¹, hence are definitely supersonic. Their angular distance to the star is of the order of a few tens of arc sec. The clouds are interpreted as being produced by the shock-interaction of the stellar wind with gas in the H⁺ region; the existence of such clouds may be due to variations in the pressure of the shocked stellar wind. Dramatic long-term changes in the stellar wind of the WR star HD 193793 (WC7+O5) are described in Section 7.4.

In the same vein is an aspect first noted by Lamers *et al.* (1978) in high-resolution *BUSS* spectra of β Ori (B8 Ia): there are indications that mass-loss is not a continuous process, at least not in all supergiants, – but may take place in 'puffs'. Indications of such events are the presence of shells of matter, as inferred from the appearance of one or two satellites to some spectral lines like the UV resonance lines of Fe II. Such satellite lines are not a permanent feature but may disappear and reappear with characteristic time-scales of the order of months or shorter. A shell with a velocity of 100 km s⁻¹ – a typical value – would travel over 10 stellar radii in one month. Similar shells were also found in α Cyg (A2 Ia) and η Leo (A0 Ia). The highly insufficient coverage of the observational sequences makes it impossible to draw conclusions about the character and scope of this effect.

Variability at short time-scales in the winds of O-type stars was discovered by York *et al.* (1977) who investigated the spectral region around 1025 Å containing H I Lβ and an O VI resonance line for the stars δ Ori A (O9.5 II), ι Ori (O9 III) and ζ Pup (O4 ef), and found variations with time-scales of the order of one to a few hours. The data show changes in the profile of the O VI line, and these occur for velocities more negative than −900 km s⁻¹. Later investigations have confirmed the existence of short-term variations in winds of these stars. In ζ Pup Wegner and Snow (1978) and Snow *et al.* (1980) detected variations in λ4686 He II and Hα, in the P Cygni-profiles of Si IV λλ1393, 1402, and O VI λ1030 with timescales of the order of one to a few hours. The observations indicate fluctuations in the stellar-wind density up to a factor 2; they are larger for the visual spectral lines than for the ultraviolet ones, so they refer to the basis of the stellar wind rather than to the more remote parts of it. In δ Ori A fluctuations in the UV P Cygni profiles were observed with a typical time scale of 1 to 2 h (Snow and Hayes, 1978).

The hypothesis that is conventionally brought forward to explain such variations is that they are caused by the outward propagation of density perturbations, such as those described by Martens (1979) and MacGregor *et al.* (1979). If this assumption is correct one should see the perturbations successively in spectral lines formed at higher levels. But while this latter statement is correct in principle, the obvious difficulty is that the wind-region of luminous early type stars and supergiants is so extended that it would take days or longer for a perturbation to travel the distance between the layers of formation of two UV resonance lines. During that time the disturbance may have damped out. Apparently this problem should be attacked by studying many UV lines at the same time, and with a high observing frequency, during as long a period of possible, and at least during several days.

In the spirit of these considerations De Jager *et al.* (1979) observed short-term variations of the P Cygni profiles of a few of the strongest UV resonance lines in the spectrum of α Cam (O9.5 Ia) with time scales of hours to days – Figure 118. These spectral changes represent considerable variations in the terminal velocities of the stellar wind. As shown in the figures several aspects are noteworthy in these observations, which refer to the outermost stellar layers: (a) variations occur only in the short-wavelength part of the line-profiles (Figure 118a), and indicate changes in the terminal velocity v_∞ of the wind; (b) there are *gradual* changes in the course of the three days of the observations; for the C IV line these would correspond to changes of $\sim7\%$ in the stellar wind acceleration; (c) there are also indications of shorter-periodic variations on a time scale of hours (Figures 118b and c); (d) the variations are *un*correlated in the various lines – which is understandable because in this very extended atmosphere the layers of formation of the various lines are far apart; (e) the *gradual* variations are strongest in the line that is formed at the largest distance to the star (C IV), while (f) the *short-periodic* changes are largest in the line with the highest ionization potential (N v).

8.6. Supergiant Atmospheric Pusations

For understanding the light and velocity variations in atmospheres of very early type stars we refer to the investigations by Schwarzschild and Härm (1959) and by Stothers and Simon (1970), described in Section 4.2 of the present volume. Early-type supergiants of masses exceeding $\sim90\mathfrak{M}_\odot$ are vibrationally unstable and may vibrate in various modes, not only in the fundamental and radial ones. This is reflected in the fact that *all* luminous early-type stars are variable – as shown in Figures 114a and 116. But how about the cooler supergiants? These are also variable – Figure 114a. It seems allowed to assume that the *intermediate-term variations* in brightness, colour and velocity are related to coherent mass-motions of large parts of the stellar atmospheres.

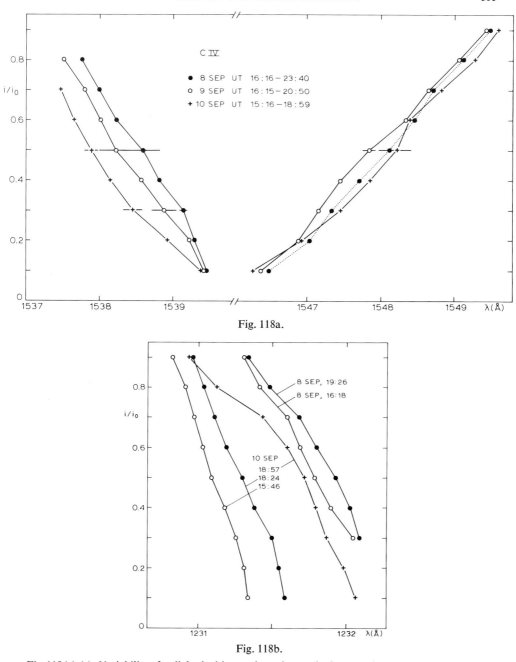

Fig. 118a.

Fig. 118b.

Fig. 118 (a)-(c). Variability of radial velocities on short-time scales in an early-type supergiant. (a) Variation of the two wings on the profile of C IV λ1543 in α Cam (O9.5 Ia): The short-wavelength wing shows a systematic displacement, indicating increase of v_∞ in the course of the three days in September 1978. (b) Very short-periodic changes in the short-λ wing profile of N V λ1235 in α Cam. Particularly the change on 10 September 1978 is very impressive. The long-wavelength-wing did not change! (c) Changes of λ (i/i_0=0.5) for four resonance lines of α Cam on three days in September 1978. For details see the text.

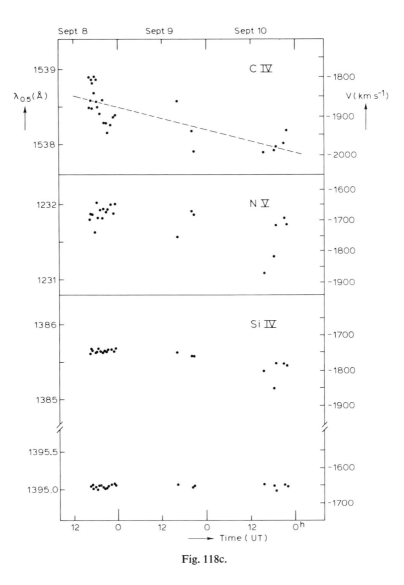

Fig. 118c.

These motions may find their origin in large-scale convective motions or other instabilities. The atmosphere responds to these disturbances with its own characteristic resonance frequency, and this leads then to semi-periodic variations of the atmospheric properties. This assumption is supported by the data shown in Figures 114b and 115, but also by systematic studies of the radial-velocities of stars, such as those presented for α Cyg, and by studies of line-widths of supergiants. But it should be stressed in addition that these very extended atmospheres will practically never pulsate as a whole in the fundamental mode: α Cyg has a radius of 10^8 km, nearly one Astronomical Unit! The time for a disturbance to propagate over such a distance is so long that it is understood already intuitively that different parts of the atmosphere of such stars will pulsate incoherently and independently: A disturbance propagating with sound-velocity traverses 10^8 km in nearly half a year. Translated in mathematical terms, this means that the pulsations are best described by the concept of *non-radial pulsations*. Let us examine this assumption!

Abt (1975b) and later Lucy (1975) have examined whether the quasi-periodicity in early-type supergiants might be ascribed to pulsations. Support for this assumption is given by the observations that the Semi-Period-Luminosity relations for supergiants run more or less parallel to the P-L relation for classical Cepheids (Figures 114b and 115). We refer to Eddington's (1926) relation for adiabatically and radially pulsating stars:

$$P \langle \varrho \rangle^{1/2} = Q \text{ or } P(\langle \varrho \rangle / \langle \varrho_\odot \rangle)^{1/2} = Q_1 = \text{constant}; \; Q = 1.19 Q_1. \quad (8.6; 1)$$

Abt found $Q_1 = 0.062d$. Sterken (1976b) has determined Q_1 for a dozen of supergiants for which he had derived semi-periods (Section 8.2), and he found $Q_1 = 0.077 \pm 0.025d$.

A more detailed analysis is from Burki (1978) who rewrote Equation (8.6; 1):

$$\log P_s = \log Q_1 - 0.5 \log(\mathfrak{M}/\mathfrak{M}_\odot) + 1.5 \log(R/R_\odot). \quad (8.6; 2)$$

With Equations (1.2; 1) and (1.2; 2), Equation (8.6; 2) transforms into

$$\log Q_1 = \log P_s + 0.5 \log(\mathfrak{M}/\mathfrak{M}_\odot) + 0.3 M_b + 3 \log T_e - 12.71. \quad (8.6; 3)$$

This equation should enable one to derive Q_1 from observational data. With assumptions for the spectral type-T_e relation (Johnson, 1966), while deriving the masses from the theoretical evolutionary tracks by Stothers and Chin (1977) – with mass-loss, however, not included so $\mathfrak{M}$ may be too high! – the $\log Q$-values are derived from Equation (8.6; 3). The $\log Q_1$-values appear to range from -0.76 to -1.49 with an average -1.14 ($Q = 0.07$), in fine agreement with Sterken (1976b). It is to be noted that the Q_1-values found are well higher than the value ($Q_1 = 0.04$) one would expect for pulsations in the fundamental radial mode (Schwarzschild, 1941; King and Cox 1968; Maeder, 1980b); see Equation (4.7; 1). This difference may be another indication that the pulsations are *non-radial*. There is a slight dependence of Q_1 on M_b, according to

$$\log Q_1 = -0.08 M_b - 1.78. \quad (8.6; 4)$$

Combining Equations (8.6; 3) and (8.6; 4) one obtains

$$\log P_s = 10.93 - 0.50 \log(\mathfrak{M}/\mathfrak{M}_\odot) - 0.38 M_b - 3 \log T_e,$$

which relation represents the best fit of the relation (8.6; 1) to the observations.

8.7. Causes of Long-Period Variability

A first question is whether the long-term and very-long-term variations (see the definitions in Section 8.2) can be caused by some form of pulsations. To this Stothers and Leung (1971) answer that pulsational overtones can develop in cool supergiants, with characteristic times of a few thousands of days (5 to 10 years). While this possibility is worth a more thorough investigation, we should reiterate our previous remark that the real state of motions in these extended atmospheres must be that of independent responses of different parts of the atmosphere to individual disturbances. The fairly erratic character of the variations indicates that the 'individual disturbances' must be of a rather irregular kind. The very large convective elements suggested by Stothers and Leung (1971) and by Schwarzschild (1975) and detected by De Jager and Vermue (1979), as described in our Section 5.10, may be responsible for part of the *long-term* variations: the life times of these elements may be of the order of a year.

CHAPTER 9

STARS OF TRANSIENT EXTREME BRIGHTNESS;
NOVAE, SUPERNOVAE

9.1. Introduction

Among the eruptive variable stars the novae and supernovae reach large brightness in their maxima. Novae have maximum absolute Blue magnitudes between ~ -6 and -8, and supernovae have -18 to -20. The first values are not far below those of the brightest stable stars; the latter are about a factor 10^5 brighter. In this chapter we describe mainly the atmospheric phenomena; evolutionary aspects are treated in Chapter 4.

First we describe the *novae*. An important monograph on that subject was written by Payne–Gaposchkin (1957); McLaughlin (1960) wrote a classical review on the spectra of novae; a colloquium was edited by M. Friedjung (1977). See also Payne–Gaposchkin's very readable reviews (1977a, b), and that of Gallagher and Starrfield (1978). A special issue of *Astron. Zh.* [**54**, 457ff=*Soviet Astron.* **21** (3)] was devoted to the very fast nova V1500 Cyg 1975.

9.2. Types, Light-Curves and Absolute Magnitudes

It is customary to divide the novae in five classes, according to the rate of light variation with time (Figures 119 and 120).

Na: *Fast novae*; a rapid increase in brightness is followed by a decrease >3 mag in less than 100 days. Examples: GK Per 1901; V603 Aql 1918; V476 Cyg 1920; NQ Vul 1976; exceptionally fast was V1500 Cyg 1975.

Nb: *Slow novae*; the brightness decrease of 3 m after the maximum takes more than 100 days: V841 Oph 1848; T Aur 1891; RR Pic 1925; DQ Her 1934; HR Del 1967 (Figure 120); Nova Vul 1979.

Nc: *Very slow novae* keep their maximum brightness for years and decay only slowly thereafter: RT Ser 1909; RR Tel 1944–49.

Nd: *Recurrent novae* have shown at least two brightness eruptions: RS Oph (1898, 1933, 1958, 1933), T Pyx (1890, 1902, 1920, 1944, 1967). There are seven recurrent novae known, the five others are VY Aqr, TCrB, V616 Mon, U Sco and V1017 Sgr. The mean period is 33 yr. (T Pyx is sometimes placed among the dwarf-novae.)

There is no sharp division line between the fast and the slow novae, and from that point of view it is an arbitrary matter how to distinguish between them. In another subdivision the groups Na, Nb and Nc are taken together to form the *classical novae* (Ne), while the *recurrent novae* are called RNe.

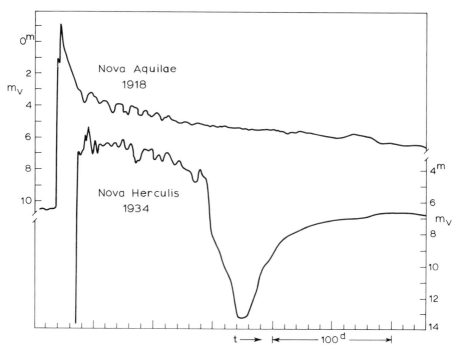

Fig. 119.   Light curves of a fast nova (Nova V603 Aql 1918 – type Na), and of a slow one (Nova DQ
Her 1934 – type Nb).

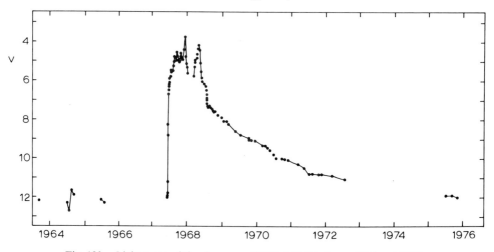

Fig. 120.   Light curve of the slow nova HR Del 1967 between 1964 and 1976.

In addition to the groups described above there are the *dwarf novae* (DNe) with
two subgroups: the Z Cam stars and the U Gem (or SS Cyg) stars. Further, there are
the nova-like variables (NL). These groups will not be discussed here. A last group,
the X-ray novae (XNe) will briefly be touched in this volume (Section 9.5).

Novae, particularly their absolute magnitudes and light-curves, can often be better studied in extra-galactic systems, like M31, than in our own Galaxy. The annual rate of occurrence of novae in our Galaxy is difficult to determine because of obscuration effects, but it is perhaps of the same order as the rate observed in M31, the Andromeda Nebula. That rate is 21 to 33 yr^{-1} (Ford, 1978).

Absolute photographic magnitudes: The average M_{pg}-value of novae at maximum brightness depends on the class, as follows:

Class	Na	Nb	Nc	Nd
M_{pg} at max.	−8.3	−6.4	−6.0	−7.5
M_{pg} at min.	+4	+4.5	+2	+0.5

Mc Laughlin (1960), who gave for $\langle M_{pg} \rangle$ of classical novae at minimum the value +4 to +5, correlated the absolute *visual* magnitude at maximum ($M_{t,v}$) with the rate of decline t_3 (in days) over 3 visual mag after maximum light: $M_{t,v} = -11.5 + 2.5 \log t_{3v}$. For *Blue* absolute magnitudes Pfau (1976) derived $M_{t,B} = -10.67 + 1.80 \log t_{3B}$.

The average absolute photographic – or Blue – magnitude of all novae at maximum is: $\langle M_{max} \rangle = -7.6$. The scatter between the four average values is large; the bolo-metric correction badly known. Yet, it may be very *fundamental* that this absolute magnitude is close to the Eddington limit for stars of about one solar mass! This suggests that nova envelopes at maximum brightness present themselves as opaque radiation-driven stellar envelopes – since novae have about one solar mass (cf. Section 9.3 and Table XLIII).

It seems also significant that at minimum magnitude recurrent novae are brighter by a few magnitudes that the other groups of novae. We return to this observation when dealing with models of the nova phenomenon – Section 9.10. Also inside the groups of novae the scatter in amplitude is large: Nova V1500 Cyg 1975, the fastest and most luminous galactic nova on record had a visual brightness amplitude of *at least* 19 mag! Derived values for $(M_v)_{max}$ for this nova include −9 (Gallagher and Ney, 1976); −9.7 (Kiselev and Narizhnaya, 1977); −9.9 (Barnes, 1976); −10 (Young *et al.*, 1976); −10.1 (McLean, 1976; Wolf, 1977); −10.2 (Shenavrin *et al.*, 1977), and even −10.3 (De Vaucouleurs, 1975). The 'best' value seems to be −10.1 mag. Reference is made to an excellent review by Wolf (1977).

9.3. Pre- and Post-Novae; Novae as Binaries; are Classical Novae Recurrent?

It is a fundamental question to ask for the properties of pre-novae, because after all it is these objects that eventually become unstable. The *photometric behaviour* of 33 pre-novae was investigated by Robinson (1975), chiefly on the basis of routine patrol-plates. Of 18 stars magnitudes were known before and after the explosion.

The following regularities appear:

– The mean apparent magnitudes of pre- and post-novae are in virtually all cases equal. Hence, as a rule, the nova-eruption does not appreciably affect the stars involved.

– About half the number of stars with well-defined pre-eruption light curves show a change in the brightness during a certain period before the eruption. This period ranges between 1 and 5 yr and this suggests some kind of 'building-up' of the eruption-configuration. Very remarkable is the very slow nova RR Tel: about seven yrs before the eruption it started to become a long-period variable ($P \approx 387d$) while also the minimum brightness increased gradually, by $\sim$1 mag, in these seven years. Mayall (1949) found evidence for this $387d$ periodicity even *during* the eruption.

– For one star only: V 466 Her 1960 the pre-eruption light curve differed from the post-eruption one. While the former had rapid intrinsic variations with a amplitude of $\sim$4 mag, the latter had an amplitude of at most $\sim$0.4 mag.

A few more observations: Pre-outburst observations of V 1500 Cyg 1975 showed that at the time of beginning of the outburst (August 28.5) the star had already obtained a luminosity $m_{ph} \approx 14$: its pre-nova brightness had hence increased by $\gtrsim 7$ mag, with an average rate of $\sim$0.3 mag per day (Wolf, 1977).

A similar behaviour was noted for the slow nova Vulpeculae 1979, which had $B > 14$ mag before November 1977, $B = 12.5$ in January 1978, and thereupon increased in brightness to $B \approx 9.5$ in May 1979, whereafter the outburst proper started. Prior to that period, during the whole of the 20th century, it showed many irregular brightness variations and 'mini-eruptions' between $B = 14$ and 16 (Liller and Liller, 1979).

For Nova V533 Her 1963 there are (indirect) indications that it experienced a pre-outburst brightening of $\sim$2 mag over $\sim$40 yr (Patterson, 1979b).

– An incidental observation that needs verification is the detection of X-radiation of DQ Her 1934, evidently at minimum light: the post-nova appears to emit a flux of $(1.1 \pm 0.38) \times 10^{-5}$ photons cm^{-2} s^{-1} keV^{-1} over the energy range 260 to 1200 eV (Coe *et al.*, 1977). Such an observation indicates that even at minimum energetic processes occur in the nova system – such as mass transfer between the components of a binary (?).

The *spectral characteristics of postnovae* have been determined for sixteen cases by Humason (1938); a modern review was given by Warner (1976). The common feature is a blue continuum with spectral types alike O, WR, or B, generally with emission lines of H, He I, He II, and Ca II. It is understandable that in a late phase postnovae are often surrounded by a expanding shell.

Mainly by work of Walker (1956), Kraft (1963, 1964a), Herbig *et al.* (1965), Paczynski, (1965), Mumford, (1967a, b) it became clear that many old novae – or perhaps all of them? – and nearly all dwarf- or sub-novae are binaries. The first well-studied case among the novae, Nova Aql 1918, is a binary consisting of a late-type main sequence star filling its Lagrangian lobe, and the nova proper being a white dwarf exhausted of hydrogen. The period is 0^d139 (Walker, 1956; Kraft, 1964; Weaver, 1974). Robinson (1976) lists orbital periods for a dozen of novae and recurrent novae. The period of revolution is mostly of the order of hours. For instance the period of Nova DQ Her 1934 is 0.194 days.

Table XLIII lists data of known binary-post-novae.

TABLE XLIII

Orbital periods and white-dwarf-masses of binary stars among novae, after Arkhipova and Mustel (1975), Robinson (1976), Warner (1976), with some additions

Star	Class	Orbital period	$\mathfrak{M}/\mathfrak{M}_\odot$	Type*	Companion's type	Additional references
RR Tel	Nc	?		VB	M5III	
T CrB	Nd	227.6d	2.1	DSB	gM3	Kraft (1958)
GK Per	Na	16^{h}26^m	>0.6	DSB	K2 IV–Vp	
T Aur	N	4^{h}54^m		EB		
DQ Her	Nb	4^{h}39^m	0.8	SB; EB	>M4	Schneider and Greenstein (1979)
			1.0			Hutchings et al. (1979)
HR Del	Nb	4^{h}05^m	1.0	SB		Hutchings (1979c)
RR Pic	Nb	3^{h}29^m		EB(?)		
V1500 Cyg	Na	3^{h}28^m		EB(?)		
V603 Aql	Na	3^{h}19^m	0.8	SB		

* DSB: double-lined spectroscopic binary;
  SB: single-lined spectroscopic binary;
  EB: eclipsing binary;
  VB: visual binary.

It is now clear [see the reviews by Robinson (1976) and Payne-Gaposhkin (1977)] that all the binary systems contain a low-luminosity white-blue star, mostly interpreted as a white dwarf surrounded by an accretion disc. The spectrum of the companion is only observable for the few stars with longest orbital periods ($>0.25d$) but this need not mean that systems with shorter or undetectable periods would not have companions. In these cases companions can simply not be large enough to contribute sufficiently to the total luminosity of the system. In all well-investigated cases the non-degenerate companion of *classical* novae is a late-type main-sequence star (or in any case: a star not more than 2 mag above the main sequence) as in the case of Nova Aql 1918. The companions of *recurrent* novae are late-type giants (Eachus *et al.*, 1976; Warner, 1976).

We now dare to *extrapolate* these results to all novae and, following Robinson (1976), we *assume* that post-novae are as a rule close binaries consisting of a cool star, in many cases filling its Lagrangian lobe, and a white dwarf – or, more generally, a star of low luminosity and high temperature. We also assume that the larger star is loosing (hydrogen-rich) material, part of which is accreted by the dwarf. This latter star seems to be the seat of the outburst.

It is indisputable that the mass-transfer from the cool star to the white dwarf must be the clue to understanding the nova phenomenon. See Section 9.10. We note in passing that the orbital period is not correlated with the range of the outburst, neither with the outburst cycle.

Once we assume that all novae are binaries containing a degenerate star we may make one further step on our way to understand nova models and mechanisms, and ask whether classical novae are *recurrent*. That this may be the case is suggested by the known recurrent novae: RS Oph, T Pyx, VY Aqr, T CrB, V616 Mon, U Sco and V1017 Sgr. (WZ Sge is not listed here because it rather seems to be a dwarf nova – Bath and Shaviv, 1979). These have a mean period of 33 yr, but since the total period during which a sufficiently complete variability patrol was made is presently marginally a century, the significance of this value may be slightly doubtful.

The answer to the question whether classical novae are also recurrent can be given on the basis of statistical data on the number-density N_N of possible novae in a well-studied stellar system. If the recurrence period is T, and if N_0 nova eruptions are observed per unit volume during an observing period Δt, then (Ford, 1978):

$$T = \Delta t \, N_N / N_0. \tag{9.3; 1}$$

A difficulty in applying Equation (9.3; 1) is that N_N is unknown. But the calculation of an *upper limit* is nevertheless possible on the basis of the assumption that all novae are white-dwarf-binaries. Let us now assume (following Bath and Shaviv, 1978) that a is the fraction of all white dwarfs that occur in nova systems. Obviously $a \lesssim 1$. With Warner's (1974) space density of nova eruptions of $N_0 = 10^{-7}$ pc^{-3}, and Weidemann's (1967) space density of white dwarfs $N_W = 10^{-2}$ pc^{-3}, and assuming $\Delta t = 100$ yr, one has with Equation (9.3; 1): $a = N_N / N_W = T N_0 / \Delta t \, N_W = 10^{-7} T$. With $a < 1$, $T < 10^7$ yr.

A *lower limit* on T can be derived if one assumes that all nova-like systems are novae in stages between eruptions. With Warner's (1974) value for the space density for such systems

$$N_{NL} = 10^{-6} \text{ pc}^{-3},$$

one derives $a = N_{NL} / N_{NW} > 10^{-4}$, and $T > 10^3$ yr.

Hence $10^3 < T < 10^7$ yr. The precise value of T is not important at this moment, but an important *conclusion* is that classical novae are recurrent, because we think that the life-time of the binary exceeds 10^7 yr. Bath and Shaviv (1978) give arguments that $T \approx 10^5$ yr but more research seems necessary to justify this value.

9.4. The Main Observational Parameters of Novae

The *bolometric luminosity* of a nova is still uncertain since ultraviolet and infrared observations are rare. But it is fundamental that ultraviolet and infrared photometry of the nova FH Ser 1970 (Gallagher and Code, 1974; Gallagher and Starrfield, 1976) showed that for more than 40 days the decline in visual light was to a large extent compensated by an increase of the ultraviolet light caused by a shift of the spectral luminosity maximum to the ultraviolet (Figure 121). So, the bolometric magnitude remained almost constant during at least forty days. This suggestive observation provoked Gallagher and Starrfield (1976) to investigate the temperature variation for a few other – fast and slow – novae, and they found that in the cases investigated an

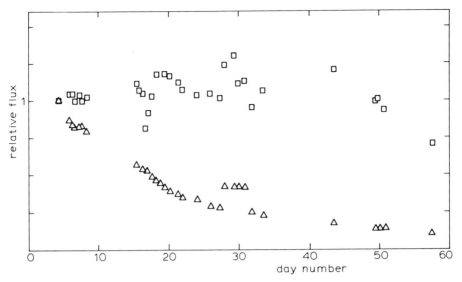

Fig. 121.   Visual (triangles) and bolometric (squares) light curve of FH Ser 1970; the bolometric magnitude remains practically constant during some 40 days. (Gallagher and Code, 1974).

increase of the temperature paralleled the visual decline so that there is a tendency
for the bolometric luminosity to remain fairly constant during a few months after
maximum. For the fast nova V603 Aql the bolometric magnitude during that phase
would be about -6 to -7 mag and the energy radiated during the first 100 days
would be $\sim 2 \times 10^{45}$ erg. This behaviour was somewhat followed by V1500 Cyg 1975
for which ultraviolet observations were available 100 days after maximum (Wu and
Kester, 1977). The luminosities at maximum and 100 days later were determined at
$5 \times 10^5 L_\odot$ and $3 \times 10^4 L_\odot$ respectively. Hence, while in these hundred days the visual
luminosity decreased by a factor of 6600, the bolometric luminosity decreased by a
factor of 20 only.

These observations lead to interesting conclusions. In Section 9.2, where we discussed absolute visual magnitudes of novae, we concluded that nova-envelopes at
maximum are to be considered as opaque radiation-driven stellar winds; or in other
words: expanding stellar photospheres at or above the Eddington limit. We are now
able to state that this is even valid during a period of a few months, roughly 10^2 days
after maximum. This has further consequences: The violet-ward shift of the wavelength of maximum radiation implies an increase of T_e. Since M_b remains approximately constant the photospheric radius decreases with time, which signifies that the
rate of mass loss $(-\mathfrak{M})$ must decrease too, to explain that the level where the continuous optical depth is ≈ 1 moves inward. Actually, precise determinations of M_b
and T_e as functions of time would enable one to determine $R(t)$ and hence $\mathfrak{M}(t)$!

Infrared observations make it possible to refine this picture. Photometry of Nova
V1500 Cyg 1975 shows that maximum brightness comes later for increasingly long
effective wavelength (~ 2 days difference for wavelengths of 0.5 and 5 μm). This is

explained as follows: the particle density in the envelope (the wind density) decreases in the course of time, so that the optical thickness τ_λ of the envelope decreases. If maximum brightness at a wavelength λ is assumed to occur when τ_λ has decreased to unity, and if the infrared radiation is assumed to be due to thermal bremsstrahlung, the observations can be explained quantitatively (Wamsteker, 1979).

The *energetics of the nova phenomenon* is not yet very well quantized. The *emission* integrated over time and wavelength in the visual and near infrared spectral region (but with an application of a bolometric correction) ranges over $E \sim 2 \times 10^{45}$ erg for a slow nova (HR Del), 3×10^{44} to 2×10^{45} for a fast one (V603 Aql) (Pottasch, 1959b; Gallagher and Starrfield, 1976) and 5×10^{44} for the very fast nova V1500 Cyg (Gallagher and Ney, 1976).

Another part of the energy output goes into the *kinetic energy* of the nova ejecta. For these Kopylov (1957) estimated masses between 5×10^{28} and 10^{31} g, inversely dependent on the maximum absolute brightness. His values were still fairly crude. The first systematic determination of nova shell-masses was made by Pottasch (1959b, c) who determined the degree of ionization and n_e-values for five novae with well-observed spectra, from a discussion of intensities measured in the Balmer and [O III] spectra and thus found the amount of matter in the shells. The following values were derived:

$$V603 \text{ Aql } 1918: \mathfrak{M}_{shell} = 37 \times 10^{-5} \, \mathfrak{M}_\odot,$$
$$DQ \text{ Her } 1934 \qquad\qquad 2 \times 10^{-5},$$
$$CP \text{ Lac } 1935 \qquad\qquad 5 \times 10^{-5},$$
$$GK \text{ Per } 1901 \qquad\qquad 7 \times 10^{-5},$$
$$RR \text{ Pic } 1925 \qquad\qquad 24 \times 10^{-5}.$$

For the slow nova HR Del 1976 a surprisingly large amount of outflown mass: $0.03\mathfrak{M}_\odot = 6 \times 10^{31}$ g was found by Antipova (1977), but a smaller value, more in agreement with current data, viz. $\lesssim 2.9 \times 10^{30}$ g ($0.0015\mathfrak{M}_\odot$) was derived by Robbins and Sanyal (1978). Ney and Hatfield (1978) found for the fast nova NQ Vul 1976: $\mathfrak{M}_{sh} = 10^{-4} \, \mathfrak{M}_\odot$. The shell-mass of the slow nova DQ Her 1934 was determined by Ahnert (1960) from a comparison of the periods of stellar revolution from the photometric eclipse curve before and after the nova explosion – an interesting but crude method. From $\Delta P/P = -2\Delta\mathfrak{M}_1/(\mathfrak{M}_1 + \mathfrak{M}_2)$ one derives $-\Delta\mathfrak{M}_1 = 2 \times 10^{-3} \, \mathfrak{M}_\odot$, albeit with a fairly large and unknown uncertainty. We note the enormous difference with Pottasch's value! For the fast nova V1500 Cyg 1975 the mass of the envelope is determined at $7 \times 10^{-5} \, \mathfrak{M}_\odot$ (Boyarchuk *et al.*, 1977a), at $(1 \ldots 5) \times 10^{-5} \, \mathfrak{M}_\odot$ by Kiselev and Narizhnaya (1977), and at $10^{-4} \, \mathfrak{M}_\odot$ by Ennis *et al.* (1977). As an average one may assume: $6 \times 10^{-5} \, \mathfrak{M}_\odot$. De Freitas Pacheco and Sodré (1979) related the hydrogen emissions of novae (assumed to be emitted by the shell) to the ionization due to $L\alpha$ emission from the central star. This is a rather indirect method, and results are somewhat uncertain. The fractional masses of the shells, thus derived for six novae range between $10^{-5} \, \mathfrak{M}_\odot$ and $8 \times 10^{-4} \, \mathfrak{M}_\odot$. Actually, the most reliable determination, involving the least number of assumptions, consists of using the multifrequency radio-observations

of nova-winds or -shells. The observations, interpreted according to the method described in Section 6.13 yield the following results (Hjellming *et al.*, 1979):

$$HR\ Del\ 1967: 8.6 \times 10^{-5}\ \mathfrak{M}_{\odot},$$
$$FH\ Ser\ 1970: 4.5 \times 10^{-5}\ \mathfrak{M}_{\odot},$$
$$V1500\ Cyg\ 1975: 2.4 \times 10^{-4}\ \mathfrak{M}_{\odot}.$$

Summarizing all data we find typical shell-masses of ~ 0.5 to $2 \times 10^{-4}\ \mathfrak{M}_{\odot}$, with
no clear difference between fast and slow novae. Assuming these values and taking
expansion velocities of 300 to 1500 km s^{-1} for slow and $-$ average $-$ fast novae, one finds
kinetic energy losses of $W = 1 \times 10^{44}$ and 2×10^{45} erg for slow and fast novae, respectively; 'average': 10^{45} erg. Similar values are derived by Pottasch (1959b). The above
values for the kinetic energy output are, in order of magnitude, comparable to the integrated visual radiation output. Hence $E : W \approx 1$.

For the recurrent novae the total output energy is smaller, of the order of 10^{43}
to 10^{44} erg (Robinson, 1976), although M_v is not different from those of other classes
of novae.

Next, the *radii of the photospheres* and their development. First a remark about
nomenclature. The terms 'photosphere' and 'shell' and often used to denote the same
object. This gives rise to confusion. Let us agree about the following. The nova emits
a *wind*; that part of the wind that is opaque for (visual) continuous radiation is called
the *photosphere*. Further away from the star is a part that is only opaque for wavelengths of certain lines. This part could be called a chromosphere, in view to our
definition in Section 6.1, but conventionally and not wholly correct the name *shell*
is used. A shell, therefore, is always outside the photosphere.

Radii of the photospheres can be determined from T_e- and M_b-values, with Equation (1.2; 2). Often these values are only very approximately known. Typically, at
maximum $R \approx 150 R_{\odot} \approx 0.7$ AU.

The *slow nova* HR Del 1967 had in the pre-nova phase a photospheric radius of
$0.6 R_{\odot}$. The maximum radius was $170 R_{\odot}$ and seven years after the outburst $R \approx 0.8 R_{\odot}$
(Drechsel *et al.*, 1977). These values agree as to the order of magnitude with those
of Antipova (1977) who finds that at maximum light (Dec. 13, 1967): $R = 65 R_{\odot}$,
$T_e = 7500$ K. These values are comparable to those of photospheres of supergiants. In
later phases the *photosphere* decreases in radius because of the decreasing rate of wind-
flux, while the *shell* is transparent for continuous radiation. While expanding further,
the shell's radius eventually exceeds the diminishing photospheric one by many orders
of magnitude. In 1974 the ratio of radii shell (:) photosphere for HR Del 1967 was
$\sim 2 \times 10^5$ (Drechsel *et al.*, 1977).

The development of the photosphere could be well studied for the *fast nova* V1500
Cyg 1975. Communicated data are not all in agreement, which is related to the physical assumptions used in deriving the radii. At maximum the photosphere had a radius
of $390 R_{\odot}$ (Kinselev and Narizhnaya, 1977) or $570 R_{\odot}$ (Wolf, 1977) to $800 R_{\odot}$
(Shenavrin *et al.*, 1977).

Thereafter the radius decreased rapidly: Boyarchuk *et al.*, (1977a) find that on the

23rd, 88th, and 100th day after maximum the radius was $6.0R_\odot$, $3.1R_\odot$, and $1.1R_\odot$, respectively.

The broad Mg II emission at 2800 Å observed during a few days after maximum by means of the *Copernicus* satellite indicated a shell electron temperature of $T_{el} \approx$ 4000 K (Jenkins *et al.*, 1977), while Voloshina *et al.* (1977) from a spectrophotometric analysis in the visual spectral range found a shell temperature of 10000 to 20000 K. The difference may indicate that the outer shell has a lower T_{el} than the inner shell. On the tenth day after maximum the spectrum became blueish. Photometric brightness changes were seen for the first time on September 1 (Rosino and Tempesti, 1977) and noted with certainty around the 10th day, showing regular brightness variations with a period of $0.^d14$ (Ambruster *et al.*, 1977) – both observations are manifestations that in the visual spectral range the central object became visible through the shell. As the expanding envelope became fainter the relative contribution of the central object increased: in the visual region its contribution increased from 10% after 50 days to 60% after a year. From infrared photometric observations (1 to 10 μm) during a period of one year, starting $2d$ before maximum, Ennis *et al.* (1977) found that the shell had a fairly constant temperature (assumed 10^4 K) during the whole period, but it changed from optically thick to thin after four days after the initial shell expulsion. After some $300d$ the photometric data at 10 μm show deviations from the expected behaviour on the basis of this simple model. This is – very tentatively – interpreted as due to the formation of dust particles.

It may be assumed – as a general rule – that the shell, while gradually becoming transparent looses most of its total thermal energy content by radiative cooling during the expansion.

9.5. Line Spectra, Description and Analysis; Abundances; 'Coronal' Lines; X-Ray Emission and X-Ray Novae

First a few words about classification of the spectra. The spectra of novae near maximum are comparable to those of supergiants or hypergiants; for example Boyarchuk *et al.* (1977b) find the spectrum of the fast nova V1500 Cyg 1975 changing from B2 Ia$^+$ to A2 Ia$^+$ around maximum light. For the slow nova HR Del 1967 the spectrum at maximum was that of an F8 Ia star (Rafanelli and Rosino, 1978). See also Table XLIV. Around and after maximum, the general time-variation of the spectra shows increasing excitation and ionisation.

The following sequence of spectra was suggested by McLaughlin (1960) as a common pattern in the evolution of novae spectra – see also Figure 122:
– *near pre-maximum halt*: early-type (B or A) absorption spectra, strongly displaced towards shorter wavelengths;
– *near maximum*: strong shortward displaced absorption spectra of type A or F with conspicuous emission lines;
– *during first fading from maximum*: the 'typical nova spectrum', being a pattern of emission lines of the P Cygni-type. Later, also emission lines of O I and still later

TABLE XLIV

Physical parameters in the photospheres of Nova (DQ) Herculis 1934, Nova (HR) Delphini 1967, and the F-type supergiant ε Aur (Antipova, 1974, 1977, and Raikova, 1977)

		Nova Del 1967			
	DQ Her	pre-max.	max.	ε Aur	units
v_t (Fe I)	15	7	7	19	km s^{-1}
v_t (ions)	21	33	26	21	km s^{-1}
T_{ex} (0–3 eV)	4500	–	4850	4800	K
T_{ex} (3–5 eV)	7500	7400	–	7400	K
T_{ex} (≈ 7 eV)	6300	–	–	6200	K
T_i	7000	–	6900	6800	K
$\log n_e$	13.4	12.5	12.8	12.7	cm^{-3}
M_v	– 5.7		– 4.6	– 7.1	mag

N II appear. This type of spectrum is often called the '*principal absorption spectrum*'; it is formed mainly in the neutral part of the envelope.

– *after having faded by 1 mag*: a still more strongly shortward displaced absorption-line spectrum. Lines are initially wide and diffuse and develop later into separate components. Most prominent are the Balmer lines; lines of Fe II and other ionized metals occur also. This is the *diffuse enhanced spectrum*. This spectrum is also formed in the neutral part of the envelope.

– *after having faded by about 2 mag* a new phase occurs, the '*Orion phase*' with wide emission and absorption lines of neutral (H I, He I) and singly ionized lines (N II, O II), which may oscillate in position. They are formed in the ionized part of the envelope.

– during the later decline (this was for the slow nova HR Del 15 months after discovery, but for the very fast V1500 Cyg it started already 9 days after maximum!) then follows the '*nebular phase*' with initially forbidden lines of [O II], [N II], ... Later, forbidden lines of higher degree of ionization like [O III], [Fe VI], [Fe VII], ..., [Fe X], [A X] ... are observed (Rosino, 1978). Figure 122 gives the visual and infrared spectra of Nova V1500 Cyg 1975 during the first hundred days (Rosino and Tempesti, 1977). The rule that the degree of ionization increases with time was nicely confirmed by ultraviolet observations of Nova Cygni 1978 (Cassastella *et al.*, 1979a).

The spectral types of novae are conventionally designated by the letter Q; a subdivision was given by McLaughlin (1946). This division, ranging from Q0 through Q9, is based on the occurrence and relative strength of the various observed systems of absorption and emission lines and is essentially a time classification of the above described variation of the spectral type during the nova outburst; the type Q0 occurs just before the maximum, Q1 at the maximum and Q9 in the very end of the declining phase. Thereupon the spectral type becomes similar to early-type spectra (see Section 9.3).

It took the very rapid nova V1500 Cyg 1975 only 28 days to pass through the sequence Q0–Q8 (Karetnikov and Medvedev, 1977). The McLaughlin system is clearly succesful; yet, not *all* novae pass through all the stages.

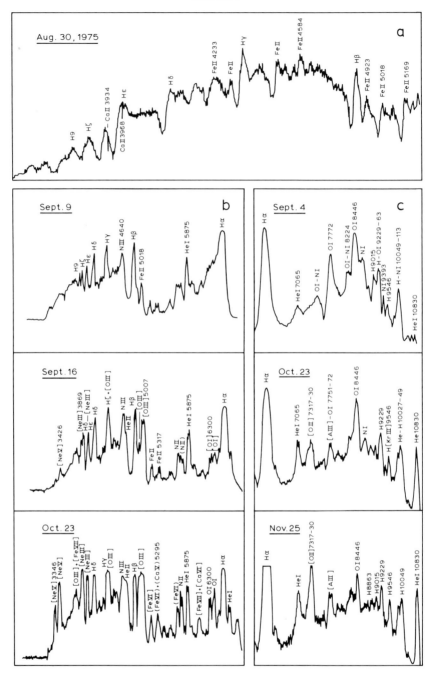

Fig. 122.   Tracings of spectra of Nova V1500 Cyg 1975. (a) From the UV to 5200 Å, on August 30–31, close to maximum light. (b) The 'visible' part of the spectrum and its evolution during September–October 1975. Note the increase of strength of [Ne v] $\lambda 3426$, the formation of nebular lines, and the appearance of [Fe vii] $\lambda 6087$. (c) The near-infrared part of the spectrum during the same period as that covered in (b) (other observing dates). Interesting is the rapid strengthening of He i $\lambda 10830$, and of [O ii] $\lambda\lambda 7317$–30 and the simultaneous fading of O i $\lambda 8446$. (After Rosino and Tempesti, 1977).

Line shapes in novae spectra: typically there are broad emission lines, often with blue-shifted absorptions, indicating outstreaming motions. P Cygni-type profiles oc-cur mostly in the early phases of the nova outburst. Average values for the outward velocities of novae as determined from absorption and/or emission lines range between very wide limits, 100 and 3500 km s^{-1}. Usually, different velocity components rep-resent multiple expanding shells. The measured values do not always represent the real velocity of expansion but can be smaller because they result from an integration over the disk. The precise correction factor depends on the limb darkening and the mechanism of line formation and can thus range between unity (no correction) and 1.4 (absorption line in *thin* atmosphere with full limb darkening); normally the emis-sion-line width gives the real expansion velocity.

The velocities change in the course of time: acceleration during the pre-maximum phase, from 1300 to 1900 km s^{-1}, has been observed for the fast nova V1500 Cyg 1975 (Fehrenbach, 1975b). A Hβ absorption at -3850 km s^{-1} occurred 1.2 days after the optical maximum; it developed into emission at 2.2d (Ferland, 1977). Near max-imum and in the first weeks thereafter the spectrum was dominated by wide emissions with shortward broad diffuse absorption features, with v_R, as derived from the emis-sions: -1700 km s^{-1} on day 0, and -2400 km s^{-1} on day 24 (23 September), when the absorption component disappeared. The expansion velocity, as derived from the widths of the emission bands, increased from 860 km s^{-1} to 1500 km s^{-1} on 3 Sep-tember (day 4), and stayed constant afterwards.

Usually there are several 'expanding shells' (or in any case: 'separately moving elements'), as observed in multiple emission- or absorption-line components. The velocities and multiplicities vary during the eruption process. The Balmer lines in the spectrum of the fast Nova V1500 Cyg 1975, as well as other emission lines, observed in the few weeks after maximum light, show periodic variations in which the 0.14d period is detectable (cf. Rush and Thompson, 1977). But the phase of the variations differs for individual wavelength points in a line. This can be explained (Campbell, 1976) by attributing these variations to a combination of periodic variations in the central ionizing source, and the light travel time towards an expanding shell. Campbell thus finds a mean expansion velocity of 1870 km s^{-1}. This model was elaborated in more detail by Hutchings *et al.* (1978b) who found best agreement for a model with an equatorial ring expanding with $\sim$1700 km s^{-1} and polar blobs expanding with $\sim$620 km s^{-1}. (This model does not account for the observed changes of the period; the consequences of this effect are described in Section 9.7).

Other spectroscopic measurements, taken with less high time resolution than those described above (cf. Kupo and Leibowitz, 1977; Aab *et al.*, 1978) did not allow the authors to build a detailed model of the event but indicate in any case that the ejec-tion of matter took place in distinct events and perhaps in different directions: a main shell was ejected with an expansion velocity of 1900 km s^{-1}, and in addition one or two cloud pairs, in which the members of each pair were ejected in opposite direc-tions.

The *slow nova HR Del* had during its maximum phase – which lasted for about a

year – at least nine different absorption systems in the spectrum, with velocities ranging from -170 to -1790 km s^{-1} (Rafanelli and Rosino, 1978) indicating the presence
of various shells of low excitation. In the course of seven years the expansion velocity
of the main shell decreased from 500 to 400 km s^{-1} (Antipova, 1978). Multiplicity
of absorption lines was also observed for the fast nova V1500 Cyg 1975, for which,
in the post-maximum phase, the emission lines became cut through by a number of
absorption lines, suggesting – as in the case of HR Del – the presence of shells of low
excitation in front of the line-emitting envelope.

For the *fast nova NQ Vul 1976* the velocities measured in the absorption lines were
-950 km s^{-1} during the initial phase ($\sim5d$ after maximum). One year after maximum
there were four components in the lines with average velocities between $+80$ km s^{-1}
and -550 km s^{-1} (Klare and Wolf, 1978). Similar values were reported by Galkina
(1977).

In the range of atmospheric depths of the envelope a *velocity gradient* may be expected, and this is also suggested by the fact that radial velocities derived from spectra
taken at the same moment show different v_R-values for lines of different excitation/
ionisation and strength. For V1500 Cyg 1975 the lines of N III, Fe II, and [O III] ex-
panded in the first months after maximum with a velocity of 600 km s^{-1} relative to
the Balmer lines (Weiler and Bahng, 1976).

Photospheric and shell abundances are derived from spectrochemical analyses. Such
a research on the basis of the curve-of-growth method was performed for several
novae by Mustel and co-workers, see Mustel and Baranova (1965), Raikova (1977),
and Antipova (1974, 1977). It may be significant that there is a certain degree of overexcitation of the high-excitation levels, perhaps due to a temperature gradient in the
part of the atmosphere where the lines are formed. The effect is quite similar in a
supergiant like ε Aur. McLaughlin (1937) already drew attention to the similarity
between the spectral types of pre-maximum novae and F-type supergiants like ε Aur
and ι^1 Sco. This similarity was further illustrated by Antipova (1974) from which Table
XLIV is derived. Also the chemical abundances are virtually equal, apart from the
elements C, N, and O for which the relative over-abundances in DQ Her 1934 range
from between factors 1 to 10 (C) to ~10 to 100 (O and N). Similar values were found
for this nova by Williams *et al.* (1978), and for the slow Nova HR Del 1967 by Antipova (1977) and Raikova (1977). The values slowly increased tenfold over a period
of approximately a month. The overabundance of these elements is also shown by
the appearance of very intense CN-bands just after the light maximum. Similar
studies, based on the emission lines, and yielding virtually the same result were carried
out by Pottasch (1967), Ruusalepp and Luud (1970), and de Freitas Pacheco (1977).
The latter found slightly larger overabundances than the others. In contrast to these
results Robbins and Sanyal (1978) found solar abundances for O for HR Del 1967;
although an attempt was made to explain the differences between this and the earlier
conflicting results no convincing arguments could so far be advanced (Williams *et
al.*, 1978).

In V1500 Cyg 1975 C, N, O, Ne are overabundant with regard to solar values by factors 25 to 30; 100 to 220; 20 and 20 respectively, whereas the Fe abundance is solar (Ferland and Shields, 1978; Boyarchuk, 1977).

A significant observation was made by Adams and Joy (1933) and later by McLaughlin (1953) who observed lines of *forbidden transitions in highly ionized atoms* ([Fe x], [Fe xiv]) in the visible spectra of novae. See also Figure 122. These 'coronal lines' are particularly important in slow novae and are observed after the nova maximum. They are obviously indicative for the origin of a hot plasma in addition to the relatively cool plasma responsible for the continuous spectrum and most of the other lines. Grasdalen and Joyce (1976) identified $\sim$ ten coronal lines in the infrared spectrum (1.9 to 4.2 μm) of V1500 Cygni 1975. They are attributed to forbidden transition in Si vi, vii and ix; Al v, vi, viii, and ix; Ca iv and viii; and Mg viii. In addition [Fe xi] 6374 Å, [Fe x] 7892 Å and [S viii] 9911 Å were observed. All these transitions, but for Ca iv would indicate ionization temperatures of the order of one or two million K. These lines were studied in some detail by Ferland *et al.* (1977). They appeared about 35 days after the outburst and persisted for some four months. As the coronal lines weakened the [Fe vii] 6087 Å line appeared. The very slow nova RR Tel (maximum in 1944–1949) studied by Thackeray (1977b) had initially a cool sharp nebular spectrum which developed through a WR phase into a 'coronal' spectrum with increasing stages of ionisation: [Fe iii] to [Fe vii], [Ni vii], [Ca vii]; e.g. [Fe iv] reached maximum intensity in 1955; [Fe vi] in 1960; [Fe vii] around 1966, but is was already visible as early as 1959. The lines of highest ionization show double structure indicating outward streaming matter. In the spectrum of the slow nova HR Del 1967 Rafanelli and Rosino (1978) found that in the 'nebular stage', which started fifteen months after discovery, and three months after the final decline, wide emission bands occurred of H, He i, N ii, and forbidden lines over a large range of degrees of ionisation: [O i], [O iii], [Ne iii], [Ne v], [Fe vi], [Fe vii], [Fe x]. All lines had complex structure. The concurrent occurrence of lines of ions with so greatly different degrees of ionisation demands a multi-temperature model for the source of photoionisation. Thus, a model with two sources, with $T \approx 4 \times 10^4$ K and $T \approx 2.5 \times 10^5$ K, each with about the same luminosity was proposed to explain the observations (Tylenda, 1978).

The customary model for explaining the hot envelopes is the following: assume there are two nova envelopes, an outer and an inner one. The inner envelope expands faster than the outer and impacts onto the outer from below. Two shock waves propagate away from the collision site in either direction. In the – optically fairly thin – regions at the shock fronts the temperature may attain several million of degrees (Bychkova and Bychkov, 1976). Shields and Ferland (1978) forwarded a slightly different suggestion: supersonic random motions in the ejected gas would 'shock' ejecta elements to temperatures of $\sim 10^6$ K.

The formation of coronal lines in a thus heated medium was discussed by Gorbatskii (1972, 1973), who included in his model calculations a computation of the temperature-relaxation behind the shocks and thus derived the time-dependence of T, and of the coronal emissions.

The explanation by shock heating would not apply to the observation of the X-ray burst of the old nova GK Per. That burst lasted for nearly two months and was tentatively ascribed to accretion of a wave of material from the primary via Roche-lobe overflow to the white dwarf of the system (King *et al.*, 1979; cf. also Brecher *et al.* (1979) for a general discussion of X-ray emission by novae).

In this relation we mention the *X-ray novae*. A well-known example is Nova Mon X-1. The most likely explanation is that such objects are comparable to the common novae, but with the white dwarf replaced by a neutron star. With that model they are very similar to the X-ray binaries (Section 4.12 and Table XXXV). The difference between the two groups of objects would be that X-ray binaries are continuous sources of X-radiation (by accretion?), while the X-ray novae are occasionally so (by a nuclear instability? – cf. Section 9.10). In that connection there is an interesting observation by Eachus (1975) who identified the X-ray source A0620–00 with the recurrent nova V616 Mon, that also erupted in 1917 and reached $M_v = -5.9 \pm 0.5$ mag at maximum (Eachus *et al.*, 1976). The X-ray spectrum is characterized by $kT = 1.3$ keV, and the increase in the X-ray flux was at least a factor 10^4 (Doxsey *et al.*, 1976).

One of the members of this system, the X-ray source, of mass $\sim 0.9 \mathfrak{M}_\odot$, is a compact object (Duerbeck and Walter, 1976); the other component is an M5 IV star. The object had maximum X-ray brightness (3–6 keV) on August 11, 1975 (Elvis *et al.*, 1975; Doxsey *et al.*, 1976). The decline of X-ray intensity was virtually exponential but was interrupted by appreciable ($\sim 2 \times$) intensity increases in October 1975 and February 1976. A more rapid decline followed in March 1976 (Kaluzienski *et al.*, 1977). Further references to this object are in Friedjung (1977).

Another – actually the first discovered – X-ray nova is the object Cen X-4 which was observed as an X-ray source with the Vela satellites in July 1969 (Conner *et al.*, 1969), reached a flux ~ 35 times the Crab nebula and then faded. It re-appeared as A1524–61 in May 1979 (Kaluziensky and Holt, 1979); reached a flux ~ 4 times the Crab on May 17, started decline on May 20 and became invisible on June 8. This object is identified with a star with 1950 coordinates $\alpha = 14^h 55^m 19\overset{s}{.}63$; $\delta = -31°28'09\overset{''}{.}0$ (Canizares *et al.*, 1980). The spectra show a strong blue continuum with absorption lines and weak Balmer emission, He II $\lambda 4686$ emission, and perhaps N III, and strong emission complexes near $\lambda 3775$, 3850, and ~ 3940.

9.6. The Continuous Spectra; Dust Shells

In Section 9.4 we referred already to the observation that – broady speaking – after maximum the photospheric temperature of a nova tends to increase. Let us now examine this in more detail! Because of the broad lines the continuous spectra are hard to determine. In the period when the photospheric temperature is not too high, it is therefore advisable to take recourse to the infrared spectral region where lines are less numerous than in the visible and ultraviolet. Gallagher and Ney (1976) made broad-band observations of Nova V1500 Cyg 1975 in 11 bands ranging from 0.5 to 12.5 μm, and found that during the *first* four days the spectrum was that of a black-body with T_c decreasing from 9000 to 5500 K, indicating an optically thick expanding envelope.

Qualitatively, this was confirmed by Stone (1977) who found that the spectrum resembled B9 Ia about one-half day before maximum light, while the continuum changed to a 5000 K blackbody about half a day after maximum. Gallagher and Ney stated further that after the fourth day post maximum the spectrum became similar to one defined by a constant monochromatic flux, F_ν=constant, which means that in that phase radiation may have been scattered at free electrons (Thomson). The mass of the envelope was estimated at 10^{28-29} g, which leads to a kinetic energy of $\sim 10^{45}$ erg. Apparently the *conclusion* is that in the period *around maximum* the *temperature decreases*; this is understandable, because the radius of the photosphere increases while the flux remains at most constant or decreases slightly. In the immediately following phase the photospheric temperature *increases*, while the radius decreases.

From spectroscopy of V1500 Cyg in the visual region (0.3 to 0.7 μm) Boyarchuk *et al.* (1977a) find the stellar temperature to increase from 43000 K (day No. 23) to 46000 K (day No. 38). Voloshina *et al.* (1977) find T=30000 K at day 67. Hundred days after maximum the continuous spectrum, as judged from ultraviolet measurements made with the Astronomical Netherlands Satellite (Wu and Kester, 1977), was like that of a black body with T=65000 K.

This all fits into the picture of a strongly decreasing radius (Section 9.4) with slowly decreasing flux.

Summarizing, the observations of the continuous spectrum confirm and extend our earlier conclusions (Sections 9.2 and 9.4): the time history of nova photospheres as inferred from simultaneous observations of M_b and T_e [e.g. by using an expression like (2.5; 1)] shows a linear increase of the diameter and a decrease of T_e till after light maximum, whereafter the *photospheric* level recedes (Gallagher and Ney, 1976; Barnes, 1976), apparently because of a decrease of the rate of mass-loss $-\mathfrak{M}$. Concurrently the photospheric temperature first decreases and later increases strongly.

The formation of a dust shell is characteristic for certain – but not for all – novae.

One of the best studied, and actually the first well-investigated case was that of Nova FH Ser 1970 (cf. review by Gallagher, 1977). Nothing 'unusual' was apparent in the spectrum and light-curve of this nova during the first 55 days after outburst, but then, in a few days only, a hot infrared radiation component appeared (Hyland and Neugebauer, 1970; Geisel *et al.*, 1970). Hundred days after maximum virtually all the luminosity of this nova had been shifted to the infrared. The obvious interpretation is that of the formation of a dust shell. Such a shell maye even become opaque to visual radiation. Nova FH Ser 1970 and V1229 Aql 1970 have become luminous infrared sources emitting more than 10^{45} erg at wavelengths >2 μm. Similar infrared emission was observed from HR Del 1967, NQ Vul 1976 (Ney and Hatfield, 1978), V1301 Aql 1975 (Vrba *et al.*, 1977), Nova Ser 1978 (Szkody *et al.*, 1979). The formation of the dust shell of NQ Vul 1976 is shown in a very convincing way by Figure 123, after Ney and Hatfield (1978), who measured the infrared intensity in ten bands between 0.9 and 18 μm. Indications of the formation of the shell are the

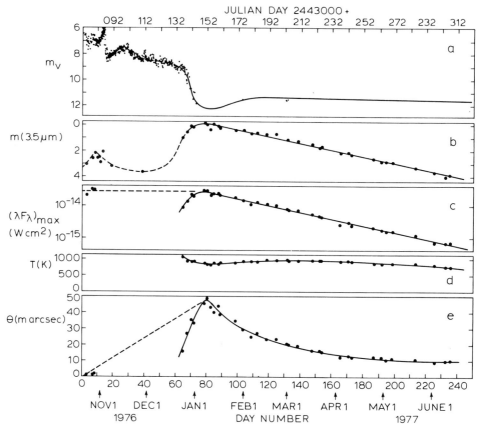

Fig. 123. Evolution of Nova NQ Vul 1976. (a) visual light curve. (b) light curve at 3.5 μm: as the 3.5 μm brightness increases, the visual luminosity decreases, indicating the formation of a dust shell. (c) variation of $(\lambda F_\lambda)_{max}$ for the blackbody phases: at day 80 the luminosity was equal to that of the initial expanding 'photosphere'. (d) blackbody temperature of the dust shell, remarkably constant during the period of observation. (e) blackbody angular diameter, determined from $(\lambda F_\lambda)_{max}$ and T. (After Ney and Hatfield, 1978).

simultaneous drop of visual brightness and the increase of infrared luminosity. The simultaneous drop in shell temperature from 1100 to 800 K is attributed to an increased efficiency in re-radiating as the grain size increases. No silicate emission was observed; this is interpreted by the assumption that the grains are mainly carbon. It is also interesting to note that the total stellar flux, characterized by $(\lambda F_\lambda)_{max}$ remains about constant – in accordance with findings by other observers (Section 9.4).

It is then tempting to associate the characteristic drop in luminosity shown by slow novae of the DQ Her type – about 2 months after maximum – with the infrared emission, and to explain it through the formation of dust particles. The observations of Nova Del 1967 were explained as due to the formation of dust particles in an expanding carbon-rich shell of gas (Malakpur, 1973; Clayton and Hoyle, 1975; Clayton and Wicramasinghe, 1976). After an initial period of rapid growth of the grain particles, in which also the infrared luminosity increased rapidly, a period of slower growth set in when the expanding gas thinned out. At the same time the temperature of the dust particles decreases from 2000 to 900 K. This was also observed for Nova Ser 1978 for which infrared observations were made between 36 and 127 days after maximum. Already in the first observations a dust-shell was evident. It persisted during the whole period of observations, while T_{dust} decreased from 1100 K to 900 K (Szkody et al., 1979). For Novae HR Del 1967, FH Ser 1970, V1301 Aql 1975, NQ Vul 1976 and Ser 1978 the masses of the dust shells were respectively $10^{-6} \, \mathfrak{M}_\odot$, $4 \times 10^{-7} \, \mathfrak{M}_\odot$, 0.3 to $3 \times 10^{-8} \, \mathfrak{M}_\odot$, $2 \times 10^{-7} \, \mathfrak{M}_\odot$, and $5 \times 10^{-7} \, \mathfrak{M}_\odot$. At the time of maximum infrared flux the dust shells had approximate radii of $6 \times 10^9 \, km \approx 40 \, AU$.

For Nova FH Ser 1970 the observations fit best to an expansion velocity of the gas shell in which the dust particles are formed, of 1200–2000 km s^{-1} (dependent on the assumed luminosity of the star). The grain temperature of 900 K is reached after ~90 days.

Linear polarization of the continuous spectrum may also be an indication for the occurrence of dust, as was found for Nova Cyg 1978 by Piirola and Korhonen (1979).

Concluding this section we *summarize* that dust particles may form in the winds of certain novae during the declining phase of the nova process. Generally, the temperature of the dust particles decreases in the course of time. This may be related to the decreasing stellar flux. It is not yet clear why dust particles do form around certain novae and not around other.

We refer also to the review paper by Gallagher and Starrfield (1978).

9.7. Rapid Brightness Fluctuations of Novae

Rapid variations in the brightness of a nova with very short periods (minutes!) must be due to changes in the central object, they cannot be due to the envelope. Such variations have so far only been detected unambiguously for a few cases and mostly at minimum light, when the degenerate star is visible. The first was DQ Her 1934. The period is 71.07 s (Walker, 1958; Herbst et al., 1974). A slow rate of change of the period has been observed $\dot{P} = (-2.69 \pm 0.08) \times 10^{-5}$ s yr^{-1}. For RR Pic the period is of the order of 30 s (values between 26 and 373 s are quoted – Warner, 1976). Nova V533 Her 1963 has brightness oscillations with a period of 63.63 s and a mean amplitude of 1% (Patterson, 1979b).

For Nova V1500 Cyg 1975 flickering has been observed with amplitudes up to 0.05 mag and characteristic periods at least as short as 100 s, but this flickering appears not to be very persistent (Kemp et al., 1977; Cacciari et al., 1977; Ambruster et al., 1977). These variations may be alike those observed in some dwarf novae, and must most probably be considered as a random process. An interesting feature that may be a clue to the understanding of the rapid oscillations is the phase-shift observed. While the oscillations are absent at mid-eclipse they appear to undergo a phase shift of $+360°$ during the eclipse. The phase shift is smooth, begins at the start of the eclipse and terminates at its end (Robinson, 1976).

Very interesting are the *very rapid brightness fluctuations* ('flickering') observed for Nova Cyg 1978 at *maximum*! The flickering occurs at frequencies below ~5 Hz, with no stable oscillations detectable. Short-lived oscillations at frequencies of 15 and 81 Hz were seen for short periods, lasting 100 to 200 s (Giuricin, *et al.* 1979).

An unique feature, discovered for Nova V1500 Cyg 1975 is the three-hours brightness oscillation. Different periods had been reported by various authors, until it was understood that the period changed during the period of observation: a *fundamental* discovery! Some of the periods communicated are e.g. $P=0\overset{d}{.}1382$ (Kemp *et al.*, 1977); $0\overset{d}{.}14096$ (Rosino and Tempesti, 1977); $0\overset{d}{.}1490\pm0.0001$ (Chia *et al.*, 1977). Semeniuk *et al.* (1977) found that the period decreased from $0\overset{d}{.}1490$ in September 1975 to $0\overset{d}{.}1380$ in autumn 1976. This was confirmed by Patterson (1978) who found furthermore that it *increased* again, by 1% in 1977; the same is found for the associated v-values. The light amplitude varied greatly, in the range of 0.03 to 0.7 mag. After a first maximum in the amplitude, 0.15 mag in mid-September 1975 it decreased, but it reached 0.7 mag in October 1976. The period and amplitude stabilized in early 1977 at $P=0.139617\pm0.000002d$, and $\Delta M\approx0.65$ mag (Kemp *et al.*, 1980; Patterson, 1979a).

Against the interpretation of this variation by the binary hypothesis are the facts that the change of period has a sign, opposite to the normal case, and is larger than any possible binary-related value by at least an order of magnitude (Fabian and Pringle, 1977). One of the proposed explanations is that the variations would be due to some 'searchlight' mechanism, related to a radiant spot stellar spot (Campbell, 1976; Hutchings and McCall, 1977). The star should be non-spherical in shape and characterized by an equatorial ring and polar blobs (Hutchings *et al.*, 1978b, c). The most likely hypothesis is that the gas, outflowing from the star assumes a spiral shape around the star, and since this gas is fairly dense it gives the star an asymmetric shape (Fabian and Pringle, 1977; Ivanov, 1978; Patterson, 1978). The observed changes in the period might be related to changes in the shape and density of the outflowing gas spiral, rotating with the star. The model needs a total mass ejected of $\sim10^{-3}\,\mathfrak{M}_{\odot}$, during a period of about 500 days, which is *much* more than the observed mass-loss! Basically the two proposed explanations are similar in that both assume an asymmetric rotating radiating body.

9.8. Morphology of Nova Envelopes and Surrounding Nebulosities

In slitless high-resolution spectra the emission lines of the novae envelopes observed several months after the outburst show a remarkable fine structure which indicates a certain spatial distribution function of the expanding gas. In Section 9.5 we mentioned that the envelopes of most novae consist of various shells. In the case of the slow nova HR Del it was thus found that the activity of the central star in emitting shells of gas persisted for at least one year (Rafanelli and Rosino, 1978). Important is that the outburst is not spherically symmetric, as was shown for the first time by Wright for Nova V603 Aquilae 1918. Exposures obtained with different position angles of the refracting edge of the spectrograph with regard to the expanding shell show the variation of the shape of the monochromatic images. For Nova V603 Aql 1918, the images were symmetric in position angle 112°, they were anti-symmetric in position angle 202° (=112°+90°!). Such observations were later used by Weaver (1974, with earlier references) to determine in a fairly detailed way the three-dimensional distribution of the emitting mass around the nova. A similar treatment is possible for other novae. While details in the resulting models may change from one case to the other and may even not always be significant, it is consistently found that equatorially symmetric rings and polar caps characterize the ejecta of various novae. [DQ Her: Mustel and Boyarchuk (1970); HR Del 1967, LV Vul 1968, JH Ser 1970: Hutchings (1972); V1500 Cyg 1975: Boyarchuk and Gershberg (1977); Hutchings et al., (1978b)]. Differences refer to the existence or not of spherical shells, and the number of ejected blobs or rings. The nebulosities of nova HR Del 1967 had $T_{el} \approx 13\,000$ K around 1970, and polar caps and equatorial rings with $n_e \approx 5 \times 10^6$ in 1969 and 9×10^5 in 1970 (Malakpur, 1979). The image showed equatorial rings at 2″ in 1976 (≈ 1600 AU if the distance is 800 pc – Sanyal (1974)) and equatorial blobs at 0″.6, – which would be 480 AU (Soderblom, 1976).

Postnovae are not only surrounded by distant, expanding shells: From the intensity variation of various emission lines in Nova DQ Her 1934 during its photometric binary cycle of $0^d.194$ it is apparent that the postnova is also surrounded by a fairly close shell emitting He II $\lambda 4686$, which follows the eclipses of the star, and a more extended shell of higher ionisation (C III/N III $\lambda 4640$) which shows only a shallow, broad eclipse (Margon et al., 1977).

9.9. Theories and Models of the Development of the Photospheres of Novae

Two kinds of hypotheses have been advanced to explain the light-curves of novae. Grotrian (1937) suggested *continuous emission of matter*, as in P Cygni and Wolf–Rayet stars. The distance of the photosphere of the nova to the star's center is in that case a function of the assumed velocity and the flux of the matter. If the expelled shell is optically thick its radius determines the increased luminosity. Actually, the luminosity can be derived from these data, and – in principle – any light curve can be explained by a suitable choice of the assumed basic data such as the expelled mass $-\mathfrak{M}(t)$ and velocity $v(t)$.

Pottasch (1959a) has dealt with the other extreme, the case of *instantaneous ejection of matter*. The inclusion of radiation transfer in the theory was due to Sparks (1969). Thereafter, Nariai (1974) has further elaborated the case of instantaneous emission of matter. After the very first phase, when the expansion process is governed by hydrodynamical processes, the accelerating processes are gravitation and radiation forces. Nariai could show that the latter process is unimportant as compared to the former; hence the photospheric $R(t)$-variation just depends on the gravitational acceleration, and can be computed. The nova's brightness can then be derived as soon as the bolometric correction is known.

It then appears, as was shown for the first time by Friedjung (1966a, b) that instantaneous ejection models are incapable of explaining the observed light-curves. Particularly the fundamental observation of Gallagher and Code (1974) that the bolometric luminosity remains approximately constant during a certain period after maximum forces us to accept that mass-loss occurs during this period, but with a rate, declining with time. Cf. Section 9.2 and particularly 9.4. This supports Grotrian's model.

In elaborating this model Bath (1978a) attempted to describe the various phases of the nova-process in a model based on one parameter only, *viz.* the mass-loss rate $-\dot{\mathfrak{M}}$. It then appears that the initial eruption – the origin of which is left open, see Section 9.10 – generates a luminosity close to and sometimes even in excess of the Eddington limit. This was already stressed by Friedjung (1966c) and appears also from our Table I (Chapter 1). As a result a radiatively driven wind originates, optically thick till a distance $> 10^7$ km. An initial mass-loss rate of $\gtrsim 10^{22}$ g s^{-1} appears to be needed. In the subsequent optical intensity decline the total luminosity stays close to the Eddington value; the temperature increases, shifting the main emission towards the ultraviolet, with a slow decrease of $-\dot{\mathfrak{M}}$.

An attractive aspect of this model is that it yields a relation between the rate of mass-loss $-\dot{\mathfrak{M}}(t)$ and the absolute Blue magnitude $M_B(t)$ for an average (fast) nova. This is illustrated in Figure 124. In addition, the increase of T_e with time, as described

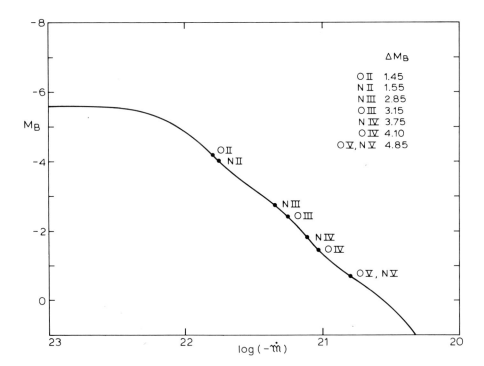

Fig. 124.   Relation between M_B and $-\dot{\mathfrak{M}}$ for a typical (fast) nova, also showing the increasing shell ionization with time. $\dot{\mathfrak{M}}$ is in g s^{-1}. (From Bath, 1978b).

in Section 9.4, produces an increasing flux of increasingly-high-energy photons, so that by radiative ionization the shell becomes more and more ionized. This explains the observed occurrence of ions of increasing degree of ionization in later phases of the nova-phenomena, as illustrated in the figure.

Quantitatively the optically thick wind-model has been worked out by Ruggles and Bath (1979). While they confirmed that the photospheric luminosity must be close to the Eddington limit, another very important aspect is that they could show that only models with a stellar radius comparable to those of white dwarfs can well reproduce the observations. This confirms the current nova model.

It is *concluded* that the nova process can be described by the ejection of an envelope with mass $\sim 10^{-4}\,\mathfrak{M}_\odot$ from a white dwarf in about 10^2 days.

9.10. The Origin of the Nova Instability

The previous sections have provided arguments for the following model of a nova. In a semi-detached binary system a late-type main-sequence star (for classical novae) or giant (for recurrent novae) is accompanied by a white dwarf. The degenerate companion is so near to the other object that in most cases the latter fills its Lagrangian lobe, and a gas-stream flows from it to the dwarf. Probably, there is a centrifugally-supported accretion disk around the dwarf; probably there is a hot spot where the infalling gas hits the disk. The latter aspect is supported by asymmetry in the light-curve – cf. the study of DQ Her by Schneider and Greenstein (1979). The model is illustrated in Figure 125.

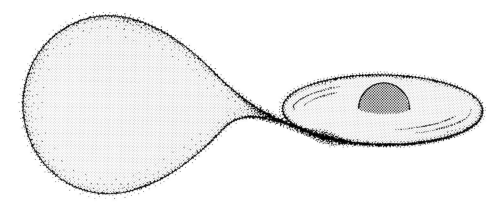

Fig. 125.   Model of novae (Bath, 1978b).

It is at this point that we have to notice the similarity between novae and X-ray binaries (Section 4.11). But in novae binaries the degenerate companion is a white dwarf, while in X-ray binaries it is a neutron star. Also these binaries can develop into novae: the X-ray novae are described in Section 9.5.

The instability that produces the nova outburst seems related to the binary character of the nova and to the transfer of material from the main-sequence star to the dwarf. To explain this, several theories have been advanced. In one group of theories the instability is intrinsic to the accretion, in another group it has a thermonuclear character. In both groups of theories the difference between classical and recurrent novae is that the rate of mass-transfer from the giant stars (that are the companions of the recurrent novae) is assumed to be much larger than for the classical novae – where the companion is a main sequence star. This higher rate of mass-transfer has two consequences: (a) a larger luminosity of recurrent novae at minimum (see the observational data in Section 9.2), and a faster repetition rate of the outbursts. Let us next consider the two groups of nova (c.q. dwarf-nova) theories!

In the first group (Bath, 1977) one considers the luminosity L_a due to liberation of potential energy of accreted material (mass accretion rate $\dot{\mathfrak{M}}$) onto a body with mass $\mathfrak{M}_1$ and radius R_1:

$$L_a = \frac{G\mathfrak{M}_1}{R_1} \dot{\mathfrak{M}} \text{ erg s}^{-1}.$$

For white dwarf masses, radii and luminosity ($\sim 10^{32}$ erg s^{-1}) one finds $\dot{\mathfrak{M}} \approx 10^{15}$ g s$^{-1} = 10^{-11} \mathfrak{M}_\odot$ yr^{-1}, a reasonable value for a binary system. Since during outburst $L_a \approx 10^{34}$ to 10^{35} erg s^{-1} for a dwarf nova, and 10^{38} to 10^{39} erg s^{-1} for a classical one this would correspond to a sudden increased accretion of 10^{-9} to $10^{-8} \mathfrak{M}_\odot$ yr^{-1} for a dwarf nova, and to a mass-flow of 10^{-5} to $10^{-4} \mathfrak{M}_\odot$ yr^{-1} for classical ones. This is perhaps caused by an instability in the outer parts of the accretion disk. The character of this instability is still obscure. In view of the duration of the maximum phase of the nova outburst (~ 0.01 to 0.1 yr) the total mass involved would be $\sim 10^{-6} \mathfrak{M}_\odot$ for an average nova, smaller than the observed envelope masses. An obvious difficulty with this theory is how a sudden increase of accretion would occur. No *satisfactory* explanation has so far been given. A positive aspect is that the pre-outburst luminosity increase of certain novae is understandable with this model (cf. Section 9.3).

Another theory assumes that hydrogen-rich material, accreted by the white dwarf from the cool and larger companion that fills its Roche lobe (see Section 9.3) will be accumulated by the accretion disk and finally deposited, as a gradually thickening layer, in the dwarf's photosphere. As accretion continues this material will gradually be compressed, and heated until it reaches ignition temperatures for hydrogen reactions (Starrfield et al., 1972) to occur in the envelopes. For low temperatures ($T < 15$ MK) the p–p chain closes via ^{3}He (e^-) $^3T(p, \gamma)$ ^{4}He (Harrison, 1978). However, since the proton–proton reaction chain involves very long reaction times it is the CNO-cycle, with its much shorter time scales, that starts ignition earlier, and at higher temperatures. For that reason the number of CNO nuclei in the envelope is a crucial parameter, and CNO-enhancement is needed to produce fast novae (Starrfield et al., 1977). But also this cycle never reaches equilibrium because the time scale of the nova process is very short as far as peak temperatures are considered. Then the abundances of the four β^+ unstable nuclei ^{13}N ($\tau_{1/2} = 598$ s), ^{14}O (71 s), ^{15}O (122 s) and ^{17}F (66 s) increase till the point where they become the most abundant nuclei in the envelope, so that the rate of the CNO-reactions will be determined by the half-lives of these four β^+ unstable nuclei, which do not depend on temperature or density. A related aspect is that for very high ignition temperatures ($> 10^8$ K) the 'hot CNO-cycle' (Table XXXIII) may be considered (Lazareff et al., 1979).

After initial computations by Starrfield et al. (1972) more detailed work by Sparks et al. (1977, 1978), Colvin et al. (1977), Lazareff et al. (1979), dealt with the details of the accretion and the subsequent outburst. Starting data are an assumed pure ^{12}C white dwarf of mass $1\mathfrak{M}_\odot$; the initial luminosity is $\log L/L_\odot = -3.4$. An accretion rate is assumed of 10^{-12} $\mathfrak{M}_\odot$ yr^{-1}. The assumption of a fair overabundance of C is essential for the process to operate; a pure He white dwarf would not give rise to the nova phenomenon. Another important aspect is that hot white dwarfs have extended convection zones (Böhm, 1968, 1970), so that carbon from lower layers can be convected upwards into the accreted material. Following the thermonuclear runaway in the accreted mantle, the outer edge of the core will become convectively unstable, and will mix nuclei into the envelope. The time of thermonuclear runaway appears to depend very strongly on the initial luminosity of the white dwarf.

If the accretion is at such a rate that heating is adiabatic, then

$$\frac{\dot{T}}{T} \approx \nabla_{\text{ad}} \frac{\dot{P}}{P},$$

with the pressure increase due to the increased weight of the overlying layers. With

$$P \approx \varrho G \mathfrak{M} \Delta r / r, \quad \Delta \mathfrak{M} \approx 4\pi r^2 \varrho \Delta r,$$

we have

$$\dot{P} \approx G \mathfrak{M} \dot{\mathfrak{M}} / 4\pi r^4.$$

Hence the rate of heat input into the envelope, due to compression, is:

$$L_{\text{comp}} \approx \varepsilon \frac{2.5 k \dot{T}}{\mu H} \mathfrak{M}_{\text{env}},$$

$$\approx \varepsilon \frac{2.5 kT}{\mu H} \nabla_{\text{ad}} \dot{\mathfrak{M}},$$

$$\approx 10^{-4} \, \varepsilon T_6 \, (\dot{\mathfrak{M}} / 10^{-10} \, \mathfrak{M}_\odot \, \text{yr}^{-1}) L_\odot.$$

The computations by Colvin et al. (1977) demonstrate that significant carbon enhancement by convective mixing occurs only over an extent approximately equal to the region occupied by the initial carbon convection zone. The extent of the carbon convection zone exceeds $\sim 10^{-5}$ $\mathfrak{M}_\odot$ for a $0.4\mathfrak{M}_\odot$ white dwarf, which is sufficient to have enough mass involved to produce a nova outburst. The outburst itself is due to the thermal ignition and subsequent runaway which is a matter of seconds in some of the cases studied (e.g. in a model studied by Starrfield et al. (1972) with $L=0.2L_\odot$, $T_e=35000$ K). Due to the pressure increase in the shell involved, a shock will form, which moves outward, and transports outward the energy deposited in the deepest layers. When hitting the surface, some 10^{26} g of matter ($\sim 10^{-5}\,\mathfrak{M}_{\odot}$) is expelled. The brightness increase (an amplitude is computed of ~ 18 mag) is assumed to be due to the arrival of the shock at the surface. The initial expansion velocity is calculated to be $\sim 10^{3}$ km s^{-1}, but it decreases fast by gravitational action of the white dwarf. Abundances computed to occur in the ejected material show overabundances for C ($4\times$), N ($50 \ldots 60\times$), and O ($3\times$) as compared with solar values. This agrees reasonably well with the spectral observations.

An interesting extension of these calculations was given by Sparks *et al.* (1978) who could show that the assumption of an initial overabundance of CNO-nuclei, although necessary to produce fast novae, is not needed to explain the development of a slow nova like HR Del 1967.

How, finally, to explain in the frame of such theories the occurrence of long-period regular brightness variations, as observed in RR Tel in the years before the outburst? Tentatively this phenomenon is ascribed to convection-zone instability of late-type stars (having deep convection zones) that fill the Roche lobe in a binary system (cf. Robinson, 1975).

9.11. Supernovae; Types, Lightcurves and Energies

Supernovae are rather rare phenomena in the Universe; there is one supernova for about 10^{3} novae. The amount of energy emitted in the supernova event is large and corresponds approximately to the total sub-nucleonic energy content of a solar type star, $\approx \mu \mathfrak{M} c^{2}$, where μ is the mass fraction that is transformed in nuclear reactions.

The proceedings of conferences on supernovae were edited by Cosmovici (1974) and Schramm (1977). A series of papers, partly reviews, on this topic appeared in *Mem. Soc. Astron. Ital.* **49** (2–3), 1978. A translation of a book by Shklovski: *Supernovae*, is in press.

There are mainly two types, called I and II. Initially, this subdivision was made on the basis of spectral criteria. Minkowski (1941) defined type II supernovae as those in which lines of the Balmer series are clearly present, the others are called type I. (See also the review by Oke and Searle, 1974). Nowadays the subdivision is made principally according to the shapes of the light curves: the *increase* of brightness is very fast, and seems fairly similar in both types, with about 0.2 to 0.5 mag per day, (but available data are scanty). For type I supernovae the rise in brightness before maximum can be represented by $m - m_{0} = -5 \log(1 - t/t_{0}) + pt$, where t and t_{0} are times after maximum light, and m_{0} the apparent maximum brightness. For most type I supernovae $t_{0} \approx -16d$, $p \approx 0.08$ which means, extrapolating, that brightness increase would start $\sim 16d$ before maximum (Pskovskij, 1977). There are clear differences in the brightness *decrease* after maximum: in supernovae of type I it goes initially fairly fast (3 mag in 25 to 40 days), thereupon rather slowly and smoothly by about 1 mag per 65 days – Figure 126. There is a relation between properties such as brightness, colour, expansion velocity, absolute photographic magnitude and the rate of bright-

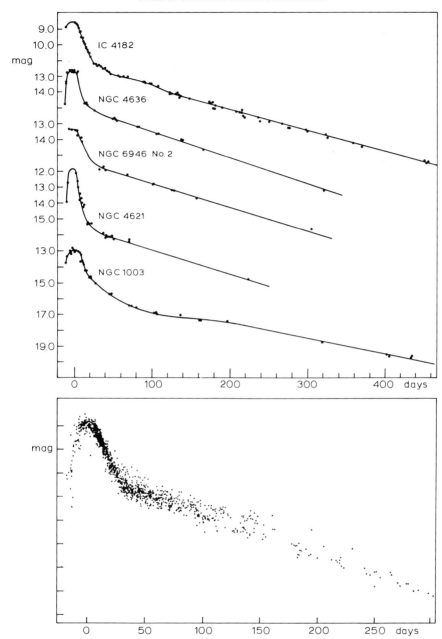

Fig. 126. *Upper diagram:* photographic light curves of type I supernovae. (From compilation by Arp, 1974). *Lower diagram:* composite blue light curve from observations of 38 type I supernovae. (From Barbon *et al.*, 1973, 1974).

ness decline, which shows that in broad outline the supernovae of type I form a physical class, dependent on one parameter only (Pskovskij, 1977). Still, Barbon *et al.* (1973) find minor differences between 'fast' and 'slow' type I's, which they call types

Ia and Ib. For these types the width of the light-curve, 2 mag below maximum brightness is 32 and 38d, respectively; the inflection point of the light-curve is 3.25 and 2.50 mag below maximum, the rate of further decline is 0.016 and 0.012 mag d^{-1}. The supernovae of type II decrease initially smoothly but slowly, with about 1 mag in 20 days, thereupon slower until 90 days after the maximum when they are about 2 mag below the maximum. Thereafter the brightness decreases faster and with a rate comparable to that of the type I supernovae; see Figure 127. Unlike the type I SN's, the light-curves of the type II's form a much less homogeneous group (Barbon *et al.*, 1974b, 1979), in which two basic types can be recognized. Those of type SN IIP have a *plateau* in their light-curve between about 35 and 80d after maximum; those of type IIL have a nearly *linear decline* after maximum. There are further differences, and there are transition-type supernovae. Actually, Zwicky introduced the additional types III, IV, and V (the latter representing η Car-like objects) but these are of less importance for the present chapter and will not be discussed here. From 129 extragalactic supernovae, for which the light-curve and spectrum could be classified, 81 were of type I and 38 of type II, 10 were of other types.

The *frequency of occurrence* depends on the type of the extra-galactic system in which the supernova occurs. The highest rate of occurrence is for spiral galaxies like our own for which several determinations from the last decade (Katgert and Oort, 1967; Caswell, 1970; Milne, 1970; Downes, 1971; Ilovaisky and Lequeux, 1972) initially led to an average rate of occurrence τ of one per 50 ± 25 year. However Tammann (1977a, b, 1978) finds along two different considerations $\tau=15(+16, -5)$ yr or $\tau=27(+23, -9)$ yr, values that apply to Sb and Sc galaxies, and thus assumes as a suitable average $\tau=15$ to 20 yr within a factor of two. This does not disagree with Clark and Stephenson (1977) who studied the historical sightings of galactic supernovae, and, including considerations on the influence of selection effects, found $\tau \lesssim 30$ yr. If this high rate of occurrence is correct it is remarkable that the last supernova in M31 was observed as long ago as 1885! There is a fair chance that the next galactic supernova will be discovered by the present generation as a faint, heavy reddened object (Van den Bergh, 1975).

The average *absolute photographic magnitudes at maximum* are approx. -19.8 for type I supernovae and -18 to -18.5 for the type II supernovae (see Table XLV). The physical scatter in these values seems of the order of 0.6 mag, perhaps smaller for type I's. No supernova has been observed less bright than -15 mag at maximum (Tammann, 1974, 1977a). The amplitude of the light variation is hard to determine since pre-outburst observations are absent. With certain assumptions on the masses of supernova progenitors (see below) one may derive amplitudes of about 25 mag (a factor 10^{10}).

An estimate of the *total emitted radiation flux* is easily obtained: Assuming for supernovae of types I and II at maximum: $M_B=-19.8$ and -18.2, taking BC$=-0.5$ and -0.6 (for $T_c=10000$ and 12000 K respectively, using our Table VII), and assuming an exponential decline $e^{-t/\tau}$ with $\tau=10d$ and $20d$ respectively, one derives a total emission $L_{\max}$. $\tau=4\times10^{49}$ and 2×10^{49} erg, respectively. The true values may be larger

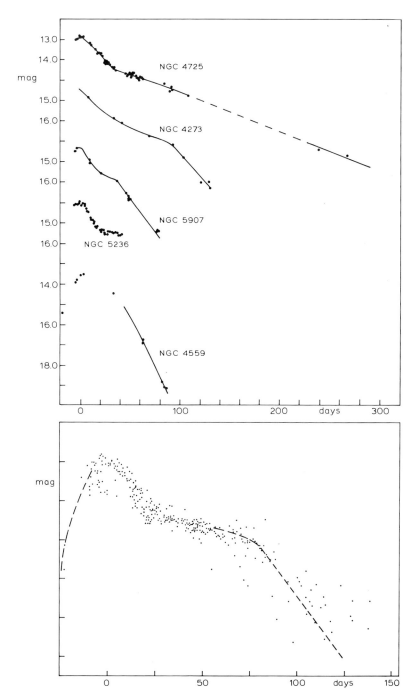

Fig. 127.   *Upper diagram:* photographic light curves of some type II supernovae. (From compilation by Arp, 1974). *Lower diagram:* composite light curve from 13 type II supernovae. (From Barbon *et al.*, 1974b).

TABLE XLV

Absolute photographic (or B) magnitudes of supernovae at maximum

Type I	Type II	References
−19.9	−18.0	Kowal (1968), Sandage and Tammann (1976)
−19.1	−17.2	Tammann (1977a,b)
−19.6±0.6		Branch (1977a,b, 1979)
−19.0		Wyckoff and Wehinger (1977)
−19.45[a]		Van den Bergh and Kamper (1977)
−21.3	−18.3	Pskovskii (1977, 1978)
−19.8±0.2	−18.7±0.2	Tammann (1978)
	−18.8	Barbon (1978)
	−16.45[b]	Barbon et al. (1979)

[a] Derived from the communicated $M_v = -19.3$, and $B-V = -0.15$, at t_{max} (Pskovskii, 1971).
[b] Assuming a Hubble constant of 100 km s^{-1} Mpc^{-1}.

because the brightness-decline is not strictly exponential. There are, however, cases in which much larger values are communicated: For the bright type I SN 1972e Chiu *et al.* (1975) estimated an integrated (over time and wavelength) radiation energy of approx. 10^{52} erg; 1974g (in NGC 4414) emitted 10^{51} erg (Wycoff and Wehinger, 1977). These numbers all suffer from lack of observational data in the ultraviolet spectral range.

From studies of the expanding shells of nebular matter around post-supernovae the *total emitted kinetic energy* W can be estimated. The mass of the expelled matter is the crucial quantity; it seems much larger for type II supernovae than for those of type I, while the velocity of expansion differs at most by a factor of about two. For type II's: $W \approx 10^{51}$ erg, for type I's it is of the same order. For the best studied remnant, the Crab (SN 1054) – often classified as type I, but this is not certain! – the mass of ejected material can be estimated; the filamentary system has $\mathfrak{M} \approx \mathfrak{M}_\odot$ (Shklovskii, 1976). This yields $W \approx 10^{51}$ erg. For the type II supernova 185: $\mathfrak{M} > 5\mathfrak{M}_\odot$ (Winkler, 1978); with $v_{exp} \approx 5000$ km s^{-1} one obtains $W > 10^{51}$ erg. The ratio between the emitted radiation E and the emitted kinetic energy W, hence, seems to be approximately equal in type I and in type II supernovae.

The differences between the two types of supernovae are certainly related to the mass, evolution, and chemical composition of the pre-supernovae. The progenitors of *type I* supernovae must be stars of fairly small, solar type masses. They occur along the major axes of elliptical extragalactic systems (which are known to have ages $\gtrsim 10^{10}$ yr) as well as in the disks of spiral systems, but not *in* the spiral arms and neither in the halos. Oemler and Tinsley (1979) found that they show preference for irregular galaxies of Hubble's class Irr II (also called I0). They used this observation to suggest that type I supernovae come from short-lived stars. However, to the present author it seems that most evidence indicates that type I progenitors belong to the old disk (intermediate) stellar population. The galactic orbit of Kepler's supernova (SN 1604), which was of type I, shows that the star has a population II-type orbit (Van den Bergh and Kamper, 1977). They are apparently fairly evolved stars; the amount of ejected matter is relatively small, they seem fairly helium-rich while their shells contain hardly any hydrogen or not at all. The assumption that they belong to the disk population sets the upper limit of their masses at about $2-3\mathfrak{M}_\odot$. Considerations on the physics of the supernova process determine the lower limit of their masses at about $1\mathfrak{M}_\odot$. Hence it seems reasonable to assume that type I progenitors have a mass of $2\mathfrak{M}_\odot \pm 1\mathfrak{M}_\odot$. Wheeler (1978a) suggests to identify their progenitors with the hydrogen-deficient C stars (Hd C), perhaps with the R Coronae Borealis stars.

The *supernovae of type II* belong to the stellar population of type I. They are concentrated in spiral arms (Maza and Van den Bergh, 1976) of Sb and Sc galaxies. Estimates of the masses of type II progenitors are derived from equalizing the rate of occurrence of such objects to the birth rate of stars of various masses. The latter rate is equal to the former for stars of approx. $5\mathfrak{M}_\odot$ (Tammann, 1977b) to $8\mathfrak{M}_\odot$ (Tinsley, 1977).

Another approach is based on the estimated value $\dot{Z}$ for the metal enrichment of the Galaxy, the assumptions that $\dot{Z}(t)=$const. and that this enrichment is due to ejection of newly synthized elements by exploding type II supernovae, and the approximation that the remnant mass is negligible with respect to the ejected mass. Thus, Trivedi (1978) derives a type II supernova mass $\mathfrak{M}=1 \ldots 10\mathfrak{M}_\odot$. Taking all evidence together it seems that one may assume $\mathfrak{M}\approx 5 \ldots 10\mathfrak{M}_\odot$ for the type II supernovae progenitor.

That some supernova *remnants* have masses exceeding the values given here is not in conflict with our assumptions about progenitor masses: a remnant sweeps up matter in interstellar space during its expansion.

9.12. The Continuous Spectra of Supernovae

Emission or absorption lines? It has been assumed during a long time that the spectra of supernovae consist for the greater part only of very broad emission bands displaced over wavelength distances corresponding to radial velocities of the order of 10^4 km s^{-1}. The difficulty of a coherent identification of the spectral lines has gradually convinced astrophysicists that the solution had to be looked for in another direction. Nowadays ideas are converging that the spectra, at least in the initial phase of the supernova phenomenon are essentially early-type (B or A) continua with broad displaced absorption lines. Later, the spectral type should approximately be of the solar type.

The continuous spectra. We refer to review papers by Mustel (1974, with references) and Kirshner (1974), also Kirshner *et al.* (1973b, 1976). The continuous spectrum can be traced better for type II supernovae than for type I's; the latter are more disturbed by emission lines (Figure 128). The variation of the spectral energy distribution of the type I SN 1972e during twenty days is shown in Figure 129. The colour temperatures can be found from measurements of absolute spectral energy distributions. Ac-

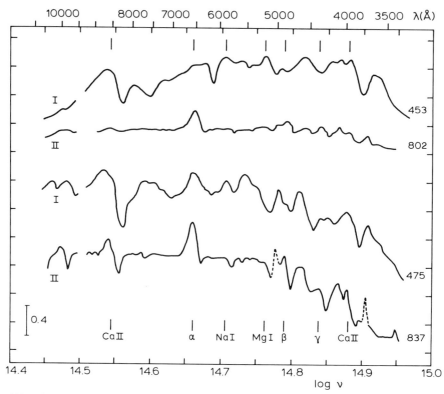

Fig. 128. Comparison of spectral energy distributions of type I and type II supernovae at two selected phases, identified by the last three digits of the Julian Day (right). The type I observations are from SN 1972e, the type II's from SN 1970g. The vertical scale gives $\log f_\nu$. (From Kirshner, 1974).

tually, Kirshner *et al.* (1973a, b, 1976) found that the continuous spectrum of a type I supernova between 0.4 and 2.2 μm can be well approximated by a black body. The derived T_c-values are rather low: about 12000 K (type I) to 10000 K (type II) in the early phase, and they decrease in 2 to 4 weeks to values of ∼6000 K for type II supernovae, and ∼7000 K for type I's. Pskovskij (1970) has determined average colours for type I supernovae, e.g. at $t=0$ (maximum light): $B-V=-0.15$; at $t=20d$: $B-V=+0.5$, and for $t>30d$: $B-V\approx$ const. $=0.9$. From these data Branch and Patchett (1973) derived colour temperatures: $T_c=20000$ $(t=0)$; 7200 $(t=20d)$; 5300 $(t>30d)$. Assuming that the spectrum is of the absorption line-type the bulk of the energy is carried in the continuous spectrum at least during the first few months. Absolute scans, made at the Hale Observatories (Oke, 1969) and compared with an adopted black body flux at the given distance determine radii of the emitting shell for a type II supernova which appear to be of the order of 10^{10} km, one month after the maximum. For a type I supernova (1960f in NGC 4496) Mustel (1971) found a radius of the order of 2×10^4 $R_\odot$, being 1.4×10^{10} km. This value applies to the (continuum emitting) photosphere in the maximum phase; from a study of the expansion velocity the radius of the emitted envelope may be derived by integration and is 2 or 3 times the photospheric value (Mustel, 1974). See also Section 9.13. Kirshner *et al.* (1973a) found for the type I supernova 1972e that the radius of the photosphere was 6×10^{15} km one month after maximum. Branch (1977b) determined radii for an 'average' type I supernova, assuming the colours of Pskovskij (1970): temperatures are found assuming a black-body energy distribution. With the average light curve of Barbon *et al.* (1973) one thus derives the relative variation of radius with time. During the period between ten days before maximum through twenty days after the radius appears to increase linearly with time $R/R(0) = \alpha t + \text{const.}$, with $\alpha^{-1} = 16 \pm 1$ days. On the other hand, spectral line observations near maximum yield a velocity of expansion of 10900 ± 700 km s^{-1}. These two data together yield $R(0) = (1.54 \pm 0.14) \times 10^{10}$ km. This, then, also gives $M_B(0) = -20.25 \pm 0.3$ mag.

9.13. Line Spectra of Type I Supernovae

Although supernovae of type I are remarkably homogeneous in their spectroscopic data, the matter of the interpretation of their line profiles has for a long time been controversial, but it now seems clear, particularly after work of Pskovskij (1968), Mustel *et al.* (1972, 1974), Mustel and Chugay (1975), and Kirshner (1974) that the spectrum contains broad displaced absorption lines with little or no hydrogen. At maximum light the excitation corresponds to that in B-type supergiants; it steadily decreases with time to correspond with an F-type supergiant $\sim$1 month after maximum. See e.g. Figures 129 and 130. The expansion velocities are of the order of 10^4 km s^{-1}. The equivalent widths are large, up to 100 Å. This can not be due to damping broadening since this would yield improbably high masses of the expanding shell ($500 \mathfrak{M}_\odot$). The most probable hypothesis seems to be that of a large velocity gradient in the expanding envelope. This is illustrated by a model-based investigation by Branch (1977a, b) who compared the observed absorption line at $\sim$6040 Å, ascribed to the Si II doublet at 6347, 6371 Å with calculations. With the expansion law $v(:)r$, and assuming an r^{-10} density law, he found reasonable agreement with an expansion velocity at the base of the envelope $v(R_e) = 10900$ km s^{-1}.

Spectrochemical analyses of type I spectra show large abundances of metals, and an apparent overabundance of N with regard to C and O.

In relation to this problem an interesting attempt was made by Assousa *et al.* (1976) who studied the spectra of the type I Supernova 1970l (six days before, one day before and 25 days after maximum). It appears (Figure 131) that the spectra can be simulated fairly well by assuming them to be due to Fe II emission lines, broadened over 3000 ± 250 (estimated) km s^{-1}. Small deviations between synthetic spectra thus obtained, and the observations can be ascribed to the addition of He II 4686 Å, and to weak emission lines of Hα and Hβ. Hence, the essential conclusion was that the spectra are mainly due to Fe II with small admixtures of He II and H I.

This result is, however, far from unambiguous: Mustel (1978) who studied the same

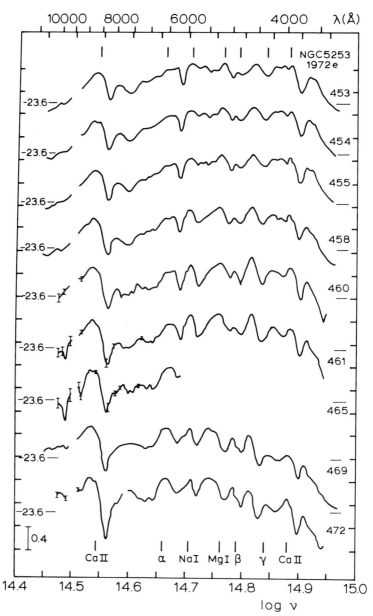

Fig. 129.   Spectral energy distribution of SN 1972e in NGC 5253 (type I). The description is as in Figure 128; f_ν is in erg cm^{-2} s^{-1} Hz^{-1}. The absolute flux level is indicated at the left of each energy distribution with a short horizontal line, repeated at the left. The wavelengths of Hα, Hβ, and Hγ are indicated, but this does not imply that the lines are present. (From Kirshner, 1974).

spectra could interpret them as a continuous spectrum with absorption lines of mostly singly ionized lines with no or negligible Balmer emission. Other authors confirm that the hydrogen lines are either absent or weak and/or masked by other emissions.

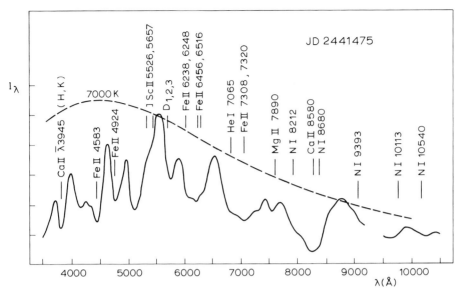

Fig. 130.   Interpolated continuous spectrum and suggested line identifications of SN 1972e (type I).
(From Mustel, 1974).

Mustel and Chugay (1975), who analysed the spectrum of SN 1972e assumed that
the hydrogen lines were invisible in the spectra of this object, and thus concluded
that the ratio (Si/H) was at least two or three orders larger in this type I supernova
than in solar-type stars. This is not the last word, because later Mustel (1978) found
indications for a weak Hα *absorption* line in the spectrum of SN 1972e while Kirshner
et al. (1973b) noted *emission* features at the expected locations of the Hα and Hγ
lines in their energy distribution curves of SN 1972e. Because of the many blends in
the spectrum they were unable to identify these features with H-emissions. Assuming
that the features near the Hα wavelength would be due to a blend of Fe II lines at
6516 and 6433 Å, Friedjung (1975) performed a detailed analysis of the expected intensities of these Fe II lines, assuming supernova model A (see Section 9.14, Figure
132). He failed to explain the observed intensities and concluded that "the emission
and absorption features near Hα are not easily explained by a blend of lines of mul-
tiplet 40 of Fe II". Assuming, then, that the emission is due to Hα, by recombination,
the total mass of hydrogen in the emitting envelope could be estimated as approx.
$0.3 \, \mathfrak{M}_{\odot}$.

A non-LTE discussion of the hydrogen level populations ($n=2$ and 3) by Bychkova
and Bychkov (1977) has confirmed the hydrogen deficiency in type I supernovae. In
a re-analysis of the spectra of SN 1972e Mustel (1978) found that the abundances
of the elements O, C, N, and S are higher than the standard – solar – values, but it
did not appear possible to give quantitative data for the abundance anomalies. It
also seems that the C I and O I absorptions appear later in the supernova event than
those of He I and N I: this may indicate stratification in the star's envelope, such

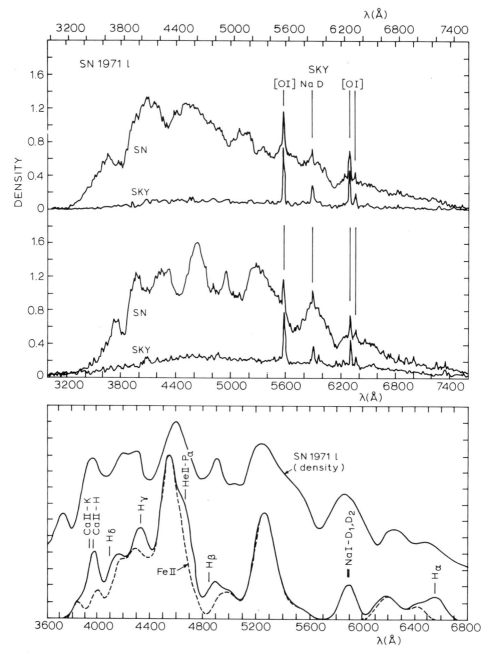

Fig. 131. *Upper diagram:* spectra of the type I supernova 1971l in NGC 6384. *Upper spectrum:* 29 July 1971, at maximum light. *Lower spectrum:* obtained 17 July 1971. The data are presented in photographic densities. *Lower diagram:* comparison of observed and calculated spectra of SN 1971l. *Upper curve:* the density recordings of 17 July 1971, smooted. *Lower curve:* synthetic emission spectrum in intensity units, with gaussian broadening assuming a velocity speed of 3000 km s^{-1}. The dashed curve gives the synthetic spectrum as due to Fe only, the full-drawn one includes H lines. (After Assousa *et al.*, 1977).

as predicted by supernova theories (ref. Figure 61 in Section 4.9). Similarly one may explain the remarkable observation that the remnants of a fairly recent type I supernovae (Kepler's SN, 1602) seem to have a normal H abundance in contrast to those of older supernovae. The explanation is that the remnants of recent supernovae still consist mainly of the stellar wind expelled *before* the supernova outburst. In older supernovae this material should then have been overtaken by material from deeper stellar layers, expelled at the outburst, with perhaps low H abundance.

Conclusion: The observational evidence is sometimes conflicting and the spectra not always easy to interpret, but the spectra of type I supernovae seem dominated by metals with little or no hydrogen. The expansion velocities are $\sim 10^4$ km s^{-1}. The line widths are due to a velocity gradient in the envelope.

9.14. Structural and Dynamical Models of Type I Supernovae

Although spectral analysis of type I supernovae has not yet advanced very far, while the available investigations are chiefly based on the supernova 1972e, it is possible to establish a crude picture of a type I supernova.

The abundances in a supernova envelope, for which indications are summarized in the previous section, indicate that type I supernovae are fairly evolved stars. Following Mustel and Chugay we consider two alternative models for the supernova structure – see Figure 132. In model (a) the envelope is detached from the photosphere and in model (b) the two are connected. Several arguments have been brought forward that should lead to adopting model (a), in which the ejected matter is in a thick envelope, well separated from the photosphere (see Mustel and Chugay, 1975; Friedjung, 1975); none of these arguments is strong, but together they may count:

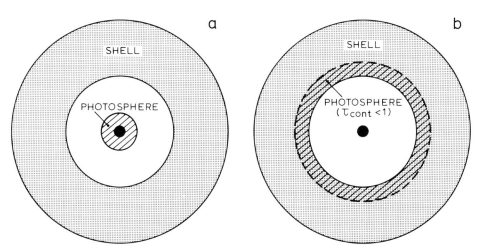

Fig. 132. Two alternative models of type I supernova shells close to light maximum: (a) photosphere not connected with the shell; (b) photosphere connected with the shell. The extent of the photosphere is larger in (b) than in (a). The black dot is the central high temperature body. (After Mustel and Chugay, 1975).

– near maximum the continuous spectrum greatly resembles that of a black body; so that the continuous radiation is mainly (quasi-)thermal;
– the expansion velocities derived from shifts of different absorption lines are nearly equal; also the velocity remains nearly constant – during 230 days in SN 1972e – over a long period of time;
– the widths of the absorption features of SN 1972e remained of the same order as the Doppler shifts for a period of three months after maximum light. While there are at the same time strong arguments for a velocity gradient across the shell (see discussion in Section 9.13) it is obvious that this gradient can only be maintained if the fast moving gas is at the outside and the slowly moving gas at the inner side;
– the radius of the densest part of the shell (6×10^{10} km two months after maximum) is considerably larger than that of the photosphere (1 to 1.6×10^{10} km), after Kirshner et al. (1973a), Branch (1977b). Wyckoff and Wehinger (1977), from the expansion velocity of SN 1972e and the reciprocal of the slope of the radius curve find $R_{ph} = 1.5 \times 10^{10}$ km ($2 \times 10^4 R_\odot$);
These arguments, put together, suggest accepting model (a).

After having thus established a structural model for a type I supernova, the next problem is that of defining *the dynamic model*: a supernova is a process rather than a structure! The two aspects can be linked via the process of *energy injection*. While it is often claimed that ejection does not occur at different discrete times (Friedjung, 1975), there are however indications that the whole of the injection process takes some time, while *IUE*-observations of the SN 1978 in M100 show that there were in this object *two* explosions: a first at low expansion velocity, followed by a more energetic one (Barbon, private communication). In fair accord with this observation Nadyozhin and Utrobin (1977) studied models based on slow injection of energy – however with a sudden start – (characteristic times of the order of one to a few tens of days) in stars with masses of $1.4\mathfrak{M}_\odot$ and core masses of $0.5\mathfrak{M}_\odot$. The dynamic model is illustrated in Figure 133, which should be considered together with the structural model of Figure 132a. In the model of Figure 133 there is a collapsing central remnant (1) surrounded by an envelope consisting of three parts: first a slowly expanding extended atmosphere (2) – the photosphere – responsible for the exponential increase of the object's brightness; thereafter a thin shell (3) containing the bulk of the mass of the ejected envelope; and finally (4) the expanding envelope with a distinct velocity gradient.

Quantitatively this model is derived and illustrated as follows. The sudden start should give rise to a shock wave progressing outward as a thin shell, while the longer lasting injection of energy would produce expanding motion of a larger part of the star. Quantitatively the basic equation is that of energy

$$\frac{\partial E}{\partial t} + P \frac{\partial}{\partial t}\left(\frac{1}{\rho}\right) = \varepsilon - \frac{\partial L}{\partial m},$$

where m is the mass variable in the theory of stellar structure, and ε the rate of energy injection. For $\varepsilon(t)$ which is the critical variable in the theory several trial functions

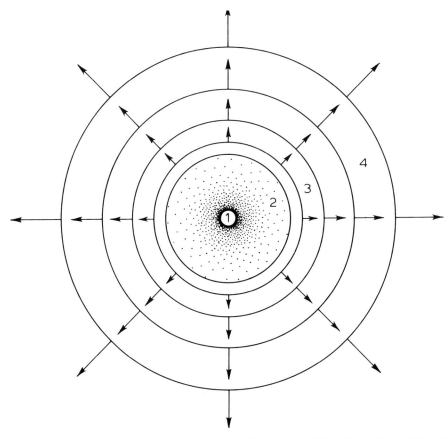

Fig. 133. A working model for a type I supernova. See the text. (After Nadoyzhin and Utrobin, 1977).

may be assumed. Part of the energy liberated at time t in the shell m will be used for accelerating the expanding envelope, the remainder will increase the thermal energy (energy losses by radiation are neglected in the expansion phase prior to maximum light), according to

$$(\varepsilon)\ t = \tfrac{1}{2}\mathfrak{M}_e\, u^2 + \tfrac{4}{3}\pi R^2\, \frac{P}{\gamma - 1},$$

where $\mathfrak{M}_e$ the mass of the envelope, R, u, and P its radius and mean velocity and pressure. The equation of motion yields

$$4\pi R^2 P \approx \mathfrak{M}_e\, \frac{\partial u}{\partial t}.$$

A comparison of the various model calculations with the average observational results shows that the best agreement is given by the model shown in Figure 133: The initial shock would produce a fairly thin outer shell ($\sim 0.01 \mathfrak{M}_\odot$) with a steep velocity gradient; from it emerges the line spectrum. The longer lasting energy injection would make a larger part of the stellar mass ($\sim 0.5 \, \mathfrak{M}_\odot$) expanding. It is this part that is the 'photosphere' proper, and which defines the stellar radius, as long as it is not yet transparent in continuous radiation.

The best agreement with observations is found for an energy release in a typical type I supernova of 2 to 5×10^{50} erg, a value that appears to be comparable to the 'observed' average kinetic energy of supernova ejecta (see Section 9.11).

On the basis of these considerations an original envelope mass of 0.3 to 0.7 $\mathfrak{M}_\odot$ is found for type I supernova remnants of SN 1006, 1572 and 1604 (Utrobin, 1978).

The most fundamental question comes next: where does the energy come from? The star of fairly low mass, considered to be the type I progenitor, must consist of a developed core, mainly of C and O (Wheeler, 1978b) with a He-mantle (see the discussions in Sections 4.8 and 4.9). The collapse of the core would form a neutron-core. Van Riper and Arnett (1979) have shown that a 'reflected shock' would thus originate, and Arnett (1979) gave arguments that this shock would process ^{56}Ni in the main part of the density gradient between the core and the more tenuous He envelope. The region of this gradient would contain a few tenths of a solar mass, and it seems that the subsequent expansion of this region and the more outer material would explain the observed envelopes with masses of 0.3 to 0.7 $\mathfrak{M}_\odot$. Arnett (1979), following Colgate and McKee (1969), assumes that the peak of the light curve would be due to ^{56}Ni decay, while the expanding He-rich envelope would receive its radiation energy from the core.

9.15. Type II Supernovae; Line Spectra, Light Curves, Scenario of Events

Unlike the type I supernovae, Balmer emission lines are visible in spectra of these objects (Figure 134). Furthermore, emission lines of [O I] and [Ca II] have been detected and – in a later phase, a few hundred days after maximum in SN 1970g – fairly strong absorption lines of $\lambda 5178$ Mg I, and $\lambda 5890$ Na I (Pronik et al., 1976). An aspect only shown by *some* type I spectra is that several of the 'emission' lines (like Hα) but also an 'absorption' line like Na I $\lambda 5890$ show P Cygni profiles, as has been remarked particularly by the Hale research team (Kirshner et al., 1973b, 1974), and later by Pronik et al. (1976); see Figure 135. This would lead, contrary to the type I case, to the assumption of a more tenuous shell, but also expanding with a velocity gradient $dv/dr \approx t_e^{-1}$, where t_e is the time since the outburst. The velocity range in type II SN is large: 3000 to 14000 km s^{-1}. The line identifications point to a solar-type spectrum with excitation temperatures of the order of 5000 K.

Radial velocity determinations can only be relied upon if based on good line identifications. Pronik et al. (1976) find for the SN 1970g in M101 that the radial velocities, determined from Hα, $\lambda 5890$ Na I and Hβ are ~ 6000 to 8000 km s^{-1} close to the supernova maximum, and decrease to about 3500 km s^{-1} in a hundred days after that. Thereafter v_R seems to remain fairly constant. It may be significant that

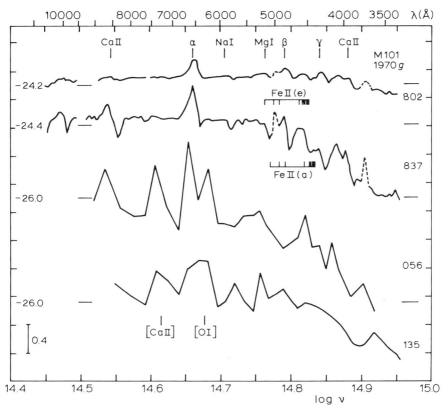

Fig. 134.   Energy distributions for the type II SN 1970g in NGC 5457 (M101). (After Kirshner, 1974).
The curves are labeled with the last three digits of the Julian Day.

the 'red half-width' of the emission lines, Δv, shows approximately the same variation with time as v_R, if one neglects a smaller value measured for Hα close to maximum. Such measurements are consistent with the assumption that the lines are formed in a freely expanding envelope whose density decreases monotonically with increasing velocity over a wide range. The smaller half-widths measured in Hα close to maximum may indicate that at that time the 'envelope' where this emission line is formed was close to the 'photosphere', thus causing a certain 'screening', reducing the total line width.

From absolute spectrophotometric data of SN 1970g near maximum and assuming a brightness temperature of 10^4 K Kirshner finds an *emission measure* $n^2 V \approx 3.5 \times 10^{64}$ cm^{-3}. Assuming a radius of 10^{10} km one obtains a shell electron density $n_e \approx 3 \times 10^9$ cm^{-3}. About a month after maximum n_e may have reduced by a factor of ~ 10.

A qualitative *spectro-chemical analysis* of the envelopes of type II supernova shows the fair abundance of hydrogen; further there are lines of He, Ca, N, Fe etc. The O I triplet at 7774 Å is identified in an absorption feature at 7660 Å (Chugai, 1977). Additional information is derived from the study of type II supernova remnants, although these may be contaminated with interstellar matter swept up by the ex-

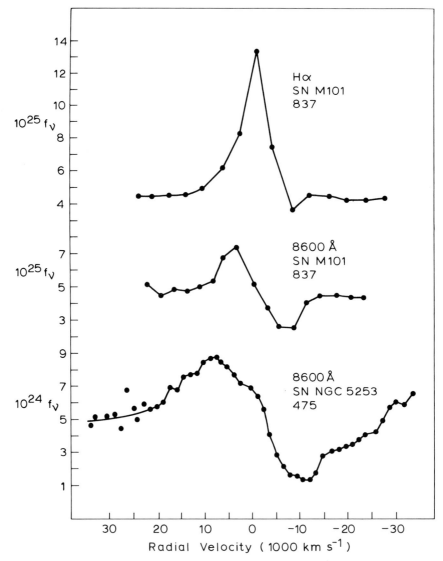

Fig. 135.   Line profiles of Hα and Ca II λ8600 in the type II SN 1970g in M101, and for the type I SN
1972e in NGC 5253. Note the P Cyg shape of the profiles. (From Kirshner, 1974).

panding envelope. An investigation of the spectra of the Cas A nebulosities by Chevalier and Kirshner (1978) yielded for the 'quasi-stationary flocculi' – supernova material that has lost its momentum by interaction with the interstellar medium – an
overabundance of N with respect to H by about an order of magnitude. This may
partly be due to H-depletion, for another part to thermonuclear production of N.
The 'fast moving knots' have considerable overabundances of various metals, for instance the ratio of number densities of S:Ar:Ca is 71:4:1, compared to the 'cosmic'
ratio 7:3:1. This may imply that the gas constituting the quasi-stationary flocculi

was originally part of the H burning core so that the pre-supernova stellar wind penetrated down to that region. The fast-moving material must consist of uncontaminated material from the core of a massive star, for which the mass is estimated from a comparison with model computations by Arnett and Schramm (1973), who suggest masses $>20\mathfrak{M}_\odot$ to $70\mathfrak{M}_\odot$, apparently in contradiction with other mass estimates (see Section 9.11!). Falk and Arnett (1973) have suggested that the $\sim$20 day peak in the light curves of type II supernovae would imply the presence of an extended circumstellar dust shell surrounding the star. The reasoning behind this assumption is evidently that pre-supernovae of type II are assumed to be red giants or supergiants around the atmospheres of which dust could occur. Simons and Williams (1976), following earlier work by Hoyle and Wickramasinghe (1970) investigated the mechanism for dust formation, and suggested that molecules, formed at $T=2000$ K, would form clusters of molecules with a radius of $\sim10^{-7}$ cm. Further grain growth would take place by the adhering together of such clusters brought together by their Brownian motion. Quantitative computations yield $r\approx2\times10^{-6}$ cm, about five times smaller than the 'observed' particle diameter. This result conflicts apparently with an investigation by Falk and Scalo (1975) who could show that radiative heating would rapidly destroy any grains but for pure quarz grains, which, however would be eliminated by sputtering and fragmentation, apart from the smallest. This problem is clearly still open.

The most probable *scenario* of the events taking place in a supernova II outburst may be outlined as follows (Chevalier, 1976; Falk and Arnett, 1977) (Figure 136): the collapse to a dense *core* region of mass $\sim1.4\mathfrak{M}_\odot$ (most probably to a neutron star) liberates energy. The core is surrounded by a *mantle* which is found interior to the helium-burning shell. This mantle with a mass $\sim0.1\mathfrak{M}_\odot$ to $5\mathfrak{M}_\odot$ contains the nucleosynthesis material. This mantle is surrounded by an optically thick *envelope* with a radius of 10^8 to 10^9 km (1–10 AU). This is surrounded by a circumstellar *shell*, a H^+ region, previously ejected from the star by mass-loss, with a radius $r\approx10^{10}$ km (≈100 AU) and $T\approx10^4$ K. An energy of 10^{51} energy must be deposited in the mantle-envelope region to yield an expansion velocity of the mantle of $\sim10^4$ km s^{-1}. The collision of the mantle with the envelope produces two effects: part of the mantle material falls back to the center, while the envelope turns into a shell expanding with high velocity. After $\sim$90 days the envelope becomes optically thin to continuous radiation, and after that time the photospheric light is due to the mantle material. This scenario has been worked out in some quantitative detail by Falk and Arnett (1977). The supernova energy release is assumed to be communicated to the stellar material in a time, short as compared with the dynamical time scale, i.e. the time for a sound wave to travel from the core to the surface ($\sim2\times10^7$ s). A reasonable but far from excellent agreement with observed light curves of type II supernovae is found for a model with initial mass $\sim5\mathfrak{M}_\odot$, and an envelope mass of $1.5\mathfrak{M}_\odot$. The integrated luminosity over a period of approx. 7×10^6 s (80 d) is 10^{50} erg; the emitted kinetic energy 2×10^{50} erg, the maximum expansion velocity 2000 km s^{-1}.

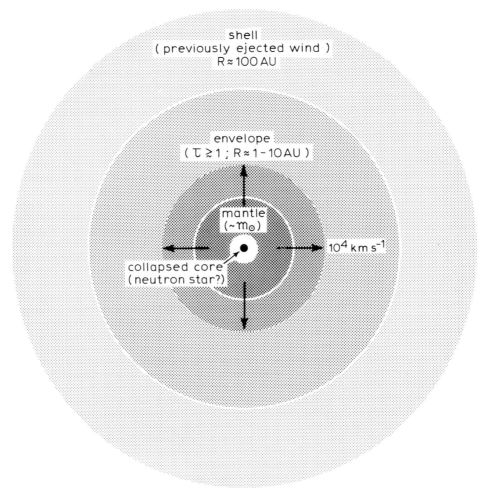

Fig. 136.   Model of a type II supernova.

9.16. High Energy Photon and Particle Emissions from Supernovae

The violence of the phenomena occurring in supernovae atmospheres makes it reasonable to assume that strong radiation will also be emitted in the X-ray and radio
ranges. No X-ray emission has yet been observed from supernovae in the first phase
of their evolution. Radio emission had been detected from the type II supernova 1970g
in M101 (Goss *et al.*, 1973). X-ray radiation had so far only been detected from remnants of old supernovae; cf. Section 9.17.

   An important unsolved problem in astrophysics is that of the origin of the galactic
cosmic ray particles. Supernovae are often suggested as a possible source. The particles could either be accelerated during the primary part of the stellar explosion, or
they may obtain their energy in the turbulent magnetic plasma of the young super-

nova-remnants, and finally they could be formed in magnetospheres of pulsars by some mode of wave-particle interaction. It is helpful to remark that these three mechanisms for particle acceleration are regularly considered in solar flare and terrestrial magneto-spheric (substorm) problems, and a critical comparision with supernova conditions may be very useful. A tool for a successful attack to this problem is by examing those cosmic-ray particles that can hardly be accelerated elsewhere than in supernovae.

It is not impossible that the observed positrons in the cosmic radiation, for which Cline and Hones (1969) measured a flux, for $E \leq 1$ MeV, of 100 $(\text{m}^2 \text{ sr s MeV})^{-1}$ or a number density of $\sim 5 \times 10^{-12}$ positrons cm^{-3}, have been injected into interstellar space by supernova outbursts. Positrons could be formed during the decay-process: $^{56}\text{Ni} \rightarrow {}^{56}\text{Co} \rightarrow {}^{56}\text{Fe}$ (see also Bychkov, 1977), or by $\gamma\gamma$ collisions. This suggestion by Burger *et al.* (1970) has been reconsidered by Colgate (1970) who predicted that 8×10^{52} positrons are emitted per supernova, while the observed positron flux density, with one supernova in our Galaxy per 20 yr would require 0.7×10^{52} positrons per supernova. (Colgate found 3.4×10^{52} but still used the rate of one supernova per 100 yr.) This is a factor 12 less than predicted, a discrepancy that may be removed by relatively small changes in the model assumptions. The theory of the physical processes in the supernova shock waves, with emphasis on high-energy processes in the shock waves was discussed quantitatively by Weaver (1976). It was found that radiative heat transport is an important shock-dissipation mechanism. So, the peak electron temperatures remain below ~ 70 keV for shock energies $\lesssim 30$ to 50 MeV nucl^{-1}. The – preliminary – conclusion of this investigation is that chances are fairly small for the production of a sufficient number of spallation cosmic ray particles from supernovae in interstellar space. The investigation leaves uncertainly on Burger's and Colgate's hypothesis on the precise number of positrons produced in supernova shock waves.

The alternative possibility that cosmic ray particles are accelerated by Fermi-acceleration in the turbulent plasma of young supernova remnants was investigated by Morfill and Scholer (1979) with the result that such acceleration is not impossible.

9.17. Supernova Remnants

More than 400 supernova remnants are known so far, mainly from their radio-emission. The best studied of them are described by Woltjer (1972); his Table 1 lists 24 galactic supernovae, detected according to their radio flux. Tables 2 and 3 of Woltjer's paper give 16 additional galactic probable supernova remnants, either being objects of fairly intense radioflux for which the association with a supernova is less certain, or faint sources, for which distance information is available. As a fine addition to Woltjer's review Van den Bergh *et al.* (1973) published an optical atlas of the 24 galactic supernova remnants of Woltjer's Table 1. Super-nova remnants are as a rule identified by their *morphology* – a more or less filamentary roundish shell – while additional support to the identification can be given by the radio and X-ray radiation distribution, by their expansion, and by their optical, radio and X-ray spectra. Van den Bergh *et al.* (1973) have attempted – but succeeded hardly!

– to set up a morphological classification. With later additions from other sources there may now be at least five classes:

– the Cygnus Loop type (3 to 4 cases) – an imcomplete shell outlined by very sharp filaments of intermediate length; the brightest filaments occurring where the super-nova shell runs into relatively dense interstellar clouds;

– the S147 type (2 cases) – characterized by the presence of very long delicate filaments, perhaps to be identified with luminious sheets seen edge-on;

– the diffuse shell type (4 examples), containing a few sharp filaments in addition to a larger number of diffuse filaments;

– the W28 type (2 cases) – shells that are filled with filamentary nebulosity.

– The Crab nebula may be grouped with 3C58 and G2 1.5–0.9 (Weiler and Wilson, 1977; confirmed for 3C58 by Van den Bergh, 1978) and with eight of nine other remnants (Caswell, 1979) in one class characterized by a centrally condensed brightness distribution, the diameter-radio-emission relation, and by the curious property that the major axes of the radio isophotes lie roughly parallel to the galactic plane!

– In addition some remnants like those of Kepler's supernova, Pup A, Car A, the γ Cyg nebula can not be classified in any of the above groups.

The explanation of the shapes of these various objects has certainly relevance to the primordial explosion and the interaction of the expanding gas with the interstellar medium. It is in this interaction that the optical, radio and X-ray emission is produced. In historical times eight to ten galactic supernovae seem to have been observed and recorded – see Table XLVI (Stephenson, 1974; Clark and Stephenson, 1975, 1976; Stephenson, 1976). Sufficient optical brightness estimates were made of the Crab supernova (1054), and those of Tycho (1572) and Kepler (1604) so that their light curves are roughly known. The Cas A supernova must have occurred in 1667, with an uncertainly of a few years, as determined from the expansion of the observed nebular filaments. However, no observations of a supernova of the expected magnitude ($\gtrsim 0$) were reported at that time (see e.g. Van den Bergh, 1974). The shape of the remnants of Cas A, as shown by optical observations with long exposure times, is virtually a round shell. This shell, with an outer diameter of 5′, is also seen in radio- and in X-ray observations (Zarnecki *et al.*, 1974; Charles *et al.*, 1977; Murray *et al.*, 1979).

The *optical spectra* of supernova remnants contain a number of forbidden transitions, such as [O II] (3726–29 Å), [S II] (6716–31 Å), [N II] (6548–84 Å), and [S II] (4959–5007 Å), and Hα (6563 Å). Hence, there is a clear difference between spectra of supernova remnants and H$^+$ regions. Many optical pictures, such as those contained in the Atlas by Van den Bergh *et al.* (1973) are therefore taken on red-sensitive plates with a broad filter around Hα, [N II] and [S II]. Similarly, D'Odorico *et al.* (1978) detect supernova remnants in extragalactic systems by comparing photographs taken in Hα, [S II] and in continuum radiation. The 'optical filaments' should have temperatures $\sim 10^5$ K (Kirshner and Chevalier, 1978). A hot plasma characterized by optical [Fe XIV] emission at 5303 Å suggesting $T_{el} \approx 2 \times 10^6$ K has been detected for SNR N49 (Murdin *et al.*, 1978), for the Cygnus Loop (Woodgate *et al.*, 1974)

TABLE XLVI

Some of the most important supernovae observed in historical times, and their remnants in so far as identified, and those other supernova remnants for which the soft X-radiation has been detected and analysed (Clark and Stephenson, 1975; Gorenstein and Tucker, 1976; Winkler and Laird, 1976; Culhane, 1977; Epstein, 1977; Winkler, 1978; Clark, 1978; Gronenschild, 1980; Murray et al, 1979; see also individual references below).

Remnant	Year of supernova	Duration of visibility (<1.5 mag.)	V_{max}	Type	$l(°)$ (for galactic remnants)	$b(°)$	Outer angular diameter (arc min)	(pc)	Distance kpc	X-ray temperatures (10^6 K)
Centaurus	+185	20 mo	−8	II	315.4	−2.3	40	30	2.5	2.5; >60
Sagittarius	386	3 mo	+1	−	11.2	−0.3	−	−	−	−
Scorpius	393	8 mo	−1	−	348.7±1	+0.3±1	−	−	−	−
Lupus	1006	1 yr	−9[1,2]	I?	327.6	14.5	270	90?	1.2?	2.5; 15[3]
Crab	1054	2 yr	−4	I−II?	184.6	−5.8	5	3	2.0	6
Cassiopeia	1181	6 mo	0?[4]	II?	130.7	+3.1	6	15	8	−
Cyg X-1	1408[10]	≥2 mo	<−3		72	+4				
Tycho	1572	16 mo	−4	I	120.1	1.4	7.0	6	3	4.6; 40
Kepler	1604	1 yr	−2.5	I	4.5	6.8	3	7?	8?	−
Cas A[11]	1667±5		>0	II	111.7	−2.1	6	4	2.8	15; 40
	age (yr)									
Pup A	42000				260.4	−3.4	55	35	2.2	3−8
Vela	13000				263	−3	200	30?	0.5?	2.5−4.3
Cygnus Loop	20000				74	−9	168	40	0.8	2−4
IC 443	?				189.1	2.9	45	20?	1.5?	10−17
LMC X-1	7000				$\alpha=5^h20^m$	$\delta=-68°$	0.6	10	55	12
W44[5]	10000				35	− 5	30	24	3	4
PKS 1209−52[6]	20000				296	+10	70	40	2	2
S 91−94[7]	10^4−10^5				79	+ 1	280	75	0.9	1−10
North Polar Spur[8]	250000								2	~3

Additional references:

1. Minkowski (1966).
2. Utrobin (1978).
3. Winkler et al. (1979).
4. with $A_v=4.3$, Searle (1971).
5. Grononeschild et al. (1978).
6. Tuohy et al. (1979).
7. Snyder (W. A.) et al. (1978).
8. Davelaar (1979).
9. Davidson et al. (1977).
10. Qi-bin (1978).
11. Murray et al. (1979).
12. Fabbiano et al. (1980).

for IC443 (Woodgate *et al.*, 1979), and was suggested for Vela (Woodgate *et al.*, 1975) but this was not confirmed by Murdin *et al.* (1978). Pup A even shows [Fe x] and [Fe xiv] (Lucke *et al.*, 1979).

The *radios images and spectra* of supernova remnants are their most characteristic property. In actual fact many of them were discovered by their radio emission.

Many supernova remnants are expected to emit an *X-ray flux*. Woltjer (1972) lists six early discoveries. At the end of 1978 soft X-ray emission had been detected from nine remnants (Gorenstein and Tucker, 1976; Culhane, 1977; Epstein, 1977) see Table XLVI. Thereafter the *Einstein-Observatory* (HEAO-2) discovered many more – Murray *et al.* (1979).

It seems that in most cases, but for the youngest remnants the X-ray spectrum can be approximated by a fairly soft thermal spectrum, indicating kinetic temperatures of the order of a few million degrees. In addition, the *Einstein Observatory* has traced line emissions from Mg, Al, Ca, and Fe in the X-ray spectrum of Cas A (Becker *et al.*, 1979). Hard emission from hot ($\sim 10^8$ K) elements is radiated by the young remnants Cas A and Tycho (Pravdo and Smith, 1979). Bychkov and Sitnik (1975) have given quantitative arguments that the 'nonsteady filaments' in Cas A and Tycho are the sources of the observed X-radiation in the one to hundreds of keV range, by the emission by swept-up interstellar gas (small energies) and by hot non-relativistic gas in the envelope (large energies).

The SN 1006 is interesting: it seems to have been the brightest supernova recorded in historic times. Its remnants are the only ones for which only X-ray and radio emissions were detected, while optical filaments are very faint (Schweizer and Lasker, 1978). Winkler (1977) found that the X-ray emission can be described best by the reverse shock model in which an inward-propagating shock, accompanying the deceleration of an ejected supernova shell by the ambient medium heats gas to such temperatures that thermal soft X-radiation is emitted. The discussion yield the following parameters.

supernova energy	$1\text{–}3 \times 10^{50}$ erg
ejected mass	$0.4\text{–}1.0\ \mathfrak{M}_\odot$
swept-up mass	$0.4\text{–}0.6\ \mathfrak{M}_\odot$
blast-wave velocity	$3000\text{–}5000$ km s^{-1}
kT for blast wave	$15\text{–}30$ keV

There exists a crude relation between the X-ray-flux and the radioflux of supernova remnants, as was found by Seward *et al.* (1976). The ratio F_X (0.2 to 10 keV, in keV cm^{-2} s^{-1}) to F_R (10 to 10^4 MHz, in erg cm^{-2} s^{-1}) appears to lie in the range 20 to 2000. The same authors found also a rough dependence of the X-ray luminosity on the radio-diameter: the latter being larger if the X-ray flux is smaller.

9.18. Supernova Remnants in the Large Magellanic Cloud

It is of interest to compare the data on galactic supernova remnants with those for the Large Magellanic Cloud (LMC), one of the nearest extragalactic systems. Mathewson and Healy (1964) discovered three supernova remnants in this system, using the Australian Parkes 64-m radio telescope. In a later search Mathewson and Clarke (1973) used the Mills cross of the Mongolo Radio Observatory at 408 MHz, supported by narrow-band image tube photography in Hα and other wavelengths at the 102 cm reflector at Siding Spring Observatory. Thus nine additional supernova remnants, were discovered, all in the extreme stellar population I region. There are, in addition, two uncertain identifications. From ten of these sources X-rays have been detected with the *Einstein Observatory* (Long and Helfand, 1979) with fluxes of 3×10^{35} to 4×10^{37} erg s^{-1} in the photon energy range 0.5 to 3.0 keV. The morphology and intensity relations [O III]/[S II] of three remnants have been described in some detail by Lasker (1977). The relation between radio-surface brightness Σ and diameter D is $\Sigma \sim D^{-3}$, as is the case for galactic supernova remnants. An initially found difference in the zero-points of the Σ, D relations for the Galaxy and the LMC (Mathewson and Clarke, 1973) is not confirmed by a later determination by Clark and Caswell (1976) based on a large homogeneous set of material: they find good agreement between the two sets of data.

9.19. The Interaction of Supernova Ejecta with the Interstellar Medium

We briefly describe the theoretical aspects of an explosion into a uniform medium. It is the transformation of the kinetic energy of the shock wave into thermal energy that explains the observed remnant structures. Shklovskii (1962) was the first to identify three phases in the dynamic evolution of supernova remnants: the early, adiabatic and radiative phases. Reviews were given by Woltjer (1972) and by Gorenstein and Tucker (1976); analytic procedures for the hydrodynamics were elaborated by Gaffet (1978).

At a time $t=0$ a mass $\mathfrak{M}_0$ with velocity v_0 and kinetic+thermal energy E_0 is injected into a rather cold medium with density ϱ_0, particle density n_{is}. The shock separating the expanding matter from the undisturbed surroundings has a radius $R(t)$; the radiative energy losses per unit time are $(dE/dt)_{rad}$.

We then distinguish between four phases:

Phase I: $\mathfrak{M}_0 \gg (4\pi/3) \varrho_0 R^3$: the relative importance of the swept-up matter is negligible; all aspects of the expansion depend on the primary explosion. This phase evidently ends when $R \approx 2(\mathfrak{M}_0/n_{is})^{1/3}$ pc ($\mathfrak{M}_0$ is here in solar units).

Phase II: $\mathfrak{M}_0 \ll (4\pi/3) \varrho_0 R^3$; $\int (dE/dt)_{rad} \, dt \ll E_0$. Swept-up matter dominates the remnant, but energy is conserved since radiative losses are still unimportant. This case is called the *adiabatic case*. For a gas with specific heat $\gamma = \frac{5}{3}$, with $\mathfrak{M}_H/\mathfrak{M}_{He} = 65:35$ one obtains for the pressure P^* and temperature T^* behind the shock (Woltjer, 1972):

$$P^* = \tfrac{3}{4}\varrho_0 v^2,$$

$$T^* = 1.45 \times 10^{-9} v^2 \text{ [K]} \quad \text{or} \quad kT^* = 1.25 \times 10^{-16} v^2 \text{ [keV]}.$$

The temperature applies to the ion gas, the electron component is slowly heated later, by Coulomb interaction. This inequality between T_i and T_{el} influences the X- and radio-fluxes calculated for model remnants. For old remnants such as the Cygnus Loop $T_i \approx T_{el}$ (Itoh, 1978). Assuming further that $T_i = T_{el}$, and by taking into account the detailed energy losses in the gas as deduced by Cox and Tucker (1969), integration of the radiative losses since $t=0$ yields

$$E_{\text{loss}} = \int_0^t (\mathrm{d}E/\mathrm{d}t)_{\text{rad}} \, \mathrm{d}t = 0.32 n_{is}^{9/5} \, E_0^{1/5} \, t^{17/5}.$$

One next assumes that the end of the adiabatic phase occurs at the time t_{rad}, when $E_{\text{loss}} = \tfrac{1}{2} E_0$, then $t_{\text{rad}} = 1.1 E^{4/17} n_{is}^{-9/17}$, while it is also possible to derive $v(t=t_{\text{rad}}) \equiv v_{\text{rad}} = 2 \times 10^7 \, n_0^{2/17} (E_0/10^{50})^{1/17}$ cm s^{-1}. At that time $R = 25(E_0/n_{is} \, 10^{51})^{5/17} \, n_{is}^{-2/17}$ pc.

Phase III: $t > t_{\text{rad}}$. Radiative losses become important. The matter cools quickly down to temperatures of $10^4 \ldots 10^5$ K. This phase is called the *radiative phase*. Hence pressure forces become relatively unimportant; the shell therefore moves with constant momentum *i.e.* $(\tfrac{4}{3})R^3 \varrho_0 v = \text{const.}$, yielding upon integration

$$R = R_{\text{rad}} \left(\frac{8}{5} \frac{t}{t_{\text{rad}}} - \frac{3}{5}\right)^{1/4},$$

$$E = E_0 \left(\frac{v}{v_{\text{rad}}}\right).$$

Phase IV: The expansion velocity becomes of the same order as the random motions of the interstellar gas; the remnants lose their identity and merge with the interstellar medium. To illustrate the relations: for $E_0 = 10^{50}$ erg, $v_0 = 10^4$ km s^{-1}, $\mathfrak{M}_0 = 0.1 \mathfrak{M}_\odot$ and $n_{is} = 1$ cm^{-3} the age t, radius R, velocity v and mass $\mathfrak{M}$ at the transitions between phases I–II, II–III, and III–IV take values as given in Table XLVII.

On the basis of essentially the above physics, and using a calibrated relation between the radio surface brightness and the linear diameter of SN remnants, Goudis (1976) has established useful graphs relating quantities such as the distance, linear and angular diameter, flux density at 1 GHz. In addition, based on the above described

TABLE XLVII

Age, radius, velocity and mass of typical supernova remnants at the interphases between the four main phases of their development. Assumed parameter values: $E_0 = 10^{50}$ erg, $v_0 = 10^4$ km s^{-1}, $\mathfrak{M}_0 = 0.1 \mathfrak{M}_\odot$; $n_{is} = 1$ cm^{-3}. (After Woltjer, 1972).

	I–II	II–III	III–IV
age t	90	22000	750000 yr
radius R	0.9	11	30 pc
velocity v	10000	200	10 km s^{-1}
mass $\mathfrak{M}$	0.2	180	3600 $\mathfrak{M}_\odot$

physical processes Clark and Culhane (1976), and Parkes *et al.* (1977), from a comparison of radio and X-ray data of four supernova remnants and using optical data on size and extent, derive temperatures of the gas behind the shock. Typical values are 2 to 5 MK for Cygnus, Puppis A and Vela, and 10 MK for IC 443. For the initial explosion energy E_0 released in the outburst one finds typically $E_0/n \approx 4 \ldots 10 \times 10^{51}$ erg cm^3; for the particle density at the shock $n=0.15$–0.3 cm^{-3}, and for the time elapsed since the outburst one finds (comparing the results of the above authors with those of Itoh (1978): for the

Cygnus loop: $(21 \pm 4) \times 10^3$ yr
Puppis A: 10 ± 3
IC 443: 1.6
Vela: 13 ± 2,

where the 'errors' give a crude indication of the uncertainly. The above treatment is only a global outline and describes the case of an isotropic explosion into a strictly homogeneous medium. No account is taken of deviations from these assumptions and the theory is therefore not able to explain the many detailed structures, such as the 'lacy structures' in the Cygnus Loop, the Vela remnant (Jenkins *et al.*, 1976) and the structures in S147 (see Van den Bergh *et al.*, 1973).

That the supernova remnants must even be more inhomogeneous than as shown by pictures taken in the visual spectral region is shown by X-ray studies such as those of the Cygnus loop: the X-ray plasma is considerably hotter than the visual remnants. Also the spectra of stars behind the Vela supernova remnant (Jenkins *et al.*, 1976) have multiple sharp lines most probably of interstellar origin with radial velocities ranging from $+90$ to -410 and -460 km s^{-1} while stars at some distance ($\gtrsim 1°$) away from the remnant do not show the high-velocity components. The physical conditions, as derived from the spectra line data are very inhomogeneous, with temperatures ranging from $\sim 10^4$ K to $\sim 3 \times 10^5 \ldots 10^6$ K necessary to explain the O VI ionisation, and densities from $[(10 \text{ to } 100) \times T/10^4]^{1/2}$ cm^{-3}, while X-ray observations indicate temperatures even above 10^6 K. Jenkins *et al.* (1976) conclude that the Vela supernova remnant is expanding into an inhomogeneous medium where the local densities exceed the average values ($\approx 10^{-1}$ cm^{-3}) by one to two orders of magnitude. The high temperatures found from X-ray observations may imply large blast-wave velocities, of the order of 400 to 500 km s^{-1}.

Apart from the fact that already existing inhomogeneities in the interstellar medium in front of the shock manifest themselves in the observed inhomogeneities, these may be further *amplified* by *thermal instabilities* in the cooling region behind the shock, as was shown, on the basis of work by Field (1965), by McCray *et al.* (1975). In a one-dimensional analysis these authors start from the basic equations (continuity, pressure, energy), neglecting thermal conductivity and the upstream pressure. Application to the shock in a homogeneous medium shows the normal picture of density increase and temperature decrease after the shock. However density *fluctuations* appear to be amplified, thus producing local condensations of lower temperature (since

pressure equilibrium should be maintained). A linear analysis shows that these condensations will continue to grow until the temperature is reached for which radiative
cooling shuts off. This happens at approx. $T_{min}=8000$ K. The scale length of the
finally reached condensations is smaller than the scale of the original inhomogeneities
by the product of the (ratio of scale lengths before and behind the shock) and (the
density increase in all condensations). McCray *et al.* were thus able to suggest that
sheets in the Cygnus loop, if originating as assumed above, would have thicknesses
of the order of an arcsec as indeed observed.

In similar investigations Mufson (1974, 1975) who considered shock-waves in the
interstellar medium found that shocks in the intercloud region can convert intercloud
gas to cloud gas, and fragment this gas to form separate clouds. It is the thermal
instability in a shocked gas that may produce the fine structures seen in supernova
remnants. Applied to the Cygnus loop it was found that the condensations may form
in about 10^4 yr; filamentary structures will originate with masses of a few percents
of a solar mass, and aligned parallel to the shock. Bychkov and Pikel'ner (1975),
further along these ideas, found that the condensed clouds emitting optical radiation
should have densities about 5 to 10 times that of the intercloud gas.

REFERENCES

Aab, A. E., Voikhanskaya, N. F., Kartasheva, T. A., and Morozova, S. M.: 1978, *Astron. Zh.* **55**, 553 (*Soviet Astron.* **22**, 318).

Abbott, D. C.: 1978, *Astrophys. J.* **225**, 893.

Abbott, D. C., Bieging, J. H., Curchwell, E., and Cassinelli, J. P.: 1980, preprint.

Abt, H. A.: 1957a, *Astrophys. J.* **126**, 138.

Abt, H. A.: 1957b, *Astrophys. J.* **126**, 547.

Acker, A.: 1978, *Astron. Astrophys. Suppl.* **33**, 367.

Adams, W. S. and Joy, A. H.: 1933, *Publ. Astron. Soc. Pacific* **45**, 301.

Adams, W. S. and MacCormack, E.: 1935, *Astrophys. J.* **81**, 119.

Ahern, F. J., FitzGerald, M. P., and Marsh, K. A.: 1977, *Astron. Astrophys.* **58**, 35.

Ahnert, P.: 1960, *Astron. Nachr.* **285**, 191.

Aitken, D. K., Jones, B., Bregman, J. D., Lester, D. F., and Rank, D. M.: 1977, *Astrophys. J.* **217**, 103.

Aizenman, M. L., Hansen, C. J., and Ross, R. R.: 1975, *Astrophys. J.* **201**, 387.

Ake, T. B.: 1979, *Astrophys. J.* **234**, 538.

Alcock, C. and Paczynski, D.: 1978, *Astrophys. J.* **223**, 244.

Alexander, D. R. and Johnson, H. R.: 1972. *Astrophys. J.* **176**, 629.

Allen, C. W.: 1973, *Astrophys. Quantities*, 3rd ed., Athlone Press, London.

Allen, D. A. and Swings, J. P.: 1976, *Astron. Astrophys.* **47**, 293.

Allen, D. A., Hyland, A. R., Longmore, A. J., Caswell, J. L., Goss, W. M., and Haynes, R. F.: 1977, *Astrophys. J.* **217**, 108.

Aller, L. H. and Faulkner, D. J.: 1914, *Astrophys. J.* **140**, 167.

Altenhoff, W. J. and Wendker, H. J.: 1973, *Nature* **241**, 37.

Altenhoff, W. J., Braes, L. L. G., Olnon, F. W., and Wendker, H. J.: 1976, *Astron. Astrophys.* **46**, 11.

Ambartsumian, V. A., Mirzoyan, L., and Snow, T. P.: 1979, *Astrophys. J.* **227**, 519.

Ambruster, C. W., Blitzstein, W., Hull, A. B., and Koch, R. H.: 1977, *Astron. Zh.* **54**, 592 (*Soviet Astron.* **21**, 336).

Amnuel, P. R. and Guseinov, O. H.: 1977, *Astron. Astrophys.* **54**, 23.

Ando, H.: 1976, *Publ. Astron. Soc. Japan* **28**, 517.

Andrews, P. J.: 1977, *Monthly Notices Roy. Astron. Soc.* **178**, 131.

Andriesse, C. D.: 1976, *Astron. Astrophys.* **48**, 137.

Andriesse, C. D.: 1979, *Astrophys. Space Sci.* **61**, 205.

Andriesse, C. D.: 1980a, *Astrophys. Space Sci.*, **67**, 461.

Andriesse, C. D.: 1980b, *Astrophys. Space Sci.* **70**, 173.

Andriesse, C. D., Donn, B. D., and Viotti, R.: 1978, *Monthly Notices Roy. Astron. Soc.* **185**, 771.

Andrillat, Y.: 1976, *20th Int. Coll. Liège*, p. 355.

Andrillat, Y. and Swings, J. P.: 1978, *Bull. Soc. Roy. Sci. Liège* **47**, 229.

Andrillat, Y. and Vreux, J. M.: 1975, *Astron. Astrophys.* **41**, 133.

d'Angelo, N.: 1969, *Solar Phys.* **7**, 321.

Antipova, L. I.: 1974, *Highlights of Astronomy* **3**, 501.

Antipova, L. I.: 1977, *Astron. Zh.* **54**, 68 (*Soviet Astron.* **21**, 38).

Antipova, L. I.: 1978, *Pisma Astron. Zh.* **4**, 177 (*Soviet Astron. Letters* **4**, 95).

Appenzeller, I.: 1970a, *Astron. Astrophys.* **5**, 355.

Appenzeller, I.: 1970b, *Astron. Astrophys.* **9**, 216.

Appenzeller, I.: 1972, *Publ. Astron. Soc. Pacific* **79**, 500.

Appenzeller, I.: 1974, *Astron. Astrophys.* **32**, 469.

Appenzeller, I. and Wolf, B.: 1979, *Astron. Astrophys. Suppl. Series* **38**, 51.

Arkhipova, V. P. and Mustel, E. R.: 1975, in V. E. Sherwood and L. Plaut (eds.), 'Variable Stars and Stellar Evolution', *IAU Symp.* **67**, 305.

Arkhipova, V. P., Dokuchaeva, O. D., and Esipov, V. F.: 1979, *Astron. Zh.* **56**, 313 (*Soviet Astron.* **23**, 174).

Arnett, W. D.: 1969, *Astrophys. Space Sci.* **5**, 180.

Arnett, W. D.: 1972, *Astrophys. J.* **176**, 699.

Arnett, W. D.: 1974a, *Astrophys. J.* **193**, 169.

Arnett, W. D.: 1974b, *Astrophys. J.* **194**, 373.

Arnett, W. D.: 1977, *Astrophys. J. Suppl. Series* **35**, 145.

Arnett, W. D.: 1979, *Astrophys. J. Letters* **230**, L37.

Arnett, W. D. and Schramm, D. N.: 1973, *Astrophys. J. Letters* **184**, L47.

Arp, H. C.: 1958, *Encyclopedia of Physics* **51**, 75, Springer, Berlin.

Arp, H. C.: 1961, *Astrophys. J.* **133**, 874.

Arp, H. C.: 1974, in C.B. Cosmovici (ed.), *Supernovae and Supernova Remnants*, D. Reidel Publ. Co., Dordrecht, Holland, p. 89.

Assousa, G. E., Peterson, C. J., Rubin, V. C., and Ford, W. K.: 1976, *Publ. Astron. Soc. Pacific* **88**, 828.

Athay, R. G.: 1970, *Astrophys. J.* **161**, 713.

Athay, R. G.: 1976, *The Solar Chromosphere and Corona: Quiet Sun*, D. Reidel Publ. Co., Dordrecht, Holland.

Athay, R. G. and White, O. R.: 1978, *Astrophys. J.* **226**, 1135.

Auer, L. H.: 1971, *J. Quant. Spectrosc. Radiat. Trans.* **11**, 573.

Auer, L. H. and Demarque, P.: 1977, *Astrophys. J.* **216**, 791.

Auer, L. H. and Mihalas, D.: 1972, *Astrophys. J. Suppl.* **24**, 193.

Auman, J. R.: 1969, *Astrophys. J.* **157**, 299.

Avni, Y.: 1976, *Communication General Assembly International Astronomical Union*, Grenoble.

Avni, Y. and Bahcall, J. N.: 1975a, *Astrophys. J.* **197**, 675.

Avni, Y. and Bahcall, J. N.: 1975b, *Astrophys. J. Letters* **202**, L131.

Avni, Y. and Milgrom, M.: 1977 *Astrophys. J. Letters* **212**, L17.

Aydin, C.: 1971, reference in Stalio (1972).

Aydin, C.: 1972, *Astron. Astrophys.* **19**, 369.

Aydin, C.: 1979, *Astrophys. Space Sci.* **64**, 481.

Ayres, T. R.: 1979, *Astrophys. J.* **228**, 509.

Azzopardi, N. and Breysacher, J.: 1979, *Astron. Astrophys.* **75**, 120 and 243.

Bahng, J. D. R.: 1971, *Astrophys. J. Letters* **167**, L75.

Bahng, J. D. R.: 1976, in A. Slettebak (ed.), 'Be and Shell Stars', *IAU Symp.* **70**, 41.

Balega, Yu. Yu. and Tikhonov, N. A.: 1977, *Pis'ma Astron. Zh.* **3**, 497. (*Soviet Astron. Letters* **3**, 272).

Baliunas, S. L. and Guinan, E. F.: 1976, *Publ. Astron. Soc. Pacific* **88**, 10.

Ballister, M., Batchelor, R. A., Haynes, R. F., Knowless, S. H., McCullock, M. G., Robinson, B. J., Wellington, K. J., and Yabsley, D. E.: 1977, *Monthly Notices Roy. Astron. Soc.* **180**, 415.

Bappu, M. K. V. and Sahade, J. (eds.): 1973, 'Wolf–Rayet and High Temperature Stars', *IAU Symp.* **49**.

Bappu, M. K. V., Ganesh, K. S., and Scaria, K. K.: 1977, *Kodaikanal Obs. Bull. Series A* **2**, 28.

Barbaro, G., Giovannone, P., Giannuzzi, M. A., and Summa, C.: 1969, in M. Hack (ed.), *Mass Loss from Stars*, D. Reidel Publ. Co., Dordrecht, Holland, p. 217.

Barbaro, G., Chiosi, C., and Nobili, L.: 1972, in M. Hack (ed.), *Colloquium on Supergiant Stars*, Oss. Astron. Trieste, p. 313.

Barbier, D. and Chalonge, D.: 1939, *Astrophys. J.* **90**, 627.

Barbier, R., Swings, J. P., Delcroix, A., Hornack, P., and Rogerson, J. B.: 1978, *Astron. Astrophys. Suppl. Series* **32**, 69.

Barbon, R.: 1978, *Mem. Soc. Astron. Ital.* **49**, 331.

Barbon, R., Ciatti, F., and Rosino, L.: 1973, *Astron. Astrophys.* **25**, 24.

Barbon, R., Ciatti, F., and Rosino, L.: 1947a, in C. B. Cosmovici (ed.), *Supernovae and Supernova Remnants*, D. Reidel Publ. Co., Dordrecht, Holland, p. 99.

Barbon, R., Ciatti, F., and Rosino, L.: 1974b, in C. B. Cosmovici (ed.), *Supernovae and Supernova Remnants*, D. Reidel Publ. Co., Dordrecht, Holland, p. 115.

Barbon, R., Ciatti, F., and Rosino, L.: 1979, *Astron. Astrophys.* **72**, 287.

Barkat, Z., Rakavy, G., and Sack, N.: 1967, *Phys. Rev. Letters* **18**, 379.

Barlow, M. J.: 1979, in P. S. Conti and C. W. H. de Loore (eds.), 'Mass Loss and Evolution of O-type Stars', *IAU Symp.* **83**, 119.

Barlow, M. J. and Cohen, M.: 1977, *Astrophys. J.* **213**, 737.

Barlow, M. J., Smith, L. J., and Willis, A. J.: 1980, submitted *Monthly Notices Roy. Astron. Soc.*

Barnes, T. G.: 1976, *Monthly Notices Roy. Astron. Soc.* **177**, 53P.

Barnes, T. G. and Evans, D. S.: 1976, *Monthly Notices Roy. Astron. Soc.* **174**, 489.

Barnes, T. G., Lambert, D. L., and Potter, A. E.: 1974, *Astrophys. J.* **187**, 73.

Barnes, T. G., Evans, D. S., and Parsons, S. B.: 1976, *Monthly Notices Roy. Astron. Soc.* **174**, 503.

Barnes, T. G., Evans, D. S., and Moffett, T. J.: 1978, *Monthly Notices Roy. Astron. Soc.* **183**, 285.

Baschek, B.: 1975, in B. Baschek, W. H. Kegel, and G. Traving (eds.), *Problems in Stellar Atmospheres and Envelopes*, Springer, Berlin, p. 118.

Baschek, B. and Scholz, M.: 1971a, *Astron. Astrophys.* **11**, 83.

Baschek, B. and Scholz, M.: 1971b, *Astron. Astrophys.* **15**, 285.

Basri, G. S. and Linsky, J. L.: 1980, preprint.

Bath, G. T.: 1977, in M. Friedjung (ed.), *Novae and Related Stars*, D. Reidel, Publ. Co., Dordrecht, Holland, p. 41.

Bath, G. T.: 1978a, *Monthly Notices Roy. Astron. Soc.* **182**, 35.

Bath, G. T.: 1978b, *Quart. J. Roy. Astron. Soc.* **19**, 442.

Bath, G. T. and Shaviv, G.: 1978, *Monthly Notices Roy. Astron. Soc.* **183**, 515.

Batten, A. H.: 1967, *Sixth Catalogue of the Orbital Elements of Spectroscopic Binary Systems*, Publ. Dominion Astrophys. Obs., XIII, (8).

Batten, A. H.: 1968, *Astron. J.* **73**, 1.

Batten, A. H.: 1976, in P. Eggleton *et al.* (eds.), 'Structure and Evolution of Close Binary Systems', *IAU Symp.* **73**, 303.

Baud, B., Habing, H. J., Matthews, H. E., and Winnberg, A.: 1979a, *Astron. Astrophys. Suppl.* **36**, 99.

Baud, B., Habing, H. J., Matthews, H. E., and Winnberg, A.: 1979b, *Astron. Astrophys. Suppl.* **36**, 193.

Baudry, A., Le Squérin, A. M., and Lépine, J. R. D.: 1977, *Astron. Astrophys.* **54**, 593.

Baumbach, S.: 1937, *Astron. Nachr.* **263**, 121.

Baumert, J. H.: 1972, Thesis, Ohio State Univ. (unpublished).

Beals, C. S.: 1938, *Trans. IAU* **6**, 248.

Beals, C. S.: 1950, *Publ. Dominion Astrophys. Obs.* **9**, 1.

Beardsly, W. R.: 1961, *Astrophys. J. Suppl.* **5**, 381.

Becklin, E. E., Frogel, J. A., Hyland, A. R., Kristian, J., and Neugebauer, G.: 1969, *Astrophys. J. Letters* **158**, L153.

Becker, R. H., Holt, S. S., Smith, B. W., White, N. E., Boldt, E. A., Mushotsky, R. F., and Serlemitsos P. J.: 1979, *Astrophys. J. Letters* **234**, L73.

Behr, A.: 1959, *Veröffentl. Göttingen* **126**.

Bell, R. A., Eriksson, K., Gustafsson, B., and Nordlund, Å.: 1976, *Astron. Astrophys. Suppl.* **23**, 37

Belyakina, T. S., Gershberg, R. E., and Shakovskaya, N. I.: 1978, *Pis'ma Astron. Zh.* **4**, 406 (*Soviet Astron. Letters* **4**, 219).

Bensammar, S., Kandel, R., Assus, P., and Journet, A.: 1978, *Astron. Astrophys.* **70**, 585.

Benson, J. M. and Mutel, R. L.: 1979, *Astrophys. J.* **233**, 119.

Benson, J. M., Mutel, R. L., Fix, J. D., and Clausen, M. J.: 1979, *Astrophys. J. Letters* **229**, L87,

Bergeat, J., Sibille, F., Lunel, M., and Lefevre, J.: 1976a, *Astron. Astrophys.* **52**, 227.

Bergeat, J., Lefevre, J., Kandel, R., Lunel, and M. Sibille, F.: 1976b, *Astron. Astrophys.* **52**, 245.

Bergeat, J., Lunel, M., Sibille, F., and Lefevre, J.: 1976c, *Astron. Astrophys.* **52**, 263.

Bergeat, J., Sibille, F., and Lunel, M.: 1978, *Astrophys.* **64**, 423.

Bernat, A. P.: 1977, *Astrophys. J.* **213**, 756.

Bernat, A. P. and Lambert, D. L.: 1967a, *Astrophys. J.* **204**, 830.

Bernat, A. P. and Lambert, D. L.: 1976b, *Astrophys. J.* **210**, 395.

Bernat, A. P. and Lambert, D. L.: 1978, *Monthly Notices Roy. Astron. Soc.* **183**, 17P.

Bernat, A. P., Barnes, T. G., Schupler, B. R., and Potter, A. E.: 1977, *Publ. Astron. Soc. Pacific* **89**, 541.

Bernat, A. P., Bruhweiler, F. C., Kondo, Y., and Van der Hucht, , K. A.: 1978a, *Publ. Astron. Soc. Pacific* **90**, 318.

Bernat, A. P., Honeycutt, R. K., Kephart, J. E., Cow, C. E., Sandford, M. T., and Lambert, D. L.: 1978b, *Astrophys. J.* **219**, 532.
Berthomieu, G., Provost, J., and Rocca, A.: 1976, *Astron. Astrophys.* **47**, 413.
Betz, A. L., McLaren, R. A., and Spears, D. L.: 1979, *Astrophys. J. Letters*, **229**, L97.
Bidelman, W. P.: 1954, *Astrophys. J. Suppl. Series* **1**, 175.
Bidelman, W. P.: 1976, in A. Slettebak (ed.), 'Be and Shell Stars', *IAU Symp.* **70**, 457.
Bidelman, W. P. and McKellar, A.: 1957, *Publ. Astron. Soc. Pacific* **69**, 31.
Bidelman, W. P. and Weitenbeck, A. J.: 1976, in A. Slettebak (ed.), 'Be and Shell Stars', *IAU Symp.* **70**, 29.
Bierman, L.: 1946, *Naturwiss.* **33**, 118.
Biermann, P. and Kippenhahn, R.: 1972, *Astron. Astrophys.* **14**, 32.
Bisnovatyj-Kogan, G. S.: 1970, *Astron. Zh.* **47**, 813.
Bisnovatyj-Kogan, G. S., Popov, Yu. P., and Samokhin, A. A.: 1976, *Astrophys. Space Sci.*, **41**, 321.
Blackwell, D. E. and Shallis, M. J.: 1977, *Monthly Notices Roy. Astron. Soc.* **180**, 177.
Blackwell, D. E. and Willis, R. B.: 1977, *Monthly Notices Roy. Astron. Soc.* **180**, 169.
Blair, G. N. and Dickinson, D. F.: 1977, *Astrophys. J.* **215**, 552.
Blazit, A., Bonneau, D., Koechlin, L., and Labeyrie, A.: 1977, *Astrophys. J. Letters* **214**, L79.
Bless, R. C., Code, A. D., and Fairchild, E. T.: 1976, *Astrophys. J.* **203**, 410.
Blumenthal, G. R. and Tucker, W. H.: 1974, *Ann. Rev. Astron. Astrophys.* **12**, 23.
Bodansky, D., Clayton, D. D., and Fowler, W. A.: 1968, *Phys. Rev. Letters* **20**, 161.
Boesgaard, A. M.: 1979, *Astrophys. J.* **232**, 485.
Boesgaard, A. M. and Magnan, C.: 1975, *Astrophys. J.* **198**, 369.
Bohannan, B. and Conti, P. S.: 1976, *Astrophys. J.* **204**, 797.
Bohannan, B. and Garmany, C. D.: 1978, *Astrophys. J.* **223**, 908.
Böhm, K. H.: 1968, *Astrophys. Space Sci.* **2**, 375.
Böhm, K. H.: 1970, *Astrophys. J.* **162**, 919.
Böhm, K. H.: 1973, in A. H. Batten (ed.), 'Extended Atmospheres and Circumstellar Matter in Spectroscopic Binary Systems', *IAU Symp.* **51**, 148.
Böhm, K. H. and Cassinelli, J.: 1971, *Astron. Astrophys.* **12**, 21.
Böhm-Vitense, E.: 1958, *Z. Astrophys.* **46**, 108.
Böhm-Vitense, E.: 1972, *Astron. Astrophys.* **17**, 335.
Boksenberg, A., Evans, R. G., Fowler, R. G., Gardner, I. S. K., Houziaux, L., Humphries, C. M., Jamar, C., Macan, D., Macan, J. P., Malaise, D., Monfils, A., Nandy, K., Thomson, G. I., Wilson, R., and Wroe, H.: 1973, *Monthly Notices Roy. Astron. Soc.* **163**, 291.
Boksenberg, A., Kirkham, B., Fowson, W. A., Venus, T. E., Bates, B., Carson, P. P. D., and Courts, G. R.: 1974, *Space Res.* **XIV**, 553.
Bolgova, G. T., Strel'nitskii, V. S., and Shmeld, I. K.: 1977, *Astron. Zh.* **54**, 828 (*Soviet Astron* **21**, 468).
Bolton, C. T. and Rogers, G. L.: 1978, *Astrophys. J.* **222**, 234.
Bond, H. E., Liller, W., and Mannery, E. J.: 1978, *Astrophys. J.* **223**, 252.
Bond, H. E., Luck, R. E., and Newman, M. J.: 1979, *Astrophys. J.* **233**, 205.
Bondi H. and Hoyle, F.: 1944, *Monthly Notices Roy. Astron. Soc.* **104**, 273.
Bonneau, D. and Labeyrie, A.: 1973, *Astrophys. J. Letters* **181**, L1.
Borsenberger, J. and Gros, M.: 1978, *Astron. Astrophys. Suppl. Series* **31**, 291.
Bossi, M., Guerrero, G., and Mantegazza, L.: 1976, *Inf. Bull. Variable Stars* **1095**.
Bouigue, R.: 1954, *Ann. Astrophys.* **17**, 104.
Bouw, G. D. and Parsons, S. B.: 1972, in M. Hack (ed.), *Colloquium on Supergiant Stars*, Osservatorio Astronomico Trieste, p. 22.
Bowers, P. F. and Kerr, F. J.: 1977, *Astron. Astrophys.* **57**, 115.
Bowers, P. F. and Kerr, F. J.: 1978, *Astron. J.* **83**, 487.
Bowers, P. F. and Kundu, M. R.: 1979, *Astron. J.* **84**, 791.
Bowers, P. F. and Sinha, R. P.: 1978, *Astron. J.* **83**, 955.
Boyarchuk, A. A.: 1968, *Astrofizika* **4**, 289.
Boyarchuk, A. A.: 1973, in A. H. Batten (ed.), 'Extended Atmospheres and Circumstellar Matter in Spectroscopic Binary Systems', *IAU Symp.* **51**, 81.
Boyarchuk, A. A.: 1977, in R. Kippenhahn *et al.* (eds.), *The Interaction of Variable Stars with their Environment*, Veröff. Remeis Sternwarte Bamberg XI, No. 121, p. 274.
Boyarchuk, A. A. and Gershberg, R. E.: 1977, *Astron. Zh.* **54**, 488.

Boyarchuk, A. A. and Kopylov, J. M.: 1958, *Soviet Astron.* **2**, 752.

Boyarchuk, A. A., Burnashev, V. I., and Gershberg, R. E.: 1977a, *Astron. Zh.* **54**, 477 (*Soviet Astron.* **21**, 269).

Boyarchuk, A. A., Galkina, T. S., Gershberg, R. E., Krasnobabtsev, V. I., Rachkovskaya, T. M., and Shakovskaya, N. I.: 1977b, *Astron. Zh.* **54**, 458.

Bracher, K.: 1966, Thesis, Indiana Univ.; and *Astron. J.* **71**, 156.

Braginskii, S. I.: 1965, in A. M. Leontovich (ed.), *Review of Plasma Physics*, Consultants Bureau, New York, p. 205.

Branch, D.: 1977a, in D. N. Schramm (ed.), *Supernovae*, D. Reidel Publ. Co., Dordrecht, Holland, p. 21.

Branch, D.: 1977b, *Monthly Notices Roy. Astron. Soc.* **179**, 401.

Branch, D.: 1979, *Monthly Notices Roy. Astron. Soc.*, **186**, 609.

Branch, D. and Patchett, B.: 1973, *Monthly Notices Roy. Astron. Soc.* **161**, 71.

Brandt, J. C.: 1970, *Introduction to the Solar Wind*, Freeman, San Francisco.

Brandt, J. C., Stecher, T. P., Crawford, D. L., and Maran, S. P.: 1971, *Astrophys. J. Letters* **163**, L99.

Branduardi, G., Mason K. O. and Sanford, P. W.: 1978, *Monthly Notices Roy. Astron. Soc.* **185**, 137.

Brecher, K., Ingham, W. H., and Morrison, P.: 1977, *Astrophys. J.* **213**, 492.

Bregman, J. D., Goebel, J. H., and Strecker, D. W.: 1978, *Astrophys. J. Letters* **223**, L45.

Breysacher, J. and Azzopardi, M.: 1979a, *Astron. Astrophys.* **75**, 243.

Breysacher, J. and Azzopardi, M.: 1979b, in P. S. Conti and C. W. H. de Loore (eds.), 'Mass Loss and Evolution of O-type Stars', *IAU Symp.* **83**, 51.

Breysacher, J. and Westerlund, B. E.: 1978, *Astron. Astrophys.* **67**, 261.

Briot, D.: 1977, *Astron. Astrophys.* **54**, 599.

Briot, D.: 1978, *Astron. Astrophys.* **66**, 197.

Brosch, N., Leibowitz, E. M., and Spector, N.: 1978: *Astron. Astrophys.* **65**, 259.

Brucato, R. J.: 1971, *Monthly Notices Roy. Astron. Soc.* **153**, 435.

Bruenn, S. W.: 1971, *Astrophys. J.* **168**, 203.

Bruenn, S. W.: 1972, *Astrophys. J. Suppl. Ser.* **24**, 283.

Bruhweiler, F. C., Morgan, T. M., and Van der Hucht, K. A.: 1978, *Astrophys. J. Letters* **225**, L74.

Brune, W. H., Mount, G. H., and Feldman, P. D.: 1979, *Astrophys. J.* **227**, 884.

Bruner, E. C.: 1978, *Astrophys. J.* **226**, 1140.

Brunet, J. P. and Prévot, L.: 1972, in M. Hack (ed.), *Colloquium on Supergiant Stars*, Osservatorio Astronomico Trieste, p. 119.

Buchkov, K. V. and Stinik, T. G.: 1975, *Soviet Astron.* **19**, 309 (*Astron. Zh.* **52**, 505).

Buhl, D., Snijder, L. E., Lovas, F. J., and Johnson, D. R.: 1974, *Astrophys. J. Letters* **192**, L97.

Buhl, D., Snijder, L. E., Lovas, F. J., and Johnson, D. R.: 1975, *Astrophys. J. Letters* **201**, L29.

Burbridge, G. R. and Burbridge, E. M.: 1954, *Astrophys. J.* **120**, 76.

Burger, J. J., Stephens, S. A., and Swanenburg, B. N.: 1970, *Astrophys. Space Sci.* **8**, 20.

Burger, M.: 1976, *Studie van het Ultraviolette Spektrum van B Sterren en Ap Sterren*, Thesis, Vrije Universiteit, Brussels.

Burger, M., de Jager, C., and Sato, N.: 1980, in preparation.

Burki, G.: 1978, *Astron. Astrophys.* **65**, 357.

Burki, G., Maeder, A., and Rufener, F.: 1978, *Astron. Astrophys.* **65**, 363.

Burnichon, M. L.: 1976, *Astron. Astrophys.* **45**, 383.

Burton, W. M., Evans, R. G., and Griffin, W. G.: 1975, *Phil. Trans. Roy. Soc. London* **A279**, 355.

Burton, W. M., Evans, R. G., and Griffin, W. G.: 1976, *Monthly Notices Roy. Astron. Soc.* **176**, 29P.

Burton, W. M., Evans, R. G., Patchett, B., and Wu, C. C.: 1978, *Monthly Notices Roy. Astron. Soc.* **183**, 605.

Bychkov, K. V.: 1977, *Astron. Zh.* **54**, 85 (1977, *Soviet Astron.* **21**, 48).

Bychkov, K. V. and Panchuk, V. E.: 1977, *Astron. Zh.* **54**, 340.

Bychkov, K. V. and Pickel'ner, S. B.: 1975, *Pis'ma Astron. Zh.* **1**, 29 (*Soviet Astron. Letters* **1**, 14).

Bychov, K. V. and Sitnik, T. G.: 1975, *Astron. Zh.* **52**, 505.

Bychkova, V. S. and Bychkov, K. V.: 1976, *Astron. Zh.* **53**, 1145 (1977, *Soviet Astron.* **20**, 675).

Bychkova, V. S. and Bychkov, K. V.: 1977, *Astron. Zh.* **54**, 772 (*Soviet Astron.* **21**, 435).

Cacciari, C., Fusi Pecci, F., and Guarnieri, A.: 1977, *Astron. Zh.* **54**, 583 (*Soviet Astron.* **21**, 331).

Cahn, J. H. and Elitzur, M.: 1979, *Astrophys. J.* **231**, 124.

Cahn, J. H. and Wyatt, S. P.: 1976, *Astrophys. J.* **210**, 508.

Cahn, J. H. and Wyatt, S. P.: 1978a, *Astrophys. J.* **221**, 163.
Cahn, J. H. and Wyatt, S. P.: 1978b, *Astrophys. J. Letters* **224**, L79.
Campbell, B.: 1976, *Astrophys. J. Letters* **207**, L41.
Campusano, L. E.: 1977, in R. Kippenhahn *et al.* (eds.), *The Interaction of Variable Stars with their Environment*, Veröffentl. Remeis Sternwarte, Bamberg, XI, No. 121, p. 128.
Canizares, C. R., McClintock, J. E., and Grindlay, J. E.: 1980, in press.
Carbon, D. C.: 1979, *Ann. Rev. Astron. Astrophys.* **17**, 513.
Carbon, D. F. and Gingerich, O.: 1969, in O. Gingerich (ed.), *3rd Harvard-Smithsonian Conference on Stellar Atmospheres*, MIT Press, Cambridge, Mass., p. 377.
Cardona-Núñez, O.: 1978, 'Analysis of the λ5696 Carbon III Line in O-stars', Thesis, Univ. of Colorado, Boulder.
Carruthers, G. R.: 1971, *Astrophys. J.* **166**, 571.
Carson, T. R.: 1976, *Ann. Rev. Astron. Astrophys.* **14**, 95.
Carty, T. F., Perkins, H. G., Warren-Smith, R. F., and Scarrott, S. M.: 1979, *Monthly Notices Roy. Astron. Soc.* **189**, 299.
Cassatella, A. and Viotti, R.: 1979, in A. J. Willis (ed.), *The First Year of IUE*, Univ. College of London, p. 117.
Cassatella, A., Benvenuti, P., Clavel, J., Heck, A., Penston, M., Selvelli, P. L., and Machetto, F.: 1979a, *Astron. Astrophys. Letters* **74**, L18.
Cassatella, A., Beeckmans, F., Benvenuti, P., Clavel, J., Heck, A., Lamers, H. J. G. L. M., Machetto, F., Penston, M., Selvelli, P. L., and Stickland, D. J.: 1979b, *Astron. Astrophys.* **79**, 223.
Cassatella, A., Gilra, D., Reimers, D., and Stickland, D.: 1979c, *IAU Circ.* **3425**.
Cassinelli, J. P.: 1971a *Astrophys. Letters* **8** 105.
Cassinelli, J. P.: 1971b, *Astrophys. J.* **165**, 265.
Cassinelli, J. P.: 1979a, in P. S. Conti and C. W. H. de Loore (eds.), 'Mass Loss and Evolution of O-type Stars', *IAU Symp.* **83**, 201.
Cassinelli, J. P.: 1979b, *Ann. Rev. Astron. Astrophys.* **17**, 275.
Cassinelli, J. P. and Castor, J. I.: 1973, *Astrophys. J.*, **179**, 189.
Cassinelli, J. P. and Hartmann, J. L.: 1975, *Astrophys. J.* **202**, 178.
Cassinelli, J. P. and Hartmann, L.: 1977, *Astrophys. J.* **212**, 488.
Cassinelli, J. P. and Olson, G.: 1978, *Astrophys. J.* **229**, 304.
Cassinelli, J. P., Castor, J. I., and Lamers, H. J. G. L. M.: 1978a, *Publ. Astron. Soc. Pacific* **90**, 496.
Cassinelli, J. P., Olson, G. L., and Stalio, R.: 1978b, *Astrophys. J.* **220**, 573.
Castellani, V. and Sacchetti, M.: 1978, *Astrophys. Space Sci.* **53**, 217.
Castelli, F.: 1978, *Astron. Astrophys.* **69**, 23.
Castor, J. I.: 1970, *Monthly Notices Roy. Astron. Soc.* **149**, 111.
Castor, J. I.: 1972, *Astrophys. J.* **178**, 779.
Castor, J. I.: 1974a, *Astrophys. J.* **189**, 273.
Castor, J. I.: 1974b, *Monthly Notices Roy. Astron. Soc.* **169**, 279.
Castor, J. I.: 1979, in P. S. Conti and C. W. H. de Loore (eds.), 'Mass Loss and Evolution of O-type Stars', *IAU Symp.* **83**, 175.
Castor, J. I. and Lamers, H. J. G. L. M.: 1979, *Astrophys. J. Suppl.* **39**, 481.
Castor, J. I. and Nussbaumer, H.: *Monthly Notices Roy. Astron. Soc.* **155**, 293.
Castor, J. I., Abbott, D. C., and Klein, R. I.: 1975, *Astrophys. J.* **195**, 157.
Castor, J. I., Abbott, D. C., and Klein, R. I.: 1976, in R. Cayrel and M. Steinberg (eds.), *Physique des Mouvements dans les Atmosphères Stellaires*, Paris, C.N.R.S., p. 363.
Caswell, J. L.: 1970, *Astron. Astrophys.* **7**, 59.
Caswell, J. L.: 1979, *Monthly Notices Roy. Astron. Soc.* **187**, 431.
Catchpole, R. M.: 1975, *Publ. Astron. Soc. Pacific* **87**, 397.
Catchpole, R. M. and Feast, M. W.: 1971, *Monthly Notices Roy. Astron. Soc.* **154**, 197.
Catura, R. C., Acton, L. W., and Johnson, H. M.: 1975, *Astrophys. J. Letters* **196**, L47.
Caughlan, G. R. and Fowler, W. A.: 1962, *Astrophys. J.* **136**, 453.
Cayrel, R.: 1963, *Compt. Rend. Acad. Sci. Paris* **257**, 3309.
Chalonge, D., Divan, L., and Mirzoyan, L. V.: 1977, *Astrofizika* **13**, 437.
Chamberlain, J. W.: 1963, *Planetary Space Sci.* **11**, 901.
Chambliss, C. R. and Leung, K. C.: 1979, *Astrophys. J.* **228**, 828.
Chandrasekhar, S.: 1934, *Monthly Notices Roy. Astron. Soc.* **94**, 444.

Chandrasekhar, S.: 1950, *Radiative Transfer*, Oxford Univ. Press.

Chapman, R. D.: 1966, *Astrophys. J.* **143**, 61.

Charles, P. A., Culhane, J. L., and Fabian, A. C.: 1977, *Monthly Notices Roy. Astron. Soc.* **178** 307.

Charles, P. A., Mason, K. O., White, N. E., Culhane, J. L., Sanford, P. W., and Moffat, A. F. J.: 1978, *Monthly Notices Roy. Astron. Soc.* **183**, 813.

Chechetkin, V. M., Gehrstein, S. S., Imshennik, V. S., Ivanova, L. N., and Khoplov, M. Yu.: 1980, *Astrophys. Space Sci.* **67**, 61.

Chevalier, C. and Ilovaisky, S. A.: 1977, *Astron. Astrophys.* **59**, L9.

Chevalier, R. A.: 1976, *Publ. Astron. Soc. Pacific* **88**, 588.

Chevalier, R. A. and Kirshner, R. P.: 1978, *Astrophys. J.* **219**, 931.

Chevalier, R. A. and Raymond, J. C.: 1978, *Astrophys. J. Letters* **225**, L27.

Chia, T. T., Milone, E. F., and Robb, R.: 1977, *Astrophys. Space Sci.* **48**, 3.

Chiosi, C. and Nasi, E.: 1978, *Astrophys. Space Sci.* **56**, 431.

Chiosi, C. and Summa, C.: 1970, *Astrophys. Space Sci.* **8**, 478.

Chiosi, C., Nasi, E., and Sreenivasan, S. R.: 1978, *Astron. Astrophys.* **63**, 103.

Chiu, B. C., Morrison, P., and Sartori, L.: 1975, *Astrophys. J.* **198**, 617.

Chugai, N. N.: 1977, *Pis'ma Astron. Zh.* **3**, 17 (*Sov. Astron. Letter* **3**, 9).

Ciatti, F., Mammano, A., and Rosino, L.: 1975, in V. E. Sherwood and L. Plaut (eds.), 'Variable Stars and Stellar Evolution', *IAU Symp.* **67**, 389.

Ciatti, F., Mammano, A., and Vittone, A.: 1977, *Astron. Astrophys.* **61**, 459.

Ciatti, F., Mammano, A., and Vittone, A.: 1978, *Astron. Astrophys.* **68**, 251.

Ciatti, F., Mammano, A., and Vittone, A.: 1979, *Astron. Astrophys.* **79**, 247.

Cimerman, M.: 1979, *Astrophys. J. Letters* **228**, L79.

Cimino, M., Gianone, P., Gianuzzi, M. A., Masani, A., and Virgopia, N.: 1963, *Nuovo Cimento X* **28**, 621.

Clark, D. H.: 1978, *Mem. Soc. Astron. Ital.* **49**, 307.

Clark, D. H. and Caswell, J. L.: 1976, *Monthly Notices Roy. Astron. Soc.* **174**, 267.

Clark, D. H., Culhane, J. L.: 1976, *Monthly Notices Roy. Astron. Soc.* **175**, 573.

Clark, D. H. and Stephenson, F. R.: 1975, *Observatory* **95**, 190.

Clark, D. H. and Stephenson, F. R.: 1976, *Quart. J. Roy. Astron. Soc.* **17**, 290.

Clark, D. H. and Stephenson, F. R.: 1977, *Monthly Notices Roy. Astron. Soc.* **179**, 87P.

Clayton, D. D.: 1968, *Principles of Stellar Evolution and Nucleosynthesis*, McGraw-Hill, New York.

Clayton, D. D. and Hoyle, F.: 1975, *Astrophys. J.* **203**, 490.

Clayton, D. D. and Wicramasinghe, N. C.: 1976, *Astrophys. Space Sci.* **42**, 463.

Climenhaga, J. L., Harris, B. L., Holts, J. T., and Smolinski, J.: 1977, *Astrophys. J.* **215**, 836.

Cline, T. L. and Hones, E. W.: 1969, *Proc. Canadian J. Phys.* **46**, 527.

Code, A. D. and Maede, M. R.: 1979, *Astrophys. J. Suppl.* **39**, 195.

Code, A. D., Houck, T. E., McNall, J. F., Bless, R. C., and Lillie, C. F.: 1970, *Astrophys. J.* **161**, 377.

Code, A. D., Davis, J., Bless, R. C., and Hanbury Brown, R.: 1974, *Wisconsin Astrophys.* **2**.

Code, A. D., Davis, J., Bless, R. C., and Hanbury Brown, R.: 1976, *Astrophys. J.* **203**, 417.

Coe, M. J., Engel, A. R., and Quenby, J. J.: 1977, *IAU Circ.* **3054**.

Cohen, M.: 1979, *Monthly Notices Roy. Astron. Soc.* **186**, 837.

Cohen, M. and Barlow, M. J.: 1973, *Astrophys. J. Letters* **185**, L37.

Cohen, M. and Kuhi, L. V.: 1977, *Monthly Notices Roy. Astron. Soc.* **180**, 37.

Cohen, M. and Vogel, S. N.: 1978, *Monthly Notices Roy. Astron. Soc.* **185**, 47.

Cohen, M. and Woolf, N. J.: 1971, *Astrophys. J.* **169**, 543.

Cohen, M., Barlow, M. J., and Kuhi, L. V.: 1975, *Astron. Astrophys.* **20**, 291.

Coleman, P. L., Bunner, A. N., Kraushaar, W. L., McCammon, D., and Williamson, F. D.: 1973, *Astrophys. J. Letters* **185**, L121.

Colgate, S. A.: 1970 *Astrophys. Space Sci.* **8**, 457.

Colgate, S. A. and McKee, C.: 1969 *Astrophys. J.* **157**, 623.

Colvin, J. D., Van Horn, J. D., Starrfield, S. G., and Truran, J. W.: 1977, *Astrophys. J.* **212**, 791.

Conner, J. P., Evans, W. D., and Belian, R. D.: 1969, *Astrophys. J. Letters* **157**, L157.

Conti, P. S.: 1973a, *Astrophys. J.* **179**, 161.

Conti, P. S.: 1973b, *Astrophys. J.* **179**, 181.

Conti, P. S.: 1973c, in M. K. V. Bappu and J. Sahade (eds.), 'Wolf-Rayet and High Temperature Stars', *IAU Symp.* **49**, 95.

Conti, P. S.: 1974, *Astrophys. J.* **187**, 589.

Conti, P. S.: 1975, in T. L. Wilson and D. Downes (eds.), *H II Regions and Related Objects*, Springer, Berlin, p. 207.

Conti, P. S.: 1976, *Mem. Soc. Roy. Sci. Liège, 6e Série* **IX**, 193.

Conti, P. S.: 1978a, *Astron. Astrophys.* **63**, 225.

Conti, P. S.: 1978b, *Ann. Rev. Astron. Astrophys.* **16**, 371.

Conti, P. S.: 1978c, in A. D. Philip and D. S. Hayes (eds.), 'The HR Diagram', *IAU Symp.* **80**, 369.

Conti, P. S. and Alschuler, W. R.: 1971, *Astrophys. J.* **170**, 325.

Conti, P. S. and Burnichon, M. L.: 1975, *Astron. Astrophys.* **38**, 467.

Conti, P. S. and de Loore, C. W. H. (eds.), 1979, 'Mass Loss and Evolution of O-type Stars', *IAU Symp.* **83**.

Conti, P. S. and Ebbets, D.: 1977, *Astrophys. J.* **213**, 438.

Conti, P. S. and Frost, S. A.: 1974, *Astrophys. J. Letters* **190**, L137.

Conti, P. S. and Frost, S. A.: 1977, *Astrophys. J.* **212**, 729.

Conti, P. S. and Garmany, C. D.: 1979, in A. J. Willis (ed.), *The First Year of IUE*, Univ. College of London, p. 285.

Conti, P. S. and Garmany, C. D.: 1980, *Astrophys. J.*, in press.

Conti, P. S. and Leep, E. M.: 1974, *Astrophys. J.* **193**, 113.

Conti, P. S. and Massey, P.: 1980, *Astrophys. J.*, in press.

Conti P. S. and Niemelä, V. S.: 1976, *Astrophys. J. Letters* **209**, L37.

Conti, P. S. and Smith, L. F.: 1972, *Astrophys. J.* **172**, 623.

Conti, P. S. and Vanbeveren, D.: 1980, *Astrophys. J.*, prepint.

Conti, P. S. and Van der Hucht, K. A.: 1979, in A. J. Willis (ed.), *The First Year of IUE*, Univ. College of London, p. 280.

Conti, P. S. and Walborn, N. R.: 1976, *Astrophys. J.* **207**, 502.

Conti, P. S., Cowley, A. P., and Johnson, G. B.: 1975, *Publ. Astron. Soc. Pacific* **87**, 327.

Conti, P. S., Leep, E. M., and Lorre, J. J.: 1977a, *Astrophys. J.* **214**, 759.

Conti, P. S., Garmany, C. D., and Hutchings, J. B.: 1977b, *Astrophys. J.* **215**, 561.

Conti, P. S., Niemelä, V. S., and Walborn, W. R.: 1979, *Astrophys. J.* **228**, 206.

Cook, A. H.: 1977, *Celestial Masers*, Cambridge, Univ. Press.

Cook, A. H.: 1978, *Quart. J. Roy. Astron. Soc.* **19**, 255.

Cosmovici, C. B.: 1974, *Supernovae and their Remnants*, D. Reidel Publ. Co., Dordrecht, Holland.

Cousins A. W. J. and Stoy, R. H.: 1963, *Roy. Obs. Bull.* **64**.

Cowley, A. P.: 1965 *Astrophys. J.* **142**, 299.

Cowley,, A. P.: 1969 *Publ. Astron. Soc. Pacific* **81**, 297.

Cowley, A. P. and Hutchings, J. B.: 1979, *Publ. Astron. Soc. Pacific* **90**, 636.

Cowley, A. P., Rogers, L., and Hutchings, J. B.: 1976, *Publ. Astron. Soc. Pacific* **88**, 911.

Cowley, A. P., Crampton, D., Hutchings, J. B., Liller, W., and Sanduleak, N.: 1977, *Astrophys. J. Letters* **218**, L3.

Cowley, A. P., Hutchings, J. B., and Popper, D. M.: 1978, *Publ. Astron. Soc. Pacific* **89**, 882.

Cox, A. N., King, D. S., and Hodson, S. W.: 1978, *Astrophys. J.* **224**, 607.

Cox, D. P. and Tucker, W. H.: 1969, *Astrophys. J.* **157**, 1157.

Cox, G. G. and Parker, E. A.: 1978, *Monthly Notices Roy. Astron. Soc.* **183**, 111.

Cox, G. G. and Parker, E. A.: 1979, *Monthly Notices Roy. Astron. Soc.* **186**, 197.

Cox, J. P. and Giuli, R. T.: 1961, *Astrophys. J.* **133**, 755.

Coyne, G. V.: 1976, in A. Slettebak (ed.), 'Be and Shell Stars', *IAU Symp.* **70**, 233.

Coyne, G. V. and McLean, I. S.: 1979, prepint.

Coyne, G. V. and Magalhães, A. M.: 1979, *Astron. J.* **84**, 1200.

Crabtree, D. R., Richer, H. B., and Westerlund, B. E.: 1976, *Astrophys. J. Letters* **203**, L81.

Cram, L. E.: 1978, *Astron. Astrophys.* **67**, 301.

Crampton, D. and Georgelin, Y. M.: 1979, prepint.

Crampton, D. and Redman, R. O.: 1975, *Astron. J.* **80**, 454.

Crampton, D., Grygar, J., Kohoutek, L., and Viotti, R.: 1970, *Astrophys. Letters* **6**, 5.

Crivellari, L. and Stalio, R.: 1977, *Mem. Soc. Astron. Ital.* **48**, 773.

Culhane, J. L.: 1977, in D. N. Schramm (ed.), *Supernovae*, D. Reidel Publ. Co., Dordrecht, Holland, p. 29.

Dallaporte, N.: 1972, in M. Hack (ed.), *Colloquium on Supergiant Stars*, Osservatorio Astronomico Trieste, p. 250.

Danks, A. C. and Houziaux, L.: 1978, *Publ. Astron. Soc. Pacific* **90**, 453.

Davelaar, J.: 1979, preprint.

Davidsen, A. F., Henry R. C., Snyder, W. A., Friedman, H., Fritz, G., Naranan, S., Shulman, S., and Yentis, D.: 1977, *Astrophys. J.* **215**, 541.

Davidson, K.: 1971, *Monthly Notices Roy. Astron. Soc.* **154**, 415.

Davidson, K. and Ostriker, J. P.: 1973, *Astrophys. J.* **179**, 585.

Davidson, P. J. N., Watson, M. G., and Pye, J. P.: 1977, *Monthly Notices Roy. Astron. Soc.* **181**, 73P.

Davies, R. D., Masheder, M. R. W., and Booth, R. S.: 1972, *Nature* **237**, 21.

Davis, J. and Shobbrook, R. R.: 1977, *Monthly Notices Roy. Astron. Soc.* **178**, 651.

Davis, L. E., Seaquist, E. R., and Purton, C. R.: 1979, *Astrophys. J.* **230**, 434.

Day, K. L.: 1976a, *Astrophys. J.* **210**, 614.

Day, K. L.: 1976b, *Astrophys. J. Letters* **203**, L99.

Dearborn, D. S. P. and Blake, J. B.: 1979, in P. S. Conti and C. W. H. de Loore (eds.), 'Mass Loss and Evolution of O-type Stars', *IAU Symp.* **83**, 349.

Dearborn, D. S. P. and Eggleton, P. P.: 1977, *Astrophys. J.* **213**, 448.

Dearborn, D. S. P., Lambert, D. L., Tomkin, J.: 1975, *Astrophys. J.* **200**, 675.

Dearborn, D. S. P., Eggleton, P. P., and Schramm, D. N.: 1976, *Astrophys. J.* **203**, 455.

Dearborn, D. S. P., Blake, J. B., Hainebach, K. L., and Schramm, D. N.: 1978, *Astrophys. J.* **223**, 552.

De Freitas Pacheco, J. A.: 1977, *Monthly Notices Roy. Astron. Soc.* **181**, 421.

De Freitas Pacheco, J. A. and Sodré, L.: 1979, *Astrophys. Space Sci.* **61**, 91.

De Greve, J. P.: 1976, *Evolutie van Zware Dubbelsterren*, Thesis, Vrije Universiteit, Brussels.

De Greve, J. P. and De Loore, C.: 1976a, in P. Eggleton, S. Mitton, and J. Whelan (eds.), 'Structure and Evolution of Close Binary Systems', *IAU Symp.* **73**, 27.

De Greve, J. P. and De Loore, C.: 1976b, *Astrophys. Space Sci.* **43**, 35.

De Greve, J. P. and De Loore, C.: 1977, *Astrophys. Space Sci.* **50**, 75.

De Greve, J. P., De Loore, C., and Van Dessel, E. L.: 1978, *Astrophys. Space Sci.* **53**, 105.

De Groot, M.: 1969, *Bull. Astron. Inst. Neth.* **20**, 225.

De Groot, M.: 1973, in M. K. V. Bappu and J. Sahade (eds.), 'Wolf-Rayet and High Temperature Stars', *IAU Symp.* **51**, 108.

De Jager, C.: 1958, in D. Bosman–Crespin (ed.), 'Etoiles à raies d'émission', *Congrès et Colloque Liège* **7**, 172.

De Jager, C.: 1959, in R. N. Bracewell (ed.), 'Paris Symposium on Radio Astronomy', *IAU Symp.* **9**, 89.

De Jager, C.: 1960, in P. Ledoux (ed.), 'Modèles d'étoiles et évolution stellaire', *Mem. in 8° Soc. Roy. Sci. Liège, (5ᵃ Série)* **16**, 280.

De Jager, C.: 1972, *Solar Phys.* **25**, 71.

De Jager, C.: 1978, *Astrophys. Space Sci.* **55**, 147.

De Jager, C. and Kuperus, M.: 1961, *Bull. Astron. Inst. Neth.* **16**, 71.

De Jager, C. and De Loore, C.: 1970, *Solar Phys.* **13**, 126.

De Jager, C. and Neven, L.: 1961, in 'Les spectres des astres dans l'ultraviolet lointain' *Congrès et Colloque Liège* **20**, 552.

De Jager, C. and Neven, L.: 1967, *Bull. Astron. Inst. Neth.* **2**, 125.

De Jager, C. and Neven, L.: 1975 *Astrophys. Space Sci.*, **33**, 295.

De Jager, C. and Vermue, J.: 1979, *Astrophys. Space Sci.* **62**, 245.

De Jager, C., Hoekstra, R., van der Hucht, K. A., Kamperman, T. M., Lamers, H. J., Hammerschlag, A., Werner, W., and Emming, J. G.: 1974, *Astrophys. Space Sci.* **26**, 207.

De Jager, C., Kondo, Y., Hoekstra, R., Van der Hucht, H. A., Kamperman, T. M., Lamers, H. J. G. L. M., Modisette, J. L., and Morgan, T. H.: 1979, *Astrophys. J.* **230**, 534.

De Jager, C., Lamers, H. J. G. L. M., Machetto, F., and Snow, T. P.: 1979, *Astron. Astrophys.* **79**, L28.

De Jong, T., and Maeder, A. (eds.): 1977, 'Star Formation', *IAU Symp.* **77**.

De Loore, C.: 1970, *Astrophys. Space Sci.* **6**, 60.

De Loore, C.: 1977, *Highlights in Astronomy*, **4** (Part II), 155.

De Loore, C.: 1980, *Space Sci. Rev.* **26**, 133.

De Loore, C. and De Grève, J. P.: 1975, *Astrophys. Space Sci.* **35**, 241.

De Loore, C., De Grève, J. P., and De Cuyper, J. P.: 1975, *Astrophys. Space Sci.* **36**, 219.

De Loore, C., De Grève, J. P., and Lamers, H. J. G. L. M.: 1977, *Astron. Astrophys.* **61**, 251.

De Loore, C., De Grève, J. P., and Vanbeveren D.: 1978a, *Astron. Astrophys.* **67**, 373.

De Loore, C., De Grève, J. P., and Vanbeveren, D.: 1978b, *Astron. Astrophys. Suppl.* **34**, 363.

De Loore, C., De Grève, J. P., and Vanbeveren, D.: 1979a, preprint.

De Loore, C., and eight others: 1979b, *Astron. Astrophys.* **78**, 287.

Deming, D.: 1978, *Astrophys. J. Letters* **223**, L31.

Den Boggende, A. J. F., Lamers, H. J. G. L. M., and Mewe, R.: 1979, *Astron. Astrophys.* **80**, 1.

Deutsch, A. J.: 1956, *Astrophys. J.* **123**, 210.

Deutsch, A. J.: 1960, in J. L. Greenstein (ed.), *Stellar Atmospheres*, Univ. of Chicago Press, p. 543.

Deutsch, A. J.: 1966, in R. F. Stein and A. G. W. Cameron (eds.), *Stellar Evolution*, Plenum Press, New York, p. 377.

Deutsch, A. J.: 1969, in M. Hack (ed.), *Mass Loss from Stars*, Trieste, p. 1.

De Vancouleurs, G.: 1975, *IAU Circ.* **2839**.

Deutsch, A. J.: 1970, in D. C. Morton (ed.), 'Ultraviolet Spectra and Ground Based Observations of Stars', *IAU Symp.* **36**, 317.

Dickinson, D. F. and Kleinmann, S. G.: 1977, *Astrophys. J. Letters* **214**, L135.

Doazan, V.: 1976, in A. Slettebak (ed.), 'Be and Shell Stars', *IAU Symp.* **70**, 37.

D'Odorico, S., Benvenuti, P., and Sabbadin, F.: 1978, *Astron. Astrophys.* **63**, 63.

Dominy, J. F.: 1978, *Astrophys. J.* **223**, 949.

Dominy, J. F. and Smith, M. A.: 1977, *Astrophys. J.* **217**, 494.

Dominy, J. F., Hinkle, K. H., Lambert, D. L., Hall, D. N. B., and Rodgway, S. T.: 1978, *Astrophys. J.* **223**, 949.

Dorschner, J., Friedemann, C., and Gürtler, J.: 1977, *Astrophys. Space Sci.* **48**, 305.

Dorschner, J., Friedemann, C., and Gürtler, J,: 1978, *Astrophys. Space Sci.*, **54**, 181.

Dorschner, J., Friedemann, C. and Gürtler J.: 1978b, *Astron. Nachr.* **299**, 269.

Doxsey, R., and fourteen others: 1976, *Astrophys. J. Letters* **203**, L9.

Downes, D.: 1971, *Astrophys. J.* **76**, 305.

Drechsel, H., Rahe, J., Duerbeck, H. W., Kohoutek, L., and Seitter, W. C.: 1977, *Astron. Astrophys. Suppl. Series* **30**, 323.

Drobyshevski, E. M. and Resnikov, B. I.: 1974, *Acta Astron.* **24**, 29.

Duerbeck, H. W. and Walter, K.: 1976, *Astron. Astrophys.* **48**, 141.

Dufton, P. L.: 1972, *Astron. Astrophys.* **16**, 301.

Dufton, P. L.: 1973, *Astron. Astrophys.* **28**, 267.

Dufton, P. L.: 1979, *Astron. Astrophys.* **73**, 203.

Duley, W. W., Maclean, S., and Millar, T. J.: 1978, *Astrophys. Space Sci.* **53**, 223.

Duncan, J. C.: 1922, *Publ. Astron. Soc. Pacific* **34**, 290.

Dupree, A. K.: 1976, in R. Cayrel (ed.), *Physique des mouvements dans les atmosphères stellaires*, Centr. Nat. Rech. Sci., Paris, p. 439.

Dupree A. K.: 1980, *Highlights of Astronomy* **5**, 263.

Dupree, A. K., Baliunas, S. L., and Lester, J. B.: 1977, *Astrophys. J. Letters* **218**, L71.

Dupree, A. K., Black, J. H., Davis, R., Hartmann, L., and Raymond J. C.: 1979 in A. H. Willis (ed.), *The First Year of IUE*, Univ. College of London, p. 217.

Dupree, A. K., Hartmann, L., and Raymond, J. C.: 1980, in M. J. Plavec and R. K. Ulrich (eds.), 'Close Binary Stars', *IAU Symp.* **88**, 39.

Durand, R. A., Eoll, J. G., and Schlesinger, B. M.: 1976, *Monthly Notices Roy. Astron. Soc.* **174**, 671

Durrant, C. J.: 1979, *Astron. Astrophys.* **73**, 137.

Dyck, H. M. and Jennings, M. C.: 1971, *Astron. J.* **76**, 431.

Dyck, H. M. and Sandford, M. T.: 1971, *Astrophys. J.* **76**, 43.

Dyck, H. M., Lockwood, G. W., and Caps, R. W.: 1974, *Astrophys. J.* **189**, 89.

Eachus, L. J.: 1975, *IAU Circ.* **2823**.

Eachus, L. J., Wright, E. L., and Liller, W.: 1976, *Astrophys. J. Letters* **203**, L99.

Eaton, J. A.: 1978, *Acta Astron.* **28**, 63.

Ebbets, D.: 1979, *Astrophys. J.* **227**, 510.

Ebbets, D.: 1980, *Astrophys. J.*, preprint.

Eddington, A. S.: 1921, *Z. Phys.* **7**, 351.

Eddington, A. S.: 1926, *The International Constitution of the Stars*, Cambridge.

Edlén, B.: 1941, *Arkiv Math. Astron. Phys.* **B28**, 1.

Edlén, B.: 1942, *Z. Astrophys.* **22**, 30.

Edmunds, M. G.: 1978, *Astron. Astrophys.* **64**, 103.

Edwards, D. L.: 1944, *Monthly Notices Roy. Astron. Soc.* **104**, 283.

Edwards, D. L.: 1956, *Vistas Astron.* **2**, 1470.

Eggen, O. J.: 1971, *Astrophys. J.* **163**, 331.

Eggen, O. J.: 1973, *Publ. Astron. Soc. Pacific* **85**, 289.

Eggen, O. J.: 1975a, *Astrophys. J.* **195**, 661.

Eggen, O. J.: 1975b, *Astrophys. J.* **198**, 131.

Elitzur, M., Goldreich, P., and Scoville, N.: 1976, *Astrophys. J.* **205**, 384.

Elmegreen, B. G. and Morris, M.: *Astrophys. J.* **229**, 593.

Elvis, M., Page, C. G., Pounds, K. A., Ricketts, M. J., and Turner, M. L. J.: 1975, *Nature* **257**, 656.

Elvius, A.: 1974, *Astron. Astrophys.* **34**, 371.

Emden, R.: 1907, *Gaskugeln*, Teubner, Leipzig.

Endler, F., Hammer, R., and Ulmschneider, P.: 1979, *Astron. Astrophys.* **73**, 190.

Engels, D.: 1979, *Astron. Astrophys. Suppl.* **36**, 337.

Engvold, O. and Rygh, B. O.: 1978, *Astron. Astrophys.* **70**, 399.

Ennis, D., Becklin, E. E., Beckwith, S., Elias, J., Gatley, I., Matthews, K., Neugebauer, G., and Willner, S. P.: 1977, *Astrophys. J.* **214**, 478.

Epstein, A.: 1977, *Astrophys. J. Letters* **218**, L49.

Ergma, E. and Vilhu, O.: 1978, *Astron. Astrophys.* **69**, 143.

Ergma, E. V. Kudryashov, A. D., and Shcherbatyuk, V. A.: 1976, *Astron. Zh.* **53**, 983.

Evans, N. J., Beckwith, S., Brown, R. L., and Gilmore, W.: 1979, *Astrophys. J.* **227**, 450.

Evans, R. G., Jordan, C., and Wilson, R.: 1975, *Monthly Notices Roy. Astron. Soc.* **172**, 585.

Fabbiano, G., Doxsey, R. E., Griffiths, R. E., and Johnston, M. D.: 1980, *Astrophys. J. Letters*, preprint.

Fabian, A. C. and Pringle, J. E.: 1977, *Montly Notices Roy. Astron. Soc.* **180**, 749.

Fahlham, C. G., Carlberg, R. G., and Walker, G. A. H.: 1977, *Astrophys. J. Letters* **217**, L35.

Falk, S. W. and Arnett, W. D.: 1973, *Astrophys. J. Letters* **180**, L65.

Falk, S. W. and Arnett, W. D.: 1977, *Astrophys. J. Suppl. Series* **33**, 515.

Falk, S. W. and Scalo, J. M.: 1975, *Astrophys. J.* **202**, 690.

Fath, E. A.: 1935, *Lick. Obs. Bull.* **17**, 115.

Fawley, W. M.: 1977, *Astrophys. J.* **218**, 181.

Feast, M. W.: 1976, *Monthly Notices Roy. Astron. Soc.* **174**, 9P.

Feast, M. W. and Thackeray, A. D.: 1956, *Monthly Notices Roy. Astron. Soc.* **116**, 587.

Feast, M. W., Thackeray, A. D., and Wesselink, A. J.: 1960, *Monthly Notices Roy. Astron. Soc.* **121**, 337.

Feautrier, P.: 1968, *Ann. Astrophys.* **31**, 257.

Fehrenbach, Ch.: 1975a, *Compt. Rend. Acad. Sci. Paris Série B* **281**, 169.

Fehrenbach, Ch.: 1975b, *Compt. Rend. Acad. Sci. Paris Série B* **281**, 365.

Fehrenbach, Ch., Duflot, M., and Acker, A.: 1976, *Astron. Astrophys. Suppl.* **24**, 379.

Feinstein, A. and Marraco, H. G.: 1974, *Astron. Astrophys.* **30**, 271.

Feinstein, A. and Marraco, H. G.: 1979, *Astron. J.* **84**, 1713.

Feldman, P. A., Purton, C. R., and Marsh, K. A.: 1973, *Nature Phys. Sci.* **245**, 7.

Ferland, G. J.: 1977, *Astrophys. J. Letters* **212**, L21.

Ferland, G. J. and Wootten, H. A.: 1977, *Astrophys. J. Letters* **214**, L27.

Ferland, G. J., Lambert, D. L., and Woordman, J. H.: 1977, *Astrophys. J.* **213**, 132.

Ferlands, G. J. and Shields, G. A.: 1978, *Astrophys. J.* **226**, 172.

Fernie, J. D. and Brooker, A. H.: 1961 *Astrophys. J.* **133** 1088.

Fessenkov, V. G.: 1949, *Astron. Zh.* **26**, 67.

Field, G. B.: 1965, *Astrophys. J.* **142**, 531.

Finzi, A. and Yahel, R.: 1978, *Astron. Astrophys.* **68**, 173.

Finzi, A., Finzi, R., and Shaviv, G.: 1974, *Astron. Astrophys.* **37**, 325.

Firmani, C., Koenigsberger, G., Bisiacchi, G. F., Ruiz, E., and Solas, A.: in P. S. Conti and C. W. H. de Loore (eds.), 'Mass-Loss and Evolution of O-type Stars', *IAU Symp.* **83**, 421.

FitzGerald, M. P. and Pilavaki, A.: 1974: *Astrophys. J. Suppl. Series* **28**, 147.

Fix, J. D.: 1978, *Astrophys. J. Letters* **223**, L25.

Florkowski, D. R. and Gottesman, S. T.: 1977, *Monthly Notices Roy. Astron. Soc.* **179**, 105.

Flower, D. R., Nussbaumer, H., and Schild, H.: 1979, *Astron. Astrophys.* **72**, L1.

Flower, P. J.: 1975, *Astron. Astrophys.* **41**, 391.

Flower, P. J.: 1977, *Astron. Astrophys.* **54**, 31.

Ford, H. C.: 1978, *Astrophys. J.* **219**, 595.

Fowler, W. A. and Hoyle, F.: 1964, *Astrophys. J. Suppl.* **9**, 201.

Foy, R., Chelli, A., Sibille, F., and Léna, P.: 1979, *Astron. Astrophys. Letters* **79**, L5.

Fraley, G. S.: 1968, *Astrophys. Space Sci.* **2**, 96.

Freire, R.: 1979, *Astron. Astrophys.* **78**, 148.

Friedemann, C., Gürtler, J., and Dorschner, J.: 1979, *Astrophys. Space Sci.* **60**, 297.

Friedjung, M.: 1966a, *Monthly Notices Roy. Astron. Soc.* **131**, 447.

Friedjung, M.: 1966b, *Monthly Notices Roy. Astron. Soc.* **132**, 143.

Friedjung, M.: 1966c, *Monthly Notices Roy. Astron. Soc.* **132**, 317.

Friedjung, M.: 1975, *Astron. Astrophys.* **44**, 431.

Friedjung, M.: 1977, *Novae and Related Stars*, D. Reidel Publ. Co., Dordrecht, Holland.

Frost, S. A., Conti, P. S.: 1976, in A. Slettebak (ed.), 'Be and Shell Stars', *IAU Symp.* **70**, 139.

Frisch, H.: 1966, *J. Quant. Spectrosc. Radiat. Transfer* **6**, 629.

Fujimoto, M. Y.: 1977a, *Publ. Astron. Soc. Japan* **29**, 331.

Fujimoto, M. Y.: 1977b, *Publ. Astron. Soc. Japan* **29**, 537.

Fujimoto, M. Y., Nomoto, K., and Sugimoto, D.: 1976, *Publ. Astron. Soc. Japan* **28**, 89.

Fujita, Y.: 1966, *Vistas Astron.* **7**, 71.

Fujita, Y.: 1970, *Interpretation of Spectra and Atmospheric Structure in Cool Stars*, Univ. Press, Tokyo.

Fujita, Y.: 1980, *Space Sci. Rev.* **25**, 89.

Fujita, Y. and Tsuji, T.: 1976, *Proc. Japan Acad.* **52**, 296.

Fujita, Y. and Tsuji, T.: 1977, *Proc. Astron. Soc. Japan* **29**, 711.

Fusi-Pecci, F. and Renzini, A.: 1975, in P. Ledoux (ed.), 'Problèmes de l'hydrodynamique stellaire', *Mém. Soc. Roy. Sci. Liège, Coll. 8°, 6e Série* **8**, 383.

Gabriel, M. and Noels, A.: 1976, *Astron. Astrophys.* **53**, 149.

Gabriel, M. and Noels, A.: 1977, *Astron. Astrophys.* **54**, 634.

Gaffet, B.: 1978, *Astrophys. J.* **225**, 442.

Galkina, T. S.: 1977, *Pis'ma Astron. Zh.* **3**, 536 (*Soviet Astron. Letters* **3**, 536).

Gallagher, J. S.: 1977, *Astron. J.* **82**, 209.

Gallagher, J. S. and Code, A. D.: 1974, *Astrophys. J.* **189**, 303.

Gallagher, J. S. and Ney, E. P.: 1976, *Astrophys. J. Letters* **204**, L35.

Gallagher, J. S. and Starrfield, S.: 1976, *Monthly Notices Roy. Astron. Soc.* **176**, 53.

Gallagher, J. S. and Starrfield, S..: 1978, *Ann. Rev. Astron. Astrophys.* **16**, 171.

Ganesh, K. S. and Bappu, M. K. V.: 1967, *Kodaikanal Obs. Bull.* **A183**.

Garmany, C. D. and Conti, P. S.: 1980, *Highlights in Astronomy* **5**.

Garrison, L. M.: 1978, *Astrophys. J.* **224**, 535.

Gascoigne, S. C. B.: 1972, *Quart. J. Roy. Astron. Soc.* **13**, 274.

Gautier, T. N., Thompson, R. I., Fink, U., and Larson, H. P.: *Astrophys. J.* **205**, 841.

Gaviola, E.: 1950, *Astrophys. J.* **111**, 408.

Geballe, T. R., Wollman, E. R., Lacy, J. H., and Rank, D. M.: 1977, *Publ. Astron. Soc. Pacific* **89**, 840.

Geballe, T. R., Lacy, J. H., and Beck, S. C.: 1979, *Astrophys. J. Letters* **230**, L47.

Gebbie, K. B.: 1968, *J. Quant. Spectr. Radiat. Transfer* **8**, 265.

Gebbie, K. B. and Thomas, R. N. (eds.): 1968, *Wolf-Rayet Stars*, Nat. Bureau of Standards, Special Publ. No. 307.

Gehrz, R. D. and Woolf, N. J.: 1971, *Astrophys. J.* **165**, 285.

Gehrz, R. D., Ney, E. P., Becklin, E. E., and Neugebauer, G.: 1973, *Astrophys. Letters* **13**, 89.

Gehrz, R. D., Hackwell, J. A., and Jones, T. W.: 1974, *Astrophys. J.* **191**, 675.

Geisel, S. L., Kleinmann, D. E., and Low, F. J.: 1970, *Astrophys. J. Letters* **161**, L101.

Genova, F. and Schatzman, E.: 1979, *Astron. Astrophys.* **78**, 323.

Genzel, R. and Downes, D.: 1977a, *Astron. Astrophys.* **61**, 117.

Genzel, R. and Downes, D.: 1977b, *Astron. Astrophys. Suppl.* **30** 145.

Genzel, R., Downes, D. and eleven others: 1978, *Astron. Astrophys.* **66**, 13.

Gezari, D. Y., Labeyrie, A., and Stachnil, R. V.: 1972, *Astrophys. J. Letters* **173**, L1.

Giacconi, R., and many others: 1979, *Astrophys. J.* **230**, 540.

Giangrande, A., Viotti, R., and Cassatella, A.: 1979, *Mem. Soc. Astron. Ital.* **50**, 217.
Gillett, F. C., Stein, W. A., and Solomon, P. M.: 1970, *Astrophys. J. Letters* **160**, L173.
Gilman, R. C.: 1972, *Astrophys. J.* **178**, 423.
Gilman, R. C.: 1974, *Astrophys. J.* **188**, 87.
Gingerich, O., Latham, D. W., Linsky, J., and Kumar, S. S.: 1967, in M. Hack (ed.), *Colloquium on Late Type Stars*, Trieste, Oss. Astron., p. 291.
Giuricin, G., Mardirossian, F., Mezzetti, M., Pucillo, M., Santin, P., and Sedmak, G.: 1979, *Astron. Astrophys.* **80**, 9.
Goldreich, P. and Kwan, J.: 1974, *Astrophys. J.* **191**, 93.
Gorbatskii, V. G.: 1972, *Astron. Zh.* **49**, 42 (*Soviet Astron.* **16**, 32).
Gorbatskii, V. G.: 1973, *Astron. Zh.* **50**, 19 (*Soviet Astron.* **17**, 11).
Gordon, C.: 1968, *Publ. Astron. Soc. Pacific* **80**, 597.
Gorenstein, P. and Tucker, W. H.: 1976, *Ann. Rev. Astron. Astrophys.* **14**, 373.
Goss, W. N., Allen, R. J., Ekers, R. D., and De Bruyn, G.: 1973, *Nature Phys. Sci.* **243**, 42.
Goss, W. M., Nguyen-Quang-Rieu, and Winnberg, A.: 1974a, *Astron. Astrophys.* **29**, 435,
Goss, W. M., Winnberg, A., and Habing, H. J.: 1974b, *Astron. Astrophys.* **30**, 349.
Gottlieb, E. W. and Liller, W.: 1978, *Astrophys. J.* **225**, 488.
Goudis, C.: 1976, *Astrophys. Space Sci.* **45**, 133.
Gow, C. E.: 1977, *Publ. Astron. Soc. Pacific* **89**, 510.
Grasdalen, G. L. and Joyce, R. R.: 1976, *Nature*, **259**, 187.
Greenberg, M., De Jager, C., and Tinbergen, J.: 1980, *Astron. Astrophys.*, in press.
Greenstein, J. L.: 1948, *Astrophys. J.* **108**, 78.
Gregory, P. C. and Seaquist, E. R.: 1976, *Astrophys. J.* **204**, 626.
Grieg, W. E.: 1971, *Astron. Astrophys.* **10**, 161.
Gronenschild, E. H. B. M.: 1980, *Astron. Astrophys.*, in press.
Gronenschild, E. H. B. M., Mewe, R., Heise, J., den Boggende, A. J. F., Schrijver, J., and Brinkman, A. C.: 1978, *Astron. Astrophys.* **65**, L9.
Groth, H. G.: 1961, *Z. Astrophys.* **51**, 231.
Groth, H. G.: 1972, *Astron. Astrophys.* **21**, 337.
Grotrian, W.: 1937, *Z. Astrophys.* **13**, 215.
Grotrian, W.: 1939 *Naturwiss.* **27**, 214.
Guerrero, G. and Mantegazza, L.: 1979, *Astron. Astrophys. Suppl.* **36**, 471.
Guibert, J.: 1979, in B. E. Westerlund (ed.), *Stars and Star Systems*, D. Reidel, Publ. Co., Dordrecht, Holland, p. 85.
Gürther, J. Dorschner, J., and Friedemann, C.: 1979, *Astron. Nachr.* **300**, 17.
Gustafsson, B., Bell, R. A., Eriksson, K., and Nordlund, Å.: 1975, *Astron. Astrophys.* **42**, 407.
Habing, H. J., Israel, F. P., and de Jong, T.: 1972, *Astron. Astrophys.* **17**, 329.
Hack, M.: 1976, *Mem. Soc. Astron. Ital.* **47**, 417.
Hack, M.: 1979, in A. J. Willis (ed.), *The First Year of IUE*, Univ. College London, p. 81.
Hack, M. and Selvelli, P. L.: 1979, *Astron. Astrophys.* **75**, 316.
Hackwell, J. A., Gehrz, R. D., and Smith, J. R.: 1974, *Astrophys. J.* **192**, 383.
Haefner, R., Metz, K., and Schoembs, R.: 1977, *Astron. Astrophys.* **55**, 5.
Hagen, W.: *Astrophys. J. Letters* **222**, L37; *Astrophys. J. Suppl. Series* **38**, 1.
Hagen, W., Black, J. H., Dupree, A. K., and Holm, A. V.: 1980, *Astrophys. J.*, in press.
Haisch, B. M.: 1979, *Astron. Astrophys.* **72**, 161.
Hammer, R.: 1979, private communication.
Hammerschlag-Hensberge, G., Henrichs, H. F., and Shaham, J.: 1979, *Astrophys. J. Letters* **228**, L75.
Hammerschlag-Hensberge, G., Van den Heuvel, E. P. J., and Lamers, H. J. G. L. M.: 1979a, in A. J. Willis (ed.), *The First Year of IUE*, p. 331.
Hammerschlag-Hensberge, G., Van den Heuvel, E. P. J., and 11 others: 1979b, *Astron. Astrophys.*, in press.
Hanbury Brown, R., Davis, J., Herbison-Evans, D., and Allen, L. R.: 1970, *Monthly Notices Roy. Astron. Soc.* **148**, 103.
Hanbury Brown, R., Davis, J., and Allen, L. R.: 1974, *Monthly Notices Roy. Astron. Soc.* **167**, 121.
Handbury, M. J. and Williams, I. P.: 1976, *Astrophys. Space Sci.* **45**, 439.
Hansen, C. J.: 1978, *Ann. Rev. Astron. Astrophys.* **16**, 15.

Härm, R. and Schwarschild, M.: 1975, *Astrophys. J.* **200**, 324.

Harmanec, P. and Kříž, S.: 1976, in A. Slettebak (ed.), 'Be and Shell Stars', *IAU Symp.* **70**, 383.

Harmanec, P., Horn, J., Koubsky, P., Kříž, S., Zdarsky, F., Papousek, J., Doazan, V., Bourdonneau, B., Baldinelli, L., Ghedini, S., and Pavlovsky, K.: 1978, *Bull. Astron. Inst. Czech.* **29**, 278.

Harmer, D. L., Lawson, P. A., and Stickland, D. J.: 1978, *Observatory* **98**, 250.

Harnden, F. R., Branduari, G., Elvis, M., Gorenstein, P., Grindlay, J., Pye, R. Rosner R., Topka, K., and Vaiana, G. S.: 1979, *Astrophys. J. Letters* **234**, L51.

Harris, D. L., Stand, K. A., and Worley, C. E.: 1963, in K. A. Strand (ed.), *Basic Astronomical Date*, Chicago, Chapter XV.

Harrison, T. G.: 1978, *Astrophys. J. Suppl. Series* **36**, 199.

Hartmann, L.: 1978, *Astrophys. J.* **224**, 520.

Hartmann, L. and Cassinelli, J. P.: 1977, *Astrophys. J.* **215**, 155.

Harvey, G. M.: 1972, *Monthly Notices Roy. Astron. Soc. South Africa* **31**, 81.

Harvey, P. M., Hoffman, W. F., and Campbell, M. F.: 1978, *Astron. Astrophys.* **70**, 165.

Harvey, P. M., Thronson, H. A. and Gatley, I.: 1979, *Astrophys. J.*, **231**, 115.

Hashimoto, J., Maihara, T., Okuda, H., and Sato, S.: 1970, *Publ. Astron. Soc. Japan* **22**, 335.

Hatchett, S. and McCray, R.: 1977, *Astrophys. J.* **211**, 552.

Hayashi, C.: 1961, *Publ. Astron. Soc. Japan* **13**, 450.

Hayashi, C.: 1966a, in R. F. Stein and A. G. W. Cameron (eds.) *Stellar Evolution*, Plenum, New York, p. 253.

Hayashi, C.: 1966b, *Ann. Rev. Astron. Astrophys.* **4**, 171.

Hayashi, C. and Cameron, R.: 1962, *Astrophys. J.* **136**, 166.

Hayashi, C., Hoshi, R., and Sugemoto, D.: 1962, *Prog. Theor. Phys. Supl.* **22**, 1.

Hayes, D. P.: 1975, *Astrophys. J. Letters* **197**, L55.

Hayes, D. P.: 1978, *Astrophys. J.* **219**, 952.

Heap, S.: 1972, *Astrophys. Letters* **10**, 49.

Heap, S.: 1976a, in A. Slettebak (ed.), 'Be and Shell Stars', *IAU Symp.* **70**, 165.

Heap, S.: 1976b, in A. Slettebak (ed.), 'Be and Shell Stars', *IAU Symp.* **70**, 315.

Heap, S.: 1977, *Astrophys. J. Letters* **218**, L17.

Hearn, A. G.: 1972, *Astron. Astrophys.* **19**, 417.

Hearn, A. G.: 1973, *Astron. Astrophys.* **23**, 97.

Hearn, A. G.: 1975a, *Astron. Astrophys.* **40**, 277.

Hearn, A. G.: 1975b, *Astron. Astrophys.* **40**, 355.

Hearn, A. G.: 1979, *Astron. Astrophys. Letters* **79**, L1.

Hearn, A. G.: 1980, *Highlights in Astornomy* **5**, 591.

Heintze, J. R. W.: 1973, in B. Hauck and B. E. Westerlund (eds.), 'Problems of Calibration of Absolute Magnitude and Temperature of Stars', *IAU Symp.* **54**, 231.

Helt, B. E. and Gyldenkerne, K.: 1975, *Astron. Astrophys. Suppl.* **22**, 171.

Henize, K. G.: 1961, *Publ. Astron. Soc. Pacific* **73**, 159.

Henize, K. G., Wray, J. D., Parsons, S. B., Benedict, G. F., Bruhweiler, F. C., Rubski, P. M., and O'Callaghan, F. G.: 1975, *Astrophys. J. Letters* **199**, L119.

Henize, K. G., Wray, J. D., Parsons, S. B., and Benedict, G. F.: 1976a, in M. J. Rycroft (ed.), *Space Research* **XVI**, 923.

Henize, K. G., Wray, J. D., Parsons, S. B., and Benedict, G. F.: 1976b, in A. Slettebak (ed.), 'Be and Shell Stars', *IAU Symp.* **70**, 191.

Hensberge, G.: 1974, *Astron. Astrophys.* **36**, 295.

Hensberge, G., Van den Heuvel, E. P. J., and Paes de Barros, M. H.: 1973, *Astron. Astrophys.* **29**, 69.

Herbig, G. H.: 1960, *Astrophys. J. Suppl. Series* **4**, 337.

Herbig, G. H.: 1969, *Mém. Soc. Roy. Sci. Liège*, *Série* **8**, 13.

Herbig, G. H.: 1972, *Astrophys. J.* **172**, 375.

Herbig, G. H. and Boyarchuk, A. A.: 1968, *Astrophys. J.* **153**, 297.

Herbig, G. H. and Zappala, R. R.: 1970 *Astrophys. J. Letters* **162**, L15.

Herbig, G. H., Preston, G. W., Smak, J., and Paczynski, B.: 1965, *Astrophys. J.* **141**, 617.

Herbst, W., Hesser, J. E., and Ostriker, J. P.: 1974, *Astrophys. J.* **193**, 679.

Herbst, W., Racine, R., and Richer, H. B.: 1977, *Publ. Astron. Soc. Pacific* **89**, 663.

Herczeg, T.: 1963, in H. H. Voigt (ed.), *Landolt Bornstein VI*, Springer Verlag, Berlin, p. 487.

Herzberg, G.: 1948, *Astrophys. J.* **107**, 94.

Higurashi, T. and Hirata, R.: 1978, *Publ. Astron. Soc. Japan*, **30**, 615.

Hill, G., Crawford, D. L., and Barness, J. V.: 1974, *Astron. J.* **79**, 1271.

Hill, S. J. and Willson, L. A.: 1979, *Astrophys. J.* **229**, 1029.

Hillebrandt, W.: 1978, *Space Sci. Rev.* **21**, 639.

Hiltner, W. A. and Schild, R. E.: 1966, *Astrophys. J.* **134**, 770.

Hinkle, K. H.: 1978, *Astrophys. J.* **220**, 210.

Hinkle, K. H. and Barnes, T. G.: 1979, *Astrophys. J.* **227**, 923.

Hinkle, K. H., Lambert, D. L., and Snell, R. L.: 1976, *Astrophys. J.* **210**, 684.

Hirata, R. and Kogure, T.: 1977, *Publ. Astron. Soc. Japan* **29**, 477.

Hirata, R. and Kogure, T.: 1978, *Publ. Astron. Soc. Japan* **30**, 601.

Hjellming, R. M.: 1968, *Astrophys. J.* **154**, 533.

Hjellming, R. M. and Wade, C. M.: 1971, *Astrophys. J. Letters* **168**, L115.

Hjellming, R. M., Wade, C. M., Vandenberg, N. R., and Newell, R. T.: 1979, *Astron. J.* **84**, 1619.

Hollars, D. R. and Beebe, H. A.: 1976, *Publ. Astron. Soc. Pacific* **88**, 934.

Holm, A. V. and Cassinelli, J. P.: 1977, *Astrophys. J.* **211**, 432.

Holzer, T. E.: 1977, *J. Geophys. Res.* **82**, 23.

Höppner, W., Kähler, H., Roth, M. L., and Weyert, A.: 1978, *Astron. Astrophys.* **63**, 391.

Horn, J. Kriz, S., and Plavec, M.: 1969, *Bull. Astron. Inst. Czech.* **20**, 193.

Houziaux, L. and Andrillat, Y.: 1976, in A. Slettebak (ed.), 'Be and Shell Stars', *IAU Symp.* **70**, 87.

Hoyle, F. and Wickramasinghe, N. C.: 1970, *Nature* **226**, 62.

Huang, S.-S: 1976, *Publ. Astron. Soc. Pacific* **88**, 448.

Hubble, E. and Sandage, A.: 1953, *Astrophys. J.* **118**, 353.

Humason, M. L.: 1938, *Astrophys. J.* **88**, 228.

Hummer, D. G.: 1976, in A. Slettebak (ed.), 'Be and Shell Stars', *IAU Symp.* **70**, 281.

Hummer, D. G. and Mihalas, D.: 1970, *Monthly Notices Roy. Astron. Soc.* **147**, 339.

Hummer, D. G. and Rybicki, G. B.: 1971, *Monthly Notices Roy. Astron. Soc.* **152**, 1.

Humphreys, R. M.: 1970, *Astrophys. J. Letters* **6**, L1.

Humphreys, R. M.: 1974, *Astrophys. J.* **188**, 75.

Humphreys, R. M.: 1975, *Astrophys. J.* **200**, 426.

Humphreys, R. M.: 1978, *Astrophys. J. Suppl.* **38**, 309.

Humphreys, R. M.: 1979, *Astrophys. J. Suppl.* **39**, 389.

Humphreys, R. M. and Lockwood, G. W.: 1972, *Astrophys. J. Letters* **172** L59.

Humphreys, R. M., Strecker, D. W., and Ney, E. P.: 1971, *Astrophys. J. Letters* **167**, L35.

Humphreys, R. M., Strecker, D. W., and Ney, E. P.: 1972, *Astrophys. J.* **172**, 75.

Hundt, E., Kodaira, K., Schmidt-Burgk, J., and Scholtz, M.: 1975, *Astron. Astrophys.* **41**, 37.

Hutchings, J. B.: 1967a, *Observatory* **87**, 273.

Hutchings, J. B.: 1967b, *Observatory*, **87**, 289.

Hutchings, J. B.: 1968a, *Monthly Notices Roy. Astron. Soc.* **141**, 219.

Hutchings, J. B.: 1968b, *Monthly Notices Roy. Astron. Soc.* **141**, 329.

Hutchings, J. B.: 1969, *Monthly Notices Roy. Astron. Soc.* **144**, 235.

Hutchings, J. B.: 1970a, *Monthly Notices Roy. Astron. Soc.* **147**, 161.

Hutchings, J. B.: 1970b, *Monthly Notices Roy. Astron. Soc.* **150**, 55.

Hutchings, J. B.: 1972, *Monthly Notices Roy. Astron. Soc.* **158**, 177.

Hutchings, J. B.: 1974, *Astrophys. J.* **192**, 685.

Hutchings, J. B.: 1975a, *Publ. Astron. Soc. Pacific* **87**, 245.

Hutchings, J. B.: 1975b, *Publ. Astron. Soc. Pacific* **87**, 529.

Hutchings, J. B.: 1976a, *Astrophys. J.* **203**, 438.

Hutchings, J. B.: 1976b, *Publ. Dom. Astrophys. Obs.* **14**, 355.

Hutchings, J. B.: 1976c, *Astrophys. J. Letters* **204**, L99.

Hutchings, J. B.: 1976d, in A. Slettebak (ed.), 'Be and Shell Stars', *IAU Symp.* **70**, 13.

Hutchings, J. B.: 1976e, *Publ. Astron. Soc. Pacific* **88**, 5.

Hutchings, J. B.: 1976f, in P. Eggleton, S. Mitton, and J. Whelan (eds.), 'Structure and Evolution of Close Binary Stars', *IAU Symp.* **73**, 9.

Hutchings, J. B.: 1977, *Publ. Astron. Soc. Pacific* **89**, 668.

Hutchings, J. B.: 1978a, *Earth Extraterr. Sci.*, **3**, 123.

Hutchings, J. B.: 1978b, *Astrophys. J. Letters* **225**, L63.

Hutchings, J. B.: 1979a, *Monthly Notices Roy. Astron. Soc.* **187**, 53P.

Hutchings, J. B.: 1979b, in P. S. Conti and C. W. H. de Loore (eds.), 'Mass Loss and Evolution of O-type Stars', *IAU Symp.* **83**, 3.

Hutchings, J. B.: 1979c, *Astrophys. J.* **232**, 176.

Hutchings, J. B.: 1979d, *Astrophys. J.* **233**, 913.

Hutchings, J. B.: 1980, *Space Sci. Rev.* **26**, 331.

Hutchings, J. B. and Cowley, A. P.: 1976, *Astrophys. J.* **206**, 490.

Hutchings, J. B. and McCall, M. L.: 1977, *Astrophys. J.* **217**, 775.

Hutchings, J. B. and Redman, R. O.: 1973, *Monthly Notices Roy. Astron. Soc.* **163**, 209.

Hutchings, J. B. and Sanyal, A.: 1976, *Publ. Astron. Soc. Pacific* **88**, 279.

Hutchings, J. B. and Stoeckley, T. R.: 1977, *Publ. Astron. Soc. Pacific* **89**, 19.

Hutchings, J. B. and Wright, K. O.: 1971, *Monthly Notices Roy. Astron. Soc.* **155**, 203.

Hutchings, J. B., Auman, J. R., Gower, A. C., and Walker, G. A. H.: 1971, *Astrophys. J. Letters* **170**, L73.

Hutchings, J. B., Crampton, D., Cowley, A. P., and Osmer, P., 1977: *Astrophys. J.* **217**, 186.

Hutchings, J. B., Bernard, J. E., and Margetish, L.: 1978, *Astrophys. J.* **224**, 899.

Hutchings, J. B., Cowley, A. P., and Crampton, D.: 1979, *Astrophys. J.* **232**, 500.

Hyland, A. R. and Bessell, M. S.: 1975, *Proc. Astron. Soc. Australia* **2**, 353.

Hyland, A. R. and Neugebauer, G.: 1970, *Astrophys. J. Letters* **160**, L177.

Hyland, A. R., Becklin, E. E., Neugebauer, G., and Wallerstein, G.: 1969, *Astrophys. J.* **158**, 619.

Hyland, A. R., Becklin, E. E., Frogel, J. A., and Neugebauer, G.: 1972, *Astron. Astrophys.* **16**, 204.

Hyland, A. R., Robinson, G., Mitchell, R. M., Thomas, J. A., and Becklin, E. E.: 1979, *Astrophys. J.* **233**, 145.

Iben, I.: 1974, *Ann. Rev. Astron. Astrophys.* **14**, 215.

Iben, I., Jr.: 1975, *Astrophys. J.* **196**, 525.

Iben, I.: 1976, *Astrophys. J.* **208**, 165.

Ilovaisky, S. A. and Lequeux, J.: 1972, *Astron Astrophys.* **20** 347.

Innes, R. T.: 1903, *Ann. Cape Obs.* **9**, 75B.

Inoue, M. N.: 1979, *Publ. Astron. Soc. Japan* **31**, 11.

Isserstedt, J.: 1978, *Astron. Astrophys.* **65**, 57.

Itoh, H.: 1978, *Publ. Astron. Soc. Japan* **30**, 489.

Ivanov, L. N.: 1978, *Pis'ma Astron. Zh.* **4**, 260 (*Soviet Astron. Letters* **4**, 141).

Ivanova, L. N., Imshennik, V. S., and Chechetkin, V. M.: 1977, *Astron. Zh.* **54**, 1009 (*Soviet Astron.* **21**, 571).

Jamar, C., Macau-Hercot, D., Monfils, A., Thomson, G. I., Houziaux, L., and Wilson, L.: 1976a, *ESA Sci. Rep.* **27**.

Jamar, C., Macau-Hercot, D., and Praderie, F.: 1976b, *Astron. Astrophys.* **52**, 373,

Jaschek, C. and Jaschek, M.: 1963, *Publ. Astron. Soc. Pacific* **75**, 509.

Jaschek, M. and Jaschek, C.: 1967, *Astrophys. J.* **150**, 355.

Jaschek, M. and Jaschek, C.: 1974, *Astron. Astrophys.* **36**, 401.

Jeffers, H. M., Van den Bos, W. H., and Greeby, F. M.: 1963, *Lick Obs. Publ.* **21**, 49.

Jenkins, E. B., Silk, J., and Wallerstein, G.: 1976, *Astrophys. J. Letters* **209**, L87.

Jenkins, E. B., Snow, T. P., Upson, W. L., Starrfield, S. G., Gallagher, J. S., Friedjung, M., Linsky, J. L., Anderson, R., Henry, R. C., and Moos, H. W.: 1977, *Astrophys. J.* **212**, 198.

Jernigan, J. G.: 1976, *IAU Circ.* **2900**.

Jerzykiewicz, M. M. and Sterken, C.: 1979, preprint.

Jewell, P. R., Elitzur, M., Webber, J. C., and Snyder, L. E.: 1979, *Astrophys. J. Suppl.* **41**, 191.

Johannson, S.: 1977, *Monthly Notices Roy. Astron. Soc.* **178**, 17P.

Johansson, L. E. B., Andersson, C., Goss, W. M., and Winnberg, A.: 1977, *Astron. Astrophys.* **54**, 323.

Johnson, H. L.: 1966, *Ann. Rev. Astron. Astrophys.* **4**, 193.

Johnson, H. L.: 1974, 'Model Atmospheres for Cool Stars', NCAR Techn. Note No. 95, Boulder, Colo.

Johnson, H. L. and Méndez, M. E.: 1970, *Astron. J.* **75**, 785.

Johnson, H. M.: 1976, *Astrophys. J.* **206**, 469.

Johnson, H. M.: 1978, *Astrophys. J. Suppl. Series*, **36**, 217.

Johnson, H. R.: 1978, *Astrophys. J.* **223**, 238.

Jones, T. W. and Merrill, K. M.: 1976, *Astrophys. J.* **209**, 509.

Jones, B., Merrill, K. M., Puetter, R. C., and Willner, S. P.: 1978, *Astron. J.* **83**, 1437.

Jones, T. J.: 1979, *Astrophys. J.* **228**, 787.

Joshi, S. C. and Rautela, B. S.: 1978, *Monthly Notices Roy. Astron. Soc.* **183**, 55.

Joyce, R. R.: 1975, *Publ. Astron. Soc. Pacific* **87**, 917.

Kähler, H.: 1972, *Astron. Astrophys.* **20**, 105.

Kaifu, N., Buhl, D., and Snyder, L. E.: 1975, *Astrophys. J.* **195**, 359.

Kaler, J. B.: 1978a, *Astrophys. J.* **220**, 887.

Kaler, J. B.: 1978b, *Astrophys. J.* **226**, 947.

Kaler, J. B., Iben, J., and Becker, S. A.: 1978, *Astrophys. J. Letters* **224**, L63.

Kalkofen, W.: 1970, in H. G. Groth and P. Wellmann (eds.), 'Spectrum Formation in Stars with Steady-State Extended Atmospheres', Nat. Bureau of Standards, Special Publ. 332, Washington D.C., p. 120.

Kaluzienski, L. J. and Holt, S. S.: 1979, *Astrophys. J. Letters*, preprint.

Kaluzienski, L. J., Holt, S. S., Boldt, E. A., and Serlemitsos, P. J.: 1977, *Astrophys. J.* **212**, 203.

Kamp, L. W.: 1978, *Astrophys. J. Suppl. Series* **36**, 143.

Kaplan, H. and Shapiro, M.: 1979, *Astrophys. J. Letters* **229**, L91.

Karetnikov, V. G. and Medvedev, Yu. A.: 1977, *Astron. Zh.* **54**, 580 (*Soviet Astron.* **21**, 330).

Kartasheva, T. A.: 1974, *Astrofis. Issled. Isv. Spets. Astrofis. Obs.* **6**, 11.

Katafos, M., Michalitsanos, A. G., and Vardya, M. S.: 1977, *Astrophys. J.* **216**, 526.

Katgert, P. and Oort, J. H.: 1967, *Bull. Astron. Inst. Neth.* **19**, 239.

Kato, S.: 1966, *Publ. Astron. Soc. Japan* **18**, 374.

Katz, J. I., Malone R. C., and Salpeter, E. E.: 1974, *Astrophys. J.* **190**, 359.

Keely, D. A.: 1970, *Astrophys. J.* **161**, 643.

Keenan, P. C.: 1963, in K. A. Strand (ed.), *Basic Astronomical Data*, Chicago, Chapter 8.

Keenan, P. C.: 1971, Contr. Kitt Peak Nat. Obs. No. 554, p. 35.

Keenan, P. C. and Morgan, W. W.: 1941, *Astrophys. J.* **94**, 501.

Keenan, P. C. and Morgan, W. W.: 1951, in Hynek (ed.), *Astrophysics*, New York, Chapter I.

Keenan, P. C. and McNeil, R. C.: 1976, *An Atlas of Spectra of the Cooler Stars: Types G, K, M, S, and C*, Ohio State Univ. Press.

Kemp, J. C., Sykes, M. V., and Rudy, R. J.: 1977, *Astrophys. J. Letters* **211**, L71.

Kemp, J. C., King, R., Parker, T. E., and Johnson, P. E.: 1980, preprint.

Khaliullin, Kh. F. and Cherepashchuk, A. M.: 1976, *Astron. Zh.* **53**, 327 (*Soviet Astron.* **20**, 186).

Kharadse, E. K. and Magalashvili, N. L.: 1967, *Observatory* **87**, 295.

Khromov, G. S.: 1976, *Astron. Zh.* **53**, 961.

King, A. R., Ricketts, M. J., and Warwick, R. S.: 1979, *Monthly Notices Roy. Astron. Soc.* **187**, 77P.

King, D. S. and Cox, J. P.: 1968, *Publ. Astron. Soc. Pacific* **80**, 365.

Kippenhahn, R.: 1969, *Astron. Astrophys.* **3**, 83.

Kippenhahn, R.: 1971, in A. B. Muller (ed.), *The Magellanic Clouds*, D. Reidel Publ. Co., Dordrecht, Holland, p. 144.

Kippenhahn, R.: 1974, in R. J. Tayler (ed.), 'Late Stages of Stellar Evolution', *IAU Symp.* **66**, 20.

Kippenhahn, R. and Weigert, A.: 1966, *Mitt. Astron. Ges.* **21**, 106.

Kippenhahn, R. and Weigert, A.: 1967a, *Z. Astrophys.* **65**, 251.

Kippenhahn, R. and Weigert, A.: 1967b, *Z. Astrophys.* **66**, 58.

Kippenhahn, R. and Meyer-Hoffmeister, E.: 1977, *Astron. Astrophys.* **54**, 539.

Kipper, T. A.: 1978, *Pis'ma Astron. Zh.* **4**, 280 (*Soviet Astron. Letters* **4**, 152).

Kirshner, R. P.: 1974, *Highlights of Astronomy* **3**, 533.

Kirshner, R. P. and Chevalier, R. A.: 1978, *Astron. Astrophys.* **67**, 267.

Kirshner, R. P., Willner, S. P., Becklin, E. E., Neugebauer, G., and Oke, J. B.: 1973a, *Astrophys. J.* **180**, L97.

Kirshner, R. P., Oke, J. B., Penston, M. V., and Searle, L.: 1973b, *Astrophys. J.* **185**, 303.

Kirshner, R. P., Arp, H. C., and Dunlap, J. R.: 1976, *Astrophys. J.* **207**, 44.

Kiselev, N. N. and Narizhnaya, N. V.: 1977, *Astron. Zh.* **54**, 606 (*Soviet Astron.* **21**, 344).

Kitchin, C. R.: 1970, *Astrophys. Space Sci.* **8**, 3.

Kitchin, C. R.: 1976, *Astrophys. Space Sci.* **45**, 119.

Klare, G. and Wolf, B.: 1978, *Astron. Astrophys. Suppl.* **33**, 327.

Klein, R. I. and Castor, J. I.: 1978, *Astrophys. J.* **220**, 902.

Kleinman, S. G., Dickinson, D. F., and Sargent, D. G.: 1978, *Astron. J.* **83**, 1206.

Klingesmith, D. A.: 1971, NASA Special Publ. No. 3065.

Koch, R. H. and Pfeiffer, R. J.: 1978, *Astron. J.* **83**, 183.

Kodaira, K. and Hoekstra, R.: 1979, *Astron. Astrophys.* **78**, 292.

Kondo, Y., Giuli, R. T., Modisette, J. I., and Rydgren, A. G.: 1972, *Astrophys. J.* **176**, 153.

Kondo, Y., Modisette, J. L., and Wolf, G. W.: 1975a, *Astrophys. J.* **199**, 110.

Kondo, Y., Morgan, T. H., and Modisette, J. L.: 1975b, *Astrophys. J. Letters* **196**, L125.

Kondo, Y., Morgan, T. H., and Modisette, J. L.: 1975c, *Astrophys. J. Letters* **198**, L37.

Kondo, Y., Modisette, J. L., Dufour, R. J., and Whaley, R. S.: 1976a, *Astrophys. J.* **206**, 163.

Kondo, Y., Morgan, T. H., and Modisette, J. L.: 1976b, *Astrophys. J.* **207**, 167.

Kondo, Y., Morgan, T. H., and Modisette, J. L.: 1976c, *Astrophys. J.* **209**, 489.

Kondo, Y., Morgan, T. H., and Modisette, J. L.: 1977, *Publ. Astron. Soc. Pacific* **89**, 675.

Kondo, Y., De Jager, C., Hoekstra, R., Kamperman, Th., Lamers, H. J. G. L. M., Modisette, J. L., Morgan, T., and Van der Hucht, K. A.: 1979, *Astrophys. J.* **230**, 526.

Kopal, Z.: 1959, *Close Binary Systems*, Wiley, New York, p. 136.

Koornneef, J.: 1978, *Monthly Notices Roy. Astron. Soc.* **184**, 477.

Kopylov, J. M.: 1957, *Isw. Krymsk, Astrophys. Obs.* **10**, 200.

Kosirev, N. A.: 1934, *Monthly Notices Roy. Astron. Soc.* **94**, 430.

Koubsky, P.: 1978, *Bull. Astron. Inst. Czech.* **29**, 288.

Kowal, C. T.: 1968, *Astron. J.* **73**, 1021.

Kraft, R. P.: 1958, *Astrophys. J.* **127**, 625.

Kraft, R. P.: 1963, *Adv. Astron. Astrophys.* **2**, 43.

Kraft, R. P.: 1964a, *Astrophys. J.* **139**, 457.

Kraft, R. P., 1964, *Astrophys. J.* **139**, 469.

Kraft, R. P.: 1966, *Astrophys. J.* **144**, 1008.

Kraft, R. P.: 1975, in H. Gurski and R. Ruffini (eds.) *Neutron Stars, Black Holes and Binary X-ray Sources*, D. Reidel Publ. Co., Dordrecht, Holland, p. 235.

Kraitcheva, Z. T.: 1974, *Nauchn. Inform. Astron. Soviet* **31**, 58.

Kraitcheva, Z. T.: 1978, *Nauchn. Inform. Astron. Soviet* **41**, 37.

Křiž, S.: 1979a, *Bull. Astron. Inst. Czech.* **30**, 83.

Křiž, S.: 1979b, *Bull. Astron. Inst. Czech.* **30**, 95.

Krupp, B. M., Collins, J. G., and Johnson, H. R.: 1978, *Astrophys. J.* **219**, 963.

Kruszewski, A. and Coyne, G. V.: 1976, *Astron. J.* **81**, 641.

Kruszewski, A., Gehrels, T., and Serkowski, K.: 1968, *Astron. J.* **73**, 677.

Kuan, P. and Kuhi, L. V.: 1975, *Astrophys. J.* **199**, 418.

Kudritzki, R. P.: 1979, *Astron. Astrophys.*, preprint.

Kudritzki, R. P. and Reimers, D.: 1978, *Astron. Astrophys.* **70**, 227.

Kuhi, L. V.: 1966, *Astrophys. J.* **145**, 715.

Kuhi, L. V.: 1973, in M. K. V. Bappu and J. Sahade (eds.), 'Wolf-Rayet and High-Temperature Stars', *IAU Symp.* **49**, 205.

Kuiper, G. P.: 1941, *Astrophys. J.* **93**, 133.

Kukarkin, B. V., Parenago, P. P., Efremov, Yu. I., and Khopolov: 1958, 'Obshtshiy Katalog Peremennykh Zvezd, Part I', Izdat. Akad. Nauk SSSR, p. 636.

Kulsrud, R. M.: 1955, *Astrophys. J.* **121**, 461.

Kunasz, P. B.: 1980, *Astrophys. J.*, preprint.

Kunasz, P. B., Hummer, D. G., and Mihalas, D.: 1975, *Astrophys. J.* **202**, 92.

Kunasz, P. and Van Blerkom, D.: 1978, *Astrophys. J.* **224**, 193.

Kuperus, M.: 1965, *Recherches Astron. Obs. Utrecht* **XVIII** (1).

Kupo, I. and Leibowitz, E. M.: 1977, *Astron. Astrophys.* **56**, 181.

Kurucz, R. L.: 1969, in O. Gingerich (ed.), *Theory and Observation of Normal Stellar Atmospheres*, MIT Press, Cambridge, Mass., p. 377.

Kurucz, R. L.: 1979, *Astrophys. J. Suppl.* **40**, 1.

Kurucz, R. L. and Schild, R. E.: 1976, in A. Slettebak (ed.), 'Be and Shell Stars', *IAU Symp.* **70**, 377.

Kurucz, R. L., Peytremann, E., and Avrett, E. A.: 1974, *Blanketed Model Atmospheres for Early-Type Stars*, Smithsonian Inst., Washington D.C.

Kurucz, R. L., Peytremann, E., and Avrett, E. A.: 1976, *Blanketed Model Atmospheres for Early-Type Stars*, Smithsonian Inst., Washington D.C.

Kwan, J. and Hill, F.: 1977, *Astrophys. J.* **215**, 781.

Kwan, J. and Scoville, N.: 1974, *Astrophys. J. Letters* **194**, L97.

Kwok, S.: 1975, *Astrophys. J.* **198**, 583.

Kwok, S. and Purton, C. R.: 1979, *Astrophys. J.* **229**, 187.

Kwok, S., Purton, C. R., and FitzGerald, P. M.: 1978, *Astrophys. J. Letters* **219**, L125.

Labeyrie, A.: 1970, *Astron. Astrophys.* **6**, 85.

Labeyrie, A.: 1974, *Nouv. Rev. Optique* **5**, 141.

Labeyrie, A., Koechlin, L., Bonneau, D., Blazit, A., and Foy, R.: 1977, *Astrophys. J. Letters* **218**, L75.

Lacy, C. H.: 1977, *Astrophys. J.* **212**, 132.

Lacy, C. H.: 1978, *Astrophys. J.* **226**, 138.

Lada, C. L. and Reid, M. J.: 1978, *Astrophys. J.* **219**, 95.

Lambert, D. L. and Luck, R. E.: 1978, *Monthly Notices Roy. Astron. Soc.* **184**, 405.

Lambert, D. L. and Snell, R. L.: 1975, *Monthly Notices Roy. Astron. Soc.* **172**, 277.

Lambert, D. L., Dearborn, D. S., and Sneden, C.: 1974, *Astrophys. J.* **193**, 621.

Lambert, D. L. and Tomkin, J.: 1974, *Astrophys. J. Letters* **194**, L89.

Lamers, H. J.: 1972, in M. Hack (ed.), *Colloquium on Supergiant Stars*, Osservatorio Astronomico Trieste, p. 83.

Lamers, H. J. G. L. M.: 1974, *Astron. Astrophys.* **37**, 237.

Lamers, H. J. G. L. M.: 1975, *Phil. Trans. Roy. London* **A279**, 445.

Lamers, H. J. G. L. M.: 1976, in R. Cayrel (ed.), *Physique des mouvements dans les atmosphères stellaires*, Centre Nat. Rech. Sci. Paris, p. 405.

Lamers, H. J. G. L. M. and Castor, J. C.: 1979, in P. S. Conti and C. W. H. de Loore, (eds.), 'Mass Loss and Evolution of O-type Stars', *IAU Symp.* **83**, 81.

Lamers, H. J. and De Loore C.: 1976. in R. Cayrel (ed.), *Physique des mouvements dans les atmosphères stellaires*, Centre Nat. Rech. Sci. Paris, p. 453.

Lamers, H. J. G. L. M. and Morton, D. C.: 1976, *Astrophys. J. Suppl. Series.* **32**, 715.

Lamers, H. J. G. L. M. and Snijders, M. A. J.: 1975, *Astron. Astrophys.* **41**, 715.

Lamers, H. J. G. L. M. and Snow, T. P.: 1978, *Astrophys. J.* **219**, 504.

Lamers, H. J. G. L. M., Van der Hucht, K. A. Snijders, M. A. J., and Sakhibulin, N.: 1973, *Astron. Astrophys.* **25**, 105.

Lamers, H. J. G. L. M., Van den Heuvel, E. P. J., and Petterson, J. A.: 1976, *Astron. Astrophys.* **49**, 327.

Lamers, H. J. G. L. M. Stalio, R., and Kondo, Y.: 1978, *Astrophys. J.* **223**, 207.

Lamers, H. J. G. L. M., Paerels, F., and De Loore, C.: 1980, *Astron. Astrophys.*, preprint.

Landini, M. and Monsignori Fossi, B. C.: 1973, *Astron. Astrophys.* **25**, 9.

Lang, K. R.: 1974, *Astrophysical Formulae*, Springer, Berlin.

Langer, G. E.: 1971, *Monthly Notices Roy. Astron. Soc.* **155**, 199.

Langer, G. E., Kraft, R. P., and Anderson, K. S.: 1974, *Astrophys. J.* **189**, 509.

Larimer, J. W. and Anders, E., 1967, *Geochim. Cosmochim. Acta* **31**, 1239.

Larson, R. and Starrfield, S.: 1971, *Astron. Astrophys.* **13**, 190.

Lasker, B. M.: 1977, *Publ. Astron. Soc. Pacific* **89**, 474.

Latour, J., Spiegel, E. A., Toomre, J., and Zahn, J. P.: 1976, *Astrophys. J.* **207**, 233.

Lazareff, B., Audouze, J., Starrfield, S., and Truran, J. W.: 1979, *Astrophys. J.* **228**, 875.

Latham, D. W.: 1970, Smithson. Astrophys. Obs., Special Report No. 321.

Lauterborn, D.: 1969, in M. Hack (ed.), *Mass Loss from Stars*, D. Reidel Publ. Co., Dordrecht, Holland, p. 262.

Lauterborn, D.: 1970, *Astron. Astrophys.* **7**, 150.

LeBlanc, J. M. and Wilson, J. R.: 1970, *Astrophys. J.* **161**, 541.

Lebofsky, M. J. and Rieke, G. H.: 1977, *Astron. J.* **82**, 646.

Ledoux, P.: 1941, *Astrophys. J.* **94**, 537.

Ledoux, P.: 1947, *Astrophys. J.* **105**, 305.

Ledoux, P. and Renson, P.: 1966, *Ann. Rev. Astron. Astrophys.* **4**, 293.

Lee, T. A.: 1970, *Astrophys. J.* **162**, 217.

Leep, E. M.: 1978, *Astrophys. J.* **225**, 265.

Lépine, J. R. D. and Paes de Barros, M. H.: 1977, *Astron. Astrophys.* **56**, 219.

Lépine, J. R. D., Le Squerin, A. M., and Scalise, E.: 1978, *Astrophys. J.* **225**, 869.

Lesh, J. R. and Aizenman, M. L.: 1973, *Astron. Astrophys.* **22**, 229.

Lester, J. B.: 1973, *Astrophys. J.* **185**, 253.

Lester, J. B.: 1979, *Astrophys. J.* **231**, 164.

Le Squeren, A. M., Baudry, A., Brillet, J., and Darchy, B.: 1979, *Astron. Astrophys.* **72**, 39.

Leung, K. C. and Schneider, D. P.: 1978a, *Astrophys. J.* **223**, 202.

Leung, K. C. and Schneider, D. P.: 1978b, *Astrophys. J.* **224**, 565.

Leung, K. C. and Schneider, D. P.: 1979, in P. S. Conti and C. W. H. de Loore (eds.), 'Mass Loss and Evolution of O-type Stars', *IAU Symp.* **83**, 265.

Leung, K. C., Moffat, A. F. J., and Seggewiss, W.: 1979, *Astrophys. J.* **231**, 742.

Li F., Rappaport, S., and Epstein, A.: 1978, *Nature* **271**, 37.

Lighthill, M. J.: 1952, *Proc. Roy. Soc.* **A211**, 564.

Lighthill, M. J.: 1960, *Phil. Trans. Roy. Soc. London* **A252**, 397.

Liller, M. H. and Liller, W.: 1979, *Astron. J.* **84**, 1357.

Limber, D. N.: 1960, *Astrophys. J.* **131**, 168.

Limber, D. N. and Marlborough, J. M.: 1968, *Astrophys. J.* **152**, 181.

Linsky, J. L.: 1977, in O. R. White (ed.), *The Solar Output and its Variations*, Colorado Associated Univ. Press, p. 477.

Linsky, J. L. and Ayres, T. R.: 1978, *Astrophys. J.* **220**, 619.

Linsky, J. L. and Haisch, B. M.: 1979, *Astrophys. J. Letters* **229**, L27.

Linsky, J. L., Worden, S. P., McClintock, W., and Robertson, R. M.: 1979a, *Astrophys. J. Suppl. Series* **41**, 47.

Linsky, J. L., Hunten, D. M., Sowell, R., Glackin, D. L., and Kelch, W. L.: 1979b, *Astrophys. J. Suppl. Series* **41**, 481.

Litvak, M. M.: 1969, *Astrophys. J.* **156**, 471.

Litvak, M. M.: 1972, in T. R. Carson and M. J. Roberts (eds.), *Atoms and Molecules in Astrophysics*, Academic Press, p. 201.

Lloyd Evans, T.: 1978a, *Monthly Notices Roy. Astron. Soc.* **183**, 305.

Lloyd Evans, T.: 1978b, *Monthly Notices Roy. Astron. Soc.* **183**, 319.

Lloyd Evans, T.: 1979, *Monthly Notices Roy. Astron. Soc.* **186**, 13.

Long, K. S. and Helfand, D. J.: 1979, *Astrophys. J. Letters* **234**, L77.

Loumos, G. L., Lambert, D. L., and Tomkin, J.: 1975, *Publ. Astron. Soc. Pacific* **87**, 859.

Luck, R. E.: 1975, *Astrophys. J.* **202**, 743.

Luck, R. E.: 1977a, *Astrophys. J.* **212**, 743.

Luck, R. E.: 1977b, *Astrophys. J.* **218**, 752.

Lucke, R. L. Zarnecki J. C., Woodgate, B. E., Culhane, J. L., and Socker, D. G.: 1979, *Astrophys. J.* **228**, 763.

Lucy, L. B.: 1971, *Astrophys. J.* **163**, 95.

Lucy, L. B.: 1975, *Mém. Soc. Roy. Sci. Liège, 6e Série* **VIII**, 359.

Lucy, L. B.: 1976a, *Astrophys. J.* **206**, 499.

Lucy, L. B.: 1976b, in W. S. Fitch (ed.), 'Multiple Periodic Variable Stars', *IAU Colloq.* **29**, 105.

Lucy, L. B.: 1976c, *Astrophys. J.* **205**, 482.

Lucy, L. B. and Solomon, P. M.: 1970, *Astrophys. J.* **159**, 879.

Lutz, T. E. and Pagel, B. E. J.: 1979, preprint.

Lynds, B. T.: 1979, in P. S. Conti and C. W. H. de Loore (eds.), 'Mass Loss and Evolution of O-type Stars', *IAU Symp.* **83**, 55.

Lynds, C. R., Worden, S. P., and Harvey, J. W.: 1976, *Astrophys. J.* **207**, 174.

MacGregor, K. B., Hartmann, L., and Raymond, J. C.: 1979, *Astrophys. J.* **231**, 514.

Maciel, W. J.: 1976, *Astron. Astrophys.* **48**, 27.

Maciel, W. J.: 1977, *Astron. Astrophys.* **57**, 273.

Maeder, A.: 1980a, *Highlights of Astronomy* **5**, 473.

Maeder, A.: 1980b, *Astron. Astrophys.*, preprint.

Maeder, A. and Rufener, F.: 1972, *Astron. Astrophys.* **20**, 437.

Maillard, J. P.: 1974, Highlights of Astronomy **3**, 269.

Malakpur, I.: 1973, *Astron. Astrophys.* **28**, 393.

Malakpur, I.: 1979, *Astron. Astrophys.* **78**, 7.

Mallama, A. D. and Trimble, V. L.: 1978, *Quart. J. Roy. Astron. Soc.* **19**, 430.

Mammano, A. and Righini. M. G.: 1973. *Mem. Soc. Astron. Ital.* **44**, 23.

Mammano, A. and Taffara, S.: 1978, *Astron. Astrophys. Suppl.* **34**, 211.

Mammano, A., Margoni, R., and Stagni, R.: 1978, *Astron. Astrophys.* **59**, 9.

Manduca, A.: 1979, *Astron. Astrophys. Suppl.* **36**, 411.

Manduca, A., Bell, R. A., and Gustafsson, B.: 1977, *Astron. Astrophys.* **61**, 809.

Margon, B., Kieniewicz, P., and Stone, R. P. S.: 1977, *Publ. Astron. Soc. Pacific* **89**, 300.

Marlborough, J. M.: 1977, *Publ. Astron. Soc. Pacific* **89**, 122.

Marlborough, J. M. and Roy, J. R.: 1970, *Astrophys. J.* **160**, 221.

Marlborough, J. M. and Snow, T. P.: 1976, in A. Slettebak (ed.), 'Be and Shell Stars', *IAU Symp.* **70**, 179.

Marlborough, J. M., Snow, T. P., and Slettebak, A.: 1978, *Astrophys. J.* **224**, 157.

Martens, P. C. H.: 1979, *Astron. Astrophys. Letters* **75**, L7.

Marques dos Santos, P., Lépine, J. R. D., and Gomez Balboa, A. M.: 1979, *Astron. J.* **84**, 787.

Masheder, M. R. W., Booth, R. S., and Davies, R. D.: 1974, *Montly Notices Roy. Astron. Soc.* **166**, 561.

Mason, K. O., Hawkins, F. J., Sanford, P. W., Murdin, P., and Savage, A.: 1974, *Astrophys. J. Letters* **192**, L65.

Mason, K. O. White. N. E. and. Sanford. P. W.: 1976, *Nature* **260**, 690.

Massa, D.: 1975, *Publ. Astron. Soc. Pacific* **87**, 777.

Massevitch, A. G.: 1949, *Astron. Zh.* **26**, 207; *Isw. Akad. Nauk.* **13**, 121.

Massevitch, A. G., Tutukov, A. V.: 1974, in R. J. Tayler (ed.), 'Late Stages of Stellar Evolution', *IUA Symp.* **66**, 73.

Massevitch, A. G. and Yungelson, L. R.: 1975, *Mem. Soc. Astron. Ital.* **46**, 217.

Massevitch, A. G., Tutukov, A. V., and Yungelson, L. R,: 1976, *Astrophys. Space Sci.* **40**, 115.

Massevitch, A. G., Popopa, E. I., Tutukov, A. V., and Yungelson, L. R.: 1979, *Astrophys. Sci.* **62**, 451.

Massey, P.: 1980, *Astrophys. J.* **236**, 526.

Massey, P. and Conti, P. S.: 1977, *Astrophys. J.* **218**, 431.

Mathewson, D. S. and Healy, J. R.: 1964, in F. J. Kerr and A. W. Rodgens (eds.), *The Galaxy and the Magellanic Clouds*, Canberra, Australian Acad. of Sci., p. 283.

Mathewson, D. S. and Clarke, J. N.: 1973, *Astrophys. J.* **180**, 725.

Matilsky, T., La Salla, J., and Jessen, J.: 1978, *Astrophys. J. Letters* **224**, L119.

Matthews H. E. Goss, W. M., Winnberg, A., and Habing, H. J.: 1977, *Astron. Astrophys.* **61**, 261.

Mayall, M. W.: 1949, *Harvard Bull.* **919**, 15.

Maza, J. and Van den Bergh, S.: 1976, *Astrophys. J.* **204**, 519.

McCarthy, D. W., Low, F. J., and Howell, R.: 1977, *Astrophys. J. Letters* **214**, L85.

McCarthy, D. W., Howell, R., and Low, F. J.: 1978, *Astrophys. J. Letters* **223**, L113.

McClintock, J. E.: 1976, *Astrophys. J. Letters* **206**, L99.

McClintock, W., Henry, R. C., Moos, H. W., and Linsky, J. L.: 1975, *Astrophys. J.* **202**, 733.

McClintock, W. and Henry, R. C.: 1977, *Astrophys. J.* **218**, 205.

McClusky, G. E. and Kondo, Y.: 1972, *Astrophys. Space Sci.*, **17**, 34.

McClusky, G. E., Kondo, Y., and Kondo, Y.: 1976, *Astrophys. J.* **208**, 760.

McClusky, G. E., Kondo, Y., and Morton, D. C.: 1975, *Astrophys. J.* **201**, 607.

McCray, R., Stein, R. F., and Kafatos, M.: 1975, *Astrophys. J.* **196**, 565.

McGee, R. X., Newton, L. M., and Brooks, J. W.: 1977, *Monthly Notices Roy. Astron. Soc.* **180**, 565.

McLaughlin, D. B.: 1937, *Astrophys. J.* **85**, 362.

McLaughlin, D. B.: 1946, *Astron. J.* **52**, 46.

McLaughlin, D. B.: 1953, *Astrophys. J.* **118**, 27.

McLaughlin, D. B.: 1960, in J. L. Greenstein (ed.), *Stellar Atmospheres*, Univ. of Chicago Press, p. 585.

McLaughlin, D. B.: 1961, *J. Roy. Astron. Soc. Canada* **55**, 13.

McLaughlin, D. B.: 1973, *Michigan Obs. Publ.* **6**, 103.

McLean, I. S.: 1976, *Monthly Notices Roy. Astron. Soc.* **176**, 73.

McLean, I. S.: 1979a, *Monthly Notices Roy. Astron. Soc.* **186**, 21.

McLean, I. S.: 1979b, *Monthly Notices Roy. Astron. Soc.* **186**, 265.

McLean, I. S. and Brown, J. C.: 1978, *Astron. Astrophys.* **69**, 291.

McLean, I. S., Coyne, G. V., Frecker, J. E., and Serkowski, K.: 1979a, *Astrophys. J. Letters* **231**, L141.

McLean, I. S. Coyne G. V., Frecker, J. E., and Serkowski, K.: 1979b, *Astrophys. J.* **228**, 802.

McMillan, R. S. and Tapia, S.: 1978, *Astrophys. J. Letters* **226**, L87.

McWhirtor, R. W. P., Thoneman, P. C., and Wilson, R.: 1975, *Astron. Astrophys.* **40**, 63.

Meadows, A. J.: 1960, *Astron. J.* **65**, 335.

Meier, D. L., Epstein, R. I., Arnett, W. D., and Schramm, D. N.: 1976, *Astrophys. J.* **204**, 869.

Melnick, J.: 1978, *Astron. Astrophys. Suppl.* **34**, 383.

Mendoza, E. E.: 1970, *Bol. Obs. Tonantzintla y Tacubaya* **5**, 269.

Mendoza, E. E. and Johnson, H. L.: 1965, *Astrophys. J.* **141**, 161.

Merrill, K. M. and Ridgway, S. T.: 1979, *Ann. Rev. Astron. Astrophys.* **17**, 9.

Merrill, K. M. and Stein, W. A.: 1976, *Publ. Astron. Soc. Pacific* **88**, 285.

Merrill, P. W.: 1933, *Publ. Astron. Soc. Pacific* **45**, 198.

Merrill, P. W. and Burwell, C. Q.: 1943, *Astrophys. J.* **98**, 153.

Merrill, P. W., Deutsch, A. J., and Keenan, P. C.: 1962, *Astrophys. J.* **136**, 21.

Mestel, L., Moore, D. W., and Perry, J. J.: 1976, *Astron. Astrophys.* **52**, 203.

Mewe, R.: 1972, *Solar Phys.* **22**, 459.

Mewe, R.: 1975a, *Solar Phys.* **44**, 383.

Mewe, R.: 1979, *Space Sci. Rev.* **24**, 101.

Mewe, R., Heise, J., Gronenschild, E., Brinkman, A. C., Schrijver, J., and Den Boggende, A. J. F.:
    1975a, *Nature Phys. Sci.* **256**, 711.

Mewe, R., Heise, J. Gronenschild, E., Brinkman, A. C., Schrijver, J. and Den Boggende, A. J. F.:
    1975b, *Astrophys. J. Letters* **202**, L67.

Mewe, R., Heise, J. Gronenschild, E., Brinkman, A. C., Schrijver, J., and Den Boggende, A. J. F.:
    1976, *Astrophys. Space Sci.* **42**, 217.

Mezger, P. G. and Smith, L. F.: 1977, in T. de Jong and A. Maeder (eds.), 'Star Formation', *IAU
    Symp.* **75**, 133.

Michalitsianos, A. G., Kafatos, M.: 1978, *Astrophys. J.*, **226**, 430.

Mihalas, B.: 1979, *Internal Gravity Waves in the Solar Atmospheres* Thesis Univ. of Colorado,
    Boulder.

Mihalas, D.: 1965, *Asthrophys. J.* **141**, 564.

Mihalas, D.: 1966, *Astrophys. J. Suppl.* **13**, 1.

Mihalas, D.: 1970, *Stellar Atmospheres*, Freeman, San Francisco.

Mihalas, D.: 1972a, *Astrophys. J.* **176**, 139.

Mihalas, D.: 1972b, NCAR Techn. Note No. 76, Boulder, Colo.

Mihalas, D.: 1974, *Astron. J.* **79**, 1111.

Mihalas, D.: 1978, *Stellar Atmospheres*, 2nd ed., Freeman, San Francisco.

Mihalas, D. and Hummer, D. G.: 1973, *Astrophys. J.* **179**, 287.

Mihalas, D. and Hummer, D. G.: 1974a, *Astrophys. J. Letters* **189**, L39.

Mihalas, D. and Hummer, D. G.: 1974b, *Astrophys. J. Suppl.* **28**, 343.

Mihalas, D., Hummer, D. G., and Conti, P. S.: 1972, *Astrophys. J. Letters* **175**, L99.

Mihalas, D., Frost, S. A., and Lockwood, G. W.: 1975a, *Publ. Astron. Soc. Pacific* **87**, 153.

Mihalas, D., Kunasz, P. B., and Hummer, D. G.: 1975b, *Astrophys. J.* **202**, 465.

Mihalas, D., Kunasz, P. B., and Hummer, D. G.: 1976, *Astrophys. J.* **203**, 647.

Mikami, T.: 1975, *Publ. Astron. Soc. Japan* **27**, 445.

Mikami, T.: 1978, *Publ. Astron. Soc. Japan* **30**, 191.

Milkey, R. W.: 1970, *Solar Phys.* **14**, 77.

Miller, F. D.: 1954, *Astron. J.* **58**, 222.

Milne, D. K.: 1970, *Australian J. Phys.* **23**, 425.

Milne, D. K. and Aller, L. H.: 1975, *Astron. Astrophys.* **38**, 183.

Milne, E. A.: 1926, *Monthly Notices Roy. Astron. Soc.* **86**, 459.

Minkowski, R.: 1966, *Astron. J.* **71**, 371.

Minnaert, M.: 1930, *Z. Astrophys.* **1**, 209.

Moffat, A.: 1977, *Astron. Astrophys.* **57**, 151.

Moffat, A. F. J.: 1978, *Astron. Astrophys.* **68**, 41.

Moffat, A. F. J.: 1979, preprint.

Moffat, A. F. J. and Haupt, W.: 1975, *Mem. Soc. Australian Ital.* **45**, 811.

Moffat, A. F. and FitzGerald, M. P.: 1977, *Astron. Astrophys.* **54**, 263.

Moffat, A. F. and Seggewiss, W.: 1977, *Astron. Astrophys.* **54**, 607.

Moffat, A. F. J. and Seggewiss, W.: 1979, in P. S. Conti and C. W. H. de Loore (eds.), 'Mass Loss and
    Evolution of O-type Stars', *IAU Symp.* **83**, 447.

Möllenhoff, C. and Schaifers, K.: 1978, *Astron. Astrophys.* **64**, 253.

Moran, J. M., Ball, J. A., Yen, J. L., Schwartz, P. R., Johnston, K. J., and Knowles, S. H.: 1977a, *Astrophys. J.* **211**, 160.
Moran, J. M., Johnston, K. J., Spencer, J. H., and Schwarz, P. R.: 1977b, *Astrophys. J.* **217**, 434.
Morfill, G. E. and Scholer, M.: 1979, *Astrophys. J.* **232**, 473.
Morgan, T. H.: 1975, *Astrophys. J.* **195**, 391.
Morgan, W. W. and Keenan, P. C.: 1973, *Ann. Rev. Astron. Astrophys.* **11**, 29.
Morgan, W. W., Keenan, P. C., and Kellman, E.: 1943, *An Atlas of Stellar Spectra*, Chicago.
Morgan, W. W., Abt, H. A., and Tapscott, J. W.: 1978, *Revised Spectral Atlas for Stars Earlier than the Sun*, Yerkes Obs., and Kitt Peak Nat. Obs.
Morris, M.: 1975, *Astrophys. J.* **197**, 603.
Morris, M. and Alcock, C.: 1977, *Astrophys. J.* **218**, 687.
Morris, M., Redman, R., Reid, M. J., and Dickinson, D. F.: 1979, *Astrophys. J.* **229**, 257.
Morrison, D. and Simon, T.: 1973, *Astrophys. J.* **186**, 193.
Morrison, N. D.: 1975, *Astrophys. J.* **200**, 113.
Morton, D. C.: 1967a, *Astrophys. J.* **147**, 1017.
Morton, D. C.: 1967b, *Astrophys. J.* **150**, 535.
Morton, D. C.: 1969a, *Astrophys. J.* **158**, 629.
Morton, D. C.: 1969b, *Astrophys. J.* **160**, 215.
Morton, D. C.: 1970, *Astrophys. J.* **158**, 629.
Morton, D. C.: 1973, in M. Bappu and J. Sahade (eds.), 'Wolf-Rayet and High Temperature Stars', *IAU Symp.* **49**, 54.
Morton, D. C.: 1975, *Phil. Trans. Roy. Soc. London, Ser. A.* **279**, 443.
Morton, D. C.: 1976, *Astrophys. J.* **203**, 386.
Morton, D. C.: 1979, *Monthly Notices Roy. Astron. Soc.* **189**, 57.
Morton, D. C. and Adams, T. F.: 1968, *Astrophys. J.* **151**, 611.
Morton, D. C. and Underhill, A. B.: 1977, *Astrophys. J. Suppl.* **33**, 83.
Morton, D. C. and Wright, A. E.: 1978, *Monthly Notices Roy. Astron. Soc.* **182**, 47P.
Morton, D. C. and Wright, A. E.: 1979, in P. S. Conti en C. W. H. de Loore (eds.), 'Mass Loss and Evolution of O-type Stars', *IAU Symp.* **83**, 155.
Morton, D. C., Jenkins, E. B., and Bohlin, R. C.: 1968, *Astrophys. J.* **154**, 66.
Morton, D. C., Jenkins, E. B., and Macy, W. W.: 1972, *Astrophys. J.* **177**, 235.
Motteran, M.: 1972, in M. Hack (ed.), *Colloquium on Supergiant Stars*, Osservatorio Astronomico Trieste, p. 292.
Mufson, S. L.: 1974, *Astrophys. J.* **193**, 561.
Mufson, S. L.: 1975, *Asprophys. J.* **202**, 372.
Mullan, D. J.: 1978, *Astrophys. J.* **226**, 151.
Mumford, G. S.: 1967a, *Astrophys. J. Suppl.* **15**, 1.
Mumford, G. S.: 1967b, *Publ. Astron. Soc. Pacific* **79**, 382.
Murdin, P., Clark, D. H., and Culhane, J. L.: 1978, *Monthly Notices Roy. Astron. Soc.* **184**, 79P.
Murray, S. S., Fabbiano, G., Fabian, A., Epstein, A., and Giacconi, R.: 1979, *Astrophys. J. Letters* **234**, L69.
Musielak, Z. and Sikorski, J.: 1979, *Acta Astron.* **29**, 381.
Mustel, E. R.: 1971, *Astron. Zh.* **48**, 3.
Mustel, E. R.: 1972, *Astron. Zh.* **50**, 1121.
Mustel, E. R.: 1973, *Soviet Astron.* **17**, 711.
Mustel, E. R.: 1974, *Highlights of Astronomy* **3**, 545.
Mustel, E. R.: 1978, *Astrophys. Space Sci.* **58**, 41.
Mustel, E. R. and Baranova, L. I.: 1965, *Astron. Zh.* **42**, 42.
Mustel, E. R. and Boyarchuk, A. A.: 1970, *Astrophys. Space Sci.* **6**, 183.
Mustel, E. R. and Chugay, N. N.: 1975, *Astrophys. Space Sci.* **32**, 39.
Mutel, R. L., Fix, J. D., Benson, J. M., and Webber, J. C.: 1979, *Astrophys. J.* **228**, 771.
Muthsam, H.: 1979, *Astron. Astrophys. Suppl.* **35**, 253.
Nadyozhin D. K. and Utrobin V. P.: 1976, *Astron. Zh.* **53**, 992.
Nadyozhin, D. K. and Utrobin, V. P.: 1977, *Astron. Zh.* **54**, 996 (*Soviet Astron.* **21**, 564).
Nariai, K.: 1970, *Publ. Astron. Soc. Japan* **22**, 475.
Nariai, K.: 1974, *Astron. Astrophys.* **36**, 231.

Nelson, G. D. and Hearn, A. G.: 1978, *Astron. Astrophys.* **65**, 223.

Neo, S., Miyaji, S., Nomoto, K., and Sugimoto, D.: 1977, *Publ. Astron. Soc. Japan* **29**, 249.

Neugebauer, G. and Leighton, R. B.: 1969, 'Two Micron Sky Survey – a Preliminary Catalogue', NASA SP-3047.

Neugebauer, G., Martz, D. E., and Leighton, R. B.: 1965, *Astrophys. J.* **142**, 399.

Ney, E. P. and Hatfield, B. F.: 1978, *Astrophys. J. Letters* **219**, L111.

Niemelä, V. S.: 1973, *Publ. Astron. Soc. Pacific* **85**, 220.

Niemelä, V. S.: 1976, *Astrophys. Space Sci.* **45**, 191.

Niemelä, V. S.: 1979, in P. S. Conti and C. W. H. de Loore (eds.), 'Mass Loss and Evolution of O-type Stars', *IAU Symp.* **83**, 291.

Niemelä, V. S. and Mendez, R. H.: 1974, *Astrophys. J. Letters* **187**, L25.

Niemelä, V. S. and Sahade, J.: 1979, in P. S. Conti and C. W. H. de Loore (eds.), 'Mass Loss and Evolution of O-type Stars', *IAU Symp.* **83**, 287.

Noguchi, K., Maihara, T., Okudia, H., and Sato, S.: 1977, *Publ. Astron. Soc. Japan* **29**, 511.

Noerdlinger, P. D.: 1979, in P. S. Conti and C. W. H. de Loore (eds.), 'Mass Loss and Evolution of O-type Stars', *IAU Symp.* 83, 253.

Noerdlinger, P. D. and Rybicki, G. B.: 1974, *Astrophys. J.* **193**, 561.

Nordh, H. L. and Olofsson, S. G.: 1977, *Astron. Astrophys.* **56**, 117.

Nordh, H. L., Olofsson, S. G., and Augason, G. C.: 1978, *Astron. J.* **83**, 188.

Nordlund, Å.: 1976, *Astron. Astrophys.* **50**, 23.

Nugis, T.: 1975, in V. E. Sherwood and L. Plaut (eds.), 'Variable Stars and Stellar Evolution', *IAU Symp.* **67**, 291.

Nugis, T., Kolka, I., and Luud, L.: 1979, in P. S. Conti and C. W. H. de Loore (eds.), 'Mass Loss and Evolution of O-type Stars', *IAU Symp.* **83**, 39.

Nussbaumer, H., Schmutz, W., Smith, L. J., Willis, A. L., and Wilson, R.: 1979, in A. L. Willis (ed.), *The First Year of IUE*, p. 259.

O'Brien, G. and Lambert, D. L.: 1979, *Astrophys. J. Letters* **229**, L33.

O'Connell, D. J. K.: 1956, *Vistas Astron.* **2**, 113.

O'Dell, C. R.: 1968, in D. E. Osterbrock and C. R. O'Dell (eds.), 'Planetary Nebulae', *IAU Symp.* **34**, 361.

Oemler, A. and Tinsley, B. M.: 1979, *Astron. J.* **84**, 985.

Oke, J. B.: 1969, *Publ. Astron. Soc. Pacific* **81**, 11.

Oke, J. B. and Searle, L.: 1974, *Ann. Rev. Astron. Astrophys.* **12**, 315.

Olnon, F. M.: 1975, *Astron. Astrophys.* **39**, 217.

Olnon, F. M.: 1977, 'Shells Around Stars', Thesis, Leiden Univ.

Olnon, F. M.: 1979, private communication.

Olson, G. L.: 1978, *Astrophys. J.* **226**, 124.

Olson, G. L.: 1979, in P. S. Conti and C. W. H. de Loore (eds.), 'Mass Loss and Evolution of O-type Stars', *IAU Symp.* **83**, 257.

Olson, B. I. and Richter, H. B.: 1975, *Astrophys. J.* **200**, 88.

Olson, B. I. and Richter, H. B.: 1979, *Astrophys. J.* **227**, 534.

Oppenheimer, J. R. and Snyder, H.: 1939, *Phys. Rev.* **56**, 455.

Osaki, Y.: 1974, *Astrophys. J.* **189**, 469.

Osborn, W.: 1975, *Monthly Notices Roy. Astron. Soc.* **172**, 631.

Osmer, P. S.: 1972, *Astrophys. J. Suppl.* **24**, 247.

Osmer, P. S.: 1973, *Astrophys. J.* **186**, 459.

Osterbrock, D. E.: 1961, *Astrophys. J.* **134**, 347.

Ostriker, J. P., McCray, R., Weaver, R., and Yahil, A.: 1976, *Astrophys. J. Letters* **208**, L61.

Packet, W., Vanbeveren, D., De Grève, J. P., De Loore, C., and Sreenivasan, S. R.: 1980, *Astron. Astrophys.*, preprint.

Paczyński, B.: 1965, *Acta Astron.* **15**, 305.

Paczyński, B.: 1966, *Acta Astron.* **16**, 231.

Paczyński, B.: 1967, *Acta Astron.* **17**, 355.

Paczyński, B.: 1970a, in K. Gyldenkerne and R. M. West (eds.), *Mass Loss and Evolution of Close Binary Systems*, Copenhagen Univ. Publ., p. 139.

Paczyński, B.: 1970b, *Acta Astron.* **20**, 195.

Paczyński, B.: 1970c, *Acta Astron.* **20**, 47.

Paczyński, B.: 1971a, *Ann. Rev. Astron. Astrophys.* **9**, 183.
Paczyński, B.: 1971b, *Acta Astron.* **21**, 1.
Paczyński, B.: 1971c. *Acta Astron.* **21**, 417.
Paczyński, B.: 1973, in M. K. V. Bappu and J. Sahade (eds.), 'Wolf-Rayet and High Temperature Stars', *IAU Symp.* **49**, 143.
Paczyński, B.: 1977, *Astrophys. J.* **214**, 812.
Paczyński, B. and Ziolkowski, J.: 1967, *Acta Astron.* **17**, 7.
Paddock, G. F.: 1935, *Lick Obs. Bull.* **17**, 99.
Pagel, B. E. J.: 1977, *Highlights in Astronomy* **4**, 119.
Pagel, B. E. J. and Wilkins, D. R.: 1979, preprint.
Panagia, N. and Felli, M.: 1975, *Astron. Astrophys.* **39**, 1.
Panagia, N. and Felli, M.: 1975, *Astron. Astrophys.* **39**, 41.
Panchuk, V. E.: 1978, *Pis'ma Astron. Zh.* **4**, 374 (*Soviet Astron. Letters* **4**, 201).
Panek, R. J. and Savage, B. D.: 1966, *Astrophys. J.* **206**, 167.
Papaloizou, J. C. B.: 1973, *Monthly Notices Roy. Astron. Soc.* **162**, 169.
Parker, E. N.: 1958, *Astrophys. J.* **128**, 664.
Parker, E. N.: 1965, *Space Sci. Rev.* **4**, 666.
Parker, R. A. R.: 1978, *Astrophys. J.* **224**, 873.
Parkes, G. E., Charles, P. A., Culhane, J. L., and Ives, J. C.: 1977, *Monthly Notices Roy. Astron. Soc.* **179**, 55.
Parsons, S. B.: 1967, *Astrophys. J.* **150**, 263.
Parsons, S. B.: 1971, *Monthly Notices Roy. Astron. Soc.* **152**, 121.
Parthasarathly, M.: 1978, *Monthly Notices Roy. Astron. Soc.* **185**, 485.
Patterson, J.: 1978, *Astrophys. J.* **225**, 954.
Patterson, J.: 1979a, *Astrophys. J.* **231**, 789.
Patterson, J.: 1979b, *Astrophys. J. Letters* **233**, L13.
Payne, C. H.: 1930, *The stars of High Luminosity*, McGraw-Hill, New York.
Payne-Gaposchkin, C. H.: 1957, *The Galactic Novae*, North-Holland Publ. Co., Amsterdam.
Payne-Gaposchkin, C. H.: 1977, in M. Friedjung (ed.), *Novae and Related Stars*, D. Reidel Publ. Co., Dordrecht, Holland, p. 3.
Payne-Gaposchkin, C. H.: 1977b, *Astron. J.* **82**, 665.
Pecker, J.-C.: 1962, *Space Res.* **3**, 1076.
Pecker, J.-C., Praderie, F., and Thomas, R. N.: 1973, *Astron. Astrophys.* **29**, 289.
Peraiah, A.: 1973, *Astrophys. Space Sci.* **21**, 223.
Peraiah, A.: 1976, *Astron. Astrophys.* **46**, 237.
Perry, B. F.: 1966, *Astrophys. J.* **144**, 672.
Peters, G. J.: 1976a, in A. Slettebak (ed.), 'Be and Shell Stars', *IAU Symp.* **70**, 69.
Peters, G. J.: 1976b, in A. Slettebak (ed.), 'Be and Shell Stars', *IAU Symp.* **70**, 157.
Peters, G. J.: 1979, *Astrophys. J. Suppl.* **39**, 175.
Peterson, D. M. and Scholz, M.: 1971, *Astrophys. J.* **163**, 51.
Peytremann, E.: 1974a, *Astron. Astrophys.* **33**, 203.
Peytremann, E.: 1974b, *Astron. Astrophys. Suppl.* **18**, 81.
Peytremann, E.: 1975, *Astron. Astrophys.* **38**, 417.
Pfau, W.: 1976, *Astron. Astrophys.* **50**, 113.
Pickering, E. C.: 1897, *Harvard Circ.* **19**.
Pietsch W., Voges, W., Reppin, C., Trümper, J., Kendziorra, E., and Staubert, R.: 1980, *Astrophys. J.*, prepint.
Piirola, V.: 1979, *Astron. Astrophys. Suppl.* **38**, 193.
Piirola, V. and Korhonen, T.: 1979, *Astron. Astrophys.* **79**, 254.
Plaskett, J. S. and Pearce, J. A.: 1931, *Publ. Dominion Astrophys. Obs. Victoria* **5**, 99.
Plavec, M.: 1968, *Adv. Astron. Astrophys.* **6**, 201.
Plavec, M.: 1976, in A. Slettebak (ed.), 'Be and Shell Stars', *IAU Symp.* **70**, 1.
Plavec, M., Ulrich, R. M. K., and Polidan, R. S.: 1973, *Publ. Astron. Soc. Pacific* **85**, 769.
Poeckert, R.: 1979, *Astrophys. J. Letters* **223**, L73.
Poeckert, R. and Marlborough, J. M.: 1977, *Astrophys. J.* **218**, 220.
Poeckert, R. and Marlborough, J. M.: 1978a, *Astrophys. J.* **220**, 940.
Poeckert, R. and Marlborough, J. M.: 1978a, *Astrophys. J. Suppl.* **38**, 229.

Poeckert, R., Bastien, P., and Landstreet, J. D.: 1979, *Astron. J.* **84**, 812.
Polidan, R. S.: 1976, in A. Slettebak (ed.), 'Be and Shell Stars', *IAU Symp.* **70**, 401.
Polidan, R. S. and Peters, G. J.: 1976, in A. Slettebak (ed.) ,'Be and Shell Stars', *IAU Symp.* **70**, 59.
Polidan, R. S., Oegerle, W. R., Pollard, G. S. G., Sanford, P. W., and Parmar, A. N.: 1979, *Astrophys. J. Letters*, **233**, L7.
Popova, E. I., Tutukov, A. V., and Yungelson, L. R.: 1978, *Nauchn. Inform. Astron. Soviet* **42**, 45.
Popper, D. M.: 1959, *Astrophys. J.* **129**, 647.
Pottasch, S. R.: 1959a, *Ann. Astrophys.* **22**, 310.
Pottasch, S. R.: 1959b, *Ann. Astrophys.* **22**, 394.
Pottasch, S. R.: 1959c, *Ann. Astrophys.* **22**, 412.
Pottasch, S. R.: 1967, *Bull. Astron. Inst. Neth.* **19**, 227.
Pottasch, S. R., Wesselius, P. R., Wu, C.-C., and Van Duinen, R. J.: 1977, *Astron. Astrophys.* **54**, 435.
Praderie, F.: 1973, in S. D. Jordan and E. H. Avrett (eds.), 'Stellar Chromospheres', NASA SP-317, Washington, p. 79.
Praderie, F.: 1976, *Mem. Soc. Astron. Ital.* **47**, 553.
Praderie, F., Talavera, A., and Lamers, H. J. G. L. M.: 1980, *Astron. Astrophys.*, preprint.
Pravdo, S. H., Smith, B. W.: 1979, preprint, NASA Techn. Mem., No. 80558.
Prialnik, D., Shara, M. M., and Shahiv, G.: 1978, *Astron. Astrophys.* **62**, 339.
Price, S. D. and Walker, R. G.: 1976, 'The AFGL Four Colour Sky Survey', Air Force Geophys. Lab. TR-76-0208.
Primini, F., Rappaport, S., and Joss, P. C.: 1977, *Astrophys. J.* **217**, 543.
Pronik, V. I., Chuvaev, K. K., and Chugai, N. N.: 1976, *Astron. Zh.* **53**, 1182 (1977, *Soviet Astron.* **20**, 666).
Proudman, I.: 1952, *Proc. Roy. Soc.* **A214**, 119.
Przybylski, A.: 1965, *Nature* **205**, 163.
Przybylski, A.: 1968 *Monthly Notices Roy. Astron. Soc.* **139** 313.
Pskovskij, Yu. P.: 1968, *Astron. Zh.* **45**, 942.
Pskovskij, Yu. P.: 1970, *Astron. Zh.* **47**, 994 (1971, *Soviet Astron.* **14**, 798).
Pskovskij, Yu. P.: 1977, *Astron. Zh.* **54**, 188 (*Soviet Astron.* **21**, 675).
Pskovskij, Yu. P.: 1978, *Astron. Zh.* **55**, 350 (*Soviet Astron.* **22**, 201).
Puetter, R. C., Russell, R. W., Soifer, B. T., and Willner, S. P.: 1978, *Astrophys. J. Letters* **223**, L93.
Purton, C. R.: 1976, in A. Slettebak (ed.), 'Be and Shell Stars', *IAU Symp.* **70**, 157.
Qi-Bin, L.: 1978, *Acta Astron. Sinica* **19**, 210 (*Chinese Astron.* **3**, 315).
Querci, F., Querci, M., and Kunde, V. G.: 1971, *Astron. Astrophys.* **9**, 1.
Querci, F., Querci, M., and Tsuji, T.: 1974, *Astron. Astrophys.* **31**, 265.
Rafanelli, P. and Rosino, L.: 1978, *Astron. Astrophys. Suppl.* **31**, 337.
Raikova, D. V.: 1977, *Astron. Zh.* **54**, 55 (*Soviet Astron.* **21**, 30).
Rakos, K. D., Schermann, A., Weis, W. W., and Wood, H. J.: 1977, *Astron. Astrophys.* **56**, 453.
Rao, N. K. and Mallik, G. V.: 1978, *Monthly Notices Roy. Astron. Soc.* **183**, 211.
Rappaport, S., Joss, P. C., and Stothers, R.: 1980, *Astrophys. J.* **235**, 570.
Rautela, B. S. and Joshi, S. C.: 1979, *Bull. Astron. Soc. India* **7**, 43.
Raymond, J. C., Cox, D. P., and Smith, B. W.: 1976, *Astrophys. J.* **204**, 290.
Rees, M. J.: 1974, *Highlights in Astronomy* **3**, 89.
Reid, M. J.: 1976, *Astrophys. J.* **207**, 784.
Reid, M. J. and Dickinson, D. F.: 1976, *Astrophys. J.* **209**, 505.
Reid, M. J., Muhleman, D. O., Moran, J. M., Johnston, K. J., and Schwarz, P. R.: 1977, *Astrophys. J.* **214**, 60.
Reimers, D.: 1973, *Astron. Astrophys.* **24**, 79.
Reimers, D.: 1975a, in B. Baschek, W. H. Kegel, and G. Traving (eds.) *Problems in Stellar Atmospheres and Envelopes*, Springer, Berlin, p. 229.
Reimers, D.: 1975b, *Mem. Soc. Roy. Sci. Liège, 6e Série* **VIII**, 369.
Reimers, D.: 1977a, *Astron. Astrophys.* **61**, 217.
Reimers, D.: 1977b, *Astron. Astrophys.* **54**, 485.
Reimers, D.: 1977c, *Astron. Astrophys.* **57**, 395.
Reimers, D.: 1978, in R. Kippenhahn *et al.* (eds.), 'The Interaction of Variable Stars with their Environment', *IAU Colloq.* **42**, 559.
Reimers, D.: 1979, private communication.

Renzini, A., Cacciari, C., Ulmschneider, P., and Schmitz, F.: 1977, *Astron. Astrophys.* **61**, 39.

Richter, G.: 1960, *Astron. Nachr.* **285**, 274.

Ridgway, S. T.: 1974, *Astrophys. J.* **190**, 591.

Ridgway, S. T., Hall, D. N. B., Kleinmann, S. G., Weinberger, D. A., and Wojsław, R. S.: 1976, *Nature Phys. Sci.* **264**, 345.

Ridgway, S. T., Hall, D. N. B., and Carbon, D. F.: 1977a, *Bull. Am. Astron. Soc.* **9**, 636.

Ridgway, S. T., Wells, D. C., and Joyce, R. R.: 1977b, *Astron. J.* **82**, 414.

Ridgway, S. T., Carbon, D. F., and Hall, D. N. B.: 1978, *Astrophys. J.* **225**, 138.

Ridgway, S. T., Wells, D. C., Joyce, R. R., and Allen, R. G.: 1979, *Astron. J.* **84**, 247.

Rieu, N. Q., Laury-Micoulaut, C., Winnberg, A., and Schultz, G. V.: 1979, *Astron. Astrophys.* **75**, 351.

Robbins, R. R. and Sanyal, A.: 1978, *Astrophys. J.* **219**, 985.

Robertson, J. W.: 1972, *Astrophys. J.* **177**, 473.

Robinson, B. J. Caswell J. L., and Goss, W. M.: 1970, *Astrophys. J. Letters* **7**, L79.

Robinson, E. L.: 1975, *Astron. J.* **80**, 515.

Robinson, E. L.: 1976, *Ann. Rev. Astron. Astrophys.* **14**, 119.

Robinson, G., Hyland, A. R., and Thomas, J. A.: 1973, *Monthly Notices Roy. Astron. Soc.* **161**, 281.

Robinson, G. and Hyland, A. R.: 1977, *Monthly Notices Roy. Astron. Soc.* **180**, 495.

Rodgers, A. W. and Searle, L.: 1967. *Monthly Notices Roy. Astron. Soc.* **135**, 99.

Rodriguez, L. F., Moran, J. M., Dickinson, D. F., and Gyulbudaghian, A. L.: 1978, *Astrophys. J.* **226**, 115.

Rogerson, J. B., Spitzer, L., Drake, J. G., Dressler, K., Jenkins, E. B., Norton, D. C., and York, D. G.: 1973, *Astrophys. J. Letters* **181**, L97.

Rogerson, J. B. and Lamers, H. J.: 1975, *Nature Phys. Sci.* **256**, 190.

Rosen, B. R., Moran, J. M., Reid, M. J., Walker, R. C., Burke, B. F., Johnston, K. J., and Spencer, J. H.: 1978, *Astrophys. J.* **222**, 132.

Rosendhal J. D.: 1970 *Astrophys. J.* **159**, 107.

Rosendhal, J. D.: 1973a, *Astrophys. J.* **182**, 523.

Rosendhal, J. D., 1973b, *Astrophys. J.* **186**, 909.

Rosendhal, J. D.: 1974, *Astrophys. J.* **187**, 261.

Rosendhal, J. D. and Wegner, G.: 1970, *Astrophys. J.* **162**, 547.

Rosendhal, J. D. and Snowden, M. S.: 1971, *Astrophys. J.* **169**, 281.

Rosino, L.: 1978, *Astrophys. Space Sci.* **55**, 383.

Rosino, L. and Bianchini, A.: 1973, *Astron. Astrophys.* **22**, 453.

Rosino, L. and Tempesti, P.: 1977, *Astron. Zh.* **54**, 517 (*Soviet Astron.* **21**, 291).

Rosner, R., Tucker, W. H., and Vaiana, G. S.: 1978a, *Astrophys. J.* **220**, 643.

Rosner, R., Golub, L., Coppi, B., and Vaiana, G. S.: 1978b, *Astrophys. J.* **222**, 317.

Rublev, S. V.: 1964, *Astron. Zh.* **41**, 223.

Rublev, S. V.: 1975, in V. E. Sherwood and L. Plaut (eds.), 'Variable Stars and Stellar Evolution', *IAU Symp.* **67**, 259.

Rubbra, F. T. and Cowling, T. G.: 1960, in P. Ledoux (ed.), 'Modèles d'Etoiles et Evolution Stellaire', *Mém.* 8° *Soc. Roy. Sci. Liège,* (*5e Série*) **3**, 274.

Rufener, F., Maeder, A., and Burki, G.: 1978, *Astron. Astrophys. Suppl. Series* **31**, 179.

Ruggles, C. L. N. and Bath, G. T.: 1979, *Astron. Astrophys.* **80**, 97.

Runciman, W., A., Sengupta, D, and Gourley, J. T.: 1973, *Am. Miner.* **58**, 451.

Rush, W. F. and Thompson, R, W.: 1977, *Astrophys. J.* **211**, 184.

Russell, H. N.: 1933, *J. Roy. Astron. Soc. Canada* **27**, 374 411.

Ruusalepp. M. and Luud, L.: 1970, *Tartu Obs. Publ.* **39**, 89.

Rybicki, G. B.: 1970, in H. G. Groth and P. Wellmann (eds.), 'Spsctrum Formation in Stars with Steady State Extended Atmospheres', Nat. Bur. Standards, Special Publ. No. 332, Washington, D.C., p. 87.

Rybicki, G. B. and Hummer, D. G.: 1975, *Monthly Notices Roy. Astron. Soc.* **170**, 423.

Saijo, K. and Saito, M.: 1977, *Publ. Astron. Soc. Japan* **29**, 739.

Saito, M.: 1970, *Publ. Astron. Soc. Japan* **22**, 455.

Saito, M.: 1973, *Astrophys. Space Sci.* **22**, 133.

Saito, M. and Kawabata, S.: 1976, *Astrophys. Space Sci.* **45**, 63.

Sakashita, S. and Hayashi, C.: 1961, *Progr. Theor. Phys.* **26**, 942.

Sakashita, S., Ono, Y., and Hayashi, C.: 1959, *Progr. Theor. Phys.* **22**, 830.

Salpeter, E. E.: 1971, *Ann. Rev. Astron. Astrophys.* **9**, 127.
Salpeter, E. E.: 1974, *Astrophys. J.* **193**, 585.
Sandage, E.: 1972, *Quart. J. Roy. Astr. Soc.* **13**, 302.
Sandage, A. and Tammann, G. A.: 1976, *Astrophys. J.* **210**, 7.
Sanduleak, N.: 1971, *Astrophys. J. Letters* **164**, L71.
Sanford, R. F.: 1950, *Astrophys. J.* **111**, 252.
Sanner, F.: 1976a, *Astrophys. J. Letters* **204**, L41.
Sanner, F.: 1976b, *Astrophys. J. Suppl. Series*, **32**, 115.
Sanner, F.: 1977, *Astrophys. J. Letters* **211**, L35.
Sanyal, A.: 1974, *Astrophys. J. Suppl. Series* **28**, 115.
Sargent, W. L. W.: 1961, *Astrophys. J.* **134**, 142.
Sargent, W. L. W.: 1965, *Observatory* **85**, 33.
Sargent, W. L. W. and Osmer, P. S.: 1969, in M. Hack (ed.), *Mass Loss from Stars*, D. Reidel Publ. Co., Dordrecht, Holland, p. 57.
Savonije, G. J.: 1978a, *Astron. Astrophys.* **62**, 317.
Savonije, G. J.: 1978b, *Astron. Astrophys.* **62**, 317.
Scalo, J. M.: 1976, *Astrophys. J.* **206**, 474.
Scalo, J. M.: 1977, *Astrophys. J.* **215**, 194.
Scalo, J. M. and Ulrich, R. K.: 1973, *Astrophys. J.* **183**, 151.
Scargle, J. D. and Strecker, D. W.: 1979, *Astrophys. J.* **228**, 838.
Scargle, J. D., Erickson, E. F., Witteborn, F. C., and Strecker, D. W.: 1978, *Astrophys. J.* **224**, 527.
Scarle, L.: 1971, *Astrophys. J.* **168**, 41.
Scharmer, G. B.: 1976, *Astron. Astrophys.* **53**, 341.
Schatzman, E.: 1949, *Ann. Astrophys.* **12**, 203.
Schatzman, E.: 1958, *White Dwarfs*, North-Holland Publ. Co., Amsterdam.
Schatzman, E.: 1977, *Astron. Astrophys.* **56**, 211.
Schild, R. E.: 1966, *Astrophys. J.* **146**, 142.
Schild, R. E.: 1976, in A. Slettebak (ed.), 'Be and Shell Stars', *IAU Symp.* **70**, 107.
Schild, R. E.: 1978, *Astrophys. J. Suppl. Series* **37**, 77.
Schild, R. E. and Romanishin, W.: 1976, *Astrophys. J.* **204**, 493.
Schild, R. F.: 1970, *Astrophys. J.* **161**, 855.
Schlesinger, B. M.: 1975, *Astrophys. J.* **199**, 166.
Schmalberger, D. L.: 1960, *Astrophys. J.* **132**, 591.
Schmidt, M.: 1959, *Astrophys. J.* **129**, 243.
Schmidt-Burgk, J. and Scholz, M.: 1977, *Monthly Notices Roy. Astron. Soc.* **179**, 563.
Schmidt-Kahler, Th.: 1965, in H. H. Voigt (ed.), *Landolt-Börnstein, Zahlenwerte und Funktionen*, Springer, Berlin, p. 312.
Schmidt-Kahler, Th.: 1965, in H. H. Voigt (ed.), *Landolt–Börnstein* **VI**, Band 1, 309.
Schmitz, F. and Ulmschneider, P.: 1980a, *Astron. Astrophys.* **84**, 93.
Schmitz, F. and Ulmschneider, P.: 1980b, *Astron. Astrophys.* **84**, 191.
Schneider, D. P. and Greenstein, J. L.: 1979, *Astrophys. J.* **233**, 935.
Schneider, D. P., Darland, J. J., and Leung, K. C.: 1979, *Astron. J.* **84**, 236.
Scholtz, M.: 1972, *Astron. Astrophys. Suppl.* **7**, 469.
Schramm, D. N. (ed.): 1977, *Supernovae*, D. Reidel Publ. Co., Dordrecht, Holland.
Schuerman, D. W.: 1972, *Astrophys. Space Sci.* **19**, 351.
Schultz, G. V., Sherwood, W. A., and Winnberg, A.: *Astron. Astrophys. Letters* **63**, L5.
Schuster, A.: 1879, *Monthly Notices Roy. Astron. Soc.* **40**, 35.
Schwartz, P. R.: 1977, *Publ. Astron. Soc. Pacific* **89**, 693.
Schwartz, P. R.: 1979, in P. S. Conti and C. W. H. de Loore (eds.), 'Mass Loss and Evolution of O-type Stars', *IAU Symp.* **83**, 147.
Schwartz, P. R. and Barrett, A. H.: 1970, *Astrophys. J. Letters* **159**, L123.
Schwartz, P. R. and Spencer, J. H.: 1977, *Monthly Notices Roy. Astron. Soc.* **180**, 297.
Schwarzschild, M.: 1941, *Astrophys. J.* **94**, 245.
Schwarzschild, M.: 1948, *Astrophys. J.* **107**, 1.
Schwarzschild, M.: 1975 *Astrophys. J.* **195**, 137.
Schwarzschild, M. and Härm, R.: 1958, *Astrophys. J.* **128**, 348.
Schwarzschild, M. and Härm, R.: 1959, *Astrophys. J.* **129**, 637.

Schwarzschild, M. and Härm, R.: 1967, *Astrophys. J.* **150**, 961.

Schweizer, F. and Lasker, B. W.: 1978, *Astrophys. J.* **226**, 167.

Seaquist, E. R.: 1976, *Astrophys. J. Letters* **203**, L35.

Seaquist, E. R.: 1977, *Astrophys. J.* **211**, 547.

Seaton, M. J.: 1966, *Monthly Notices Roy. Astron. Soc.* **132**, 113.

Semeniuk, I., Kruszewski, A., Schwarzenberg-Czerny, A., Chlebowski, T., Mikołajewski, M., and Wołczyk, J.: 1977, *Acta Astron.* **27**, 301.

Serkowski, K.: 1970, *Astrophys. J.* **160**, 1083.

Severny, A. B., Kurshinov, V. M., and Nikulin, N. S.: 1974, *Isv. Krimsk. Astrophys. Obs.* **50**, 3.

Seward, F., D. Burginyon, G., Grader, R., Hill, R., Palmieri, T., Stoering, P., and Toor, R.: 1976, *Astrophys. J.* **205**, 238.

Seward, F. D., Forman, W. R., Giacconi, R., Griffiths R. E., Harnden, F. R., Jones, C., and Pye, J. P.: 1979, *Astrophys. J. Letters* **234**, L55.

Shakura, N. I. and Sunyaev, R. A.: 1973, *Astron. Astrophys.* **24**, 337.

Sharov, A. S.: 1975, in V. E. Sherwood and L. Plaut (eds.), 'Variable Stars and Stellar Evolution', *IAU Symp.* **67**, 275.

Shawl, S. J.: 1969, *Astrophys. J. Letters* **157**, L57.

Shenavrin, V. I., Moroz, V. I., and Liberman, A. A.: 1977, *Astron. Zh.* **54**, 629 (*Soviet Astron.* **21**, 358).

Shields, G. A. and Ferland, G. J.: 1978, *Astrophys. J.* **225**, 950.

Shklovskii, I. S.: 1962, *Astron. Zh.* **39**, 209.

Shklovskii, I. S.: 1976, *Astron. Zh.* **53**, 750 (*Soviet Astron.* **20**, 422).

Shobbrook, R. R.: 1978a, *Monthly Notices Roy. Astron. Soc.* **184**, 43.

Shobbrook, R. R.: 1978b, *Monthly Notices Roy. Astron. Soc.* **185**, 825.

Silverglate, P., Zuckermann, B., Terzian, Y., and Wolff, M.: 1979, *Astron. J.* **84**, 345.

Simon, N. R. and Stothers, R.: 1970, *Astron. Astrophys.* **6**, 183.

Simon, R.: 1957, *Bull. Cl. Sci. Acad. Roy. Belgique (5)* **43**, 471, 610.

Simons, S. and Williams, I. P.: 1976, *Astrophys. Space Sci.* **39**, 123.

Simpson, E. E.: 1971, *Astrophys. J.* **165**, 295.

Slettebak, A.: 1976, in A. Slettebak (ed.), 'Be and Shell Stars', *IAU Symp.* **70**, 123.

Slettebak, A.: 1979, *Space Sci. Rev.* **23**, 541.

Slettebak, A. and Reynolds, R. C.: 1978, *Astrophys. J. Suppl.* **38**, 205.

Slettebak, A. and Snow, T. P.: 1978, *Astrophys. J. Letters* **224**, L127.

Smak, J. I.: 1966, *Ann. Rev. Astron. Astrophys.* **4**, 19.

Smith, A. M.: 1970, *Astrophys. J.* **160**, 595.

Smith, L. F.: 1968a, in K. G. Gebbie and R. N. Thomas (eds.), 'Wolf–Rayet Stars', *Natl. Bur. Std., Special Publ.* **307**.

Smith, L. F.: 1968b, *Monthly Notices Roy. Astron. Soc.* **138**, 109.

Smith, L. F.: 1973, in M. K. C. Bappu and J. Sahade (eds.), 'Wolf–Rayet and High Luminosity Stars', *IAU Symp.* **49**, 33 and 237.

Smolinski, J.: 1971, in M. Hack (ed.), *Colloquium on Supergiant Stars*, Osserv. Astron. Trieste, p. 68.

Smolinski, J., Climenhaya, J. L., and Kipper, T. A.: 1976, *Publ. Astron. Soc. Pacific* **88**, 67.

Smolinski, J., Feldman, P. A., and Higgs, L. A.: 1977, *Astron. Astrophys.* **60**, 277.

Snijders, M. A. J. and Lamers, H. J. G. L. M.: 1975, *Astron. Astrophys.* **41**, 245.

Snijders, M. A. J. and Underhill, A. B.: 1975, *Astrophys. J.* **200**, 634.

Snow, T. P.: 1977, *Astrophys. J.* **217**, 760.

Snow, T. P. and Hayes, D. P.: 1978, *Astrophys. J.* **226**, 897.

Snow, T. P. and Marlborough, J. M.: 1976, *Astrophys. J. Letters* **203**, L87.

Snow, T. P. and Morton, D. C.: 1976, *Astrophys. J. Suppl.* **32**, 429.

Snow, T. P. and Jenkins, E. B.: 1977, *Astrophys. J. Suppl.* **33**, 269.

Snow T. P., Peters, G. J., and Mathieu, R. D.: 1979, *Astrophys. J. Suppl.* **39**, 359.

Snow, T. P., Wegner, G. A., and Kunasz, P. B.: 1980, *Astrophys. J.*, preprint.

Snyder, L. E. and Buhl, D.: 1974, *Astrophys. J. Letters* **189**, L31.

Snyder, L. E. and Buhl, D.: 1975, *Astrophys. J.* **197**, 329.

Snyder, L. E., Dickinson, D. F., Brown, L. W., and Buhl, D.: 1978, *Astrophys. J.* **224**, 512.

Snyder, W. A., Davidsen, A. F., Henry, R. C., Shulman, S., Fritz, G., and Friedman, H.: 1978, *Astrophys. J. Letters* **222**, L13.

Sobolev, V. V.: 1947a, *Astron. Zh.* **24**, 1.

Sobolov, V. V.: 1947b, *Dvizhushchiesya Obolochki Zvezd*, Leningrad State Univ. (translated by S. Gaposhkin, *Moving Envelopes of Stars*, Harvard Univ. Press, Cambridge, Mass., 1960).

Sobolov, V. V.: 1962 *Astron. Zh.* **39**, 632 (*Soviet Astron.* **6**, 531).

Soderblom, D.: 1976, *Publ. Astron. Soc. Pacific* **88**, 517.

Solf, J.: 1978, *Astron. Astrophys. Suppl.* **34**, 409.

Souffrin, P.: 1966, *Ann. Astrophys.* **29**, 55.

Sparks, W. M.: 1969, *Astrophys. J.* **156**, 569.

Sparks, W. M., Starrfield, S., and Truran, J. W.: 1977, in M.Friedjung (ed.), *Novae and Related Stars*, D. Reidel Publ. Co., Dordrecht, Holland, p. 189.

Sparks, W. M., Starrfield, S., and Truran, J. W.: 1978, *Astrophys. J.* **220**, 1063.

Spencer, J. H., Johnston, K. J., Moran, J. M., Reid, M. J., and Walker, R. C.: 1979, *Astrophys. J.* **230**, 449.

Spinrad, H. and Vardya, M. S.: 1966, *Astrophys. J.* **146**, 399.

Spinrad, H. and Wing, R. F.: 1969, *Ann. Rev. Astron. Astrophys.* **7**, 249.

Sreenivasan, S. R. and Ziebarth, K. E.: 1974, *Astrophys. Space Sci.* **30**, 57.

Sreenivasan, S. R. and Wilson, W. J. F.: 1978a, *Astrophys. Space Sci.* **53**, 193.

Sreenivasan, S. R. and Wilson, W. J. F.: 1978b, *Astron. Astrophys.* **70**, 755.

Stalio, R.: 1972, in M. Hack (ed.), *Colloquium on Supergiant Stars*, Osservatorio Astronomico Trieste, p. 28.

Stark, A. A. and Blitz, L.: 1978, *Astrophys. J. Letters* **225**, L15.

Starrfield, S., Truran, J. W., Sparks, W. M., and Kutter, G. S.: 1972, *Astrophys. J.* **176**, 169.

Starrfield, S., Truran, J. W. and Sparks, W. M.: 1977, in J. Audouze (ed.), *CNO-Isotopes in Astrophysics*, D. Reidel Publ. Co., Dordrecht, Holland, p. 49.

Stecher, T. P.: 1970, *Astrophys. J.* **159**, 543.

Stein, R. F.: 1967, *Solar Phys.* **2**, 385.

Stein, R. F.: 1968, *Astrophys. J.* **154**, 297.

Stein, R. F. and Leibacher, J.: 1974, *Ann. Rev. Astron. Astrophys.* **12**, 407.

Stellingwerf, R. F.: 1978, *Astron. J.* **83**, 1184.

Stencel, R. E.: 1977, *Astrophys. J.* **215**, 176.

Stencel, R. E.: 1978, *Astrophys. J. Letters* **223**, L37.

Stencel, R. E.: 1978, in A. G. D. Philip and D. S. Hayes (eds.), 'The Hertzsprung–Russell Diagram', *IAU Symp.* **80**, 59.

Stencel, R. E.: 1979, in D. F. Gray (ed.), 'Turbulence in Stellar Armospheres', *IAU Colloq.* **51**.

Stencel, R. E., Kondo, Y., Bernat, A. P., and McCluskey, G. E.: 1979, *Astrophys. J.* **233**, 621.

Stephens, J. R. and Russell, R. W.: 1979, *Astrophys. J.* **228**, 780.

Stephenson, C. B.: 1973, *Publ. Warner Swasey Obs.* **1**, No. 4.

Stephenson, C. B.: 1976, *Publ. Warner Swasey Obs.* **2**, No. 2.

Stephenson, D. J.: 1979, *Monthly Notices Roy. Astron. Soc.* **187**, 129.

Stephenson, F. R.: 1974, in C. B. Cosmovici (ed.), *Supernovae and Supernovae Remnants*, D. Reidel Publ. Co., Dordrecht, Holland, p. 75.

Stephenson, F. R.: 1976, *Quart. J. Roy. Astron. Soc.* **17**, 97.

Sterken, C.: 1976a, *Astron. Astrophys.* **47**, 453.

Sterken, C.: 1976b, 'Variabiliteit van Extreme Galaktische B- en A-Superreuzen', Thesis, Vrije Universiteit, Brussels.

Sterken, C.: 1977, *Astron. Astrophys.* **57**, 361.

Sterken, C. and Wolf, B.: 1978a, *Astron. Astrophys.* **70**, 641.

Sterken, C. and Wolf, B.: 1978b, *Astron. Astrophys. Suppl.* **35**, 69.

Stickland, D. J. and Harmer, D. L.: 1978, *Astron. Astrophys. Letters* **70**, L53.

Stift, M. J.: 1978, *Astron. Astrophys. Suppl.* **32**, 343.

Stone, R. P. S.: 1977, *Publ. Astron. Soc. Pacific* **89**, 155.

Stothers, R.: 1970, *Monthly Notices Roy. Astron. Soc.* **151**, 65.

Stothers, R.: 1972a, *Astrophys. J.* **175**, 431.

Stothers, R.: 1972b, *Astrophys. J.* **175**, 717.

Stothers, R.: 1972c, *Astron. Astrophys.* **18**, 325.

Stothers, R.: 1973a, *Publ. Astron. Soc. Pacific* **85**, 363.

Stothers, R.: 1973b, *Astrophys. J.* **184**, 181.

Stothers, R.: 1974, *Astrophys. J.* **192**, 145.

Stothers, R.: 1976, *Astrophys. J.* **209**, 800.

Stothers, R. and Chin, C.: 1968, *Astrophys. J.* **152**, 225.

Stothers, R. and Chin, C.: 1973, *Astrophys. J.* **179**, 555.

Stothers, R. and Chin, C.: 1975, *Astrophys. J.* **198**, 407.

Stothers, R. and Chin, C.: 1976, *Astrophys. J.* **204**, 472.

Stothers, R. and Chin, C.: 1977, *Astrophys. J.* **211**, 189.

Stothers, R. and Chin, C.: 1978, *Astrophys. J.* **226**, 231.

Stothers, R. and Chin, C.: 1979, *Astrophys. J.* **233**, 267.

Stothers, R. and Leung, K. C.: 1971, *Astron. Astrophys.* **10**, 290,

Stothers, R. and Simon, N.: 1967a, *Astrophys. J.* **152**, 233.

Stothers, R. and Simon, N.: 1967b, *Astrophys. J.* **157**, 673.

Stothers, R. and Simon, N.: 1970, *Astrophys. J.* **160**, 1019.

Strel'nitskii, V. S.: 1977, *Astron. Zh.* **54**, 674 (*Soviet Astron.* **21**, 381).

Struve, O.: 1931, *Astrophys. J.* **73**, 94.

Struve, O. and Elvey, C. T.: 1934, *Astrophys. J.* **79**, 409.

Struve, O. and Swings, P.: 1940, *Astrophys. J.* **91**, 546.

Suchkov, A. A. and Shchekinov, Yu. A.: 1977, *Pis'ma Astron. Zh.* **3**, 546 (*Soviet Astron. Letters* **3**, 297).

Suess, H. E. and Urey, H. C.: 1956, *Rev. Mod. Phys.* **28**, 53.

Sugimoto, D. and Nomoto, K.: 1980, *Space Sci. Rev.* **25**, 155.

Surdej, J.: 1979, *Astron. Astrophys.* **73**, 1.

Sutton, E., Becklin, E. E., and Neugebauer, G.: 1974, *Astrophys. J. Letters* **190**, L69.

Sutton, E. C., Storey, J. W. V., Betz, A. L., Townes, C. H., and Spears, D. L.: 1977, *Astrophys. J. Letters* **217**, L97.

Sutton, E. C., Storey, J. W. V., and Townes, C. H.: 1978, *Astrophys. J. Letters* **224**, L123.

Sutton, E. C., Betz, A. L., Storey, J. W., and Spears, D. L.: 1979, *Astrophys. J. Letters* **230**, L105.

Sweigart, A. V.: 1974, *Astrophys. J.* **189**, 289.

Sweigart, A. V. and Gross, P. G.: 1978, *Astrophys. J. Suppl. Series* **36**, 405.

Sweigart, A. V. and Mengel, J. G.: 1979, *Astrophys. J.* **229**, 624.

Swings, J. P.: 1973, in 'Les Nébuleuses Planétaires', *Mém. Soc. Roy. Sci. Liège*, *Coll.* 8°, *6e Série* **5**, 321.

Swings, J. P.: 1976, in A. Slettebak (ed.), 'Be and Shell Stars', *IAU Symp.* **70**, 219.

Swings, P. and Struve, O.: 1941a, *Astrophys. J.* **94**, 291.

Swings, P. and Struve, O.: 1941b, *Proc. Acad. Washington* **27**, 225.

Swings, J. P. and Klutz, M.: 1976, *Astron. Astrophys* **46**, 303.

Szkody, P., Dyck, H. M., Capps, R. W., Becklin, E. E., and Cruikshank, D. P.: 1979, *Astron. J.* **84**, 1359.

Takada, M.: 1977, *Publ. Astron. Soc. Japan* **29**, 439.

Talbot, R. J., 1971a, *Astrophys. J.* **163**, 17.

Talbot, R. J., 1971b, *Astrophys. J.* **165**, 121.

Tammann, G. A.: 1974, in C. B. Cosmovici (ed.), *Supernovae and Supernova Remnants*, D. Reidel Publ. Co., Dordrecht, Holland, p. 155.

Tammann, G. A.: 1977a, in D. N. Schramm (ed.), *Supernovae*, D. Reidel Publ. Co., Dordrecht, Holland, p. 95.

Tammann, G. A.: 1977b, in *Eight Texas Symposium on Relativistic Astrophysics*, Ann. N.Y. Acad. Sci.

Tammann, G. A.: 1978, *Mem. Soc. Astron. Ital.* **49**, 315.

Tammann, G. A. and Sandage, A.: 1968, *Astrophys. J.* **151**, 285.

Tanaka, Y.: 1966a, *Publ. Astron. Soc. Japan* **18**, 47.

Tanaka, Y.: 1966b, *Progr. Theor. Phys. Kyoto* **36**, 844.

Tananbaum. H. and Tucker, W. H.: 1974, in R. Giaccoci and H. Gursky (eds.), *X-Ray Astronomy*, D. Reidel Publ. Co., Dordrecht, Holland, p. 207.

Tanzi, E. G., Treves, A., Salinari, G., and Tarenghi, M.: 1979, *Astron. Astrophys.* **78**, 226.

Tarafdar, S. P., Krishna Swamy, K. S., and Vardya, M. S.: 1980, preprint.

Taylor, K. and Münch, G.: 1978, *Astron. Astrophys.* **70**, 359.

Terzian, T., Balick, B., and Bignell, C.: 1974, *Astrophys. J.* **188**, 257.

Thackeray A. D.: 1950, *Monthly Notices Roy. Astron. Soc.* **110**, 524.

Thackeray, A. D.: 1973, *IAU Circ.* **2584**.

Thackery, A. D.: 1977, *Monthly Notices Roy. Astron. Soc.* **180**, 95.

Thackeray, A. D.: 1977b, *Mem. Roy. Astron. Soc.* **83**, 1.

Thackeray, A. D. and Emerson, B.: 1969, *Monthly Notices Roy. Astron. Soc.* **142**, 429.

Thackeray, A. D. and Velasco, R.: 1976, *Observatory* **96**, 104.

The, P. S. and Vleeming, G.: 1971, *Astron. Astrophys.* **14**, 120.

The, P. S., Tjin A. Djie, H. R. E., Kudritzki, R. P., and Wesselius, P. R.: 1980a, preprint.

The, P. S., Tjin A. Djie, H. R. E., and Wamsterker, W.: 1980b, *Astron. Astrophys.* **84**, 263.

Thomas, H. C.: 1977, *Ann. Rev. Astron. Astrophys.* **15**, 127.

Thomas, J., A. Hyland, A. R., and Robinson, G.: 1976, *Monthly Notices Roy. Astron. Soc.* **174**, 711.

Thomas, R. N.: 1973, *Astron. Astrophys.* **29**, 297.

Thompson, G. I., Nandy, K,. Jamar, C., Monfils, A., Houziaux, L., Carnochan, D. J., and Wilson, R.: 1978, *Catalogue of Stellar Ultraviolet Fluxes*, The Science Research Council, London.

Thompson, R. I.: 1977, *Astrophys. J.* **212**, 754.

Thompson, R. I., Erickson, E. F., Witteborn, F. C., and Strecker, D. W.: 1976, *Astrophys. J. Letters* **210**, L31.

Thornton, H. A. and Harvey, P. M.: 1979, *Astrophys. J. Letters* **229**, L133.

Tifft, W. G. and Snell, Ch. M.: 1971, *Monthly Notices Roy. Astron. Soc.* **151**, 365.

Tinsley, B. M.: 1975, *Publ. Astron. Soc. Pacific* **87**, 837.

Tinsley, B. M.: 1977, in D. N. Schramm (ed.), *Supernovae*, D. Reidel Publ. Co., Dordrecht, Holland, p. 117.

Tjin A Djie, R. R. E. and Thé, P. S.: 1978, *Astron. Astrophys.* **70**, 311.

Tomkin, J. and Lambert, D. L.: 1974, *Astrophys. J.* **193**, 631.

Toombs, R. I., Becklin, E. E., Frogel, J. A., Law, S. K., Porter, F. C., and Westphal, J. A.: 1972, *Astrophys. J. Letters* **173**, L71.

Toomre, J., Zahn, J. P., Latour, J., and Spiegel, E. A.: 1976, *Astrophys. J.* **207**, 545.

Torres-Peimbert, S. and Peimbert, M.: 1976, *Bull. Am. Astron. Soc.* **8**, 535.

Treffers, R. and Cohen, M.: 1974, *Astrophys. J.* **188**, 545.

Trimble, V. and Sackmann, I. J.: 1978, *Monthly Notices Roy. Astron. Soc.* **182**, 97.

Trivedi, B. M. P.: 1978, *Astrophys. J.* **225**, 209.

Troland, T. A. Heiles, C., Johnson, D. R., and Clark, F. O.: 1979, *Astrophys. J.* **232**, 143.

Tsuji, T.: 1971, *Publ. Astron. Soc. Japan*, **23**, 275.

Tsuji, T.: 1976a, *Publ. Astron. Soc. Japan* **28**, 543.

Tsuji, T.: 1976b, *Publ. Astron. Soc. Japan* **28**, 567.

Tsuji, T.: 1978a, *Publ. Astron. Soc. Japan* **30**, 435.

Tsuji, T.: 1978b, *Astron. Astrophys. Letters* **68**, L23.

Tsuji, T.: 1979a, preprint.

Tsuji, T.: 1979b, *Publ. Astron. Soc. Japan* **31**, 43.

Tuchman, Y., Sack, N., and Barkat, Z.: 1979, *Astrophys. J.* **234**,, 217.

Tuohy, I. R., Mason, K. O., Clark, D. H., Cordova, F., Charles, P. A., Walter, F. M., and Garmire, G. P.: 1979, *Astrophys. J. Letters* **230**, L27.

Tutukov, A. V. and Yungelson, R. L.: 1971, *Nauchn. Inform. Astron. Soviet* **20**, 86.

Tutukov, A. V. and Yungelson, L. R.: 1973, *Nauchn. Inform. Astron. Soviet* **27**, 58.

Tutukov, A. V., Yungelson, R. L., and Klayman, A. J.: 1973, *Nauchn. Inform. Astron. Soviet* **27**, 3.

Tylenda, R.: 1978, *Acta Astron.* **28**, 333.

Uesugi, A. and Fukuda, I.: 1970, *Contr. Inst. Astrophys. Kwasan Obs. Kyoto* **189**.

Ulmschneider, P.: 1967, *Z. Astrophys.* **67**, 193.

Ilmschneider, P.: 1971a, *Astron. Astrophys.* **12**, 297.

Ulmschneider, P.: 1971b, *Astron. Astrophys.* **14**, 275.

Ulmschneider, P.: 1974, *Solar Phys.*, **39**, 327.

Ulmschneider, P.: 1979, *Space Sci. Rev.* **24**, 71.

Ulmschneider, P., Schmitz, F., Renzini, A., Cacciari, C., Kalkofen, W., and Kurucz, R.: 1977, *Astron. Astrophys.* **61**, 515.

Ulrich, R. K. and Burger, M. L.: 1976, *Astrophys. J.* **206**, 509.

Underhill, A. B.: 1966, *The Early Type Stars*, D. Reidel, Publ. Co., Dordrecht, Holland.

Underhill, A. B.: 1968, *Ann. Rev. Astron. Astrophys.* **6**, 39.

Underhill, A. B.. 1969, in N. Hack (ed.), *Mass Loss from Stars*, D. Reidel Publ. Co., Dordrecht, Holland, p. 17.

Underhill, A. B.: 1973, in M. K. V. Bappu and J. Sahade (eds.), 'Wolf–Rayet and High Luminosity Stars', *IAU Symp.* **49**, 237.

Underhill, A. B.: 1978, in A. Reiz and T. Andersen (eds.), *Astronomical Papers Dedicated to Bengt Strömgren*, Copenhagen Univ. Obs., p. 155.

Underhill, A. B.: 1979, *Astrophys. J.* **234**, 528.

Underhill, A. B. and Faney, R. P.: 1973, *Astrophys. J. Suppl.* **25**, 463.

Underhill, A. B., Divan, L., Prévot-Burnichon, M. L., and Doazan, B.: 1979, *Monthly Notices Roy. Astron. Soc.* **189**, 601.

Unno, W. and Kondo, M.: 1978, *Publ. Astron. Soc. Japan* **29**, 693.

Unsöld, A.: 1955, *Physik der Sternatmosphären*, 2nd. ed., Springer, Berlin.

Utrobin, V. P.: 1978, *Astrophys. Space Sci.* **55**, 441.

Utsumi, K.: 1970, *Publ. Astron. Soc. Japan* **22**, 93.

Vaiana, G. and Rosner, R.: 1978, *Ann. Rev. Astron. Astrophys.* **16**, 393.

Vanbeveren, D.: 1977a, *Astron. Astrophys.* **54**, 877.

Vanbeveren, D.: 1977b, *Astron. Astrophys.* **54**, 877.

Vanbeveren, D.: 1978, *Astrophys. Space Sci.* **57**, 41.

Vanbeveren, D. and Conti, P S : 1980, preprint.

Vanbeveren, D. and De Grève, J. P.: 1979, *Astron. Astrophys.* **77**, 295.

Vanbeveren, D. and De Loore, C.: 1980, preprint.

Vanbeveren, D. and Packet, W.: 1979, *Astron. Astrophys.* **80**, 242.

Vanbeveren, D., De Grève, J. P., Van Dessel, E. L., and De Loore, C.: 1979, *Astron. Astrophys.* **73**, 19.

Van Blerkom, D.: 1971, *Astrophys. J.* **166**, 343.

Van Blerkom, D.: 1978a, *Astrophys. J.* **221**, 186.

Van Blerkom, D.: 1978b, *Astrophys. J.* **223**, 835.

Van Bueren, H. G.: 1973, *Astron. Astrophys.* **23**, 247.

Van Citters, G. W. and Morton, D. C.: 1970, *Astrophys. J.* **161**, 695.

Van de Hulst, H. C.: 1950, *Bull. Astron. Inst. Neth.* **11**, 135.

Van de Kamp, P.: 1975, *Ann. Rev. Astron. Astrophys.* **13**, 295.

Van de Kamp, P.: 1977, *Astron. J.* **82**, 750.

Van de Kamp, P.: 1978, *Astron. J.* **83**, 975.

Van den Bergh, S.: 1974, *Highlights of Astronomy* **3**, 559.

Van den Bergh, S.: 1975, *Astrophys. Space Sci.* **38**, 447.

Van den Bergh, S. 1978, *Sky Telesc.* **55**, 196.

Van den Bergh, S. and Kamper, K. W.: 1977, *Astrophys. J.* **218**, 617.

Van den Bergh, S., Marscher, A. P., and Terzian, Y.: 1973, *Astrophys. J. Suppl.* **26**, 19.

Van den Heuvel, E. P. J.: 1968, *Bull. Astron. Inst. Neth.* **19**. 309,

Van den Heuvel, E. P. J.: 1973, *Nature Phys. Sci.* **242**, 71.

Van den Heuvel, E. P. J.: 1975a, *Astrophys. J. Letters* **198**, L109.

Van den Heuvel, E. P. J.: 1975b, in *Astrophysics and Gravitation*, Proc. 16th Solvay Conf., Brussels, p. 119.

Van den Heuvel, E. P. J.: 1976, in P. Eggleton, S. Metton, J. Whelan, (eds.) 'Structure and Evolution of Close Binary Systems', *IAU Symp.* 73, 35.

Van den Heuvel, E. P. J. and Heise, J.: 1972, *Nature Phys. Sci.* **239**, 67.

Van den Heuvel, E. P. J. and De Loore, C.: 1973, *Astron. Astrophys.* **25**, 387.

Van der Hucht, K. A. Lamers, H. G. J. L. M., Faraggiana, R., Hack, M., and Stalio, R.: 1976, *Astron. Astrophys. Suppl.* **25**, 65.

Van der Hucht, K. A., Stencel, R. E., Haisch, B. M., and Kondo, Y.: 1979a, *Astron. Astrophys. Suppl.* **36**, 377.

Van der Hucht, K. A., Cassinelli, J. P., Wesselius, P. R., and Wu, C.-C.: 1979b, *Astron. Astrophys. Suppl.* **38**, 279.

Van der Hucht, K. A., Bernat, A. P., and Kondo, Y.: 1980, *Astron. Astrophys.*, preprint.

Van der Hucht, K. A., Conti, P. S., Lundström, I., and Stenholm, B.: 1980, *Space Sci. Rev.*, preprint.

Van Duinen, R. J., Aalders, J. W. G., Wesselius, P. R., Wildeman, K. J., Wu, C. C., Luinge, W., and Snel, D.: 1975, *Astron. Astrophys.* **39**, 159.

Van Genderen, A. M.: 1974, *Inf. Bull. Variable Stars* **877**.

Van Genderen, A. M.: 1979, *Astron. Astrophys. Suppl.* **38**, 151.

Van Genderen, A. M.: 1980a, *Astron. Astrophys.*, preprint.

Van Genderen, A. M.: 1980b, *Astron. Astrophys.*, preprint.

Van Helden, R.: 1972, *Astron. Astrophys.* **21**, 209.

Van Paradijs, J.: 1972, *Nature Phys. Sci.* **238**, 37.

Van Paradijs, J.: 1973a, *Astron. Astrophys. Suppl.* **11**, 25.

Van Paradijs, J.: 1973b, *Astron. Astrophys* **23**, 369.

Van Paradijs, J.: 1976, *Astron. Astrophys.* **49**, 53.

Van Paradijs, J. and De Ruiter, H.: 1972, *Astron. Astrophys.* **20**, 169.

Van Paradijs, J. and Zuiderwijk, E. J.: 1977, *Astron. Astrophys. Letters* **61**, L19.

Van Paradijs, J., Hammerschlag-Hensberge, G., Van den Heuvel, E. P. J., Takens, P. J., Zuiderwijk, E. J., and De Loore, C.: 1976, *Nature* **259**, 547.

Van Paradijs, J., Zuiderwijk, E. J., Takens, R. J., Hammerschlag, G., Van den Heuvel, E. P. J., and De Loore, C.: 1977, *Astron. Astrophys. Suppl. Series* **30**, 195.

Van Paradijs, J. A., Hammerschlag, J. A., Hammerschlag-Hensberge, G., and Zuiderwijk, E. J.: 1978, *Astron. Astrophys. Suppl. Series* **31**, 189.

Van Riper, K. and Arnett, W. D.: 1978, *Astrophys. J. Letters* **225**, L129.

Van Tend, W.: 1980, *Solar Phys.* **66**, 29.

Vardya, M. S.: 1972, in M. Hack (ed.), *Colloquium on Supergiant Stars*, Osservatorio Astronomico, Trieste, p. 207.

Vardya, M. S.: 1976, *Astrophys. Space Sci.* 41, L1.

Vardya, M. S.: 1977, *Publ. Astron. Soc. Pacific* **89**, 811.

Vardya, M. S. and Kandel, R.: 1967, *Ann. Astrophys.* **30**, 111.

Varshalovich, D. A. and Khersonskij, V. K.: 1978, *Astron.Zh* **55**, 328 (*Soviet Astron.* **22**, 192).

Varshavskyj, V. I. and Tutukov, A. V.: 1972, *Sci. Inform. Astron. Council U.S.S.R.* **32**, 32.

Vaughan, A. H. and Skumanich, A.: 1970, in H. G. Groth and P. Wellmann (eds.), *Spectrum formations with Steady State Extended Atmospheres*, N.B.S. Special Publ. No. 332, p. 295.

Vermue, J. and De Jager, C.: 1979, *Astrophys. Space Sci.* **61**, 129.

Vermury, S. K. and Stothers, R.: 1977, *Astrophys. J.* **214**, 809.

Viotti, R.: 1969, *Astrophys. Space Sci.* **5**, 323.

Viotti, R.: 1971, *Publ. Astron. Soc. Pacific* **83**, 170.

Viotti, R. and Nesci, R.: 1974, *Inf. Bull. Variable Stars* **878**.

Vogt, H.: 1926, *Astron. Nachr.* **226**, 301.

Voloshina, I. B., Glushneva, I. N., and Doroshenko, V. T.: 1977, *Astron. Zh.* **54**, 541 (*Soviet Astron.* **21**, 306).

Vorontsov-Vel'yaminov, B. A.: 1948, *Gazovye Tumannosti i novye zvezdi*, Chapter 9; Isdat. Akad, Nauk., S.S.S.R.

Vrba, F. J., Schmidt. G. D., and Burke, E. W.: 1977, *Astrophys. J.* **211**, 480.

Wackering, L. R.: 1970, *Mem. Roy. Astron. Soc.* **73**, 153.

Wade, C. M. and Hjelming, R. M.: 1971, *Astrophys. J. Letters* **163**, L105.

Walborn, N. R.: 1971a, *Astrophys. J. Letters* **164**, L67.

Walborn, N. R.: 1971b, *Astrophys. J. Suppl.* **23**, 257.

Walborn, N. R.: 1971c, *Astrophys. J. Letters* **167**, L31.

Walborn, N. R.: 1972a, *Astrophys. J. Letters* **176**, L119.

Walborn, N. R.: 1972b, *Astrophys. J. Letters* **177**, 312.

Walborn, N. R.: 1972c, *Astron. J.* **77**, 312.

Walborn, N. R.: 1973a, *Astrophys. J.* **179**, 517.

Walborn, N. R.: 1973b, *Astrophys. J. Letters* **179**, L123.

Walborn, N. R.: 1973c, *Astrophys. J. Letters* **180**, L35.

Walborn, N. R.: 1973d, *Astron. J.* **78**, 1067.

Walborn, N. R.: 1974, *Astrophys. J.* **189**, 269.

Walborn, N. R.: 1975, *Astrophys. J. Letters* **202**, L129.

Walborn, N. R.: 1976a, *Astrophys. J.* **205**, 419.

Walborn, N. R.: 1976b, *Astrophys. J. Letters* **204**, L17.

Walborn, N. R.: 1977, *Astrophys. J.* **215**, 53.

Walborn, N. R.: 1978a, in A. G. D. Philips and D. S. Hayes (eds.), 'The Hertzsprung–Russell Diagram', *IAU Symp.* **80**, 5.

Walborn, N. R.: 1978b, *Astrophys. J. Letters* **224**, L133.

Walborn, N. R. and Liller, M. H.: 1977, *Astrophys. J.* **211**, 181.

Walborn, N. R., Blanco, B. M., and Thackeray, A. D.: 1978, *Astrophys. J.* **219**, 498.

Walker, A. R., Wild, P. A. T., and Byrne, P. B.: 1979, *Monthly Notices Roy. Astron. Soc.* 189 455.

Walker, M. F.: 1956, *Astrophys. J.* **123**, 68.

Walker, M. F.: 1958, *Astrophys. J.* **127**, 319.

Walker, R. C., Johnston, K. J., Burke, B. F., and Spencer, J. H.: 1977, *Astrophys. J. Letters* **211**, L135.

Walker, R. C., Burke, B. F., Haschick, A. D., Crane, P. C., Moran, J. M., Johnston, K. J., Lo, K. Y., Yen, J. L., Broten, N. W., Legg, T. H., Greisen, E. W., and Hanssen, S. S.: 1978, *Astrophys. J.* **226**, 95.

Walker, R. G. and Price, S. D.: 1975, 'AFCRL Infrared Sky Survey', Air Force Cambridge Res. Lab. TR-0373.

Wallerstein, G.: 1973, *Ann. Rev. Astron. Astrophys.* **11**, 15.

Wallerstein, G.: 1977, *Astrophys. J.* **211**, 170.

Wallerstein, G.: 1978, *Observatory* **98**, 224.

Walraven, Th. and J.: 1971, in A. B. Muller (ed.), *The Magellanic Clouds*, D. Reidel Publ. Co., Dordrecht, Holland, p. 117.

Wamsteker, W.: 1979, *Astron. Astrophys.* **76**, 226.

Wan, F. S. and Van der Borght, R.: 1966, *Australian J. Phys.* **19**, 467.

Wannier, P. G., Leighton, R. B., Knapp, G. R., Redman, R. O., Phillips, T. G., and Huggins, P. J.: 1979, *Astrophys. J.* **230**, 149.

Wares, G. W., Ross, J. E., and Aller, L. H.: 1968, *Astrophys. Space Sci.* **2**, 344.

Warner, B.: 1969, *Monthly Notices Roy. Astron. Soc.* **144**, 333.

Warner, B.: 1974, *Monthly Notices Roy. Astron. Soc. South Africa* **33**, 21.

Warner, B.: 1976, in P. Eggleton *et al.* (eds.), 'Structure and Evolution of Close Binary Systems', *IAU Symp.* **73**, 85.

Warner, J. W. and Wing, R. F.: 1977, *Astrophys. J.* **218**, 105.

Warren, P. R.: 1973a, *Monthly Notices Roy. Astron. Soc.* **161**, 427.

Warren, P. R.: 1973b, *Monthly Notices Roy. Astron. Soc.* **163**, 337.

Warren-Smith, R. F., Scarrott, S. M., Murdin, P., and Bingham, R. G.: 1979, *Monthly Notices Roy. Astron. Soc.* **187**, 761.

Watanabe, T. and Kodaira, K.: 1978, *Publ. Astron. Soc. Japan* **30**, 21.

Watanabe, T. and Kodaira, K.: 1979, *Publ. Astron. Soc. Japan* **31**, 61.

Watson, W. D.: 1974, in R. Balian, P. Encrenaz, J. Lequeux (eds.), *Physiqie Atomique et Moléculaire et Matière Interstellaire*, North-Holland, Amsterdam, p. 117.

Weaver, H.: 1974, *Highlights of Astronomy* **3**, 509.

Weaver, T. A.: 1976, *Astrophys. J. Suppl. Series* **32**, 233.

Weaver, T. A., Zimmerman, G. B., and Woosley, S. E.: 1978, *Astrophys. J.* **225**, 1021.

Webbink, R. F.: 1975, Ph.D. Dissertation, Cambridge, England.

Wegner, G. A. and Snow, T. P.: 1978, *Astrophys. J. Letters* **226**, L25.

Weidemann, V.: 1967, *Z. Astrophys.* **67**, 286.

Weidemann, V.: 1968, *Ann. Rev. Astron. Astrophys.* **6**, 351.

Weidemann, V.: 1977, *Astron. Astrophys.* **61** L27.

Weiler, E. J. and Bahng, J. D. R.: 1976, *Monthly Notices Roy. Astron. Soc.* **174**, 563.

Weiler, E. J. and Oegerle, W. R.: 1979, *Astrophys. J. Suppl.* **39**, 537.

Weiler, K. W. and Wilson, A. S.: 1977, in D. N. Schramm (ed.), *Supernovae*, D. Reidel Publ. Co., Dordrecht, Holland, p. 67.

Wendker, H. J., Baars, J. M. W., and Altenhof, W. J.: 1973, *Nature Phys. Sci.* **245**, 118.

Wendker, H. J., Smith, L. F., Israel, F. P., Habing, H. J., and Dickel, H. R.: 1975, *Astron. Astrophys.* **42**, 173.

Wesselink, A. J.: 1969, *Monthly Notices Roy. Astron. Soc.* **144**, 297.

Weymann, R.: 1960, *Astron. J.* **65**, 503.

Weymann, R.: 1962a, *Ann. Rev. Astron. Astrophys.* **1**, 97.

Weymann, R.: 1962b, *Astrophys. J.* **136**, 844.

Wheeler, J. C.: 1978a, *Astrophys. J.* **225**, 212.

Wheeler, J. C.: 1978b, *Mem. Soc. Astron. Ital.* **49**, 349.

Whitaker, W. E.: 1963, *Astrophys. J.* **137**, 914.

White, G. J.: 1979, *Monthly. Notices Roy. Astron. Soc.* **186**, 377.

White, G. J., Macdonald, G. H.: 1979, *Monthly Notices Roy. Astron. Soc.* **188**, 745.

White, N. E.: 1978, *Nature*, **271**, 39.

White, N. E. and Davidson, P. J.: 1977, *IAU Circ.* **3095**.

White, N. E. and Pravdo, S. H.: 1979, NASA Techn. Mem., No. 80327.

White, N. M. and Wing, R. F.: 1978, *Astrophys. J.* **222**, 209.

Whitney, C. A.: 1952, *Harvard Bull.* **921**, 8.

Wickramasinghe, N. C.: 1975, *Monthly Notices Roy. Astron. Soc.* **170**, 11P.

Wildley, R. L.: 1964, *Astrophys. J. Suppl.* **8**, 439.

Wilkerson, M. S. and Worden, S. P.: 1977, *Astron. J.* **82**, 642.

Williams, I. P.: 1967, *Monthly Notices Roy. Astron. Soc.* **136**, 341.

Williams, I. P.: 1969, in M. Hack (ed.), *Mass Loss from Stars*, D. Reidel Publ. Co., Dordrecht, Holland, p. 139.

Williams, P. M. and Antonopoulou, E.: 1979, *Monthly Notices Roy. Astron. Soc.* **187**, 183.

Williams, P. M., Beattie, D. H., Lee, T. J., Steward, J. M., and Antonopoulou, E.: 1978, *Monthly Notices Roy. Astron. Soc.* **185**, 467.

Williams, R. E., Woolf, N. J., Hege, E. K., Moore, R. L., and Kopriva, D. A.: 1978, *Astrophys. J.* **224**, 171.

Willis, A. J.: 1979, *The First Year of IUE*, Univ. College of London.

Willis, A. J. and Stickland, D. J.: 1980, *Monthly Notices Roy. Astron. Soc.* **190**, 27P.

Willis, A. J. and Wilson, R.: 1978, *Monthly Notices Roy. Astron. Soc.* **182**, 559.

Willis, A. J. and Wilson, R.: 1979, in P. S. Conti and C. W. H. de Loore 'Mass Loss and Evolution of O-type Stars', *IAU Symp.* 83, 461.

Willis, A. J., Wilson, R., Machetto, F., Beekmans, F., Van der Hucht, K. A., and Stickland, D. J. (eds.): 1979, in *The First Year of IUE.*, Univ. College of London, p. 394.

Willis, A. J.: 1980, communic. Tübingen Symposium on *I.U.E.*

Wilson, O. C.: 1959, *Astrophys. J.* **130**, 499.

Wilson, O. C.: 1963, *Astrophys. J.* **138**, 832.

Wilson, O. C.: 1970a, *Astrophys. J.* **160**, 225.

Wilson, O. C.: 1970b, *Bull. Astron. Soc. Pacific* **82**, 865.

Wilson, O. C.: 1976, *Astrophys. J.* **205**, 823.

Wilson, O. C. and Bappu, M. K. V.: 1957, *Astrophys. J.* **125**, 661.

Wilson, R., and many others: 1978, *Nature* **275**, 372.

Wilson, W. J.: 1971, *Astrophys. J. Letters* **116**, L13.

Wilson, W. J. and Barrett, A. H.: 1968, *Science* **161**, 778.

Wilson, W. J. and Barrett, A. H.: 1972, *Astron. Astrophys.* **17**, 385.

Wilson, W. J., Schwarz, P. R., Neugebauer, G., Harvey, P. M., and Becklin, E. G.: 1972, *Astrophys. J.* **177**, 523.

Wing, R. F.: 1974, *Highlights of Astronomy* 3, 285.

Wing, R. F. and Yorka, S. B.: 1977, *Monthly Notices Roy. Astron. Soc.* **178**, 383.

Winkler, P. F.: 1977, *Astrophys. J.* **211**, 562.

Winkler, P. F.: 1978, *Astrophys. J.* **221**, 220.

Winkler, P. F. and Laird, F. N.: 1976, *Astrophys. J. Letters* **204**, L111.

Winkler, P. F., Hearn, D. R., Richardson, J. A., and Behnken, J. M.: 1979, *Astrophys. J. Letters* **229**, L123.

Winnberg, A.: 1977, *Infrared Phys.* **17**, 557.

Winnberg, A.: 1978, in R. Kippenhahn *et al.* (eds.), 'The Interaction of Variable Stars with their Environment', *IAU Colloq.* 42, 495.

Wolf, B.: 1972, *Astron. Astrophys.* **20**, 275.

Wolf, B.: 1973, *Astron. Astrophys.* **28**, 335.

Wolf, B.: 1975, *Astron. Astrophys.* **41**, 471.

Wolf, B.: 1977, in R. Kippenhahn *et al.* (eds.), *The Interactions of Variable Stars with their Environment*, Veröff. Remeis Sternwarte, Bamberg, Vol. XI, No. 121, p. 151.

Wolf, B. and Sterken, C.: 1976, *Astron. Astrophys.* **53**, 355.

Wolf, B., Campusano, L., and Sterken, C.: 1974, *Astron. Astrophys.* **36**, 87.

Wolf, C. J. E. and Rayet, G.: 1867, *Compt. Rend. Acad. Sci. Paris* **65**, 292.

Wolf, M.: 1923, *Astron. Nachr.* **217**, 475.

Wolf, N. J., Stein, W. A., and Strittmatter, P. A.: 1970, *Astron. Astrophys.* **9**, 252.

Wolff, S. C. and Beichman, C. A.: 1979, *Astrophys. J.* **230**, 519.

Wolff, S. C. and Preston, G. W.: 1978, *Astrophys. J. Suppl.* **37**, 371.

Woltjer, L.: 1972, *Ann. Rev. Astron. Astrophys.* **10**, 129.

Wood, F. B.: 1977, in R. Kippenhahn *et al.* (eds.), 'The Interaction of Variable Stars with their Environment', *IAU Coll.* **42**, 639.

Wood, P. R.: 1975, *Monthly Notices Roy. Astron. Soc.* **171**, 15P.

Wood, P. R.: 1976, *Monthly Notices Roy. Astron. Soc.* **174**, 531.

Wood, P. R. and Cahn, J. H.: 1977, *Astrophys. J.* **211**, 499.

Woodgate, B. E., Stockman, H. R., Angel, J. R. P., and Kirsher, R. R.: 1974, *Astrophys. J. Letters* **188**, L79.

Woodgate, B. E., Angel, J. R. P., and Kirsher, R. R.: 1975, *Astrophys. J.* **200**, 715.

Woodgate, B. E., Lucke, R. L., and Socker, D. G.: 1979, *Astrophys. J. Letters* **229**, L119.

Woolf, N. J. and Ney, E. P.: 1969, *Astrophys. J. Letters* **155**, L181.

Woolf, N. J., Stein, W. A., and Strittmatter, P. A.: 1970, *Astron. Astrophys.* **9**, 252.

Wright, A. E. and Barlow, M. J.: 1975, *Monthly Notices Roy. Astron. Soc.* **170**, 41.

Wright, K. O.: 1970, in A. Beer (ed.), *Vistas in Astronomy*, **12**, 147.

Wright, K. O.: 1972, in S. D. Jordan and E. H. Avrett (eds.), 'Stellar Chromospheres', Scient. and Techn. Inf.Office, NASA, p. 159.

Wright, K. O.: 1973, in A. H. Batten (ed.), 'Extended Atmospheres and Circumstellar Matter in Spectroscopic Binary Systems', *IAU Symp.* **51**, 117.

Wright, K. O.: 1977, *J. Roy. Astron. Soc. Canada* **71**, 152.

Wu, C. C. and Kester, D.: 1977, *Astron. Astrophys.* **58**, 331.

Wyckoff, S. and Clegg, R. E. S.: 1978, *Monthly Notices Roy. Astron. Soc.* **184**, 127.

Wyckoff, S. and Wehinger, P. A.: 1976, *Monthly Notices Roy. Astron. Soc.* **175**, 587.

Wyckoff, S. and Wehinger, P. A.: 1977, *Astrophys. Space Sci.* **48**, 421.

Yahel, R. Z.: 1977, *Astrophys. Space Sci.* **51**, 135.

Yamashita, Y.: 1975, *Publ. Dom. Astrophys. Obs.* **12**, 293.

Yamashita, Y.: 1966, in M. Hack (ed.), *Colloquium on Late Type Stars*, Trieste, p. 98.

Yamashita, Y.: 1967, *Publ. Dom. Astrophys. Obs.* **13**, 67.

Yamashita, Y.: 1972, *Ann. Tokyo Astron. Obs* **13**, 169.

Yamashita, Y.: 1975, *Publ. Astron. Soc. Japan* **27**, 459.

Yamashita, Y. and Maehara, H.: 1977, *Publ. Astron. Soc. Japan* **29**, 319.

Yamashita, Y. and Maehara, H.: 1978, *Publ. Astron. Soc. Japan* **30**, 409.

Yamashita, Y., Nishimura, S., Shimizu, M., Noguchi, T., Watanabe, E., and Okida, K.: 1977, *Publ. Astron. Soc. Japan* **29**, 731.

York, D. G., Vidal-Madjar, A., Laurent, C., and Bonnet, R.: 1977, *Astrophys. J. Letters* **213**, L61.

Yorka, S. B. and Wing, R. F.: 1979, *Astron. J.* **84**, 1010.

Young, P. J., Corwin, H. G., Bryan, J., and De Vaucouleurs, G.: 1976, *Astrophys. J.* **209**, 882.

Yungelson, L. R.: 1973, *Nauchn. Inform. Astron. Soviet* **26**, 71.

Zaikowski, A. and Knacke, R. F.: 1975, *Astrophys. Space Sci.* **37**, 3.

Zarnecki, J. C., Culhane, J. L., Fabian, A. C., Rapley, C. G., Boyd, R. L. F., Parkinson, J. H., and Silk, R.: 1974, in G. Contopoulos (ed.), *Highlights of Astronomy* **3**, 565.

Zarnecki, J. C., Stark, J. P., Charles, P. A., and Culhane, J. L.: 1975, *Monthly Notices Roy. Astron. Soc.* **173**, 103.

Zellner, B. H. and Serkowski, K.: 1972, *Publ. Astron. Soc. Pacific* **84**, 619.

Ziebarth, K.: 1970, *Astrophys. J.* **162**, 947.

Ziolkowski, J.: 1972, *Acta Astron.* **22**, 327.

Ziolkowski, J.: 1976, *Astrophys. J.* **204**, 512.

Zirin, H.: 1976, *Astrophys. J.* **208**, 414.

Zuckerman, B.: 1979, *Astrophys. J.* **230**, 442.

Zuckerman, B., Palmer, P., Morris, M., Turner, B. E., Gilra, D. P., Bowers, P. F., and Gilmore, W.: 1977, *Astrophys. J. Letters* **211**, L97.

SYMBOLS AND NOTATIONS

a	semi-major axis of binary orbit
A	Einstein transition probability for emission
B	Einstein transition probability for absorption or stimulated emission
B	Black-body radiation intensity
c	velocity of light
c_v, c_p	specific heat at constant volume, pressure
c	stellar magnitude in Strömgren's c-band
d	distance of object from observer
E	integrated total radiation output of stellar outburst (nova, supernova)
f	stellar integrated flux received at Earth's distance
f_v, f_λ	monochromatic flux at frequency v or wavelength λ received at Earth's distance
f	oscillator strength
πF	stellar radiation flux $= L/4\pi R^2$
πF_v	monochromatic flux at frequency v
πF_λ	monochromatic flux at wavelength λ
F_m	mechanical flux
F_{ac}	acoustic flux
$F(k)dk$	spectrum of turbulence at wavenumber k
g, g_{grav}	acceleration of gravity
g_{rad}	acceleration by radiation forces
g_t	acceleration by turbulent forces
g_{eff}	effective acceleration in a stellar atmosphere
g_l, g_u	statistical weight of lower, upper level
g	gaunt factor
G	constant of gravity
H_d, H_p	density, pressure scale height in an atmosphere
H	Eddington's first moment of radiation intensity
I	radiation intensity
J	average radiation intensity in an atmosphere
J	orbital angular momentum of a binary
k	Boltzmann's constant
k, k_F, k_R	opacity, flux-mean opacity, Rosseland-mean opacity
k	wavenumber $= 2\pi/l$
K	Eddington's second moment of radiation intensity

K	coefficient of thermal conduction
l	stellar luminosity expressed in solar value
l	wavelength in velocity field if decomposed according to Fourier
l	(suffix) lower
$L, L_\odot$	stellar, solar luminosity; $L_\odot = 3.83 \times 10^{33}$ erg s^{-1}
m	stellar mass expressed in solar value
M	absolute magnitude; suffixes: b – bolometric; V – V-band; B – B-band; v – visual; pg – photographic
$\mathfrak{M}, \mathfrak{M}_\odot$	stellar, solar mass
n	number density
n	exponent in the mass-luminosity equation
N_u, N_l	number of ions in upper, lower level of excitation
P	pressure; suffixes: g – gas, e – electron, r – radiation, t – turbulent, μ – micro-turbulent, M – macro-turbulent, tot – total
Q_A	heat input by dissipation of mechanical energy
Q_R	radiative energy losses
r	distance from star's center
r_s	distance of sonic point
r_c	distance of critical point
$R, R_\odot$	stellar, solar radius; $R_\odot = 6.96 \times 10^5$ km
$\mathfrak{R}$	gas constant
s	velocity of sound
S	source function
S_ν, S_λ	monochromatic source function at frequency ν, at wavelength λ
T	temperature; cf. definitions in Section 1.8
T_e	effective temperature
u	(suffix) upper
U	internal energy of a star
v	velocity; suffixes: t – turbulent; μ – micro-turbulent; M – macro-turbulent
v_z, ζ	vertical velocity component
v_{th}	thermal velocity
v_m	average Maxwellian velocity
v_{esc}	escape velocity
v_∞	asymptotic velocity of stellar wind (terminal wind velocity)
W	integrated kinetic energy output of stellar outburst (nova, supernova)
X	relative abundance of hydrogen by mass
y	stellar magnitude in Strömgren's y-band
Y	relative abundance of helium by mass
z	height ordinate in a star (positive outward)
Z	relative abundance of 'heavy' elements by mass
Z	average ionic charge
α	radio spectral index

β	P_g/P_{tot}
γ	c_P/c_v
Γ	g_r/g_{grav}; suffixes: s – calculated with Thomson electron scattering; F – with the flux-mean opacity; C – with the flux mean opacity, neglecting lines
ζ	velocity component along z-axis; suffixes: μ – micro-turbulent component; M – macroturbulent component; t – total turbulent component
θ	angular diameter of a star
θ	optical scale height in the atmosphere
Θ	$5040/T$
$\kappa, \kappa_\lambda, \kappa_\nu$	absorption coefficient, at wavelength λ or frequency ν
λ	wavelength
μ	molecular weight
μ	$\sec \vartheta$
ν	frequency
ϱ	density of matter
σ, σ_e	scattering coefficient, by free electrons
τ	optical depth; suffixes: λ, ν: monochromatic at wavelength λ or frequency ν; 0.5 or .5: at 0.5 µm; R: Rosseland
$\varphi(\nu)\,d\nu$	profile of absorption coefficient
ω	rotational angular velocity, [rad s^{-1}]
Ω	potential energy of a star
∇	$d(\log T)/d(\log P)$

Definitions regarding relative abundances.

Let E, E_1, E_2 be symbols for chemical elements then we define:

$(E) \equiv A_E \equiv N(E)/N(H)$: abundance by number, relative to hydrogen
$(E_1/E_2) \equiv N(E_1)/N(E_2)$
$[E] \equiv \log\{(E^*)/(E_\odot)$
$[E_1/E_2] \equiv \log(\{(E^*)/E_\odot)\}/\{(E^*)/(E_\odot)\}) = [E_1] - [E_2]$.

Here, an asterisk or the symbol $\odot$ denote stellar or solar values.

ALPHABETIC LIST OF STARS

The stars, mentioned in this book, are listed here; firstly according to their constellation (three-letter abbreviated notation – alphabetic). This list is followed by one containing those stars that are mentioned in the book with their HD(E)-number. When the constellation-designation and the HD(E)-number occur both, the page-references are given only with the constellation name, while appropriate cross-references are given in the HD(E) list. This latter list is followed by lists of stars that are mentioned otherwise (HR, BD, their names, or 'other designations'). When a star is mentioned under more than one designation the page-reference is always, and only, given in the first of the lists that contain that object.

(HD(E))			(HD(E))		
46223		19, 73	163181	→	V 453 Sco
47129		32, 34, 324	164353		67 Oph
48329		328	164402		47
50896		72, 76, 186	166397	→	μ Sgr
53138	→	o^2 CMa	166734		32
54662		324	167264		342
57060	→	UW CMa	167771		324
58350	→	η CMa	167838		345
65818	→	V Pup	168076		324
66811		324	168206	→	CV Ser
68273	→	γ^2 Bel	168607		342
77581	→	Vela X-1	169454		325
80077		100, 104	174237	→	CX Dra
86161		71, 74	186943		75
87737	→	η Leo	187399		326
88661		93, 94	190429		73, 324
90657		75	190603		325
91316	→	ϱ Leo	190918		75
91619		342, 349	190967	→	V 448 Cyg
92740		71, 75	191765		76
93128		268	192103		76
93129A		3, 20, 62,71,73, 80, 268, 324	192163		76, 77, 272, 322
93129B		20, 268	192881		62
93131		71, 73, 74, 80	192641		75
93162		71, 74, 80, 300, 301	193077		80
93205		32, 301	193576	→	V 444 Cyg
93206	→	QZ Car	193611	→	V 478 Cyg
93249		300	193793		78, 272, 322
93250		6, 71, 72, 268, 301, 324	194279		325
93356		300	197345	→	α Cyg
93403A		32	197406		75
96159		99	198478	→	55 Cyg
96214		99	198846	→	Y Cyg
96248		57, 99	206267		32
96548		71, 74	208606		328
96919		342, 349	210839		308, 324
102567		89	211853	→	GP Cep
112244		308	216014	→	AH Cep
113904	→	θ Mus	216946		328
119796		100, 102, 104, 345	217476		100, 101, 224, 327, 347
135591		324	218066	→	CW Cep
138481		270	218319		95
148937		324	226868	→	Cyg X-1
151804		308, 309, 324	227696	→	V 453 Cyg
151890	→	μ^2 Sco	228766		32, 80, 324
151932		7,1 72	228854		V 382 Cyg
152233		324	268757		344
152236	→	ζ^1 Sco	268907		344
152424		324	269006		342, 344
152408		64, 308, 309, 324	269700		325
152248		32	269723		346
152667	→	V 861 Sco	269810		62
153919		190, 324	269896		99, 100
159176		32	269953		346
160529		100, 103, 243, 341, 343, 345, 349, 350, 352	303308		73

HR(BS)

894	→	HD 18552
1040		327
2874		327
3165		ζ Pup
4441		328
4337		100, 103
5171	→	HD 119796
5999		93
6392		74, 103
8752	→	HD 217476

BD

+60°2522		62, 65, 287
+40°4220	→	V 729 Cyg
−14°5037		325, 342

Other Designations

A 0620 − 00		
A 1524 − 61	→	Cen X-4
ADS 782 B		97
AFGL 333		133
AS 320		78
Boss 1970		107
Boss 1985		34, 107
CIT 6		293
CRL 3068		116, 287
GL 3068	→	CRL 3068
HBV 475		273
IC 1470		287
IRC + 10011		281
IRC + 10216	→	CW Leo
IRC + 10420		284, 343

(Other Designations)

IRC 30308		120, 278
LMC-X4		191
Lk Hα 101		272, 287
Lk Hα 198		88
M1-11		273
M1-70		130, 273, 284
M 1-92		130, 273, 284
MWC 342		92
MWC 349		92, 273
MWC 957		92
MWC 1080		287
MR 82		78
NGC 2438		273
NGC 7635	→	BD + 60°2522
OH 26.5 +06		282
OH 30.1 −0.7		277
Sk 160	→	SMC X-1
SMC X-1		191
Tr 27-28		78
4 U 0115 − 37	→	SMC-X1
4 U 0352 + 30	→	X Per
4 U 0900 − 40	→	Vela X-1
4 U 1119 − 60	→	Cen X-3
4 U 1538 − 52	→	190
4 U 1700 − 37	→	HD 153919
4 U 1956 + 35	→	Cyg X-1
Var 2		340
Ve 2-45		78
Vy 2-2	→	M1-70
W 49		282

OBJECTS BY NAME

SUPERNOVAE (BY YEAR OF APPEARANCE)

185	405		
386	405		
393	405		
1006	398, 405		
1054	388, 404, 405	(Crab)	
1181	405		
1408	405		
1572	398, 404, 405	(Tycho)	

1604	395, 398, 404, 405	(Kepler)
1970g	398, 399, 400, 402	
1971l	391, 394	
1972e	388, 389, 392, 393, 395, 396, 400	
1974g	388	
1978 (M 100)	396	
Cas A	404	

GEOPHYSICS AND ASTROPHYSICS MONOGRAPHS

AN INTERNATIONAL SERIES OF FUNDAMENTAL TEXTBOOKS